Polylactide-Based Materials: Synthesis and Biomedical Applications

Polylactide-Based Materials: Synthesis and Biomedical Applications

Editors

Marek Brzeziński
Małgorzata Baśko

MDPI • Basel • Beijing • Wuhan • Barcelona • Belgrade • Manchester • Tokyo • Cluj • Tianjin

Editors
Marek Brzeziński
Centre of Molecular and
Macromolecular Studies,
Polish Academy of Sciences,
Lodz, Poland

Małgorzata Baśko
Centre of Molecular and
Macromolecular Studies,
Polish Academy of Sciences,
Lodz, Poland

Editorial Office
MDPI
St. Alban-Anlage 66
4052 Basel, Switzerland

This is a reprint of articles from the Special Issue published online in the open access journal *Molecules* (ISSN 1420-3049) (available at: https://www.mdpi.com/journal/molecules/special_issues/polylactide_material).

For citation purposes, cite each article independently as indicated on the article page online and as indicated below:

LastName, A.A.; LastName, B.B.; LastName, C.C. Article Title. *Journal Name* **Year**, *Volume Number*, Page Range.

ISBN 978-3-0365-6126-4 (Hbk)
ISBN 978-3-0365-6641-2 (PDF)

© 2023 by the authors. Articles in this book are Open Access and distributed under the Creative Commons Attribution (CC BY) license, which allows users to download, copy and build upon published articles, as long as the author and publisher are properly credited, which ensures maximum dissemination and a wider impact of our publications.

The book as a whole is distributed by MDPI under the terms and conditions of the Creative Commons license CC BY-NC-ND.

Contents

About the Editors vii

Marek Brzeziński and Malgorzata Basko
Polylactide-Based Materials: Synthesis and Biomedical Applications
Reprinted from: *Molecules* 2023, 28, 1386, doi:10.3390/molecules28031386 1

Yulia A. Kadina, Ekaterina V. Razuvaeva, Dmitry R. Streltsov, Nikita G. Sedush, Eleonora V. Shtykova, Alevtina I. Kulebyakina, et al.
Poly(Ethylene Glycol)-*b*-Poly(D,L-Lactide) Nanoparticles as Potential Carriers for Anticancer Drug Oxaliplatin
Reprinted from: *Molecules* 2021, 26, 602, doi:10.3390/molecules26030602 7

Sana Ayari-Riabi, Noureddine Ben khalaf, Balkiss Bouhaouala-Zahar, Bernard Verrier, Thomas Trimaille, Zakaria Benlasfar, et al.
Polylactide Nanoparticles as a Biodegradable Vaccine Adjuvant: A Study on Safety, Protective Immunity and Efficacy against Human Leishmaniasis Caused by Leishmania Major
Reprinted from: *Molecules* 2022, 27, 8677, doi:10.3390/molecules27248677 23

Sasivimon Pramual, Kriengsak Lirdprapamongkol, Korakot Atjanasuppat, Papada Chaisuriya, Nuttawee Niamsiri and Jisnuson Svasti
PLGA-Lipid Hybrid Nanoparticles for Overcoming Paclitaxel Tolerance in Anoikis-Resistant Lung Cancer Cells
Reprinted from: *Molecules* 2022, 27, 8295, doi:10.3390/molecules27238295 39

Katharina Wulf, Madeleine Goblet, Stefan Raggl, Michael Teske, Thomas Eickner, Thomas Lenarz, et al.
PLLA Coating of Active Implants for Dual Drug Release
Reprinted from: *Molecules* 2022, 27, 1417, doi:10.3390/molecules27041417 51

Júlia Ribeiro Garcia Carvalho, Gabriel Conde, Marina Lansarini Antonioli, Clarissa Helena Santana, Thayssa Oliveira Littiere, Paula Patrocínio Dias, et al.
Long-Term Evaluation of Poly(lactic acid) (PLA) Implants in a Horse: An Experimental Pilot Study
Reprinted from: *Molecules* 2021, 26, 7224, doi:10.3390/molecules26237224 67

Karima Belabbes, Coline Pinese, Christopher Yusef Leon-Valdivieso, Audrey Bethry and Xavier Garric
Creation of a Stable Nanofibrillar Scaffold Composed of Star-Shaped PLA Network Using Sol-Gel Process during Electrospinning
Reprinted from: *Molecules* 2022, 27, 4154, doi:10.3390/molecules27134154 83

Rodrigo Osorio-Arciniega, Manuel García-Hipólito, Octavio Alvarez-Fregoso and Marco Antonio Alvarez-Perez
Composite Fiber Spun Mat Synthesis and In Vitro Biocompatibility for Guide Tissue Engineering
Reprinted from: *Molecules* 2021, 26, 7597, doi:10.3390/molecules26247597 99

Raasti Naseem, Giorgia Montalbano, Matthew J. German, Ana M. Ferreira, Piergiorgio Gentile and Kenneth Dalgarno
Influence of PCL and PHBV on PLLA Thermal and Mechanical Properties in Binary and Ternary Polymer Blends
Reprinted from: *Molecules* 2022, 27, 7633, doi:10.3390/molecules27217633 115

Huan Hu, Ang Xu, Dianfeng Zhang, Weiyi Zhou, Shaoxian Peng and Xipo Zhao
High-Toughness Poly(lactic Acid)/Starch Blends Prepared through Reactive Blending Plasticization and Compatibilization
Reprinted from: *Molecules* **2020**, *25*, 5951, doi:10.3390/molecules25245951 . . . . **127**

Joanna Bojda, Ewa Piorkowska, Grzegorz Lapienis and Adam Michalski
Shear-Induced Crystallization of Star and Linear Poly(L-lactide)s
Reprinted from: *Molecules* **2021**, *26*, 6601, doi:10.3390/molecules26216601 . . . . **143**

Mariia Svyntkivska, Tomasz Makowski, Ewa Piorkowska, Marek Brzezinski, Agata Herc and Anna Kowalewska
Modification of Polylactide Nonwovens with Carbon Nanotubes and Ladder Poly(silsesquioxane)
Reprinted from: *Molecules* **2021**, *26*, 1353, doi:10.3390/molecules26051353 . . . . **161**

Qing Zhang, Huiyuan Liu, Junxia Guan, Xiaochun Yang and Baojing Luo
Synergistic Flame Retardancy of Phosphatized Sesbania Gum/Ammonium Polyphosphate on Polylactic Acid
Reprinted from: *Molecules* **2022**, *27*, 4748, doi:10.3390/molecules27154748 . . . . **175**

Raasti Naseem, Charalampos Tzivelekis, Matthew J. German, Piergiorgio Gentile, Ana M. Ferreira and Kenny Dalgarno
Strategies for Enhancing Polyester-Based Materials for Bone Fixation Applications
Reprinted from: *Molecules* **2021**, *26*, 992, doi:10.3390/molecules26040992 . . . . **187**

Jean Coudane, Hélène Van Den Berghe, Julia Mouton, Xavier Garric and Benjamin Nottelet
Poly(Lactic Acid)-Based Graft Copolymers: Syntheses Strategies and Improvement of Properties for Biomedical and Environmentally Friendly Applications: A Review
Reprinted from: *Molecules* **2022**, *27*, 4135, doi:10.3390/molecules27134135 . . . . **207**

Bartłomiej Kost, Marek Brzeziński, Marta Socka, Małgorzata Baśko and Tadeusz Biela
Biocompatible Polymers Combined with Cyclodextrins: Fascinating Materials for Drug Delivery Applications
Reprinted from: *Molecules* **2020**, *25*, 3404, doi:10.3390/molecules25153404 . . . . **231**

Seung Hyuk Im, Dam Hyeok Im, Su Jeong Park, Justin Jihong Chung, Youngmee Jung and Soo Hyun Kim
Stereocomplex Polylactide for Drug Delivery and Biomedical Applications: A Review
Reprinted from: *Molecules* **2021**, *26*, 2846, doi:10.3390/molecules26102846 . . . . **275**

Melania Bednarek, Katarina Borska and Przemysław Kubisa
Crosslinking of Polylactide by High Energy Irradiation and Photo-Curing
Reprinted from: *Molecules* **2020**, *25*, 4919, doi:10.3390/molecules25214919 . . . . **303**

Anna Kowalewska and Maria Nowacka
Supramolecular Interactions in Hybrid Polylactide Blends—The Structures, Mechanisms and Properties
Reprinted from: *Molecules* **2020**, *25*, 3351, doi:10.3390/molecules25153351 . . . . **331**

About the Editors

Marek Brzeziński

Marek Brzeziński obtained a PhD degree in 2014 at the Centre of Molecular and Macromolecular Studies in Lodz, investigating the novel properties of PLA-based materials. He was a Humboldt Fellow at Freie Universität Berlin and Helmholtz-Zentrum Berlin for Materials and Energy, working on a project involving the use of microfluidics to design supramolecular capsules for drug delivery. In 2021, he received habilitation for developing micro- and nanoparticles based on polylactides as potential drug delivery carriers in anticancer therapy. He is now a professor at the Centre of Molecular and Macromolecular Studies in Lodz, and his current research interests include functional polymer design and microparticles and nanoparticles for biomedical applications.

Małgorzata Baśko

Malgorzata Baśko received a Ph.D. from the Centre of Molecular and Macromolecular Studies, Polish Academy of Sciences in Lodz, (Poland) in 2003 (thesis on "Copolymers of Methyl Glyoxylate-Synthesis and properties"). Afterward, she completed postdoctoral fellowships at Ghent University (Belgium, Prof. Filip Du Prez's research group) and at Leuven University (Belgium, Prof. Ivo Vankelecom's research group) working on the synthesis of solvent-resistant nanofiltration membranes based on segmented polymer networks. She has been an adjunct at the CMMS PAS since 2008. Her research interests encompass the synthesis of biocompatible and degradable polymers for application in different areas, including biomedicine and filtration.

Editorial

Polylactide-Based Materials: Synthesis and Biomedical Applications

Marek Brzeziński * and Malgorzata Basko *

Centre of Molecular and Macromolecular Studies, Polish Academy of Sciences, Sienkiewicza 112, 90-363 Lodz, Poland
* Correspondence: marek.brzezinski@cbmm.lodz.pl (M.B.); malgorzata.basko@cbmm.lodz.pl (M.B.)

Polylactide (PLA) is a biocompatible polyester that can be obtained by polycondensation of lactic acid or the ring-opening polymerization (ROP) of lactide. It is worth noting that its unique properties are also related to the existence of two enantiomeric forms: poly(L-lactide) (PLLA) and poly(D-lactide) (PDLA), which form a supramolecular complex. This complex is called a stereocomplex, and its formation leads to enhanced thermal and mechanical properties, decreased degradation time or drug release. All these advantageous features predispose PLLA to applications in the biomedical field. Various materials are prepared from PLA matrix, such as nanoparticles (NPs), microparticles (MPs), nonwovens, implants, hydrogels, 3D-printed materials, etc. However, there is an unmet need for novel and more effective materials that can deliver their cargo to the desired site of the body. The main aim is to overcome drug resistance to antibiotics or anti-cancer drugs by appropriately designing PLA-based materials. In this regard, supramolecular materials are required since their unique structure allows them to self-heal or release the drug with slight environmental fluctuations (pH, temperature, light, etc.). It is also important to evaluate the in vivo stability of biosafety of these novel materials before their administration to humans. Due to the above-mentioned issues, there are goals that still need to be achieved, despite the tremendous applications of PLA-based materials in biomedicine; moreover, artificial intelligence may be useful in designing these materials in the near future.

This Special Issue presents the original research and reviews articles describing the synthesis and application of PLA-based materials. In this regard, Razuvaeva et al. [1] reported the preparation of NPs composed of methoxy poly(ethylene glycol)-*b*-poly(D,L-lactide) copolymers loaded with the anticancer drug: oxaliplatin. The core–corona structure was obtained after the nanoprecipitation of copolymers. It was shown that the higher amount of hydrophobic block results in a decrease in the core–corona interface and a lowering of the drug loading. The size of drug-loaded NPs ranges from 19 to 27 nm, and its highest loading is for NPs built from copolymers with shorter hydrophobic blocks, which is related to the low hydrophobicity of the drug. In addition, hybrid NPs composed of PLGA and a lipid block for the delivery of paclitaxel against lung cancer cells was proposed by Lirdprapamongkol et al. [2]. The self-assembled NPs were uniform, with a size of around 100 nm and a negative charge on their surface. Those NPs were tested against A549 attached and floating cells. Both cell lines show dose-dependent cytotoxicity; treatment of the floating cells (anoikis resistant) by NPs increased efficiency, since the 71.6-fold decrease in IC_{50} compared to the free drug was observed. Several mechanisms may be responsible for paclitaxel resistance of A459 cancer cells; however, the enhanced efficiency of nanocarriers is probably related to their ability to deliver paclitaxel into the cell interior. Ayari-Riabi [3] proposes a nanoparticle for the preparing unit vaccine that can be used against the vector-borne disease leishmaniasis. This tropical disease, and its most common form, Cutaneous leishmaniasis, causes skin infections. To overcome this problem, the poly(D,L-lactide)-based NPs were obtained by nanoprecipitation and mixed with *L. major* histone H2B (*L. major* H2B) protein, and they acquired immunogenicity. The

adsorption of proteins on the surface of NPs (adsorption capacity: 2.8 ($\pm$0.24% w/w) induces the shift in charge from negative to positive, while the size of NPs increases from 270 to 340 nm. Such particles release H2B protein in a sustained way, and this feature facilitates the prolonged activity of the encapsulated protein. To test their efficiency, BALB/c mice were infected with *L. major* parasite, and the following formulations were administered: H2B protein/CpG-ODN7909 adjuvant, H2B protein/PLA-NPs, and H2B protein alone. The humoral and cellular immune response is induced for a combination of H2B protein with CpG-ODN7909 or PLA-NPs, respectively. Most importantly, the footpads of mice were unoccupied with necrosis, with fewer parasites on their surface in the group treated with H2B protein/PLA-NPs compared to the control. This proves that such a PLA-based system can be an anti-leishmanial vaccine.

Due to their biocompatibility, PLA-based materials can also be used as implants or dressings. In this regard, Wulf et al. [4] propose to cover cochlear implants with PLLA coating (loaded with diclofenac-DCF) along with medical-grade silicone as a matrix (loaded with dexamethasone-DMS) to achieve sustained release of these anti-inflammatory drugs. This is especially important after implantation since the long-term release of those drugs may cause undesired tissue reactions. To prepare the desired material, the silicone surface was treated with O_2 plasma in the presence of (3-glycidyloxypropyl)trimethoxysilane. Subsequently, the surface was spray-coated with PLLA with amino functionalities. This material structure leads to totally different release profiles of encapsulated drugs. The PLLA layer on the silicon surface decreases the release of DMS; however, it accelerates the release of DCF. This dual-release profile will enhance the patient's comfort after implantation. Furthermore, the impedance measurements were performed, and since PLA is an insulator, the masking on the surface of the research electrode contacts should be applied during the coating process. Those promising materials are planned to be investigated in vivo. On the contrary, the in vivo investigation of PLA biocompatibility and biodegradation after subcutaneous implantation on the lateral surface of the neck of a horse was described by Ferraz et al. [5]. The implants were constructed from commercially available PLA (Ingeo 3251D, NatureWorks) by hot pressing with the size of 1 cm^2 and 1 mm thickness. The histochemical examination was performed 24, 28, 34, 38, and 57 weeks after implantation. The plasma fibrinogen level was constant, indicating a negligible inflammatory process. The foreign body response was detected until 38 weeks. Most importantly, SEM analysis indicated the occurrence of the biodegradation process since the implant's surface became rough and the pore diameter increased. Interestingly, the polymer fragmentation after 57 weeks was significant; therefore, the implant cannot be removed and analysed, indicating PLA degradation. This process is complicated, since several factors must be considered during the degradation process, such as D-monomer content, crystallinity, molar mass, and especially conditions of this process. However, the proposed approach paved the way for the use of commercial grade PLA in veterinary because the chosen PLA was not a medical grade one.

Electrospinning (ES) [6] and the air-jet spinning technique (AJS) [7] are the most relevant techniques for the preparation of PLA fibres. Pinese et al. [6] combined the ES process with chemical crosslinking to prepare nanofibers resistant to degradation. Star-shaped PDLLA was synthesised and equipped with triethoxysilylpropyl (TPES) end groups to achieve this aim. These end groups are subsequently used during the ES process to crosslink the resulting nanofibers with 360 to 780 nm diameters. It relies on the reaction of hydrolysis of TPES and the formation of silanol groups that can condense to form a network. Several parameters, such as polymer dilution, molecular weight, and the addition of low-molecular-weight PTES-functionalized PLA to prepare the material able to withstand 6 months of degradation. Most importantly, the obtained materials exhibit acceptable biocompatibility with murine and mouse fibroblasts cells lines that predispose their application in biomedicine. Alvarez-Perez et al. [7] use the AJS technique to develop nanocomposite fibre scaffolds for tissue bone regeneration. The first step towards this aim was the synthesis of ZrO_2 NPs by the hydrothermal method. Those NPs (0.1 or 0.5 g) were added to commercially available PLA (NatureWorks), and the AJS technique was suitable for obtaining composite fibres with an average diameter of 395 nm. The human foetal osteoblast cell attachment on the surface of obtained mats was at the suitable level, whereas the highest viability was for the nanocomposite with 0.5 g of nanoceramics; therefore, those fibres were used for the biomineralisation essay. It was shown that PLA/ZrO_2 exhibit a higher concentration of calcium precipitates on their surface than those composed solely of PLA. These good pro-osteogenic properties indicate their potential for bone regeneration application.

PLA also has several drawbacks, such as brittleness or low impact strength; therefore, the methods of its properties modification are desirable. In this respect, the blending of PLLA with polycaprolactone (PCL) and poly(3-hydroxybutyrate-co-3-hydroxyvalerate) (PHBV) is proposed to prepare materials for orthopaedical applications by Dalgarno et al. [8]. The main goal was to modify the creep behaviour of PLLA to fulfil the requirements for materials applied in bone regeneration. Twin screw extrusion is suitable for combining this material, and the analysis by differential scanning calorimetry (DSC) proved their effective blending. The ternary blends PLLA/PCL/PHBV (80/10/10) were ideal since the increased modulus, strength, and elongation at break can be achieved. Therefore, the biomaterials with the desired properties can be obtained by adjusting the ratio of macromolecules in the resulting composite. Alternatively, starch/PLA (NatureWorks) blends with oil polyols, polyethylene glycol (PEG), and citric acid (CA) as modifiers were proposed by Zhao et al. [9]. Reactive blending leads to the preparation of desired materials with the elongation at break, and impact strength is greatly improved compared to pure PLA. This affected the esterification reaction between CA, PEG, and starch. Interestingly, the value of tan δ in the low-frequency region shows gel-like behaviour indicating a crosslinking reaction. Moreover, the resulting material was hydrophilic in contrast to hydrophobic PLA. A different approach was shown by Bojda et al. [10], which investigated the influence of topology on shear to induce the crystallisation of PLA. Therefore, the high-molecular-weight 4-arm and 6-arm PLLAs were synthesised, and their properties were compared with linear counterparts. As expected, the influence of the macromolecular structure and molecular weight on the shear-induced crystallisation was determined. Another way to modify the properties of PLA-based materials is the addition of nanofillers such as carbon nanotubes (CNTs), as proposed by Makowski et al. [11]. To achieve this aim, the ES method was used to prepare nonwovens with 0.1 wt.% CNTs and 5 and 10 wt.% oligomeric linear ladder poly(silsesquioxane)s (LPSQ) as additives. The porous fibres with a diameter from 0.68 to 3.5 μm were obtained, and their size decreased with the addition of CNTs and LPSQ. Most importantly, mechanical parameters can be greatly improved due to the presence of additives. It was shown that a combination of 10 wt.% LPSQ-COOMe and 0.1 wt.% of MWCNT lead to a 2.4-times increase in the tensile strength and elongation at maximum stress. In contrast, the flame retardancy of PLA (4032D, NatureWorks) can also be modified by appropriate additives, as shown by Zhang et al. [12]. Firstly, 9,10-dihydro-9-oxa-10-phosphaphenanthrene-10-oxide (DOPO) was reacted with endic anhydride (EA), and the resulting product was melt blended with commercially available PLA. Using vertical combustion (UL-94), the limiting oxygen index (LOI) tests indicate that the addition of ammonium polyphosphate (APP) to the composite is required to achieve an increased LOI value of 32.2% and a UL V-0 rating. Their good dispersion causes the synergistic effect of these two additives in the PLA matrix and the formation of an intumescent protective layer on the surface of the resulting material, which results in high thermal stability and flame retardancy.

The reviews in this Special Issue cover many topics, focusing mainly on strategies for extending the range of polylactide-based materials applications in biomedicine. The properties of PLA homopolymers, their copolymers, blends, and composite materials prepared for the manufacture of devices suitable for bone fixation were reviewed by Naseem et al. [13]. In conclusion, the authors indicated that current and prospective research should focus mainly on the search for polyester materials with appropriate mechanical properties, adjustable degradation time, modifiable brittleness, modified degradation chemistry to avoid excess acid degradation products, and improved interactions between the device and native tissue. Furthermore, the contribution concerning the chemical modification of the PLA backbone by introducing reactive structures that enabled the graft copolymers synthesis in the next step was presented by Coudane et al. [14]. After presenting the methods enabling the functionalization of the PLA backbone, the results of grafting various polymer segments onto the polyester backbone were discussed. By reacting the anhydride or epoxide groups incorporated into the PLA chain, various polymers, including cellulose derivatives, polyesters, polyamides, natural rubbers and PMMA, were combined with PLA. The main applications of these PLA graft copolymers in the field of environmental protection and biomedicine are proposed. In addition, the preparation of drug carriers, hydrogels and fibres in which supramolecular interactions occur in the cyclodextrin/polymer system has been reviewed by Kost and Brzeziński et al. [15]. Cyclodextrins are a group of cyclic oligosaccharides with a specific structure that enables the formation of inclusion complexes with various molecules through non-covalent host–guest interactions. Due to these specific association abilities, cyclodextrins are used in the pharmaceutical industry to increase the solubility and stability of drugs. By combining biocompatible polymers with cyclodextrins, various multifunctional materials have been developed. The presented work is divided into four sections which describe the drug delivery systems based on poly(lactide), poly(ε-caprolactone), poly(ethyleneglycol), and poly(sacharides). The paper discusses the release of different biologically active substances, including antibiotics, vitamins, hormones, enzymes, anticancer drugs, non-steroidal anti-inflammatory drugs and physiologically active lipid compounds. In contrast, Kim [16] reviews strategies for applying stereocomplex PLA in drug delivery systems and biomedicine. Synthesis and processing methods that enabled the production of therapeutic carriers in various forms (stereocomplex micelles, self-assembly nanoparticles, emulsions, hydrogels, and 3D-printed materials) were discussed in the article. Presented fields for potential stereocomplex PLA material applications concerned drug delivery systems, anti-cancer therapy, tissue engineering, and anti-microbial activity. Moreover, systems based on crosslinked polyesters (polymeric networks) are widely used in the area of drug delivery, wound healing, tissue engineering or medical implants. Methods of PLA crosslinking by applying different types of irradiation, i.e., high-energy electron beam or gamma irradiation and UV light, were reported by Bednarek et al. [17]. Finally, the progress of research on PLA nanocomposites containing synthetic organic nucleators (arylamides, hydrazides and 1,3:2,4-dibenzylidene-D-sorbitol) and biological nucleators (orotic acid, humic acids, fulvic acids, nanocellulose, cyclodextrins) in the context of their biomedical application was discussed by Kowalewska et al. [18]. The main role of these additives was to improve the rate of PLA of crystallization kinetics (nucleation and crystal growth) through supramolecular interactions based on hydrogen bonds or host–guest effects. Supramolecular interactions operating in those blends play a very important role in the properties of those novel hybrid materials.

In summary, the Special Issue entitled "Polylactide-Based Materials: Synthesis and Biomedical Applications" compiles the most recent research works on the modification of PLA materials by supramolecular interactions or the addition of nanofillers. It highlights the biomedical applications of PLA-based materials in drug delivery, tissue regeneration, bone fixation, etc. It is anticipated that acquired knowledge will lead to better understanding of the structure–properties relationship and will help to develop novel advanced strategies in biomedical and pharmaceutical field.

Author Contributions: M.B. (Malgorzata Basko): writing—review and editing; M.B. (Marek Brzeziński): writing—review and editing. All authors have read and agreed to the published version of the manuscript.

Acknowledgments: The CMMS-PAS, Lodz, Poland, is acknowledged for the support.

Conflicts of Interest: The authors declare no conflict of interest.

References

1. Kadina, Y.A.; Razuvaeva, E.V.; Streltsov, D.R.; Sedush, N.G.; Shtykova, E.V.; Kulebyakina, A.I.; Puchkov, A.A.; Volkov, D.S.; Nazarov, A.A.; Chvalun, S.N. Poly(Ethylene glycol)-*b*-poly(D,L-lactide) nanoparticles as potential carriers for anticancer drug oxaliplatin. *Molecules* **2021**, *26*, 602. [CrossRef]
2. Pramual, S.; Lirdprapamongkol, K.; Atjanasuppat, K.; Chaisuriya, P.; Niamsiri, N.; Svasti, J. PLGA-Lipid Hybrid Nanoparticles for Overcoming Paclitaxel Tolerance in Anoikis-Resistant Lung Cancer Cells. *Molecules* **2022**, *27*, 8295. [CrossRef] [PubMed]
3. Ayari-Riabi, S.; Ben khalaf, N.; Bouhaouala-Zahar, B.; Verrier, B.; Trimaille, T.; Benlasfar, Z.; Chenik, M.; Elayeb, M. Polylactide Nanoparticles as a Biodegradable Vaccine Adjuvant: A Study on Safety, Protective Immunity and Efficacy against Human Leishmaniasis Caused by Leishmania Major. *Molecules* **2022**, *27*, 8677. [CrossRef] [PubMed]
4. Wulf, K.; Goblet, M.; Raggl, S.; Teske, M.; Eickner, T.; Lenarz, T.; Grabow, N.; Paasche, G. PLLA Coating of Active Implants for Dual Drug Release. *Molecules* **2022**, *27*, 1417. [CrossRef] [PubMed]
5. Garcia, R.; Conde, G.; Antonioli, M.L.; Santana, C.H.; Littiere, T.O.; Patroc, P.; Chinelatto, M.A.; Al, P.; Jos, F.; Ferraz, G.C. Long-Term Evaluation of Poly(lactic acid) (PLA) Implants in a Horse: An Experimental Pilot Study. *Molecules* **2021**, *26*, 7224. [CrossRef]
6. Belabbes, K.; Pinese, C.; Leon-Valdivieso, C.Y.; Bethry, A.; Garric, X. Creation of a Stable Nanofibrillar Scaffold Composed of Star-Shaped PLA Network Using Sol-Gel Process during Electrospinning. *Molecules* **2022**, *27*, 4154. [CrossRef] [PubMed]
7. Osorio-Arciniega, R.; García-Hipólito, M.; Alvarez-Fregoso, O.; Alvarez-Perez, M.A. Composite fiber spun mat synthesis and in vitro biocompatibility for guide tissue engineering. *Molecules* **2021**, *26*, 7597. [CrossRef] [PubMed]
8. Naseem, R.; Montalbano, G.; German, M.J.; Ferreira, A.M.; Gentile, P.; Dalgarno, K. Influence of PCL and PHBV on PLLA Thermal and Mechanical Properties in Binary and Ternary Polymer Blends. *Molecules* **2022**, *27*, 7633. [CrossRef] [PubMed]
9. Hu, H.; Xu, A.; Zhang, D.; Zhou, W.; Peng, S.; Zhao, X. High-Toughness Poly(lactic Acid)/Starch Blends Prepared through Reactive Blending Plasticization and Compatibilization. *Molecules* **2020**, *25*, 5951. [CrossRef] [PubMed]
10. Bojda, J.; Piorkowska, E.; Lapienis, G.; Michalski, A. Shear-induced crystallization of star and linear poly(L-lactide)s. *Molecules* **2021**, *26*, 6601. [CrossRef] [PubMed]
11. Svyntkivska, M.; Makowski, T.; Piorkowska, E.; Brzezinski, M.; Herc, A.; Kowalewska, A. Modification of polylactide nonwovens with carbon nanotubes and ladder poly(silsesquioxane). *Molecules* **2021**, *26*, 1353. [CrossRef] [PubMed]
12. Zhang, Q.; Liu, H.; Guan, J.; Yang, X.; Luo, B. Synergistic Flame Retardancy of Phosphatized Sesbania Gum/Ammonium Polyphosphate on Polylactic Acid. *Molecules* **2022**, *27*, 4748. [CrossRef] [PubMed]
13. Naseem, R.; Tzivelekis, C.; German, M.J.; Gentile, P.; Ferreira, A.M.; Dalgarno, K. Strategies for enhancing polyester-based materials for bone fixation applications. *Molecules* **2021**, *26*, 992. [CrossRef] [PubMed]
14. Coudane, J.; Van Den Berghe, H.; Mouton, J.; Garric, X.; Nottelet, B. Poly(Lactic Acid)-Based Graft Copolymers: Syntheses Strategies and Improvement of Properties for Biomedical and Environmentally Friendly Applications: A Review. *Molecules* **2022**, *27*, 4135. [CrossRef]
15. Kost, B.; Brzezinski, M.; Socka, M.; Basko, M.; Biela, T. Biocompatible polymers combined with cyclodextrins: Fascinating materials for drug delivery applications. *Molecules* **2020**, *25*, 3404. [CrossRef] [PubMed]
16. Im, S.H.; Im, D.H.; Park, S.J.; Chung, J.J.; Jung, Y.; Kim, S.H. Stereocomplex polylactide for drug delivery and biomedical applications: A review. *Molecules* **2021**, *26*, 2846. [CrossRef] [PubMed]
17. Bednarek, M.; Borska, K.; Kubisa, P. Crosslinking of Polylactide by High Energy Irradiation and Photo-Curing. *Molecules* **2020**, *25*, 4919. [CrossRef] [PubMed]
18. Kowalewska, A.; Nowacka, M. Supramolecular interactions in hybrid polylactide blends-the structures, mechanisms and properties. *Molecules* **2020**, *25*, 3351. [CrossRef] [PubMed]

Disclaimer/Publisher's Note: The statements, opinions and data contained in all publications are solely those of the individual author(s) and contributor(s) and not of MDPI and/or the editor(s). MDPI and/or the editor(s) disclaim responsibility for any injury to people or property resulting from any ideas, methods, instructions or products referred to in the content.

Article

Poly(Ethylene Glycol)-*b*-Poly(D,L-Lactide) Nanoparticles as Potential Carriers for Anticancer Drug Oxaliplatin

Yulia A. Kadina [1], Ekaterina V. Razuvaeva [1,*], Dmitry R. Streltsov [1], Nikita G. Sedush [1], Eleonora V. Shtykova [2], Alevtina I. Kulebyakina [1], Alexander A. Puchkov [1], Dmitry S. Volkov [3], Alexey A. Nazarov [3] and Sergei N. Chvalun [1,4]

[1] National Research Center "Kurchatov Institute", 123182 Moscow, Russia; yellow_jk@mail.ru (Y.A.K.); streltsov.dmitry@gmail.com (D.R.S.); nsedush@gmail.com (N.G.S.); alya.kulebyakina@gmail.com (A.I.K.); puchkov1208@gmail.com (A.A.P.); s-chvalun@yandex.ru (S.N.C.)
[2] Federal Scientific Research Centre "Crystallography and Photonics" of Russian Academy of Sciences, 119333 Moscow, Russia; viwopisx@yahoo.co.uk
[3] Department of Chemistry, Lomonosov Moscow State University, 119991 Moscow, Russia; dmsvolkov@gmail.com (D.S.V.); nazarov@med.chem.msu.ru (A.A.N.)
[4] Enikolopov Institute of Synthetic Polymeric Materials Russian Academy of Sciences, 117393 Moscow, Russia
* Correspondence: razuvaeva.kate@gmail.com

Abstract: Nanoparticles based on biocompatible methoxy poly(ethylene glycol)-*b*-poly(D,L-lactide) (mPEG$_{113}$-*b*-P(D,L)LA$_n$) copolymers as potential vehicles for the anticancer agent oxaliplatin were prepared by a nanoprecipitation technique. It was demonstrated that an increase in the hydrophobic PLA block length from 62 to 173 monomer units leads to an increase of the size of nanoparticles from 32 to 56 nm. Small-angle X-ray scattering studies confirmed the "core-corona" structure of mPEG$_{113}$-*b*-P(D,L)LA$_n$ nanoparticles and oxaliplatin loading. It was suggested that hydrophilic oxaliplatin is adsorbed on the core-corona interface of the nanoparticles during the nanoprecipitation process. The oxaliplatin loading content decreased from 3.8 to 1.5% wt./wt. (with initial loading of 5% wt./wt.) with increasing PLA block length. Thus, the highest loading content of the anticancer drug oxaliplatin with its encapsulation efficiency of 76% in mPEG$_{113}$-*b*-P(D,L)LA$_n$ nanoparticles can be achieved for block copolymer with short hydrophobic block.

Keywords: poly(lactide); poly(ethylene glycol); block copolymers; self-assembly; nanoparticles; drug delivery systems; anticancer agent

1. Introduction

In the last decades, great attention has been paid to the development of nanoscale vehicles for drug delivery [1–3]. The incorporation of drug molecules into nanocarriers allows us to overcome poor water solubility of hydrophobic drugs, as well as increase stability against hydrolytic degradation of hydrophilic ones [4]. Moreover, nanoparticulate drug formulations can act in a targeted and prolonged manner, enhancing the efficacy of treatment, e.g., cancer treatment [4]. Platinum-based complexes (cisplatin, carboplatin, oxaliplatin, etc.) are widely used chemotheraputics agents for the treatment of various types of cancer [5,6]. Cisplatin (*cis*-(eblock)dichloridoplatinum(II)) is the first-generation platinum drug, that has a therapeutic effect against breast cancer, ovarian cancer, lung and head and neck cancer, cervix carcinoma, etc. However, it produces significant side effects such as ototoxicity, hematological, and emetogenicity [7]. Carboplatin (cis-diammine(1,1-cyclobutanedicarboxylato)platinum(II)), that is the second-generation platinum complex and cisplatin analogue, was designed to reduce the dose limiting toxicity of cisplatin. Oxaliplatin ((*trans*-R,R-cyclohexane-1,2-diamine)oxalatoplatinum(II)) is the third-generation platinum complex that was designed to overcome cellular resistance to cisplatin and carboplatin. Oxaliplatin shows higher solubility and less toxicity than cisplatin. It is used as a standard treatment for colorectal cancer. Moreover, oxaliplatin can be active against refractory ovarian cancer, germ-cell cancers, non-small cell lung cancer, etc. Nevertheless, its low water solubility, short half-life in the bloodstream, and non-selective biodistribution reduces the effective dose of oxaliplatin in the targeted tissues and enhances the systemic toxicity [8].

Design of nanocarriers for oxaliplatin delivery is one of the strategies to overcome its limitations and improve the efficacy of cancer treatment [7–9]. A wide range of various types of nanocarriers including inorganic nanoparticles [10], dendrimers [11], liposomes [12], polymeric nanoparticles [13], block-copolymer micelles [14,15], nanogels [16], etc., have been investigated for drug delivery of oxaliplatin. Several liposomal formulations of oxaliplatin, e.g., Lipoxal (Regulon, Inc.) and MBP-426 (Mebiopharm Co., Ltd.), are under clinical investigation [12]. Lipoxal is based on oxaliplatin-loaded PEGylated liposomes prepared from soy phosphatidylcholine, cholesterol, dipalmitoyl phosphatidylglycerol, and mPEG-distearoyl phosphatidylethanolamine, which exhibits reduced side effects compared to free oxaliplatin [17]. MBP-426 is a transferrin (Tf)-conjugated N-glutaryl phosphatidylethanolamine liposomal formulation of oxaliplatin, which can provide selective tumor targeting by binding to transferrin receptors [18]. Nonetheless, this technology has some disadvantages, namely rapid leakage of the incorporated drug molecules from liposomes during storage and after administration [19]. The promising nanoformulations of oxaliplatin are based on various polymeric carriers. The strategies of oxaliplatin incorporation into polymeric vehicles are generally based on chemical conjugation (complexation) or physical encapsulation. Chemical conjugation is usually accomplished between block copolymers, containing poly(glutamic acid) Pglu [20–22], poly(methacrylic acid) (PMA) [14], or poly(lactide-co-2-methyl-2-carboxyl-propylene carbonate) P(LA-co-MCC) [23,24] block and active part of oxaliplatin dichloro(1,2-diaminocyclohexane) platinum(II) DACHPt. Cabral et al. prepared DACHPt-loaded poly(ethylene glycol)-b-Pglu PEG-b-Pglu micelles, consisting of Pglu core surrounded by PEG corona, through polymer-metal complex formation of DACHPt with Pglu block [20,21]. It was observed that an increase of the initial [DACHPt]/[Glu] molar ratio from 0.25 to 1.5 leads to enhancement of platinum encapsulation efficiency from 20% to 90% and a growth of hydrodynamic diameter of DACHPt-loaded micelles from 25 to 50 nm [20]. In ref. [21], the effect of PEG_{272}-b-$Pglu_n$ (n = 20, 40, 70) copolymer composition on the DACHPt loading efficacy and biodistribution in vivo of DACHPt-loaded micelles was studied. It was observed that incorporation efficacy of DACHPt was approximately 30%, regardless of Pglu block length (molar ratio [DACHPt]/[Pglu] = 1). Meanwhile, in vivo biodistribution assay performed on tumor-bearing mice showed that DACHPt-loaded micelles based on PEG_{272}-b-$Pglu_{20}$ copolymer with the shortest Pglu block exhibited the lowest non-specific accumulation in normal tissues (liver, kidney, spleen) providing the highest accumulation in tumor. The authors suggested that PEG_{272}-b-$Pglu_{20}$ copolymer allows the formation of particles with effective surface coverage by PEG leading to reduction of accumulation of particles in liver. Although chemical conjugation usually leads to higher drug loading content in micelles, polymers can coordinate with Pt-complexes in a non-specific geometry, resulting in uncontrolled crosslinking [25]. Physical encapsulation of oxaliplatin into polymeric carriers allows us to overcome this limitation. Cui et al. developed poly(lactide) PLA nanoparticles stabilized by Tween80 as potential carriers for oxaliplatin [13]. The authors studied the effect of formulation parameters on the size, stability, drug loading content, and encapsulation efficacy of the nanoparticles. Optimal oxaliplatin loading content and its encapsulation efficacy was found to be 3.52 ± 0.07 wt.% and 17.40 ± 0.47 %, correspondingly. Micelles based on a stearic acid-grafted chitosan oligosaccharide CSO-SA were also investigated as carriers for oxaliplatin [26]. The highest drug loading content in CSO-SA micelles was 3.5 wt.%, and its encapsulation efficiency was 47%. The authors reported that a chitosan-based nanoformulation of oxaliplatin demonstrates enhanced antitumor activity in vitro against several cancer cells (about 3–6 folds) compared to free oxaliplatin. Despite a number of nanoformulations developed using different platforms, none of them have received FDA approval at the moment. The design of new types of nanocarriers for delivery of oxaliplatin is of high interest. We believe that PLA-*b*-PEG nanoparticles are a promising platform due to their flexibility and successful track record as nanocarriers for development of targeted anticancer drug formulations.

Nanoparticles of amphiphilic block copolymers comprising PLA hydrophobic block and PEG hydrophilic block have been extensively studied as potential vehicles for drug delivery due to their biocompatibility and biodegradability [27–29]. In aqueous solution, PLA-*b*-PEG copolymers are able to self-organize into "core-corona" nanosized structures, where hydrophobic PLA block chains form an inner core surrounded by a corona of hydrophilic PEG block chains [27,28]. PEG chains provide a "stealth effect" for the nanoparticles, minimize their undesirable interaction with proteins and capturing by the reticuloendothelial system, that extends circulation time of the nanoparticles in the body [4]. The nanoscale size of these block copolymer particles is favorable for passive targeting due to the so-called enhanced permeability and retention (EPR) effect. It is known that tumors and inflamed tissues are characterized by increased vascular permeability and impaired lymphatic drainage, which result in selective accumulation of nanoparticles into them [30]. Depending on the characteristics of the block copolymer (composition, molecular architecture, and block lengths) as well as the preparation conditions, PLA-*b*-PEG nanoparticles of different sizes and morphologies can be obtained [30]. Garofalo et al. showed that PLA/PEG block copolymer architecture, as well as its chemical composition, strongly affect size, stability, and tumor uptake of the micelles [27]. Linear and "tree-shaped" block copolymers mPEG$_{45}$-*b*-(PLA)$_n$ and mPEG$_{113}$-*b*-(PLA)$_n$, where n = 1 and 2 or 4, with various length and tacticity of the hydrophobic PLA blocks, were synthesized. It was reported that only the two-arm mPEG$_{45}$-*b*-poly(D,L-lactide)$_2$ mPEG$_{45}$-*b*-(P(D,L)LA)$_2$ copolymer produces high-stable monodispersed micelles with a hydrodynamic diameter of about 250 nm, whereas the other block copolymers show low stability and tendency to aggregate with formation of large submicron clusters. The influence of hydrophobic block length on the size of P(D,L)LA-*b*-PEG nanoparticles was studied in ref. [31]. It was reported that the hydrodynamic diameter of P(D,L)LA-*b*-PEG$_{113}$ nanoparticles produced by nanoprecipitation increased from 27.7 to 174.6 nm, with an increase of P(D,L)LA block molecular weight from 3 to 110 kDa. The effect of the nanoprecipitation parameters on the size of mPEG$_{113}$-*b*-P(D,L)LA$_{800}$ nanoparticles was studied by Y. Dong and S.-S. Feng [32]. The increase of the polymer concentration in organic phase from 4 to 13 g/L leads to increasing hydrodynamic diameter of the particles from 77 to 111.2 nm. Meanwhile, the increase of the organic solvent volume from 5 to 25 mL with a fixed polymer concentration of 10 g/L results in reducing hydrodynamic diameter of the particles from 89.8 to 79.7 nm.

For application of block copolymer nanoparticles as carriers for drug delivery, considerable attention should be given to their structure. Small-angle X-ray (SAXS) and neutron (SANS) scattering are powerful techniques for characterization of nanostructure, which can be implemented to analyze micelles and nanoparticles in aqueous medium [33]. Important characteristics affecting physicochemical properties of core-corona nanoparticles, and their drug loading capability can be determined from small-angle scattering data, e.g., density and size of the core, thickness and surface density of the corona, etc. [34–38]. Using SANS, T. Riley et al. studied the structure of deuterated P(D,L)LA(d)-*b*-PEG nanoparticles in aqueous solution, prepared by solvent evaporation method [36]. It was found that the P(D,L)LA(d) block length affects the core size as well as conformation of the corona-forming PEG chains. Thus, P(D,L)LA(d)-*b*-PEG$_{113}$ with short P(D,L)LA block (3 kDa) forms nanoparticles with a small core and highly splayed PEG chains in corona. The increase of the P(D,L)LA molecular weight to 15 kDa at a fixed length of the PEG block leads to increasing core size and, consequently, reducing PEG chain grafting density. The authors also observed that with decreasing PEG chain grafting density, the corona becomes more radially homogeneous. Using SAXS, Ma et al. investigated the influence of PLA stereostructure on the density of core and doxorubicin loading efficacy of PLA/PEG nanoparticles, which were prepared by a precipitation/solvent evaporation method [37]. It was observed that nanoparticles formed of enantiomerically-mixed P(L)LA$_{64}$-*b*-PEG$_{113}$/P(D)LA$_{71}$-*b*-PEG$_{113}$

copolymers and P(L)LA$_{64}$-b-PEG$_{113}$ copolymer exhibit larger core density compared with nanoparticles with stereoblock copolymer core, e.g., PEG$_{113}$-b-P(L)LA$_{32}$-P(D)LA$_{34}$, and P(D,L)LA$_{58}$-b-PEG$_{113}$.

Various fabrication techniques were developed to incorporate drugs in the core of the nanoparticles or at the core-corona interface, e.g., direct dissolution, dialysis, nanoprecipitation, salting-out method, emulsification method, etc. [39]. There are many factors that affect drug loading, including drug-core compatibility [40–42], hydrophobic block length [31,43–45], its crystallization capability [37,44,45], preparation conditions [32], etc. Despite the recent progress in development of the drug nanoformulations based on nanoparticles of amphiphilic block copolymers, there is still a lack of systematic data on the relationship between the chemical structure of block copolymer and the drug loading ability of nanoparticles. X. Zhang et al. investigated nanoparticles of P(D,L)LA-b-mPEG copolymers with various P(D,L)LA/mPEG weight ratios prepared by a solution casting method as potential carriers of highly hydrophobic drug paclitaxel [43]. It was reported that the particles based on P(D,L)LA-b-mPEG copolymers with higher P(D,L)LA content exhibited enhanced paclitaxel loading efficacy. Nanoparticles prepared by a nanoprecipitation method from P(D,L)LA-b-PEG copolymers with a fixed molecular weight of PEG block (5 kDa) and a variable molecular weight of P(D,L)LA block (from 3 to 110 kDa) were studied as a delivery system for water-soluble drug procaine hydrochloride [31]. It was demonstrated that the drug incorporation efficacy was independent of the P(D,L)LA block molecular weight.

PEG-b-PLA nanoparticles demonstrate many advantages as carriers for anticancer drugs. However, to the best of our knowledge, there is only one research article dedicated to PEG-b-PLA nanoparticles as potential carriers of oxaliplatin [15]. In addition, there is no data on the relationship between the chemical structure of PEG-b-PLA copolymers and the oxaliplatin loading ability of these nanoparticles. Development of such nanoformulation is complicated due to hydrophilicity of oxaliplatin, which limits its loading into carriers. The precise control of nanoparticles' structure and properties throughout adjustment of molecular structure of amphiphilic PEG-b-PLA copolymers is a promising approach for tuning their loading capacity. In the present work, we propose a nanoformulation of Pt(II)-based complex oxaliplatin based on mPEG$_{113}$-b-P(D,L)LA$_n$ nanoparticles, where n = 62–173 monomer units. Amphiphilic block copolymers with rather short PLA blocks (< 200 monomer units) are of great interest because significant changes in structure and characteristics of nanoparticles can occur with increasing PLA block length in this range of polymerization degrees. The aim of our study was to elucidate the effect of molecular weight of the hydrophobic P(D,L)LA block on the size, structure, morphology, and drug loading of the mPEG-b-P(D,L)LA nanoparticles.

2. Results and Discussion

2.1. Synthesis of mPEG-b-P(D,L)LA Copolymers

Amphiphilic mPEG$_{113}$-b-P(D,L)LA$_n$ copolymers with fixed molecular weight of the PEG block were synthesized by ring-opening polymerization (Figure 1). It is known that hydroxyl-containing compounds act as co-initiators in a coordination-insertion polymerization of lactide in presence of SnOct$_2$ catalyst [46]. Therefore, block-copolymers can be synthesized with mPEG as a macroinitiator.

Molecular characteristics of the synthesized polymers are presented in Table 1. The residual content of monomer determined by ^{1}H NMR was less than 1% for all block copolymers.

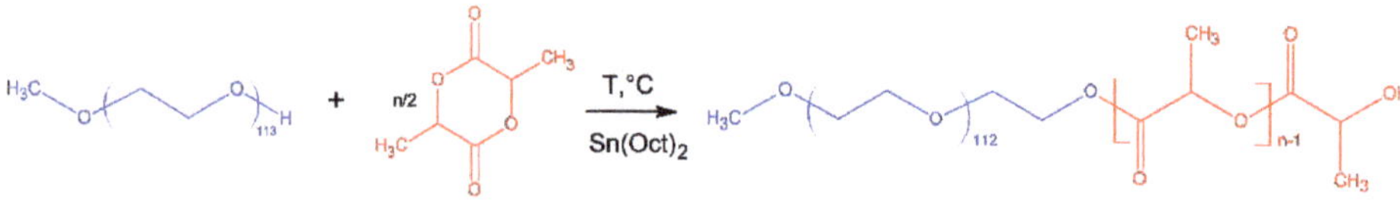

Figure 1. Scheme of synthesis of mPEG-b-P(D,L)LA copolymers by ring-opening polymerization.

Table 1. Molecular characteristics of the synthesized diblock copolymers

Sample	M_n^1, g/mol	M_n^2, g/mol	M_w^2, g/mol	PDI [2]
mPEG$_{113}$-b-P(D,L)LA$_{62}$	9500	7300	10,500	1.4
mPEG$_{113}$-b-P(D,L)LA$_{135}$	14,700	9000	14,600	1.6
mPEG$_{113}$-b-P(D,L)LA$_{173}$	17,500	10,400	18,000	1.7

[1] Determined by ^{1}H NMR. [2] Determined by GPC.

2.2. Characterization of Drug-Free mPEG-b-(D,L)LA Nanoparticles

For targeted delivery through EPR-effect hydrodynamic diameter of nanoparticles should be in the range of 10–200 nm [30]. Therefore, physicochemical characteristics of nanoparticles are important parameters that should be studied in order to consider them as potential drug carriers. Intensity size distribution curves (DLS) of aqueous suspensions of drug-free mPEG$_{113}$-b-P(D,L)LA$_n$ nanoparticles are presented in Figure 2a. Monomodal distribution with hydrodynamic radius R_h values corresponding to the peak maximum of 22 ± 10 and 28 ± 10 nm are observed for mPEG$_{113}$-b-P(D,L)LA$_{135}$ and mPEG$_{113}$-b-P(D,L)LA$_{173}$, respectively. A second peak appears on the intensity size distribution curve of mPEG$_{113}$-b-P(D,L)LA$_{62}$ nanoparticles with the shortest PLA block. One can suggest that these two peaks can be attributed to small individual nanoparticles and their large aggregates. For individual mPEG$_{113}$-b-P(D,L)LA$_{62}$ nanoparticles and their aggregates the values of R_h corresponding to the peak position were found to be 16 ± 6 and 72 ± 40 nm, respectively. Since the light scattering intensity of large objects is much stronger than that of small objects [47] and the peak intensities are almost equal, one can assume that only a minor fraction of the aggregates co-exists with the main fraction of individual nanoparticles. This assumption was also confirmed by TEM and SAXS measurements, provided below.

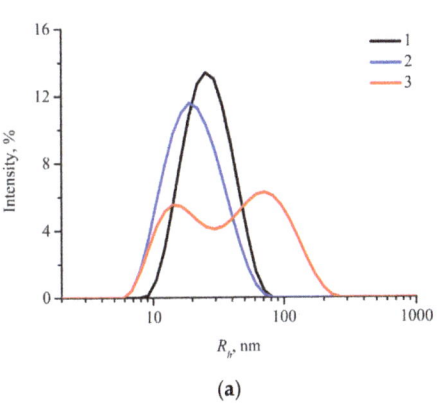

(a)   (b)

Figure 2. (a) DLS intensity size distribution curves for nanoparticles based on mPEG$_{113}$-b-P(D,L)LA$_n$ (c = 0.5 g/L): 1—mPEG$_{113}$-b-P(D,L)LA$_{173}$, 2—mPEG$_{113}$-b-P(D,L)LA$_{135}$, 3—mPEG$_{113}$-b-P(D,L)LA$_{62}$; (b) representative TEM image of spherical individual mPEG$_{113}$-b-P(D,L)LA$_{62}$ nanoparticles.

ζ-Potential of drug-free mPEG$_{113}$-b-P(D,L)LA$_n$ nanoparticles was found to be in the range from −14 to −20 mV (Table 2), which promotes their high stability in water.

For all the mPEG$_{113}$-b-P(D,L)LA$_n$ copolymers, TEM reveals only spherical individual nanoparticles (Figure 2b).

In order to evaluate the size and address the structure of nanoparticles, SAXS measurements were carried out. Scattering curves for aqueous suspensions of mPEG$_{113}$-b-P(D,L)LA$_n$ nanoparticles in Log I–Log s coordinates are presented in Figure 3a. All the SAXS curves converge to a plateau in the region of $s < 0.1$ nm^{-1}, indicating the absence of large scattering objects (aggregates) in aqueous suspensions of mPEG$_{113}$-b-P(D,L)LA$_n$

nanoparticles [48]. Apparently, the aggregates were removed during centrifugation before SAXS measurements. One also can see the secondary maximum at the SAXS curves in the region $0.3 < s < 1$ nm^{-1} (Figure 3a), suggesting that the nanoparticles have a spherical, well-defined structure with a relatively narrow size distribution.

Table 2. Characteristics of the mPEG$_{113}$-b-P(D,L)LA$_n$ nanoparticles in aqueous suspensions.

Sample	$R_h{}^1$, nm	$R_g{}^2$, nm	$R_g{}^3$, nm	$D_{max}/2^4$, nm	R^5, nm	$2R_g/D_{max}$	R_g/R	ζ^6, mV
mPEG$_{113}$-b-P(D,L)LA$_{173}$	28 ± 10	11.4 ± 0.1	12.5 ± 0.1	19.5 ± 1	15.4 ± 1	0.64 ± 0.05	0.81 ± 0.06	−14 ± 8
mPEG$_{113}$-b-P(D,L)LA$_{135}$	22 ± 10	10.1 ± 0.1	9.7 ± 0.1	17 ± 1	11.5 ± 1	0.57 ± 0.06	0.84 ± 0.09	−15 ± 4
mPEG$_{113}$-b-P(D,L)LA$_{62}$	16 ± 6	9.1 ± 0.1	8.6 ± 0.1	14 ± 1	9 ± 1	0.61 ± 0.07	0.96 ± 0.11	−20 ± 4

[1] The value of R_h corresponding to the peak position on DLS intensity size distribution curve. [2] Gyration radius of nanoparticles calculated from Guinier plots. [3] Gyration radius of nanoparticles evaluated from $P(R)$. [4] Radius of nanoparticles evaluated from $P(R)$. [5] The value of R corresponding to the maximum position on $P(R)$ function. [6] ζ-potential of nanoparticles.

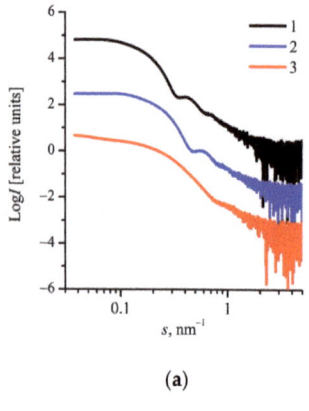
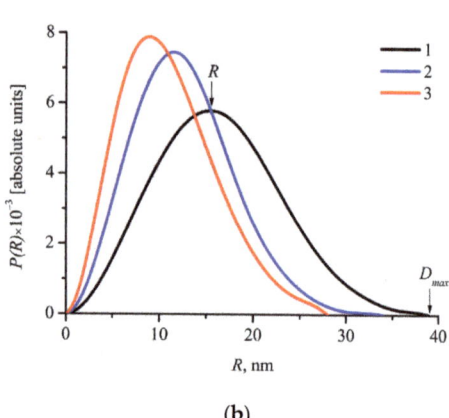

Figure 3. (a) SAXS (small-angle X-ray) curves in Log I–Log s coordinates and (b) corresponding pair distance distribution functions $P(R)$–R for mPEG$_{113}$-b-P(D,L)LA$_n$ nanoparticles: 1—mPEG$_{113}$-b-P(D,L)LA$_{173}$ (c = 7.5 g/L), 2—mPEG$_{113}$-b-P(D,L)LA$_{135}$ (c = 5 g/L), 3—mPEG$_{113}$-b-P(D,L)LA$_{62}$ (c = 5 g/L). The SAXS curves are shifted vertically for clarity.

The bell-shaped Kratky plots indicate a compact globular shape of the nanoparticles (Figure 4b) [49].

The values of gyration radius R_g were determined from the Guinier plots (Figure S1 of the Supplementary Materials). It was found that an increase in the P(D,L)LA block length results in higher R_g values (Table 2).

The pair distance distribution functions $P(R)$ are presented in Figure 3b. They are bell-shaped with a peak shifted to distances smaller than a half of maximum dimension of the scattering objects $D_{max}/2$ (Figure 3b). Since solid spherical particles display bell-shaped $P(R)$ functions with a peak at about $D_{max}/2$ [50], one can assume that the shift of $P(R)$ function maximum could be attributed to "core-corona" structure of the mPEG$_{113}$-b-P(D,L)LA$_n$ nanoparticles.

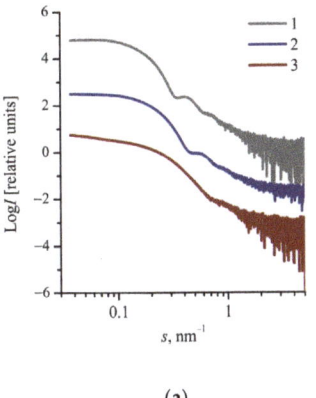 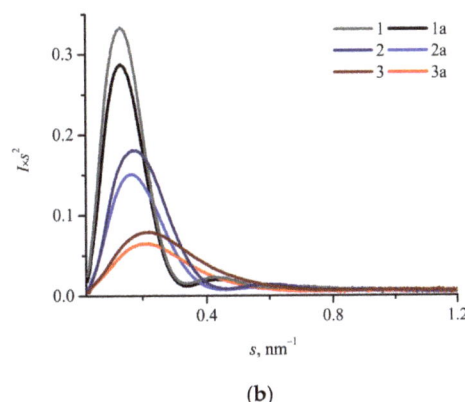

(a) (b)

Figure 4. (a) SAXS curves in Log I–Log s coordinates for oxaliplatin-loaded mPEG$_{113}$-b-P(D,L)LA$_n$ nanoparticles: 1—mPEG$_{113}$-b-P(D,L)LA$_{173}$ (c = 7.5 g/L), 2—mPEG$_{113}$-b-P(D,L)LA$_{135}$ (c = 5 g/L), 3—mPEG$_{113}$-b-P(D,L)LA$_{62}$ (c = 5 g/L). The SAXS curves are shifted vertically for clarity. (b) Kratky plots for oxaliplatin-loaded (1, 2, 3) and drug-free (1a, 2a, 3a) mPEG$_{113}$-b-P(D,L)LA$_n$ nanoparticles: 1—mPEG$_{113}$-b-P(D,L)LA$_{173}$ (c = 7.5 g/L), 2—mPEG$_{113}$-b-P(D,L)LA$_{135}$ (c = 5 g/L), 3,—mPEG$_{113}$-b-P(D,L)LA$_{62}$ (c = 5 g/L).

The values of R and $D_{max}/2$ corresponding to the peak position and the maximum dimension evaluated from $P(R)$ functions are listed in Table 2. It was found that an increase in the P(D,L)LA block length leads to higher R and $D_{max}/2$ values. R_g values evaluated from $P(R)$ functions were compared with $D_{max}/2$ and R values. The $2R_g/D_{max}$ values were found to be in the range of 0.57–0.64, whereas R_g/R was in the range of 0.81–0.96 (Table 2). It should be noted that for spherical particles with constant density $R_g/R = 0.78$. Thus, one can suppose that the mPEG$_{113}$-b-P(D,L)LA$_n$ nanoparticles have higher electron density in the inner part than that in the outer part, i.e., a "core-corona" structure [37,48,51]. It should be noted that the PLA block makes the main contribution to the X-ray scattering due to its higher electron density in comparison with less dense PEG corona. The values of $D_{max}/2$ and R are smaller than the values of R_h or evaluated by DLS (Table 2), which can be explained by higher contribution of the corona.

One can see in Table 2 that the radius of nanoparticles determined both by DSL and SAXS increases with an increase in length of amorphous P(D,L)LA block, which is in accordance with the literature [31,52,53]. Enhanced hydrophobic interactions result in higher aggregation number of the copolymer chains into a nanoparticle leading to an increase in hydrodynamic size [52]. Variation of the polymerization degree of P(D,L)LA block allows us to effectively tune the size of nanoparticles that could be important in optimizing the drug incorporation into these potential drug delivery carriers.

According to the literature the size of obtained mPEG$_{113}$-b-P(D,L)LA$_n$ nanoparticles, which is less than 100 nm, could provide delivery of the loaded anticancer agent to tumor in passive targeting manner [30], achieve high tumor extravasation efficacy, and deep tumor penetration of particles regardless of the tumor type [54].

2.3. Characterization of Oxaliplatin-Loaded mPEG-b-(D,L)LA Nanoparticles

The values of oxaliplatin loading content in mPEG$_{113}$-b-P(D,L)LA$_n$ nanoparticles after removal of free drug are listed in Table 3, which demonstrates that the length of hydrophobic PLA block in mPEG$_{113}$-b-P(D,L)LA$_n$ copolymers affects the drug loading. The highest loading content of oxaliplatin was evaluated as 3.8 wt.% for nanoparticles of mPEG$_{113}$-b-P(D,L)LA$_{62}$ copolymer with the shortest PLA block. Due to low hydrophobicity of oxaliplatin, its incorporation in the hydrophobic PLA core is unfavorable [12]. We suppose that the anticancer agent could be adsorbed at the core-corona interface of the mPEG$_{113}$-b-P(D,L)LA$_n$ nanoparticles due to the highest density of hydrophilic PEG chains there. The size of the nanoparticles decreases with a decrease of PLA block length (Table 3). The smaller size of the mPEG$_{113}$-b-P(D,L)LA$_n$ nanoparticles results in enhanced core-corona interface leading to an increase in oxaliplatin loading content.

Table 3. Characteristics of oxaliplatin-loaded mPEG$_{113}$-b-P(D,L)LA$_n$ nanoparticles in aqueous suspensions.

Sample	$R_h{}^1$, nm	$R_g{}^2$, nm	$R_g{}^3$, nm	$D_{max}/2^4$, nm	R^5, nm	$2R_g/D_{max}$	R_g/R	ζ^6, mV	DLC^7, wt.%
mPEG$_{113}$-b-P(D,L)LA$_{173}$	27 ± 10	11.1 ± 0.1	12.3 ± 0.1	19.5 ± 1.0	15.3 ± 1.0	0.63 ± 0.05	0.80 ± 0.07	−16 ± 9	1.5
mPEG$_{113}$-b-P(D,L)LA$_{135}$	25 ± 12	9.4 ± 0.1	10.4 ± 0.1	17.0 ± 1.0	12.2 ± 1.0	0.61 ± 0.06	0.85 ± 0.08	−17 ± 4	2.3
mPEG$_{113}$-b-P(D,L)LA$_{62}$	19 ± 8	9.3 ± 0.1	8.7 ± 0.1	14.0 ± 1.0	9.2 ± 1.0	0.62 ± 0.07	0.95 ± 0.11	−24 ± 8	3.8

^{1} The value of R_h corresponding to the peak position on DLS intensity size distribution curve. ^{2} Gyration radius of nanoparticles calculated from Guinier plots. ^{3} Gyration radius of nanoparticles evaluated from $P(R)$. ^{4} Radius of nanoparticles evaluated from $P(R)$. ^{5} The value of R corresponding to the maximum position on $P(R)$ function. ^{6} ζ-potential of nanoparticles. ^{7} Drug loading content determined by ICP-AES.

In order to investigate the influence of oxaliplatin loading on size, morphology, and structure of the mPEG$_{113}$-b-P(D,L)LA$_n$ nanoparticles, DLS, TEM, and SAXS were used. Intensity size distribution curves (DLS) for drug-loaded mPEG$_{113}$-b-P(D,L)LA$_{173}$ and mPEG$_{113}$-b-P(D,L)LA$_{135}$ nanoparticles reveal one peak, except for nanoparticles based on mPEG$_{113}$-b-P(D,L)LA$_{62}$ copolymer with the shortest P(D,L)LA block (Figure S2 of the Supplementary Materials). These two peaks could be attributed to individual nanoparticles with small R_h (< 100 nm) and their submicron size aggregates. It should be noted that a fraction of small particles could be undetectable in a mixture with several percent of large particles [47]. In the case of mPEG$_{113}$-b-P(D,L)LA$_{62}$, the scattering intensity of individual nanoparticles is high enough to estimate their size (Figure S2 of the Supplementary Materials). Therefore, we suppose that the investigated suspension mainly consists of individual oxaliplatin-loaded nanoparticles with a minor fraction of their aggregates.

The values of R_h of oxaliplatin-loaded mPEG$_{113}$-b-P(D,L)LA$_n$ nanoparticles are listed in Table 3. The R_h values of drug-loaded nanoparticles are almost identical (in the limits of experimental uncertainty) to the values of R_h of drug-free nanoparticles (Table 2). Thus, we assume that oxaliplatin loading does not affect the size of the mPEG$_{113}$-b-P(D,L)LA$_n$ nanoparticles.

ζ-Potential of oxaliplatin-loaded mPEG$_{113}$-b-P(D,L)LA$_n$ nanoparticles was found to be in the range from −16 to −24 mV (Table 3). The values of ζ-potential of drug-loaded nanoparticles are nearly the same as the values for drug-free nanoparticles (Table 2).

The morphology of mPEG$_{113}$-b-P(D,L)LA$_n$ nanoparticles remained unchanged with oxaliplatin loading (data not shown).

Scattering curves from oxaliplatin-loaded mPEG$_{113}$-b-P(D,L)LA$_n$ nanoparticles in Log I–Log s coordinates are presented in Figure 4a. All the SAXS profiles converge to a plateau in the region of $s < 0.1$ nm^{-1}, indicating the absence of large aggregates. The profiles also show the secondary maximum in the region $0.3 < s < 1$ nm^{-1} (Figure 4a), suggesting that the nanoparticles have a spherical well-defined structure with a relatively narrow size distribution. As one can see from Figures 3a and 4a, the drug loading does not affect the shape of SAXS curves. Thus, we suggest that oxaliplatin loading does not affect the structure and size of the mPEG$_{113}$-b-P(D,L)LA$_n$ nanoparticles.

Figure 4b shows the SAXS curves for both oxaliplatin-loaded and drug-free mPEG$_{113}$-b-P(D,L)LA$_n$ nanoparticles in $I \cdot s^2$–s coordinates. As one can see from the Kratky plots (Figure 4b), drug loading leads to an increase of the scattering intensity $I(s)$ from the mPEG$_{113}$-b-P(D,L)LA$_n$ nanoparticles. This increment could be attributed to a higher average electron density of nanoparticles with incorporated oxaliplatin compared with drug-free nanoparticles.

The R_g values of oxaliplatin-loaded mPEG$_{113}$-b-P(D,L)LA$_n$ nanoparticles determined from the Guinier plots (Figure S3 of the Supplementary Materials) are listed in Table 3. It was found that the R_g values of drug-loaded nanoparticles are almost identical (in the limits of experimental uncertainty) to the values of R_g of drug-free nanoparticles (Table 2).

The pair distance distribution functions $P(R)$ for oxaliplatin-loaded mPEG$_{113}$-b-P(D,L)LA$_n$ nanoparticles are bell-shaped with a well-defined maximum shifted to distances smaller than $D_{max}/2$ (Figure S4 of the Supplementary Materials). We assume that the shift of the $P(R)$ function maximum can be attributed to "core-corona" structure of oxaliplatin-loaded mPEG$_{113}$-b-P(D,L)LA$_n$ nanoparticles. The values of R and $D_{max}/2$ corresponding to the peak value and the maximum dimension of the scattering objects are listed in Table 3. R_g values evaluated from $P(R)$ functions were compared with $D_{max}/2$ and R values. The $2R_g/D_{max}$ values were found to be in the range of 0.61–0.63, whereas R_g/R was in the range of 0.80–0.95 (Table 3). Thus, one can suppose that oxaliplatin-loaded mPEG$_{113}$-b-P(D,L)LA$_n$ nanoparticles also have higher electron density in the inner part than that in the outer part, i.e., a "core-corona" structure [37,48,51].

Based on SAXS data, we estimated the core-corona interface area of mPEG$_{113}$-b-P(D,L)LA$_n$ nanoparticles s_{int} (Table 4). Nanoparticles of mPEG$_{113}$-b-P(D,L)LA$_{62}$ copolymer with the shortest PLA block have the largest core-corona interface area available for oxaliplatin adsorption (Table 4). In addition, as one can see from Table 4, the value of tethering density of hydrophilic PEG chains σ on the surface of PLA core is the highest for the mPEG$_{113}$-b-P(D,L)LA$_{62}$ particles (detailed description of the calculation can be found in the Supplementary Materials) that is also favorable for oxaliplatin encapsulation. Therefore, mPEG$_{113}$-b-P(D,L)LA$_{62}$ particles show the highest values of drug loading content and encapsulation efficacy (Table 4). Based on SAXS data, we suppose that oxaliplatin loading does not affect the size and structure of the mPEG$_{113}$-b-P(D,L)LA$_n$ nanoparticles (Tables 2 and 3). It seems that the amount of loaded drug is insufficient to cause significant changes in these parameters. However, a higher scattering intensity $I(s)$ observed for drug-loaded nanoparticles could be attributed to an increase of average electron density of the particles due to oxaliplatin incorporation. Higher initial oxaliplatin loading can lead to a larger drug content in nanoparticles and make it distinguishable on the SAXS curves, which will allow us to investigate the localization of drug in carrier.

Table 4. Parameters of oxaliplatin-loaded mPEG$_{113}$-b-P(D,L)LA$_n$ nanoparticles.

Sample	R_c[1], nm	s_{int}[2], nm^2/g	σ[3], nm^{-2}	DLC[4], wt.%	EE[5], %
mPEG$_{113}$-b-P(D,L)LA$_{173}$	15.3	1.5 × 10^{20}	1.0	1.5	30
mPEG$_{113}$-b-P(D,L)LA$_{135}$	12.2	2.0 × 10^{20}	0.9	2.3	46
mPEG$_{113}$-b-P(D,L)LA$_{62}$	9.2	2.7 × 10^{20}	1.6	3.8	76

[1] The value of PLA core radius evaluated from SAXS (equivalent to the maximum position R on the pair distance distribution function $P(R)$). [2] The value of core-corona interface area s_{int}. [3] The value of tethering density of PEG chains σ on the PLA core surface. [4] The value of oxaliplatin loading content. [5] The value of encapsulation efficacy.

Thus, in the present work, we showed that the highest oxaliplatin loading content at the core-corona interface of mPEG$_{113}$-b-P(D,L)LA$_n$ nanoparticles through its physical adsorption was 3.8 wt.% with initial drug loading content 5 wt.%. The obtained value is sufficiently higher compared to the value of oxaliplatin loading content in PEG-b-PLA nanoparticles reported in ref. [15], i.e., 0.053 wt.%. It should be noted that generally, physical encapsulation of oxaliplatin into polymeric carriers results in lower values of oxaliplatin loading content (3–5 wt.%) [13,26] compared to its content (tens of percent) [14,20–24] after chemical conjugation of oxaliplatin active part with polymer chains. Nevertheless, oxaliplatin-loaded chitosan based polymeric particles showed enhanced cytotoxicity activity against cancer cells compared to free oxaliplatin [26]. Moreover, chitosan-based polymeric micelles and PEG-b-PLA nanoparticles with incorporated oxaliplatin are able to selectively accumulate in tumors [15,26].

One of the strategies to enhance the encapsulation efficacy of platinum drugs into polymeric carriers is variation of their lipophilicity. Margiotta et al. investigated the effect of the carboxylate ligand chain length on the encapsulation efficacy of Pt(IV) prodrug complex in poly(lactide-co-glycolide)-PEG PLGA-PEG nanoparticles [55]. The authors reported that the amount of Pt atoms encapsulated in PLGA-PEG nanoparticles increased approximately 2 times with an increase of the length of the carboxylate ligand chain from 2 to 10 carbon atoms. In the present work, we studied the effect of mPEG$_{113}$-b-P(D,L)LA$_n$ copolymers composition on the encapsulation efficacy of oxaliplatin. Based on the obtained results, we suggest that variation of the ratio of hydrophobic and hydrophilic block lengths in mPEG$_{113}$-b-P(D,L)LA$_n$ copolymers can be another useful strategy to enhance Pt-complex loading content in polymeric nanoparticles without chemical modification of the drug (Table 4).

3. Materials and Methods

3.1. Materials

D,L-lactide (3,6-dimethyl-1,4-dioxane-2,5-dione, 99%) was purchased from Corbion (Netherlands) and recrystallized in butyl acetate before use. Poly(ethylene glycol) methyl ether (mPEG) with a molecular weight of 5000 Da, stannous (II) 2-ethylhexanoate (SnOct$_2$) were purchased from Sigma-Aldrich and used as received. All organic solvents were of analytical grade and used without further purification. Double distilled water was used for all experiments. Oxaliplatin (trans-R,R-cyclohexane-1,2-diamine)oxalatoplatinum(II)) was synthesized using a procedure described in literature [56].

3.2. Synthesis of Block Copolymers

mPEG$_{113}$-b-P(D,L)LA$_n$ diblock copolymers were synthesized by ring-opening polymerization of D,L-lactide in the presence of mPEG. Stannous (II) 2-ethylhexanoate (0.14% wt./wt. with respect to the amount of lactide) was used as a catalyst. By varying the ratio of lactide to mPEG in the reaction, it was possible to control the polymerization degree of PLA block. Before polymerization, the reactants were dried under vacuum for 30 min at room temperature.

For example, for synthesis of mPEG$_{113}$-b-P(D,L)LA$_{62}$ block-copolymer, D,L-lactide (5.04 g, 35 mmol), mPEG$_{113}$ (5 g, 1 mmol) were placed into a dried polymerization flask equipped with a magnetic stirrer. Then, 0.017 mL of 1M hexane solution of stannous (II) 2-ethylhexanoate was poured into the flask. Hexane was removed by evaporation in vacuum. The reaction flask was closed with a glass stopper and immersed into an oil bath preheated to 140 °C. The polymerization was carried out for 24 h under argon atmosphere with constant stirring. The reaction product was cooled to room temperature and dissolved in tetrahydrofuran (10 mL) and precipitated twice, first using excess of cold (+5 °C) diethyl ether and then cold hexane (200 mL). The synthesis of block copolymers with different composition was carried out analogously.

3.3. Characterization of mPEG-b-P(D,L)LA Copolymers

The degree of lactide conversion, PLA block length and number-average molecular weight of the synthesized polymers were determined by ^{1}H NMR. Spectra were recorded on a 300 MHz Bruker WP-250 SY spectrometer in 5 mm o.d. sample tubes. For measurements 30 mg of block copolymer was dissolved in 1 mL of CDCl$_3$. The integrals of the peaks corresponding to the PLA methine protons (-CH, 5.15 ppm) and PEG methylene protons (-CH$_2$-, 3.65 ppm) were used to calculate PLA block length and number average molecular weight (M_n) of the synthesized block copolymers (Figure S5 of the Supplementary Materials). The degree of conversion was found to be 90–95% for all the synthesized polymers. It was calculated using the integrals of the peaks corresponding to the unreacted lactide and PLA methine protons. The residual monomer was successfully removed by precipitation, as was confirmed by the absence of the corresponding signal (around 4.97–5.05 ppm) on the ^{1}H NMR spectrum (Figure S5 of the Supplementary Materials). Thus, the purity of synthesized polymers is not less than 99%.

Gel permeation chromatography (GPC) was performed to determine molecular weight and polydispersity index of the synthesized block copolymers (Figure S6 of the Supplementary Materials). Chromatograms were recorded on a Knauer system consisting of a pump, a refractometric detector, and Phenogel 5 µm 103 Å column. The sample concentration was 5 g/L, THF was used as the mobile phase (40 °C and 1 mL/min), and the column calibration was performed with polystyrene standards (Polymer Laboratories).

3.4. Preparation of mPEG-b-P(D,L)LA Nanoparticles

Drug-free and drug-loaded nanoparticles based on mPEG-*b*-P(D,L)LA copolymers were prepared by solvent displacement (nanoprecipitation) method with acetone as organic solvent [57]. Briefly, mPEG-*b*-P(D,L)LA (50 mg) was dissolved in acetone (10 mL). Double distilled water (10 mL) was added dropwise into the solution under stirring. The organic solvent was removed through evaporation for 4 h at room temperature.

To prepare oxaliplatin-loaded mPEG-*b*-P(D,L)LA nanoparticles, 2.5 mg of oxaliplatin (5% wt./wt. with respect to the amount of block copolymer) was preliminarily dissolved in water. Then, the nanoparticles were prepared similarly to the drug-free mPEG-*b*-P(D,L)LA nanoparticles. Finally, the aqueous suspensions were centrifuged (40000 g, 30 min) to remove the residues of the organic solvent and free drug, and the precipitated nanoparticles were dispersed in water and freeze-dried.

3.5. Characterization of mPEG-b-P(D,L)LA Nanoparticles

Measurements of size distribution and zeta-potential (ζ-potential) were performed by dynamic light scattering (DLS) on a Zetasizer Nano ZS instrument (Malvern Ltd.), equipped with a He-Ne laser with a wavelength of 633 nm at a scattering angle of 173°. All of the experiments were carried out three times. The data are presented as mean ± standard deviation.

The morphology of nanoparticles was characterized by transmission electron microscopy (TEM) using a Titan 80–300 TEM/STEM (FEI) microscope at accelerating voltage of 300 kV with a BM-Ultrascan (Gatan) camera operating in the bright field mode. Thin-carbon-film-coated copper TEM grids were glow-discharged for 10 s in the Pelco easiGlow system. A 3 µL droplet of the aqueous suspensions with concentration of 0.5 g/L was deposited on the carbon side of the grid and incubated for 1 min. Then, the carbon side of the grid was rinsed with 10 µL of distilled water, and right after that, 10 µL of uranyl acetate solution with a concentration of 1 wt.% was applied to the grid and incubated for 30 s. The excess of the solution was removed after each step by touching the grid edge with filter paper.

Synchrotron SAXS measurements of the aqueous suspensions of nanoparticles were performed at the European Molecular Biology Laboratory (EMBL) on the storage ring PETRA III (DESY, Hamburg) on the EMBL-P12 beamline equipped with a 2D photon counting pixel X-ray detector Pilatus 2 M (Dectris). The scattering intensity, $I(s)$, was recorded in the range of the momentum transfer $0.02 < s < 4.5$ nm^{-1}, where

$$s = (4\pi \sin \theta)/\lambda \quad (1)$$

2θ is the scattering angle, and $\lambda = 0.124$ nm is the X-ray wavelength. The measurements were carried out at 23 °C using continuous flow operation over a total exposure time of 1 s divided into 20 × 50 ms individual frames to monitor for potential radiation damage (no radiation effects were detected). For each sample, 20 scattering curves were captured to improve the quality of the obtained data. The data were corrected for the solvent scattering and processed using standard procedures with the program PRIMUS [58]. Data analysis was performed using the software suite ATSAS [59]. Pair distance distribution and volume size distribution functions were calculated using the program GNOM [60]. Before SAXS measurements, all suspensions were centrifuged (10000 rpm, 10 min).

3.6. Evaluation of Drug Loading

The content of oxaliplatin (weight ratio of the drug to the block copolymer) in the freeze-dried nanoparticles was determined by inductively coupled plasma atomic emission spectroscopy (ICP-AES). An axial ICP-AES 720-ES spectrometer (Agilent Technologies, USA) was used for measurements with a low flow axial quartz torch with 2.4 mm inner diameter injector tube (Glass Expansion, Australia), a double-pass glass cyclonic spray chamber (Agilent Technologies), a OneNeb nebulizer (Agilent Technologies, USA), and a Trident Internal Standard Kit (Glass Expansion). Samples were introduced manually to reduce washing volume, without preliminary digestion or dilution. A detailed description of the measurement process can be found in the Supplementary Materials. The drug loading content DLC and encapsulation efficacy EE of oxaliplatin-loaded mPEG-b-P(D,L)LA nanoparticles were calculated according to the following equations:

$$DLC = \frac{m_1^{OxPt}}{m_{NP}} \times 100\% \tag{2}$$

$$EE = \frac{m_1^{OxPt}}{m_0^{OxPt}} \times 100\% \tag{3}$$

where m_1^{OxPt} is amount of incorporated oxaliplatin in nanoparticles, m_{NP} is amount of nanoparticles, and m_0^{OxPt} is initial amount of oxaliplatin.

4. Conclusions

Drug-free and oxaliplatin-loaded biodegradable $mPEG_{113}$-b-P(D,L)LA$_n$ nanoparticles were prepared by a simple nanoprecipitation technique. The influence of hydrophobic block length on the structure, size, morphology, and drug loading content of $mPEG_{113}$-b-P(D,L)LA$_n$ nanoparticles was investigated. It was observed that in aqueous solution $mPEG_{113}$-b-P(D,L)LA$_n$ copolymers, where $n = 62$–173 monomer units, form spherical nanoparticles with hydrodynamic diameters ranging from 32 to 56 nm. The "core-corona" structure of the block copolymer nanoparticles was confirmed by SAXS. Tailoring of P(D,L)LA block length results in variation in both core-corona interface area and tethering density of hydrophilic PEG chains on the surface of P(D,L)LA core of the $mPEG_{113}$-b-P(D,L)LA$_n$ nanoparticles, which affects oxaliplatin loading content. An increase in P(D,L)LA block length from 62 to 173 monomer units results in a decrease in core-corona interface area from 2.7×10^{20} to 1.5×10^{20} nm^2/g and tethering density of PEG chains from 1.6 to 1.0 nm^{-2} and a reduction in the oxaliplatin loading content from 3.8 to 1.5% wt./wt. Thus, we suppose that oxaliplatin is adsorbed on the core-corona interface of the $mPEG_{113}$-b-P(D,L)LA$_n$ nanoparticles. SAXS measurements revealed that oxaliplatin loading does not affect the size and structure of the block copolymer nanoparticles.

The size and structure of polymeric nanoparticles are crucial characteristics that should be considered in the design of targeted nanoformulations of anticancer agents. The developed oxaliplatin formulation based on 32 nm $mPEG_{113}$-b-P(D,L)LA$_{62}$ nanoparticles loaded with 3.8 wt.% of drug with 76% encapsulation efficiency can be considered as a promising candidate for treatment of various types of cancer. In vitro and in vivo tests will be performed in order to compare its efficacy and toxicological profile with pure oxaliplatin.

Supplementary Materials: The following are available online, Figure S1: Guinier plots for the $mPEG_{113}$-b-P(D,L)LA$_n$ nanoparticles, Figure S2: DLS intensity size distribution curves for oxaliplatin-loaded $mPEG_{113}$-b-P(D,L)LA$_n$ nanoparticles, Figure S3: Guinier plots for oxaliplatin-loaded $mPEG_{113}$-b-P(D,L)LA$_n$ nanoparticles, Figure S4: Pair distance distribution functions $P(R)$–R for oxaliplatin-loaded $mPEG_{113}$-b-P(D,L)LA$_n$ nanoparticles, Figure S5: Typical ^{1}H NMR spectrum of $mPEG_{113}$-b-P(D,L)LA$_n$ (300 MHz, CDCl$_3$), Figure S6: GPC curves for synthesized $mPEG_{113}$-b-P(D,L)LA$_n$ copolymers, Table S1: Conditions of ICP-AEC measurements.

Author Contributions: Conceptualization, N.G.S. and S.N.C.; validation, Y.A.K., A.A.P., and D.S.V.; formal analysis, E.V.R., E.V.S., and A.A.N.; investigation, Y.A.K., A.A.P., E.V.R., and D.S.V.; writing—original draft preparation, E.V.R. and Y.A.K.; writing—review and editing, D.R.S., N.G.S., and A.I.K.; visualization, E.V.R. and Y.A.K.; supervision, N.G.S. and S.N.C.; project administration, N.G.S. All authors have read and agreed to the published version of the manuscript.

Funding: This research was funded by the Russian Science Foundation, grant number 18-73-10079.

Institutional Review Board Statement: Not applicable.

Informed Consent Statement: Not applicable.

Data Availability Statement: The data presented in this study are available on request from the corresponding author.

Acknowledgments: This work was carried out using the equipment of the resource centers of the National Research Center "Kurchatov Institute". Authors are grateful to European Molecular Biology Laboratory (EMBL) for SAXS experiments on the storage ring PETRA III (DESY, Hamburg).

Conflicts of Interest: The authors declare no conflict of interest.

Sample Availability: Samples of the compounds are not available from the authors.

References

1. Majumder, N.; Das, G.N. Polymeric micelles for anticancer drug delivery. *Ther. Deliv.* **2020**, *11*, 613–635. [CrossRef] [PubMed]
2. Froiio, F.; Lammari, N. Chapter 16—Polymer-based nanocontainers for drug delivery. In *Smart Nanocontainers*, 1st ed.; Nguyen-Tri, P., Do, T., Eds.; Elsevier Inc.: Amsterdam, The Netherlands, 2019; pp. 271–285.
3. Li, Z.; Tan, S. Cancer drug delivery in the nano era: An overview and perspectives (Review). *Oncol. Rep.* **2017**, *38*, 611–624. [CrossRef] [PubMed]
4. Cabral, H.; Miyata, K. Block copolymer micelles in nanomedicine applications. *Chem. Rev.* **2018**, *118*, 6844–6892. [CrossRef] [PubMed]
5. Khoury, A.; Deo, K.M. Recent advances in platinum-based chemotherapeutics that exhibit inhibitory and targeted mechanisms of action. *J. Inorg. Biochem.* **2020**, *207*, 111070. [CrossRef] [PubMed]
6. Ghosh, S. Cisplatin: The first metal based anticancer drug. *Bioorg. Chem.* **2019**, *88*, 102925. [CrossRef] [PubMed]
7. Farooq, M.A.; Aquib, M. Recent progress in nanotechnology-based novel drug delivery systems in designing of cisplatin for cancer therapy: An overview. *Artif. Cells Nanomed. Biotechnol.* **2019**, *47*, 1674–1692. [CrossRef] [PubMed]
8. Hang, Z.; Cooper, M.A. Platinum-based anticancer drugs encapsulated liposome and polymeric micelle formulation in clinical trials. *Biochem. Compd.* **2016**, *4*. [CrossRef]
9. Browning, R.J.; Reardon, P.J.T. Drug delivery strategies for platinum-based chemotherapy. *ACS Nano* **2017**, *11*, 8560–8578. [CrossRef]
10. Brown, S.D.; Nativo, P. Gold nanoparticles for the improved anticancer drug delivery of the active component of oxaliplatin. *J. Am. Chem. Soc.* **2010**, *132*, 4678–4684. [CrossRef]
11. Ho, M.N.; Bach, L.G. PEGylated PAMAM dendrimers loading oxaliplatin with prolonged release and high payload without burst effect. *Biopolymers* **2019**, *110*, e23272. [CrossRef]
12. Oberoi, H.S.; Nukolova, N.V. Nanocarriers for delivery of platinum anticancer drugs. *Adv. Drug Deliv. Rev.* **2013**, *65*, 1667–1685. [CrossRef] [PubMed]
13. Cui, Z.; Sun, Y. Preparation and evaluations in vitro of oxaliplatin polylactic acid nanoparticles. *Artif. Cells Nanomed. Biotechnol.* **2013**, *41*, 227–231. [CrossRef] [PubMed]
14. Oberoi, H.S.; Nukolova, N.V. Preparation and in vivo evaluation of dichloro(1,2-Diaminocyclohexane)platinum(II)-loaded core cross-linked polymer micelles. *Chemother. Res. Pract.* **2012**, 1–10. [CrossRef]
15. Wei, H.; Xu, S. Preliminary pharmacokinetics of PEGylated oxaliplatin polylactic acid nanoparticles in rabbits and tumor-bearing mice. *Artif. Cells Nanomed. Biotechnol.* **2015**, *43*, 258–262. [CrossRef] [PubMed]
16. Ren, Y.; Li, X. Improved anti-colorectal carcinomatosis effect of tannic acid co-loaded with oxaliplatin in nanoparticles encapsulated in thermosensitive hydrogel. *Eur. J. Pharm. Sci.* **2019**, *128*, 279–289. [CrossRef]
17. Stathopoulos, G.P.; Boulikas, T. Liposomal oxaliplatin in the treatment of advanced cancer: A phase I study. *Anticancer Res.* **2006**, *26*, 1489–1893.
18. Senzer, N.N.; Matsuno, K. Abstract C36: MBP-426, a novel liposome-encapsulated oxaliplatin, in combination with 5-FU/leucovorin (LV): Phase I results of a Phase I/II study in gastro-esophageal adenocarcinoma, with pharmacokinetics. *Mol. Cancer Ther.* **2009**, *8*, C36. [CrossRef]
19. Perez-Soler, R. Liposomes as carriers of antitumor agents: Toward a clinical reality. *Cancer Treat. Rev.* **1989**, *16*, 67–82. [CrossRef]
20. Cabral, H.; Nishiyama, N. Preparation and biological properties of dichloro(1,2-diaminocyclohexane)platinum(II) (DACHPt)-loaded polymeric micelles. *J. Control. Release* **2005**, *101*, 223–232. [CrossRef]
21. Cabral, H.; Nishiyama, N. Optimization of (1,2-diamino-cyclohexane)platinum(II)-loaded polymeric micelles directed to improved tumor targeting and enhanced antitumor activity. *J. Control. Release* **2007**, *121*, 146–155. [CrossRef]
22. Liu, G.; Gao, H. DACHPt-loaded unimolecular micelles based on hydrophilic dendritic block copolymers for enhanced therapy of lung cancer. *ACS Appl. Mater. Interfaces* **2017**, *9*, 112–119. [CrossRef]
23. Xiao, H.; Zhou, D. A complex of cyclohaxane-1,2-diaminoplatinum with an amphiphilic biodegradable polymer with pendant carboxyl groups. *Acta Biomater.* **2012**, *8*, 1859–1868. [CrossRef] [PubMed]

24. Wang, R.; Hu, X. Biological characterization of folate-decorated biodegradable polymer-platinum(II) complex micelles. *Mol. Pharm.* **2012**, *9*, 3200–3208. [CrossRef] [PubMed]
25. Kim, J.; Pramanick, S. Polymeric biomaterials for the delivery of platinum-based anticancer drugs. *Biomater. Sci.* **2015**, *3*, 1002–1017. [CrossRef]
26. Xu, Y.-Y.; Du, Y.-Z. Improved cytotoxicity and multidrug resistance reversal of chitosan based polymeric micelles encapsulating oxaliplatin. *J. Drug Target.* **2011**, *19*, 344–353. [CrossRef] [PubMed]
27. Garofalo, C.; Capuano, G. Different insight into amphiphilic PEG-PLA copolymers: Influence of macromolecular architecture on the micelle formation and cellular uptake. *Biomacromolecules* **2014**, *15*, 403–415. [CrossRef]
28. Jelonek, K.; Li, S. Self-assembled filomicelles prepared from polylactide/poly(ethylene glycol) block copolymers for anticancer drug delivery. *Int. J. Pharm.* **2015**, *485*, 357–364. [CrossRef]
29. Kapse, A.; Anup, N. Chapter 6—Polymeric micelles: A ray of hope among new drug delivery systems. In *Drug Delivery Systems*, 1st ed.; Tekade, R.K., Ed.; Elsevier Inc.: Amsterdam, The Netherlands, 2019; pp. 235–289. [CrossRef]
30. Letchford, K.; Burt, H. A review of the formation and classification of amphiphilic block copolymer nanoparticulate structures: Micelles, nanospheres, nanocapsules and polymersomes. *Eur. J. Pharm. Biopharm.* **2007**, *65*, 259–269. [CrossRef]
31. Govender, T.; Riley, T. Defining the drug incorporation properties of PLA-PEG nanoparticles. *Int. J. Pharm.* **2000**, *199*, 95–110. [CrossRef]
32. Dong, Y.; Feng, S.-S. Methoxy poly(ethylene glycol)-poly(lactide) (MPEG-PLA) nanoparticles for controlled delivery of anticancer drugs. *Biomaterials* **2004**, *25*, 2843–2849. [CrossRef]
33. Dionzou, M.; Morère, A. Comparison of methods for the fabrication and the characterization of polymer self-assemblies: What are the important parameters? *Soft Matter.* **2016**, *12*, 2166–2176. [CrossRef] [PubMed]
34. Szymusiak, M.; Kalkowski, J. Core-shell structure and aggregation number of micelles composed of amphiphilic block copolymers and amphiphilic heterografted polymer brushes determined by small-angle X-ray scattering. *ACS Macro Lett.* **2017**, *6*, 1005–1012. [CrossRef] [PubMed]
35. Kelley, E.G.; Murphy, R.P. Size evolution of highly amphiphilic macromolecular solution assemblies via a distinct bimodal pathway. *Nat. Commun.* **2014**, *5*, 3599. [CrossRef] [PubMed]
36. Riley, T.; Heald, C.R. Core-shell structure of PLA-PEG nanoparticles used for drug delivery. *Langmuir* **2003**, *19*, 8428–8435. [CrossRef]
37. Ma, C.; Pan, P. Core-shell structure, biodegradation, and drug release behavior of poly(lactic acid)/poly(ethylene glycol) block copolymer micelles tuned by macromolecular stereostructure. *Langmuir* **2015**, *31*, 1527–1536. [CrossRef] [PubMed]
38. Razuvaeva, E.V.; Kulebyakina, A.I. Effect of composition and molecular structure of poly(L-lactic acid)/poly(ethylene oxide) block copolymers on micellar morphology in aqueous solution. *Langmuir* **2018**, *34*, 15470–15482. [CrossRef] [PubMed]
39. Tyrrell, Z.L.; Shen, Y. Fabrication of micellar nanoparticles for drug delivery through the self-assembly of block copolymers. *Prog. Polym. Sci.* **2010**, *35*, 1128–1143. [CrossRef]
40. Liu, J.; Xiao, Y. Polymer-drug compatibility: A guide to the development of delivery systems for the anticancer agent, ellipticine. *J. Pharm. Sci.* **2004**, *93*, 132–143. [CrossRef]
41. Lin, W.-J.; Juang, L.-W. Stability and release performance of a series of pegylated copolymeric micelles. *Pharm. Res.* **2003**, *20*, 668–673. [CrossRef]
42. Abyaneh, H.S.; Vakili, M.R. Rational design of block copolymer micelles to control burst drug release at a nanoscale dimension. *Acta Biomater.* **2015**, *24*, 127–139. [CrossRef]
43. Zhang, X.; Jackson, J.K. Development of amphiphilic diblock copolymers as micellar carriers of taxol. *Int. J. Pharm.* **1996**, *132*, 195–206. [CrossRef]
44. Shuai, X.; Merdan, T. Core-cross-linked polymeric micelles as paclitaxel carriers. *Bioconjug. Chem.* **2004**, *15*, 441–448. [CrossRef] [PubMed]
45. Shuai, X.; Hua, A. Micellar carriers based on block copolymers of poly(epsilon-caprolactone) and poly(ethylene glycol) for doxorubicin delivery. *J. Control. Rel.* **2004**, *98*, 415–426. [CrossRef] [PubMed]
46. Kowalski, A.; Duda, A. Kinetics and mechanism of cyclic esters polymerization initiated with tin(II) octoate. 3. Polymerization of L,L-dilactide. *Macromolecules* **2000**, *33*, 7359–7370. [CrossRef]
47. Tomaszewska, E.; Soliwoda, K. Detection limits of DLS and UV-Vis spectroscopy in characterization of polydisperse nanoparticles colloids. *J. Nanomater.* **2013**, *2013*, 1–10. [CrossRef]
48. Akiba, I.; Terada, N. Encapsulation of a hydrophobic drug into a polymer-micelle core explored with synchrotron SAXS. *Langmuir* **2010**, *26*, 7544–7551. [CrossRef]
49. Kikhney, A.G.; Svergun, D.I. A practical guide to small angle X-ray scattering (SAXS) of flexible and intrinsically disordered proteins. *FEBS Lett.* **2015**, *589*, 2570–2577. [CrossRef]
50. Svergun, D.I.; Koch, M.H.J. Small-angle scattering studies of biological macromolecules in solution. *Rep. Prog. Phys.* **2003**, *66*, 1735–1782. [CrossRef]
51. Hussain, H.; Tan, B.H. Synthesis, micelle formation, and bulk properties of poly(ethylene glycol)-b-poly(pentafluorostyrene)-g-polyhedral oligomeric silsesquioxane amphiphilic hybrid copolymers. *J. Polym. Sci. Part A Polym. Chem.* **2010**, *48*, 152–163. [CrossRef]

52. Zhang, J.; Wang, L.-Q. Micellization phenomena of amphiphilic block copolymers based on methoxy poly(ethylene glycol) and either crystalline or amorphous poly(caprolactone-b-lactide). *Biomacromolecules* **2006**, *7*, 2492–2500. [CrossRef]
53. Theerasilp, M.; Nasongkla, N. Comparative studies of poly(ε-caprolactone) and poly(D,L-Lactide) as core materials of polymeric micelles. *J. Microencapsul.* **2013**, *30*, 390–397. [CrossRef] [PubMed]
54. Cabral, H.; Matsumoto, Y. Accumulation of sub-100 nm polymeric micelles in poorly permeable tumors depends on size. *Nat. Nanotechnol.* **2011**, *6*, 815–823. [CrossRef] [PubMed]
55. Margiotta, N.; Savino, S. Encapsulation of lipophilic kiteplatin Pt(IV) prodrugs in PLGA-PEG nanoparticles. *Dalton Trans.* **2016**, *45*, 13070–13081. [CrossRef] [PubMed]
56. Al-Allaf, T.A.; Rashan, L.J. Palladium(II) and platinum(II) complexes of (1R,2R)-(−)-1,2-diaminocyclohexane (DACH) with various carboxylato ligands and their cytotoxicity evaluation. *Appl. Organometal. Chem.* **2009**, *23*, 173–178. [CrossRef]
57. Fessi, H.; Puisieux, F. Nanocapsule formation by interfacial polymer deposition following solvent displacement. *Int. J. Pharm.* **1989**, *55*, 1–4. [CrossRef]
58. Konarev, P.V.; Volkov, V.V. PRIMUS: A Windows PC-based system for small-angle scattering data analysis. *J. Appl. Cryst.* **2003**, *36*, 1277–1282. [CrossRef]
59. Franke, D.; Petoukhov, M.V. ATSAS 2.8: A comprehensive data analysis suite for small-angle scattering from macromolecular solutions. *J. Appl. Cryst.* **2017**, *50*, 1212–1225. [CrossRef]
60. Svergun, D.I. Determination of the regularization parameter in indirect-transform methods using perceptual criteria. *J. Appl. Cryst.* **1992**, *25*, 495–503. [CrossRef]

Article

Polylactide Nanoparticles as a Biodegradable Vaccine Adjuvant: A Study on Safety, Protective Immunity and Efficacy against Human Leishmaniasis Caused by Leishmania Major

Sana Ayari-Riabi [1,*], Noureddine Ben khalaf [2,3], Balkiss Bouhaouala-Zahar [1,*], Bernard Verrier [4], Thomas Trimaille [5], Zakaria Benlasfar [6], Mehdi Chenik [3] and Mohamed Elayeb [1]

[1] NanoBioMedika Team, Laboratoire des Biomolécules, Venins, et Applications Théranostiques (LBVAT) LR20IPT01, Institut Pasteur Tunis, Université Tunis El Manar, BP 74, 13 Place Pasteur, Tunis 1002, Tunisia
[2] Life Sciences Department, College of Graduate Studies, Arabian Gulf University, Manama P.O. Box 26671, Bahrain
[3] Laboratory of Immunopathology, Vaccinology and Molecular Genetics, Institut Pasteur de Tunis, BP 74, 13 Place Pasteur, Tunis 1002, Tunisia
[4] Laboratoire de Biologie Tissulaire et d'Ingénierie Thérapeutique, Univ Lyon, CNRS, Université Claude Bernard Lyon 1, UMR 5305, 7 Passage du Vercors, CEDEX 07, 69367 Lyon, France
[5] Ingénierie des Matériaux Polymères, Univ Lyon, CNRS, Université Claude Bernard Lyon 1, INSA Lyon, Université Jean Monnet, UMR 5223, CEDEX, 69622 Villeurbanne, France
[6] Service des Unités Animalières, Institut Pasteur Tunis, BP 74, 13 Place Pasteur, Tunis 1002, Tunisia
* Correspondence: sana.riabii@gmail.com (S.A.-R.); balkiss.bouhaouala@fmt.utm.tn (B.B.-Z.)

Abstract: Leishmaniasis is the 3rd most challenging vector-borne disease after malaria and lymphatic filariasis. Currently, no vaccine candidate is approved or marketed against leishmaniasis due to difficulties in eliciting broad immune responses when using sub-unit vaccines. The aim of this work was the design of a particulate sub-unit vaccine for vaccination against leishmaniasis. The poly (D,L-lactide) nanoparticles (PLA-NPs) were developed in order to efficiently adsorb a recombinant *L. major* histone H2B (*L. major* H2B) and to boost its immunogenicity. Firstly, a study was focused on the production of well-formed nanoparticles by the nanoprecipitation method without using a surfactant and on the antigen adsorption process under mild conditions. The set-up preparation method permitted to obtain H2B-adsorbed nanoparticles H2B/PLA (adsorption capacity of about 2.8% (w/w)) with a narrow size distribution (287 nm) and a positive zeta potential (30.9 mV). Secondly, an in vitro release assay performed at 37 °C, pH 7.4, showed a continuous release of the adsorbed H2B for almost 21 days (30%) from day 7. The immune response of H2B/PLA was investigated and compared to H2B + CpG7909 as a standard adjuvant. The humoral response intensity (IgG) was substantially similar between both formulations. Interestingly, when challenged with the standard parasite strain (GLC94) isolated from a human lesion of cutaneous leishmaniasis, mice showed a significant reduction in footpad swelling compared to unvaccinated ones, and no deaths occurred until week 17th. Taken together, these results demonstrate that PLA-NPs represent a stable, cost-effective delivery system adjuvant for use in vaccination against leishmaniasis.

Keywords: PLA nanoparticles; *Leishmania major*; recombinant histone H2B; vaccine adjuvant; parasite' challenge; lesion swelling

1. Introduction

Leishmaniasis, a neglected tropical disease, is the third most challenging vector-borne disease after malaria and lymphatic filariasis. According to the World Health Organization's (WHO) published data, it has been reported that leishmaniasis is endemic to approximately 100 countries of the world [1]. Cutaneous leishmaniasis (CL) is the most common form of leishmaniasis. The annual registered cases are around 0.7–1.2 million across the globe. The disease is caused by the protozoan *Leishmania* parasite [2]. The principal vector responsible for mammal-to-human transmission is the Phlebotomine sandfly. Two distinct morphological forms of *Leishmania* have been identified in its life cycle: the promastigote is present in the vector, and the amastigote moves into the monocytes/macrophages of the human or mammalian host [3]. CL is a skin infection that reduces patients' quality of life and imparts psychological problems and social stigmatism [4].

When BALB/c mice, an inbred strain of mice, are infected with the *L. major* parasite, the maturation of a Th2 immune response is triggered by the interleukin-4 (IL-4) produced during the first two days. This IL-4 rapidly renders parasite-specific $CD4^+$ T cell precursors unresponsive to interleukin-12 (IL-12) [5].

Effective regulation requires macrophage activation and nitric oxide (NO) in response to the Th1-produced cytokine IFN-γ. Disease prognosis can be improved by overcoming problems such as low efficacy, systemic toxicity, insufficient drug within macrophages, poor antigen presentation to cells and expensive care. Vaccination may be the most appropriate strategy in this context.

Adjuvants for *Leishmania* vaccines to date are categorized into two types: (i) immunostimulatory molecules and (ii) nanoparticulate and/or delivery systems. The first ones include Bacille Calmette–Guerin (BCG) emulsified with antigens, e.g., *Leishmania amazonenzis* isolate IFLA/BR/1967/pH 8 [6], autoclaved *Leishmania major* ALM or GP63 [7,8], the monophosphoryl lipid (MPL) with LEISH-F1 or LEISH-F2 peptides [9,10], the interleukin-12 (IL-12) with pSP Leish-tec peptide [11], saponin with A2 Leish-tec [12], GLA with SMT Leish-tec [13] and CpG-ODN with KMP-11 Leish-tec [14]. However, immunostimulatory adjuvants suffer from rapid clearance and safety issues [15]. Nano-based delivery systems (NDS) applied as potential adjuvants in anti-leishmanial vaccines may be a useful alternative to conventional bacterial adjuvants and virus vectors [16]. An NDS could potentially deliver target vaccines to the site of action within the host's body, enhance immune reactions by facilitating antigens' absorption and uptake by antigen-presenting cells (APCs) [17], prevent its degradation such as peptides, proteins, or oligonucleotides [18], promote their controlled release and modulate the type of immune responses [19,20]. The particulate adjuvants, for example, liposomes, polymeric microspheres, and emulsions, have been utilized effectively to deliver *Leishmania* antigens in preclinical models of leishmaniasis, as well as other infectious diseases [21]. Liposomes (LPS) and derivatives have been widely studied. Despite the relevance of LPS, some shortcomings are also associated with these lipid-based nanocarriers, such as leakage of the entrapped moiety, oxidation, hydrolysis, and inadequate stability [22]. The polymeric particles, when used as adjuvants, could develop more potent immunogenicity against *Leishmania* antigens [23]. Several polymers, such as polylactide (PLA), polyglycolide (PGA), poly-lactide-co-glycolide (PLGA), poly-caprolactone (PCL), poly-cyanoacrylate (PCA) and natural protein polymers, such as albumin and gelatin, and polysaccharides, have been investigated as vaccine carriers. The polymer that has been explored most extensively is polylactide (PLA). The FDA has approved the use of this compound in human applications due to its biocompatibility and lack of toxicity [23,24]. Currently, PLA is commercialized as part of several nanomedicine tools. For vaccine purpose, antigens formulated with PLA-NPs induce broad and potent humoral and cellular immunity in mice, rabbit and macaque models [25,26]. Other researchers have shown that a mycobacterial antigen adsorbed on lamellar particles of lactide polymers have induced cellular immunity [27].

Another strategy to control leishmaniasis is antigen-target-specific vaccines. These are categorized into three types: (i) live *Leishmania* parasites; (ii) killed *Leishmania* or parasite fractions (first generation), and (iii) *Leishmania* recombinant proteins (second generation) or DNA molecules (third generation) [16]. DNA vaccines are of particular interest because they can effectively induce both CD8+ and CD4+ T cells and produce long-lived antigens and properly folded polypeptides [28]. However, these are still in the early phases of clinical trials. Some trials of developing first-generation vaccines have failed to provide convincing results in phase III due to standardization and safety issues, whereas others are still in earlier phases. Second-generation vaccines, e.g., leish-111f, suffer from a lack of an appropriate adjuvant [9,21]. Among them, the histone H2B protein was described as a potential candidate [7,10]. The histone protein H2B forms with H2A, H3 and H4, the major constituents of the nucleosome in the nucleus of eukaryotic cells [29]. This protein is conserved among various species of *Leishmania*, i.e., *L. major*, *L. infantum*, *L. donovani* and *L. tropica*. Previous studies investigated the immunogenicity and the protective role of the recombinant H2B protein from *L. major* [30]. H2B, in combination with CpG-ODN, confers effective protection to sensitive BALB/c mice infected with the virulent strain of the *L. major* parasite [31]. Later, Meddeb-Garnaoui et al. showed that this recombinant protein H2B induced a specific Th1-type cellular response in individuals who recovered from cutaneous leishmaniasis infection [32].

The purpose of this research was, on the one hand, to develop a particulate vaccine system based on PLA, a polymer that has received FDA approval, and on the other hand, to protect mice against *L. major* parasite infection. To do this, nanoparticle dispersion was developed using PLA polymer (PLA-NPs). Then, BL21 *Escherichia coli* strain cells were used to express the histone H2B protein, a sub-unit vaccine candidate for leishmaniasis. When animals were experimentally infected with the parasite *L. major* promastigotes, the immune protection of immunized mice was investigated and compared to CpG-ODN, a TLR agonist compound.

2. Results

2.1. Characterization of PLA NPs

The nanoparticles were obtained by the nanoprecipitation method without adding surfactant. Optimum experimental conditions were as follows: 0.2 g PLA polymer was dissolved in acetone (10 mL) and added to MilliQ water (1v/3v). A white and milky suspension was obtained after the formation of the nanoparticles. The production yield, based on recovered PLA after solvent evaporation, was in the 70–80% range. According to the DLS results, the obtained nanoparticles had a mean hydrodynamic diameter of 287.4 nm (± 10); and a polydispersity index (PDI) of 0.14 (± 0.06) (Figure 1). This indicates that NP distribution was in a quite narrow range. Zeta potential is one of the important parameters affecting the stability of nanoparticles. Their zeta potential value was -45 mV (± 5), prone to ensure colloidal stability through electrostatic repulsion.

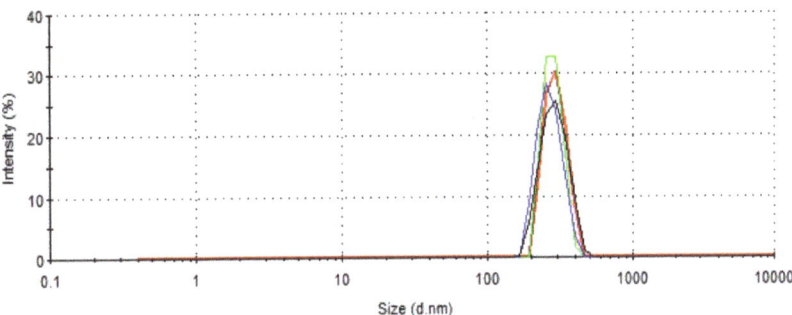

Figure 1. Size distribution of PLA-NPs using DLS technique. NPs were prepared by a nanoprecipitation process in one single step without using surfactant.

For stability analysis, nanoparticles (4 mg/mL) were kept at +4 °C for several months. No significant variation in particle size was observed until month 18.

2.2. Characterization of a Recombinant H2B Protein

Recombinant protein H2B was produced in *Escherichia coli* BL21 bacteria (*E. coli* BL21) using the pET prokaryotic expression system. Proteins were then purified by affinity chromatography over Ni-NTA resin, and purity was assessed by SDS-polyacrylamide gel electrophoresis. Staining of the gel with Coomassie blue revealed two bands of 16 kDa and

30 kDa, respectively, Figure 2. As previously described based on Western blot analysis, bands of 16 kDa and 30 kDa correspond to recombinant H2B monomeric and dimeric forms, respectively [31].

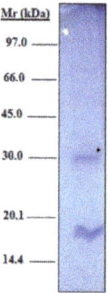

Figure 2. SDS analysis of *L. major* recombinant H2B. Protein was expressed in BL21 *Escherichia coli*, then purified by affinity chromatography over Ni-NTA resin and analyzed using SDS-PAGE (15%) followed by Coomassie blue staining. (*) indicates an additional band that corresponds to H2B dimeric form. Mr, Molecular weight markers (kDa) (GE Healthcare).

2.3. Adsorption Efficiency and Zeta Potential of H2B/PLA Particles

The characteristics of H2B-adsorbed PLA nanoparticles are reported in Figure 3. The NP concentration was fixed at 0.06 w% (i.e., 0.6 mg/mL). The adsorption process was followed by monitoring both the amount of adsorbed protein and the surface charge of H2B/PLA. The maximum adsorption was reached when an introduced protein/particle ratio ranged from 5 to 8.3 w%. Thus, the adsorption capacity was 2.8% w/w of protein particles (Figure 3a). Zeta potentials were followed as a function of introduced protein amounts under the adsorption process, as its strongly cationic character is expected to induce charge inversion (Figure 3b). Flocculation of the colloid was observed around the neutralization point (~2 w% protein/particle ratio). The colloidal stability was restored at an introduced H2B/particle ranging from 5 to 8.3% with a zeta potential of +30 mV and no longer changes, indicating the saturation of the colloid surface.

a)

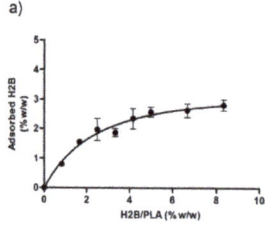

b)

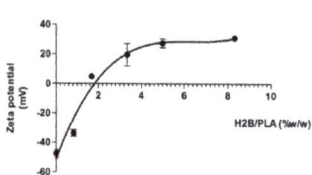

Figure 3. Adsorption curves of H2B/PLA-NPs. (**a**) Adsorption efficiency of the peptide/particles (0.06% w/v) at different protein-to-particle ratios. (**b**) Zeta potentials of H2B/PLA as a function of initial protein amount (% w/w).

In further studies, H2B-coated NPs were prepared in the presence of excess protein to avoid the flocculation process (5 w% initial protein/NP ratio) and washed from unbound protein (centrifugation/redispersion steps). The H2B/PLA exhibited a size and a zeta potential of 341 nm and +30.9 mV, respectively, Table 1.

Table 1. Main characteristics of H2B/PLA-NPs selected for animal experiments.

Adsorption Capacity % w/w	Zeta Potential mV	DLS Size nm
2.8 (±0.24)	30.95 (±0.78)	340.8 (±52)

2.4. In Vitro Desorption of Adsorbed H2B

The H2B/PLA-NPs were incubated in PBS buffer (pH 7.4) at 37 °C. The amount of protein in each supernatant was quantified at a fixed time. As shown in Figure 4, protein release followed a specific profile over time. No protein was detected in the supernatant collected during the first three days until day 7. Continuous H2B release was quantified from day 7 to day 21 with a cumulative percentage value of 20%. After that, the release continues to slow down until the end of the test and the maximum percentage value of 30% was obtained (Figure 4).

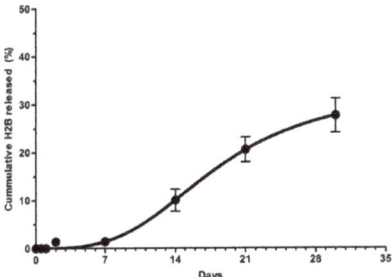

Figure 4. Release of adsorbed H2B peptide over time. A known amount of H2B/PLA was suspended in PBS buffer solution (pH~7.4) (GIBCO) at 37 °C. Protein desorption was followed at predetermined time intervals. Supernatants were collected at each point, and protein concentration was determined using a micro BCA assay kit (St. Louis, MO, USA, Sigma-Aldrich). Desorption rate was expressed as the cumulative percentage.

2.5. Assessment of the Antibody Response

To demonstrate an antigen-specific adjuvant effect, BALB/c mice were subcutaneously immunized two times on days 1 and 14, with the following vaccine formulations; H2B alone and H2B/PLA-NPs. As a positive control, H2B plus CpG-ODN7909 was also included because it triggers cytotoxic immunity. For non-specific responses, one group was given PLA-NPs, while the other was given CpG-ODN7909 only. For infection control, one group was given PBS alone. Specific IgG titers were assessed by ELISA on day 30 at the end of the immunization schedule. A significant increase in the specific IgG titer was found after one month of primary immunization for groups vaccinated with either H2B/PLA-NPs or H2B + CpG in contrast to groups that were exposed to soluble protein (Figure 5). There was an increase in the average antibody levels. These results indicate that PLA-based vaccine formulation induced humoral immunity.

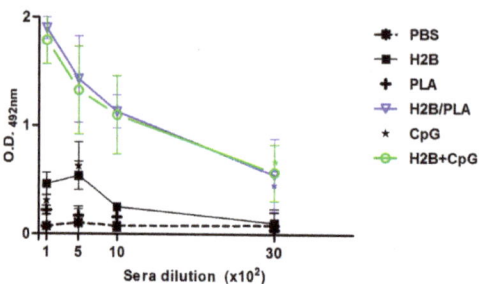

Figure 5. Specific anti-H2B antibody immune response. Pooled sera from each group were reacted in ELISA with recombinant H2B protein. Total IgG titers were calculated for each group. Titers were scored positively at the last dilution of immune sera. Results are represented as mean ± standard deviation of optical density (O.D.). (*) $p < 0.05$ between PBS control and protein-received groups.

2.6. Analysis of the Anti-H2B IgG Isotype

IgG1 and IgG2 antibody isotypes are markers of the humoral and cellular immune response, respectively. The evaluation of their titers allows them to determine by which pathway, Th2 or Th1, the immune system reacts to H2B/PLA formulation. Both adjuvants (particulate, PLA and molecular, CpG) produced a significant IgG2 isotype level. This suggested that the PLA immune response tends towards a Th1-type response, as confirmed with the CpG-ODN adjuvant (as control) (Figure 6).

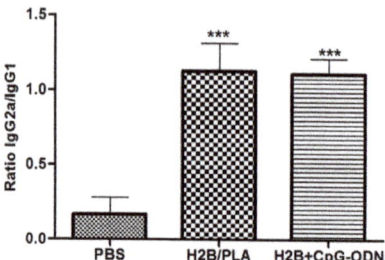

Figure 6. IgG isotype ratio in pooled sera from immunized mice. Results are represented as mean ± standard deviation (SD) of optical density (O.D.) values obtained from 5 mice per group. (***) $p < 0.001$ between PBS control and protein-received groups.

2.7. Protective Potency of H2B/PLA in Mice

The protective potential of antigen adsorbed onto PLA was compared with that of CpG-ODN7909, which has previously been shown to be a Th1 response enhancer.

The challenge tests were performed using the standard parasite strain, MHOM/TN/94/GLC94 (GLC94), isolated from a human lesion of cutaneous leishmaniasis (CL). This isolate belongs to the species of L. major, zymodeme MON25 and is the most virulent strain [31,33]. A high dose (2×10^6) of GLC94 parasites was inoculated through subcutaneous injection into the right footpad six weeks after the booster dose (week 8). Then, the course of the infection was recorded weekly for eight weeks (weeks 10 to 17) Figure 7a. The progression of lesions was similar in all groups of mice during the first three weeks post infection. Among the groups that received H2B antigen, PLA NP gave the best results, followed by CpG-ODN, applied as a control adjuvant. Since the soluble protein itself is immunogenic, a moderate protective potential has been observed. For PLA NPs, CpG-ODN, and PBS groups, lesions progressed rapidly to severe necrosis from 3 to 8 weeks post-infection. In contrast, the footpads of mice vaccinated with PLA-H2B were free of necrosis at the end of the experiment (week 8) and were 2-fold thinner than those of the

PBS group ($p < 0.001$). Compared with the CpG-ODN group, the results were significantly similar (Figure 7b).

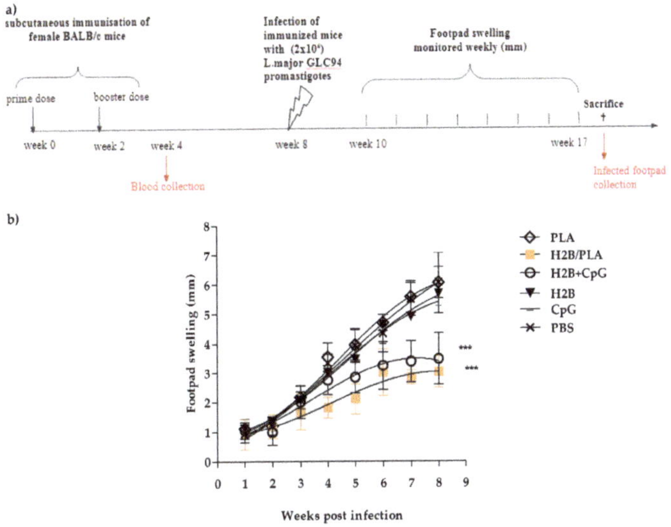

Figure 7. Immune protection of vaccinated BALB/c mice against *Leishmania major* challenge. (**a**) The timeline of the in vivo experiment. Female BALB/c mice (6–8 weeks old) were vaccinated twice on day 1 and day 14 with H2B/PLA (25 µg/protein), H2B (25 µg) + CpG or soluble H2B (25 µg). The control groups received PBS, PLANPs or CpG. Mice were infected with (2×10^6) *L. major* GLC94 promastigotes by subcutaneous injection on the right footpad. (**b**) Footpad swelling. Lesion sizes were monitored weekly for up to eight weeks. The mean lesion size ± SD is shown ($n = 8$). (***) $p < 0.001$ between PBS control and protein-received groups.

2.8. *Evaluating the Parasite Load in Mice*

Parasites disseminated in infected footpads were quantified using a limiting dilution technique. As expected, mice immunized with PLA (H2B/PLA) and CpG (H2B + CpG) induced good protection against parasites, with 2–3 log fewer parasites than with the PBS group ($p < 0.001$). Unvaccinated mice had the highest parasite load (Figure 8). The results correlated with the lesion thickness observed in respective groups. It suggests that only protein formulated with PLA nanoparticles or CpG-ODN can protect animals against parasite spread.

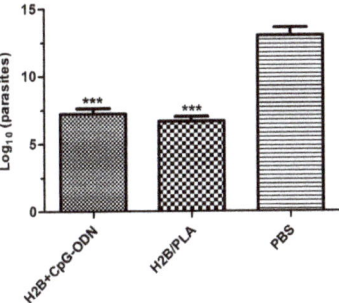

Figure 8. Parasite load in infected footpads. BALB/c mice vaccinated with H2B/PLA and H2B + CPG were challenged with *L. major* promastigotes, respectively. Eight weeks post-infection, footpads were collected. Parasites at the lesion site were counted by limiting dilutions. The results were compared with those obtained from PBS group. (***) $p < 0.001$ between PBS and protein-received group.

3. Discussion

A significant public health issue is the absence of effective treatments or preventive vaccinations for several diseases brought on by intracellular infections, such as cutaneous leishmaniasis. Strong cellular responses are needed in the vaccine formulations for these illnesses [34]. Preclinical trials are now being conducted on a variety of potential anti-leishmania vaccine candidates, including live genetically altered parasites, live attenuated parasites, recombinant proteins and DNA vaccines, and vaccinations using antigen combinations [16]. Despite the preclinical effectiveness of some vaccine candidates, their use has been restricted due to the toxicity of a living vaccine on the one hand and the minimal antigenic exposure to the antigen-presenting cells (APCs) for a subunit protein vaccine on the other hand [15,35,36]. However, a satisfactory adjuvant in vaccine formulation has yet to be approved [36]. At present, the WHO still recommends the use of alum in combination with the vaccine antigen; this encourages the investigation of new adjuvants because alum is unable to elicit cell-mediated Th1 response [37].

Some of these limitations can be resolved through the use of nanoparticulate delivery systems. Today, the administration of vaccines to organisms using PLA nanoparticles with surface-adsorbed antigens is a promising approach. The polymer is biocompatible, biodegradable, non-toxic, and non-antigenic [38]. Interestingly, PLA-based vaccines improve efficient cellular immune responses (CTL) to a number of infections [25]. Such an immune response seems crucial for the control of Leishmania, an intracellular pathogen [21].

This research aimed to develop a PLA polymer-based nanoparticle vaccine for the immunization of mice infected with *L. major* parasites. We investigated and compared the immune performance of antigen-coated PLA nanoparticles and CpG 7909, a Th1-activated adjuvant. The recombinant H2B protein, expressed in *E. coli* BL21 strain cells, is a chosen candidate for a leishmanial vaccine [31]. It is interesting to note that this protein does not interact negatively with mammalian histones, is easily expressed in the prokaryotic environment, and is not cytotoxic [39]. The first part of the study was to optimize the physico-chemical conditions of the procedure and the formulation of the vaccine since the size, zeta potential and desorption pattern are parameters that affect the particle uptake by cells. For the development of H2B-coated PLA NPs, we prepared well-dispersed nanoparticles by the nanoprecipitation method. We optimized the protocol so that the particles are obtained in a single step without the use of surface-active agents. This is a benefit in vaccine formulation that overcomes the inherent toxicity of excipients. Many studies demonstrated the emergence of skin toxicity when using surfactants in nano-delivery systems [40]. The PLA-NPs had a mean diameter of 287 nm and had a quite narrow size distribution with a polydispersity index of 0.14. Neither deposits nor agglomerates were noticed in the PLA dispersions. A zeta potential of −45 mV was high enough (in absolute value) to afford colloidal stability through electrostatic repulsion. The carboxyl groups at the particle surface outset the negative global charge. The H2B adsorption on the particle surface resulted in the inversion of the zeta potential from −45 mV to +30 mV. The carboxyl groups were neutralized by protonated amines within the protein backbone. Obviously, histones are classified as proteins with a lightweight structure and a strong basic polycation character [41]. Thus, the H2B adsorption on NPs occurred by electrostatic interactions. The positive global charge of H2B/PLA formulation could optimize cell membrane interactions. This is a crucial parameter of antigen-presenting cells [42]. Several researchers have explored the cell uptake of positively charged particles. Among them, Liu Z. et al. showed that cationic stearylamine lipid-polymer hybrid NPs (LPNPs) of ~200 nm increased the efficacy of AmB through macrophage uptake [23]. As confirmed elsewhere, the sodium alginate-glycol chitosan stearate NPs (AmB-SA-GCS-NP) showed the highest macrophagic uptake in J774.1 macrophages and the rapid in vivo localization tissues of the liver, spleen, lung, and kidney [43]. Many studies have reported that particle size modulates the assimilation of particles into cells. Among them is one concerning PLGA NPs that exhibited low toxicity and efficient uptake by APC in vitro and in vivo when particles were about 300 nm in average size [44].

Finally, the release pattern of H2B/PLA was analyzed at predetermined time intervals for 30 days. The recombinant H2B started desorption on day 7, and the total amount of 30% was reached 3 weeks later (the 30th day).

Weak desorption at pH = 7.4 could be interpreted as a more favorable interaction between protein and polymer than protein and release medium. The progressive H2B release promotes a deposit effect. For vaccination purposes, this can significantly reduce booster doses.

Secondly, we evaluated the vaccine potency of the formulation on the BALB/c mice because they are susceptible to infection with *L. major*, and if allowed to run its course, subcutaneous injection of parasites leads to uncontrolled lesion growth and eventual death [33]. The challenge tests were performed using the standard parasite strain, MHOM/TN/94/GLC94 (GLC94), isolated from a human lesion of cutaneous leishmaniasis (CL). This isolate belongs to the species of *L. major*, zymodeme MON25, and is the most virulent strain [45]. BALB/c mice were subcutaneously immunized two times with the following vaccine formulations: H2B alone, H2B/PLA-NPs and H2B/CpG-ODN7909. We recorded an increase in the average antibody level against H2B-based adjuvant formulations in contrast to groups that were exposed to free antigens. The humoral response intensity (IgG titers) was substantially similar between both formulations and, therefore, immunogenic in a specific way to adsorbed antigens. These results agree with previous work on PLA NPs/Nod ligand formulations, which induced an increase of up to 100-fold in antibody responses against Gag p24 HIV-1 antigen in comparison to alum [46]. As previously described using PLA as an adjuvant, the p24/PLA vaccine induced high antibody titers with a strong CTL response in mice, rabbits and macaque [25]. As for other infectious diseases, the pattern of the IgG subclass has been shown to play a role in the course of *Leishmania* infection. Mice, similarly to humans, show four different classes of IgGs, named IgG1, IgG2a, IgG2b and IgG3, which functionally correspond to human IgG1, IgG2, IgG4 and IgG3, respectively. In general, it is possible to conclude that in mice and humans, IgG1 (as well as IgG4 in humans) is associated with a Th2 profile, and the other subclasses are mainly associated with a Th1 profile [47]. The evaluation of the IgG2/IgG1 ratio allowed us to predict the type of immune response induced by both adjuvants. The PLA/H2B formulation tends towards a Th1-activated response due to the positive IgG2/IgG1 ratio of antibodies against H2B. Clearly, the CpG7909 preferentially activated cellular immunity, as confirmed by the prevalence of IgG2, by directing the differentiation of LB towards the secretion of specific IgG2a antibodies. Additionally, for LCL-infected individuals, PBMC proliferation and IFNγ levels increased significantly in the presence of a recombinant H2B [32]. The prevalence of the cellular response with H2B/PLA could be explained based on preliminary experiments suggesting that the H2B protein has the ability to induce a specific Th1 response in BALB/c mice [31]. When BALB/c mice were vaccinated with the H2B protein alone or in the presence of CpG adjuvant (two injections), they were able, one month after the last immunization, to produce large amounts of IFNγ when their lymph nodes (injection sites) were re-stimulated in vitro with H2B protein. Furthermore, the amounts of IL-10 and IL-4 were very low or not detectable. Such a Th1-polarized response is curative, while the Th2 response exacerbates or is ineffective in controlling the disease [48–50].

The challenge assay demonstrated that animals do not develop ulcerating lesions until week 8 post infection. Both groups (PLA particles and CpG-ODN) have the lowest lesion thickness. This suggested that H2B/PLA formulation exhibited a protection immune potency. Consequently, the parasite dissemination has been slowed down or even interrupted in vaccinated animals, while control groups lead to uncontrolled lesion growth and eventual death. Despite the effectiveness of CpG-ODN in triggering a cytotoxic immune response, its use is still restricted since the clearance is rapid. A study of its pharmacokinetics and biodistribution characteristics demonstrated its rapid renal clearance when intravenously injected into animals [51].

A previous study from Khatik et al. (2014) showed that PS–coated gelatin NPs with encapsulated amphotericin B reduced parasite burden in Wistar rats (85% vs. 50%) when compared to standard AmB [52]. For application against CL, AmB nano-encapsulation in PLGA/dimercaptosuccinic acid (DMSA) NPs (nano-DMS-AmB) was developed and tested against C57BL/6 mice infected with *L. amazonensis*. The nano DMS-AmB showed a greater reduction in the number of parasites than standard D-AmB [53]. Nanoliposomes used as the nanocarriers for soluble *Leishmania* antigens (SLA) showed parasite clearance in the footpad and spleen of a mouse model injected with this formulation. Liposomal formulations of ChimeraT (a combination of 3 leishmanial proteins), when delivered subcutaneously, protected mice against *L. infantum* infection by reducing the parasite load in the spleen, liver, bone marrow, and lymph nodes [54].

In the context of leishmanial nano-vaccines, Katebi A. and collaborators (2021) loaded PLGA NPs with SLA and TLR receptor agonists PAM3CSK4 and R848 and incubated them in vitro with immune cell lines. The results showed a marked increment in the macrophages' phagocytic potential accompanied by a significant reduction in pro-inflammatory cytokines against *L. major* parasites. However, this immunogenic candidate should be tested against a *Leishmania*-infected animal model to determine the in vivo parasitic and disease reduction potential [55,56].

4. Materials and Methods

4.1. Chemicals and Reagents

Phosphate-buffered saline tablets and acetone were obtained from Sigma-Aldrich Co. (St. Louis, MO, USA). A Low Molecular Weight Calibration Kit for SDS Electrophoresis (Code: 17-0446-01) was purchased from GE Healthcare. An s29 Gx 0.5 syringe (insulin) was purchased from Terumo. The adjuvant CpG-ODN (7698 g/mol) (ODN 7909), a human TLR9 ligand, was purchased from InvivoGen. Poly (D,L-lactide) (PLA Mw = 16,000 g/mol, Mw/Mn = 1.6) was purchased from Corbion Purac Biomaterials (4206 AC, The Netherlands). Imidazole reagent was obtained from GE Healthcare-Application Note 28-4067-41 AA. A QuantiPro BCA Assay Kit was obtained from Sigma-Aldrich Co. (St. Louis, MO, USA)). Schneider's Drosophila medium was obtained from Gibco-BRL, Paisley, Scotland.

4.2. Parasites

The standard strain of the *Leishmania major* parasite is MHOM/TN/94/GLC94 (GLC94), which was isolated in a human lesion of zoonotic cutaneous leishmaniasis (ZCL). The most virulent strain of a parasite is the *L. major* zymodeme MON25 (*L. major*-MON25) [31–33]. Promastigotes were collected in a logarithmic-phase culture, counted in a fixed volume, and then incubated once more at 26 °C, as previously mentioned. After five days, the stationary phase was attained, and the parasite concentration value was equal to 4×10^7 parasites/milliliter. Then, using the density gradient centrifugation method, metacyclic promastigotes were isolated, cleaned, and administered to BALB/c mice for inoculation (2×10^6).

4.3. Mice

Female BALB/c mice (6–8 weeks old) were obtained from The Institut Pasteur de Tunis "http://www.Pasteur.tn (accessed on 12 June 2013)". The experimental protocol (registry number: 2015/13/I/LR11IPT08 V1) was approved by the Institutional Ethical Bio-Medical Committee of the Institut Pasteur de Tunis IPT, and the US registry number is IRB00005445, FWA00010074).

4.4. Preparation of PLA Nanoparticles

Surfactant-free PLA nanoparticles (PLA-NPs) were prepared by the nanoprecipitation method as previously described [57]. Briefly, 0.2 g of polymer was dissolved in 10 mL of acetone and added dropwise to 30 mL of milliQ under moderate stirring water without using a surfactant. The organic solvent and a part of the water were removed by evaporation under reduced pressure. The final NP concentration after the preparation was determined by measuring the solid content after heating the NP dispersion (known volume) to a constant weight in an oven at 70 °C for 24 h. The size distribution was determined by dynamic light scattering technique (DLS) at 25 °C using a Zeta-Sizer Nano ZS (Malvern Instruments, Malvern, UK). Zeta potential was measured by a laser Doppler with the same instrument using diluted colloidal dispersions in 1 mM NaCl.

4.5. Expression and Purification of Recombinant Histone H2B

The BL21 *Escherichia coli* (*E. coli*) strain containing the recombinant plasmid pET-H2B was kindly provided by Dr. Chenik.

Purification and expression of the H2B protein were carried out as previously described [31]. Luria broth (LB) medium supplemented with ampicillin (100 µg/mL) and 0.1% glucose was used to grow *E. coli* BL21 pLysS DE3-pET-H2B in shaking flasks until the absorbance at 600 nm reached a value between 0.6 and 0.9. After that, IPTG (1 mM) was used to stimulate protein expression for a minimum of 4 h at 37 °C. Cells were pelleted before the periplasmic proteins were osmotically shocked out and placed into a Ni-NTA column. The histidine-tagged (His-tagged) proteins were then eluted using a linear gradient of imidazole from 0.05 to 0.5 M after being cleaned with PBS. Membrane dialysis was used to desalt fractions. SDS-PAGE was used to evaluate purified proteins, and the BCA assay kit (Pierce, Rockford, IL, USA) instructions were used to calculate protein quantities.

4.6. Synthesis of H2B/PLA, a Nanovaccine Formulation

In this study, a nanovaccine formulation was synthesized by the adsorption of H2B onto PLA nanoparticles, as previously described [57]. PLA-NPs were diluted at a concentration of 1.2 mg/mL in Milli-Q water. Then, H2B solutions were prepared at different concentrations (0–100 µg/mL) in NaCl (10 mM). One volume of NPs was added to one volume of protein solution to obtain the H2B/PLA formulation. The adsorption medium has a fixed concentration of both NPs (0.6 mg/mL) and protein, which ranged from 0 to 50 µg/mL (corresponding to an H2B/PLA weight ratio from 0 to 8.3%). Protein and particle dispersions were incubated for 15 min at room temperature with gentle stirring. The samples were centrifuged for 20 min at $5000\times g$. Separate tubes were used to collect each supernatant. Each H2B/PLA pellet was cleaned further by centrifugation/redispersion ($5000\times g$ for 20 min at 25 °C). The pellets were then resuspended in saline solution (NaCl 10 mM). In accordance with the manufacturer's recommendations, the QuantiPro BCA assay Kit (Sigma, Germany) was used to quantify the concentration of non-adsorbed H2B. A Zeta-sizer Nano ZS-based DLS technique was used to measure parameters such as the particle size, polydispersity indices and zeta potentials of the H2B/PLA formulations (Malvern Instruments, Malvern, UK).

4.7. In Vitro Release Study

To study the desorption profile of H2B, the H2B/PLA pellet (which was used at a 5% w/w ratio) was resuspended in PBS (pH 7.4) and incubated on a shaker at 37 °C. At set times (12 h, 1, 2, 3, 7, 14, 21 and 30 days), samples (in triplicate) were taken and centrifuged at $5000\times g$ for 20 min. Using the QuantiPro BCA assay Kit, the concentration of protein released in supernatants was quantified.

4.8. In Vivo Assay

4.8.1. Immunization of BALB/c Mice

In vivo studies have been performed in BALB/c mice, which are the mouse models used for preclinical evaluation of *Leishmania* vaccine candidates [33]. Forty-eight females, between 6 to 8 weeks old, were used in the experiments. The animals were divided into 6 groups, including 8 mice each. They were then administered subcutaneously twice on days 1 and 14 with the following vaccine formulations: H2B alone (25 µg/injection), H2B (25 µg) plus CpG (20 µg), H2B/PLA-NPs (25 µg/0.9 mg), CpG (20 µg), PLA-NPs (0.9 mg) and PBS. In each vaccination, 200 µl of the preparation diluted in PBS was injected into the animals. On day 30, blood samples were taken to measure serum antibody levels.

4.8.2. Identifying the IGg1 and IgG2 Isotypes

To evaluate the antibody pattern generated based on the H2B/PLA formulation, the IGg1/IgG2 subtype was examined. Enzyme immunoassay (ELISA) was used to detect specific antibodies generated against H2B formulations (H2B/PLA and H2B + CpG7909) in mouse serum. Briefly, 5 µg/mL of H2B was coated in 100 µL of carbonate-bicarbonate buffer (0.1 M) and incubated overnight at 4 °C in 96-well high-binding plates (Nunc). PBS-T20 buffer was used three times to wash the plates. PBS-T20 solution, supplemented with low-fat milk (2%), was used to block the plates at 37 °C for 1 h. After the wash steps, sera were added at a dilution of 1/1000 and incubated for 2 h at 37 °C.

4.8.3. Parasite Infection and Development of Lesions

The purpose of the treatment was to demonstrate an adjuvant effect of H2B/PLA against an experimental *L. major* infection. A minimum of 6 weeks was required between vaccination and infection. Six weeks after receiving the booster dose, animals in each group were challenged with infectious *L. major* promastigotes (GLC94) in order to observe and evaluate the efficacies of both the H2B + CpG and H2B/PLA-NPs vaccine formulations. For this, mice were subcutaneously injected with an infective parasite load (2×10^6) of *L. major* promastigotes suspended in 50 µL of PBS in the right footpad. Within eight weeks, the swelling's growth was tracked once a week using a dial-gauge caliper (Mitutoyo, Japan). Lesion values (in mm) were calculated by subtracting the thickness of the infected footpad from the thickness of the contralateral footpad that was not infected.

4.8.4. Determining the Parasite Load

Parasite load was quantified by a limiting-dilution technique adapted from the work of Laskay et al. [58]. The infected pad was cut (3 groups, *n* = 8) and homogenized before serial 10-fold dilutions were plated in triplicate in 96-well flat-bottom microtiter plates (Nunc, Roskilde, Denmark) containing Schneider's Drosophila medium (Gibco-BRL, Paisley, Scotland) with the addition of 100 U of penicillin/mL, 100 µg of streptomycin/mL, 2 mM L-glutamine, and 10% heat-inactivated fetal calf serum. Plates were incubated at 26 °C. Live parasites were attested under an inverted microscope. Parasite load is expressed as the average of the log negative of the last dilution in which mobile parasites were detected.

5. Statistical Analysis

Statistical analysis was performed using GraphPad Prism software 5.1 (GraphPad Software 2365 Northside Dr. Suite 560 San Diego, CA 92108). The data are expressed as mean ± standard deviation. Statistical significance *p*-values of less than 0.05 are considered statistically significant.

6. Conclusions

The successful development of an effective vaccine needs exact data regarding the physicochemical characteristics of the antigen and adjuvant, as well as expertise in their combination to generate a safe, steady, and immunogenic vaccine.

Our study described the immunogenicity and protective capacity of H2B/PLA as a particulate vaccine in an anti-leishmanial therapeutic setting. The outcomes clearly demonstrate that the PLA nanoparticles serve as an adjuvant for the recombinant protein H2B since BALB/c mice were protected against experimental infection by the *L. major* parasite, a standard virulent strain GLC94. In contrast, mice immunized with a soluble protein were unable to stop the parasite's spread, which led the paw to completely ulcerate. The H2B/PLA's added value refers to its colloidal stability, simple manufacturing process, and established adjuvant capacity.

It will be interesting to extend these data by incorporating immunostimulatory molecules (such as imiquimod, 3 M 052) in the NP PLA core to increase its protective effect.

Author Contributions: Conceptualization, M.E.; Formal analysis, S.A.-R. and N.B.k.; Investigation, S.A.-R.; Methodology, T.T. and M.C.; Resources, B.V., Z.B., M.C. and M.E.; Supervision, B.B.-Z., B.V., T.T. and M.E.; Validation, B.V. and T.T.; Writing—original draft, S.A.-R.; Writing—review & editing, S.A.-R., N.B.k., B.B.-Z., B.V. and T.T. All authors have read and agreed to the published version of the manuscript.

Funding: This research received no external funding.

Institutional Review Board Statement: The animal study protocol was approved by the Biomedical Ethics Committee of Institut Pasteur de Tunis, Tunisia (CEBM: 2015/13/I/LR11IPT08 V1) and according to the 2010/63/EU Directive for animal experiments.

Informed Consent Statement: Not applicable.

Data Availability Statement: Data available from the corresponding author.

Acknowledgments: In this section, special thanks are due to Hechmi Louzir, from the Institut Pasteur de Tunis, for his scientific comments and support. Thanks are addressed to Marouani A, Ellefi A, and Rammeh S, for technical assistance on laboratory animals.

Conflicts of Interest: The authors declare no conflict of interest.

References

1. WHO. Leishmaniasis. Available online: https://www.who.int/news-room/fact-sheets/detail/leishmaniasis (accessed on 8 January 2022).
2. Burza, S.; Croft, S.L.; Boelaert, M. Leishmaniasis. *Lancet* **2018**, *392*, 951–970. [CrossRef] [PubMed]
3. Pace, D. Leishmaniasis. *J. Infect.* **2014**, *69* (Suppl. 1), S10–S18. [CrossRef] [PubMed]
4. Bennis, I.; Thys, S.; Filali, H.; De Brouwere, V.; Sahibi, H.; Boelaert, M. Psychosocial impact of scars due to cutaneous leishmaniasis on high school students in Errachidia province, Morocco. *Infect. Dis. Poverty* **2017**, *6*, 46. [CrossRef] [PubMed]
5. Himmelrich, H.; Parra-Lopez, C.; Tacchini-Cottier, F.; Louis, J.A.; Launois, P. The IL-4 rapidly produced in BALB/c mice after infection with Leishmania major down-regulates IL-12 receptor beta 2-chain expression on CD4+ T cells resulting in a state of unresponsiveness to IL-12. *J. Immunol.* **1998**, *161*, 6156–6163. [PubMed]
6. Armijos, R.X.; Weigel, M.M.; Calvopina, M.; Hidalgo, A.; Cevallos, W.; Correa, J. Safety, immunogenicity, and efficacy of an autoclaved Leishmania amazonensis vaccine plus BCG adjuvant against New World cutaneous leishmaniasis. *Vaccine* **2004**, *22*, 1320–1326. [CrossRef]
7. Elfaki, M.E.; Khalil, E.A.; De Groot, A.S.; Musa, A.M.; Gutierrez, A.; Younis, B.M.; Salih, K.A.; El-Hassan, A.M. Immunogenicity and immune modulatory effects of in silico predicted L. donovani candidate peptide vaccines. *Hum. Vaccin. Immunother.* **2012**, *8*, 1769–1774. [CrossRef]
8. Khalil, E.A.; Musa, A.M.; Modabber, F.; El-Hassan, A.M. Safety and immunogenicity of a candidate vaccine for visceral leishmaniasis (Alum-precipitated autoclaved Leishmania major + BCG) in children: An extended phase II study. *Ann. Trop. Paediatr.* **2006**, *26*, 357–361. [CrossRef]
9. Duthie, M.S.; Raman, V.S.; Piazza, F.M.; Reed, S.G. The development and clinical evaluation of second-generation leishmaniasis vaccines. *Vaccine* **2012**, *30*, 134–141. [CrossRef]
10. Gillespie, P.M.; Beaumier, C.M.; Strych, U.; Hayward, T.; Hotez, P.J.; Bottazzi, M.E. Status of vaccine research and development of vaccines for leishmaniasis. *Vaccine* **2016**, *34*, 2992–2995. [CrossRef]
11. Choudhury, R.; Das, P.; De, T.; Chakraborti, T. 115 kDa serine protease confers sustained protection to visceral leishmaniasis caused by Leishmania donovani via IFN-gamma induced down-regulation of TNF-alpha mediated MMP-9 activity. *Immunobiology* **2013**, *218*, 114–126. [CrossRef]
12. Grimaldi, G., Jr.; Teva, A.; Dos-Santos, C.B.; Santos, F.N.; Pinto, I.D.; Fux, B.; Leite, G.R.; Falqueto, A. Field trial of efficacy of the Leish-tec(R) vaccine against canine leishmaniasis caused by Leishmania infantum in an endemic area with high transmission rates. *PLoS ONE* **2017**, *12*, e0185338. [CrossRef] [PubMed]
13. Moafi, M.; Rezvan, H.; Sherkat, R.; Taleban, R. Leishmania Vaccines Entered in Clinical Trials: A Review of Literature. *Int. J. Prev. Med.* **2019**, *10*, 95. [CrossRef] [PubMed]
14. Agallou, M.; Margaroni, M.; Karagouni, E. Cellular vaccination with bone marrow-derived dendritic cells pulsed with a peptide of Leishmania infantum KMP-11 and CpG oligonucleotides induces protection in a murine model of visceral leishmaniasis. *Vaccine* **2011**, *29*, 5053–5064. [CrossRef] [PubMed]
15. Azmi, F.; Ahmad Fuaad, A.A.; Skwarczynski, M.; Toth, I. Recent progress in adjuvant discovery for peptide-based subunit vaccines. *Hum. Vaccin Immunother.* **2014**, *10*, 778–796. [CrossRef] [PubMed]
16. Doroud, D.; Rafati, S. Leishmaniasis: Focus on the design of nanoparticulate vaccine delivery systems. *Expert Rev. Vaccines* **2012**, *11*, 69–86. [CrossRef]
17. Xia, Y.; Fan, Q.; Hao, D.; Wu, J.; Ma, G.; Su, Z. Chitosan-based mucosal adjuvants: Sunrise on the ocean. *Vaccine* **2015**, *33*, 5997–6010. [CrossRef]

18. Black, M.; Trent, A.; Tirrell, M.; Olive, C. Advances in the design and delivery of peptide subunit vaccines with a focus on toll-like receptor agonists. *Expert Rev. Vaccines* **2010**, *9*, 157–173. [CrossRef]
19. Rice-Ficht, A.C.; Arenas-Gamboa, A.M.; Kahl-McDonagh, M.M.; Ficht, T.A. Polymeric particles in vaccine delivery. *Curr. Opin. Microbiol.* **2010**, *13*, 106–112. [CrossRef]
20. Mallapragada, S.K.; Narasimhan, B. Immunomodulatory biomaterials. *Int. J. Pharm.* **2008**, *364*, 265–271. [CrossRef]
21. Raman, V.S.; Duthie, M.S.; Fox, C.B.; Matlashewski, G.; Reed, S.G. Adjuvants for Leishmania vaccines: From models to clinical application. *Front. Immunol.* **2012**, *3*, 144. [CrossRef]
22. Daraee, H.; Etemadi, A.; Kouhi, M.; Alimirzalu, S.; Akbarzadeh, A. Application of liposomes in medicine and drug delivery. *Artif. Cells Nanomed. Biotechnol.* **2016**, *44*, 381–391. [CrossRef] [PubMed]
23. Liu, Z.; Jiao, Y.; Wang, Y.; Zhou, C.; Zhang, Z. Polysaccharides-based nanoparticles as drug delivery systems. *Adv. Drug Deliv. Rev.* **2008**, *60*, 1650–1662. [CrossRef] [PubMed]
24. Makadia, H.K.; Siegel, S.J. Poly Lactic-co-Glycolic Acid (PLGA) as Biodegradable Controlled Drug Delivery Carrier. *Polymers* **2011**, *3*, 1377–1397. [CrossRef] [PubMed]
25. Ataman-Onal, Y.; Munier, S.; Ganee, A.; Terrat, C.; Durand, P.Y.; Battail, N.; Martinon, F.; Le Grand, R.; Charles, M.H.; Delair, T.; et al. Surfactant-free anionic PLA nanoparticles coated with HIV-1 p24 protein induced enhanced cellular and humoral immune responses in various animal models. *J. Control. Release* **2006**, *112*, 175–185. [CrossRef] [PubMed]
26. Lamalle-Bernard, D.; Munier, S.; Compagnon, C.; Charles, M.H.; Kalyanaraman, V.S.; Delair, T.; Verrier, B.; Ataman-Onal, Y. Coadsorption of HIV-1 p24 and gp120 proteins to surfactant-free anionic PLA nanoparticles preserves antigenicity and immunogenicity. *J. Control. Release* **2006**, *115*, 57–67. [CrossRef] [PubMed]
27. Venkataprasad, N.; Coombes, A.G.; Singh, M.; Rohde, M.; Wilkinson, K.; Hudecz, F.; Davis, S.S.; Vordermeier, H.M. Induction of cellular immunity to a mycobacterial antigen adsorbed on lamellar particles of lactide polymers. *Vaccine* **1999**, *17*, 1814–1819. [CrossRef]
28. Handman, E. Leishmaniasis: Current status of vaccine development. *Clin. Microbiol. Rev.* **2001**, *14*, 229–243. [CrossRef]
29. Khorasanizadeh, S. The nucleosome: From genomic organization to genomic regulation. *Cell* **2004**, *116*, 259–272. [CrossRef]
30. Maalej, I.A.; Chenik, M.; Louzir, H.; Ben Salah, A.; Bahloul, C.; Amri, F.; Dellagi, K. Comparative evaluation of ELISAs based on ten recombinant or purified Leishmania antigens for the serodiagnosis of Mediterranean visceral leishmaniasis. *Am. J. Trop. Med. Hyg.* **2003**, *68*, 312–320. [CrossRef]
31. Chenik, M.; Louzir, H.; Ksontini, H.; Dilou, A.; Abdmouleh, I.; Dellagi, K. Vaccination with the divergent portion of the protein histone H2B of Leishmania protects susceptible BALB/c mice against a virulent challenge with Leishmania major. *Vaccine* **2006**, *24*, 2521–2529. [CrossRef]
32. Meddeb-Garnaoui, A.; Toumi, A.; Ghelis, H.; Mahjoub, M.; Louzir, H.; Chenik, M. Cellular and humoral responses induced by Leishmania histone H2B and its divergent and conserved parts in cutaneous and visceral leishmaniasis patients, respectively. *Vaccine* **2010**, *28*, 1881–1886. [CrossRef] [PubMed]
33. Benhnini, F.; Chenik, M.; Laouini, D.; Louzir, H.; Cazenave, P.A.; Dellagi, K. Comparative evaluation of two vaccine candidates against experimental leishmaniasis due to Leishmania major infection in four inbred mouse strains. *Clin. Vaccine Immunol.* **2009**, *16*, 1529–1537. [CrossRef] [PubMed]
34. Vijaya BJoshi, S.M.G.; Salem, A.K. Biodegradable particles as vaccine antigen delivery systems for stimulating cellular immune responses. *Hum. Vaccines Immunother.* **2013**, *9*, 2584–2590. [CrossRef]
35. Srivastava, S.; Shankar, P.; Mishra, J.; Singh, S. Possibilities and challenges for developing a successful vaccine for leishmaniasis. *Parasites Vectors* **2016**, *9*, 277. [CrossRef]
36. Gheibi Hayat, S.M.; Darroudi, M. Nanovaccine: A novel approach in immunization. *J. Cell. Physiol.* **2019**, *234*, 12530–12536. [CrossRef]
37. Yasinzai, M.; Khan, M.; Nadhman, A.; Shahnaz, G. Drug resistance in leishmaniasis: Current drug-delivery systems and future perspectives. *Future Med. Chem.* **2013**, *5*, 1877–1888. [CrossRef]
38. Shive, M.S.; Anderson, J.M. Biodegradation and biocompatibility of PLA and PLGA microspheres. *Adv. Drug Deliv. Rev.* **1997**, *28*, 5–24. [CrossRef]
39. Soto, M.; Requena, J.M.; Quijada, L.; Perez, M.J.; Nieto, C.G.; Guzman, F.; Patarroyo, M.E.; Alonso, C. Antigenicity of the Leishmania infantum histones H2B and H4 during canine viscerocutaneous leishmaniasis. *Clin. Exp. Immunol.* **1999**, *115*, 342–349. [CrossRef]
40. Lemery, E.; Briancon, S.; Chevalier, Y.; Oddos, T.; Gohier, A.; Boyron, O.; Bolzinger, M.A. Surfactants have multi-fold effects on skin barrier function. *Eur. J. Dermatol.* **2015**, *25*, 424–435. [CrossRef]
41. Ronningen, T.; Shah, A.; Oldenburg, A.R.; Vekterud, K.; Delbarre, E.; Moskaug, J.O.; Collas, P. Prepatterning of differentiation-driven nuclear lamin A/C-associated chromatin domains by GlcNAcylated histone H2B. *Genome Res.* **2015**, *25*, 1825–1835. [CrossRef]
42. Foged, C.; Brodin, B.; Frokjaer, S.; Sundblad, A. Particle size and surface charge affect particle uptake by human dendritic cells in an in vitro model. *Int. J. Pharm.* **2005**, *298*, 315–322. [CrossRef] [PubMed]
43. Gupta, P.K.; Asthana, S.; Jaiswal, A.K.; Kumar, V.; Verma, A.K.; Shukla, P.; Dwivedi, P.; Dube, A.; Mishra, P.R. Exploitation of lectinized lipo-polymerosome encapsulated Amphotericin B to target macrophages for effective chemotherapy of visceral leishmaniasis. *Bioconjugate Chem.* **2014**, *25*, 1091–1102. [CrossRef] [PubMed]

44. Margaroni, M.; Agallou, M.; Kontonikola, K.; Karidi, K.; Kammona, O.; Kiparissides, C.; Gaitanaki, C.; Karagouni, E. PLGA nanoparticles modified with a TNFalpha mimicking peptide, soluble Leishmania antigens and MPLA induce T cell priming in vitro via dendritic cell functional differentiation. *Eur. J. Pharm. Biopharm.* **2016**, *105*, 18–31. [CrossRef]
45. Kebaier, C.; Louzir, H.; Chenik, M.; Ben Salah, A.; Dellagi, K. Heterogeneity of wild Leishmania major isolates in experimental murine pathogenicity and specific immune response. *Infect. Immun.* **2001**, *69*, 4906–4915. [CrossRef]
46. Pavot, V.; Rochereau, N.; Primard, C.; Genin, C.; Perouzel, E.; Lioux, T.; Paul, S.; Verrier, B. Encapsulation of Nod1 and Nod2 receptor ligands into poly(lactic acid) nanoparticles potentiates their immune properties. *J. Control. Release* **2013**, *167*, 60–67. [CrossRef]
47. Banerjee, K.; Klasse, P.J.; Sanders, R.W.; Pereyra, F.; Michael, E.; Lu, M.; Walker, B.D.; Moore, J.P. IgG subclass profiles in infected HIV type 1 controllers and chronic progressors and in uninfected recipients of Env vaccines. *AIDS Res. Hum. Retrovir.* **2010**, *26*, 445–458. [CrossRef] [PubMed]
48. Launois, P.; Conceicao-Silva, F.; Himmerlich, H.; Parra-Lopez, C.; Tacchini-Cottier, F.; Louis, J.A. Setting in motion the immune mechanisms underlying genetically determined resistance and susceptibility to infection with Leishmania major. *Parasite Immunol.* **1998**, *20*, 223–230. [CrossRef] [PubMed]
49. Liew, F.Y.; O'Donnell, C.A. Immunology of leishmaniasis. *Adv. Parasitol.* **1993**, *32*, 161–259. [CrossRef] [PubMed]
50. Reiner, S.L.; Locksley, R.M. The regulation of immunity to Leishmania major. *Annu. Rev. Immunol.* **1995**, *13*, 151–177. [CrossRef]
51. Palma, E.; Cho, M.J. Improved systemic pharmacokinetics, biodistribution, and antitumor activity of CpG oligodeoxynucleotides complexed to endogenous antibodies in vivo. *J. Control. Release* **2007**, *120*, 95–103. [CrossRef]
52. Khatik, R.; Dwivedi, P.; Khare, P.; Kansal, S.; Dube, A.; Mishra, P.R.; Dwivedi, A.K. Development of targeted 1,2-diacyl-sn-glycero-3-phospho-l-serine-coated gelatin nanoparticles loaded with amphotericin B for improved in vitro and in vivo effect in leishmaniasis. *Expert Opin. Drug Deliv.* **2014**, *11*, 633–646. [CrossRef] [PubMed]
53. De Carvalho, R.F.; Ribeiro, I.F.; Miranda-Vilela, A.L.; de Souza Filho, J.; Martins, O.P.; Cintra e Silva Dde, O.; Tedesco, A.C.; Lacava, Z.G.; Bao, S.N.; Sampaio, R.N. Leishmanicidal activity of amphotericin B encapsulated in PLGA-DMSA nanoparticles to treat cutaneous leishmaniasis in C57BL/6 mice. *Exp. Parasitol.* **2013**, *135*, 217–222. [CrossRef] [PubMed]
54. Lage, D.P.; Ribeiro, P.A.F.; Dias, D.S.; Mendonca, D.V.C.; Ramos, F.F.; Carvalho, L.M.; Steiner, B.T.; Tavares, G.S.V.; Martins, V.T.; Machado, A.S.; et al. Liposomal Formulation of ChimeraT, a Multiple T-Cell Epitope-Containing Recombinant Protein, Is a Candidate Vaccine for Human Visceral Leishmaniasis. *Vaccines* **2020**, *8*, 289. [CrossRef] [PubMed]
55. Katebi, A.; Varshochian, R.; Riazi-Rad, F.; Ganjalikhani-Hakemi, M.; Ajdary, S. Combinatorial delivery of antigen and TLR agonists via PLGA nanoparticles modulates Leishmania major-infected-macrophages activation. *Biomed. Pharmacother.* **2021**, *137*, 111276. [CrossRef] [PubMed]
56. Noormehr, H.; Zavaran Hosseini, A.; Soudi, S.; Beyzay, F. Enhancement of Th1 immune response against Leishmania cysteine peptidase A, B by PLGA nanoparticle. *Int. Immunopharmacol.* **2018**, *59*, 97–105. [CrossRef]
57. Ayari-Riabi, S.; Trimaille, T.; Mabrouk, K.; Bertin, D.; Gigmes, D.; Benlasfar, Z.; Zaghmi, A.; Bouhaouala-Zahar, B.; Elayeb, M. Venom conjugated polylactide applied as biocompatible material for passive and active immunotherapy against scorpion envenomation. *Vaccine* **2016**, *34*, 1810–1815. [CrossRef]
58. Laskay, T.; Diefenbach, A.; Rollinghoff, M.; Solbach, W. Early parasite containment is decisive for resistance to Leishmania major infection. *Eur. J. Immunol.* **1995**, *25*, 2220–2227. [CrossRef]

Article

PLGA-Lipid Hybrid Nanoparticles for Overcoming Paclitaxel Tolerance in Anoikis-Resistant Lung Cancer Cells

Sasivimon Pramual [1], Kriengsak Lirdprapamongkol [1,2,*], Korakot Atjanasuppat [1,3], Papada Chaisuriya [1], Nuttawee Niamsiri [4] and Jisnuson Svasti [1]

1. Laboratory of Biochemistry, Chulabhorn Research Institute, Laksi, Bangkok 10210, Thailand
2. Center of Excellence on Environmental Health and Toxicology (EHT), OPS, MHESI, Thailand
3. Division of Hematology and Oncology, Department of Pediatrics, Faculty of Medicine Ramathibodi Hospital, Mahidol University, Ratchathewi, Bangkok 10400, Thailand
4. Department of Biotechnology, Faculty of Science, Mahidol University, Ratchathewi, Bangkok 10400, Thailand
* Correspondence: kriengsak@cri.or.th; Tel.: +66-2553-8555 (ext. 8356)

Abstract: Drug resistance and metastasis are two major obstacles to cancer chemotherapy. During metastasis, cancer cells can survive as floating cells in the blood or lymphatic circulatory system, due to the acquisition of resistance to anoikis—a programmed cell death activated by loss of extracellular matrix attachment. The anoikis-resistant lung cancer cells also develop drug resistance. In this study, paclitaxel-encapsulated PLGA-lipid hybrid nanoparticles (PLHNPs) were formulated by nanoprecipitation combined with self-assembly. The paclitaxel-PLHNPs had an average particle size of 103.0 ± 1.6 nm and a zeta potential value of −52.9 mV with the monodisperse distribution. Cytotoxicity of the nanoparticles was evaluated in A549 human lung cancer cells cultivated as floating cells under non-adherent conditions, compared with A549 attached cells. The floating cells exhibited anoikis resistance as shown by a lack of caspase-3 activation, in contrast to floating normal epithelial cells. Paclitaxel tolerance was evident in floating cells which had an IC_{50} value of 418.56 nM, compared to an IC_{50} value of 7.88 nM for attached cells. Paclitaxel-PLHNPs significantly reduced the IC_{50} values in both attached cells (IC_{50} value of 0.11 nM, 71.6-fold decrease) and floating cells (IC_{50} value of 1.13 nM, 370.4-fold decrease). This report demonstrated the potential of PLHNPs to improve the efficacy of the chemotherapeutic drug paclitaxel, for eradicating anoikis-resistant lung cancer cells during metastasis.

Keywords: PLA-based materials; nanoparticles; drug delivery; paclitaxel resistance; metastasis; lung cancer; anoikis

1. Introduction

The two major problems in cancer treatment are drug resistance and the spreading of cancer cells in the body or metastasis. Metastatic tumors are difficult to treat because they frequently develop drug resistance, which remains a major cause of cancer-related death. During metastasis, cancer cells float through the blood and lymphatic circulatory system until they reach targeted locations in distant organs [1]. Normal epithelial and endothelial cells require adhesion to their appropriate extracellular matrix (ECM) and neighboring cells for maintaining their survival. Detachment from ECM or adhesion to inappropriate ECM leads to activation of a programmed cell death termed anoikis in the normal epithelial and endothelial cells [2,3]. This phenomenon avoids the misplaced growth of normal cells in other sites. However, some populations of cancer cells are able to develop anoikis resistance to support their survival when they become floating cells during the journey through the circulatory systems [4]. There is considerable interest in developing approaches for managing anoikis-resistant cancer cells.

Drug resistance of cancer cells in patients can occur before receiving chemotherapy (inherent or intrinsic resistance) or emerge after treatment (acquired resistance), this classification is based on the time when resistance is developed [5]. The effectiveness of current chemotherapeutic drugs is gradually decreased by the acquired resistance after repeated treatments, while the intrinsic resistance of cancer cells causes poor drug response from the first treatment so that patients do not receive the benefit of chemotherapy [6].

New deaths worldwide from lung cancer in 2020 have been estimated to be 1.8 million or 18% of all cancer deaths, making lung cancer the most deadly cancer [7]. A major type of lung cancer is non-small cell lung cancer (NSCLC) which accounts for about 85% of all lung cancer cases [8]. More than 75% of new lung cancer cases were advanced cases at metastatic stage III or IV at the time of diagnosis [9]. Over the past decade, several new treatments for advanced NSCLC have been developed, including targeted therapy and immunotherapy [8]. However, cytotoxic chemotherapy has been the first-line standard regimen for metastatic NSCLC patients who do not have a targetable mutation [10]. Therefore, the improving efficacy of cytotoxic anticancer drugs is still essential for cancer chemotherapy.

Paclitaxel or Taxol^{TM} is a widely used anticancer agent, either as monotherapy or in combination with other drugs, for the treatment of several cancers such as lung, breast, ovarian, prostate, liver, gastric, and bladder cancer [11,12]. Several lines of evidence show that the occurrence of paclitaxel resistance is associated with metastasis in NSCLC patients who have never received paclitaxel treatment. Intrinsic resistance to paclitaxel was observed in 76–79% of the NSCLC patients with metastatic stage III/IV cancer [13,14]. So, we previously used an in vitro non-adherent culture model to mimic floating cancer cells during metastasis and showed that floating H460 lung cancer cells exhibited paclitaxel tolerance that resulted from increased expression of the βIVa-tubulin isotype [15]. Therefore, this model may be used for studying metastasis-associated paclitaxel tolerance in anoikis-resistant lung cancer cells.

Since paclitaxel-resistant cancer cells require a higher dose of the drug, it is important to explore approaches that enable the increase in intracellular drug concentrations. Paclitaxel is a drug with low solubility in water which requires assistance to allow desired therapeutic concentrations to be reached in tumors: this is a major problem in enabling paclitaxel to achieve satisfactory results [16]. With regards to this issue, a variety of nanoformulation platforms for paclitaxel and other poorly soluble bioactive compounds have been developed to improve solubility and bioavailability [17–19]. In addition, nano-sized formulations utilize the concept of enhanced permeability and retention (EPR) effect and passively extravasate through the leaky vasculature of tumor tissues [20].

Different platforms of paclitaxel nanoformulations have been investigated for NSCLC treatment such as albumin-bound nanoparticles [21], solid lipid nanoparticles [22], lipid-based nanoparticles [23], and polymeric micelle nanoparticles [24]. Poly(D,L-lactide-co-glycolide) or PLGA is an FDA-approved copolymer widely used for producing polymeric nanoparticles with biodegradability and biocompatibility. Over the past two decades, PLGA nanoparticles have been used as a carrier of paclitaxel, either as a single drug or in combination with other agents, for lung cancer treatment. Fonseca et al. prepared paclitaxel-loaded PLGA nanoparticles by interfacial deposition method (presently known as a nanoprecipitation method) and showed the enhancement of paclitaxel cytotoxicity in NCI-H69 lung cancer cells, by the nanoparticles compared to free paclitaxel [25]. Recently, Jiménez-López et al. used a modified nanoprecipitation method to prepare paclitaxel-loaded PLGA nanoparticles and demonstrated promising results of the nanoparticles by inhibiting the proliferation of several lung cancer cell lines with an average three-fold reduction in paclitaxel IC$_{50}$ values compared to free drug. Moreover, the paclitaxel nanoparticles also decreased in vitro growth of cancer stem cells and tumor spheroids, as well as resulting in the rapid accumulation of paclitaxel in various tissues including lungs of mice after intravenous administration of paclitaxel nanoparticles, compared with free paclitaxel [26].

Until now, there have been few studies evaluating paclitaxel-loaded PLGA nanoparticles in drug-resistant lung cancer cells. Yuan et al. used PLGA-Tween80 co-polymer to fabricate paclitaxel-loaded PLGA-Tween80 nanoparticles which exhibited a greater effect than paclitaxel-loaded PLGA nanoparticles for facilitating cellular up-take in paclitaxel-resistant A549 lung cancer cells. Furthermore, the IC_{50} value of paclitaxel-PLGA-Tween80 nanoparticles in the resistant cell line was three-fold and eight-fold lower than IC_{50} values of paclitaxel-PLGA nanoparticles and free drug, respectively, indicating that the presence of hydrophilic part of Tween80 on the surface of PLGA-Tween80 nanoparticles effectively improved paclitaxel delivery into the drug-resistant cells [27].

Recently, research trends suggest that PLGA-lipid hybrid nanoparticles (PLHNPs) have a wide range of therapeutic applications. The PLHNPs consist of three parts, (i) an inner PLGA core encapsulating the hydrophobic drug, (ii) a lipid monolayer shell of lecithin coating the PLGA core, and (iii) an outer PEGylated lipid hydrophilic stealth layer of 1,2-distearoyl-sn-glycero-3-phosphoethanolamine-N-[carboxy(polyethylene glycol)-2000] (DSPE-PEG-COOH) interspersed throughout the lecithin monolayer, which prolongs systemic circulation of the nanoparticles by avoiding clearance by the immune system [28]. Previous studies have shown the effectiveness of PLHNP formulations for the treatment of multiple cancers [29].

In this work, we aimed to apply PLHNPs to improve the effectiveness of the therapeutic effect of paclitaxel towards anoikis-resistant lung cancer cells during metastasis. Thus, we prepared paclitaxel-PLHNPs and demonstrated their potential for overcoming metastasis-associated paclitaxel tolerance in A549 floating lung cancer cells.

2. Results and Discussion

2.1. Preparation and Characterization of Paclitaxel-PLHNPs

First, we investigated the effect of solvent used to dissolve the PLGA polymer for the preparation of PLHNPs. Acetonitrile and tetrahydrofuran were chosen as organic solvents in this study. Empty PLHNPs were prepared using a combination of PLGA and mixed lipid through the nanoprecipitation/self-assembly method to form nanoparticles in the nanometer-sized range with a narrow polydispersity index (PDI). As shown in Figure 1a, the particle size of nanoparticles made with PLGA/acetonitrile was smaller than those made with PLGA/tetrahydrofuran. The more water-miscible solvents tended to have greater polarity which resulted in decreasing the size of nanoparticles [30]. Use of acetonitrile as organic solvent compared with tetrahydrofuran resulted in a more highly negative charged surface (Figure 1b). Therefore, acetonitrile has been chosen as an organic solvent for the formulation of paclitaxel-PLHNPs since smaller particle sizes and higher zeta potential were obtained.

The size of formulated paclitaxel-PLHNPs was 103.0 nm with uniform particle distribution (PDI of 0.11). The paclitaxel-PLHNPs possessed a strong negatively charged surface with carboxylic acid end groups of DSPE-PEG-COOH with zeta potential −52.9 mV, indicating a stable nanoparticle dispersion (Table 1).

To determine the storage stability of nanoparticles, the prepared paclitaxel-PLHNPs were kept at 4 °C and room temperature. As shown in Figure 2, the higher storage temperature induced aggregation of paclitaxel-PLHNPs, causing an increase in particle size. Paclitaxel-PLHNPs were stable in suspension when stored at 4 °C up to 28 days, compared to storage under room temperature. This was consistent with a previous study reporting that PLGA nanospheres should be stored at 4 °C in order to avoid the aggregation [31].

2.2. Evaluation of Anoikis Resistance

In vitro non-adherent culture using polyHEMA-coated plates has been used to obtain floating cancer cells with an anoikis-resistant property that mimics metastasizing cells in the blood and lymphatic circulatory system [15]. Morphological differences were found between A549 attached cells grown as monolayers (Figure 3a) and A549 floating cells cultivated in polyHEMA-coated plates (Figure 3b). The floating cells exhibited a round shape and formed aggregates similar to the aggregated floating lung cancer cells found in lymphatic vessels of lung cancer patients [32].

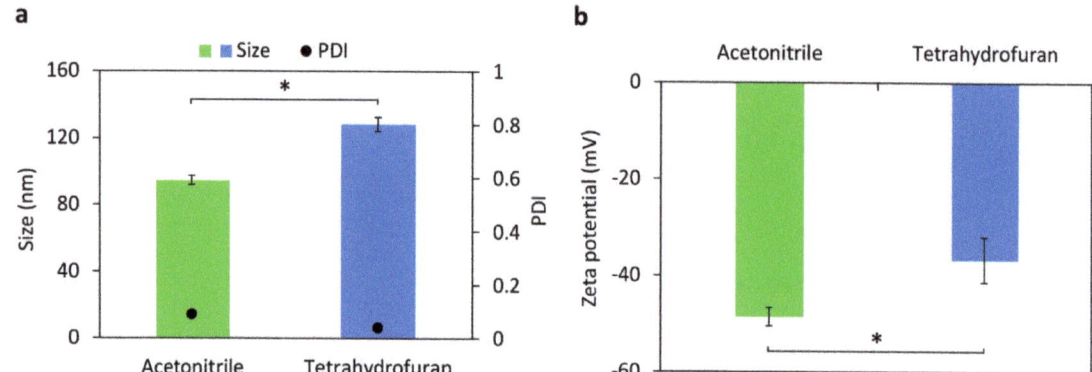

Figure 1. The effect of organic solvents on (**a**) size, PDI, and (**b**) zeta potential of PLGA–lipid hybrid nanoparticles (PLHNPs). Data are reported as mean ± SD from three independent experiments, * $p < 0.05$ compared between the two organic solvents.

Table 1. Characteristics of paclitaxel-PLHNPs. Data are reported as mean ± SD from three independent experiments. * $p < 0.05$ when comparing empty PLHNPs to paclitaxel-PLHNPs.

	Size	PDI	Zeta Potential
Empty PLHNPs	$\begin{bmatrix} 94.6 \pm 2.7 \text{ nm} \\ 103.0 \pm 1.6 \text{ nm} \end{bmatrix}$ *	0.09	* $\begin{bmatrix} -48.6 \pm 1.9 \text{ mV} \\ -52.9 \pm 2.1 \text{ mV} \end{bmatrix}$
Paclitaxel-PLHNPs		0.11	

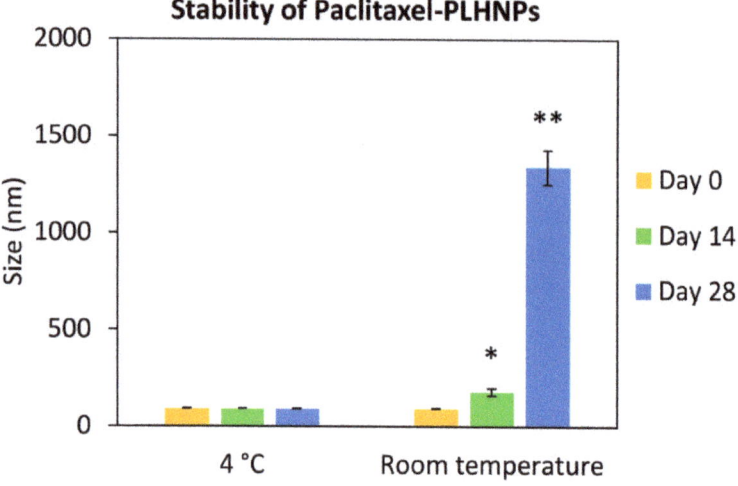

Figure 2. Storage stability of paclitaxel-PLHNPs. Data are reported as mean ± SD from three independent experiments, * and ** $p < 0.05$ compared to Day 0.

Next, we studied the activation of caspase-3 which is a hallmark of apoptotic cell death [33]. Anoikis is defined as a type of apoptosis specifically induced by loss of cell attachment, and caspase-3 activation has been used to detect anoikis in cancer cells [32]. We, therefore, used an assay of caspase-3 to evaluate anoikis resistance in our floating cancer cells. Normal epithelial cells undergo anoikis with loss of attachment. Caspase-3 activity of HMECs cultured under non-adherent conditions was significantly increased, compared with the attached cells (Figure 4a), indicating anoikis induction in these cells. In contrast, there was no significant difference between the caspase-3 activities of A549 floating cells and A549 attached cells (Figure 4b). These results demonstrated that anoikis resistance emerged in A549 floating cells when cultured under non-adherent conditions. Taken together, the results confirmed that the A549 floating cells in our in vitro model displayed morphology and anoikis resistance property similar to that found in metastasizing lung cancer cells. Therefore, this model was further used for investigating the effect of paclitaxel-PLHNPs.

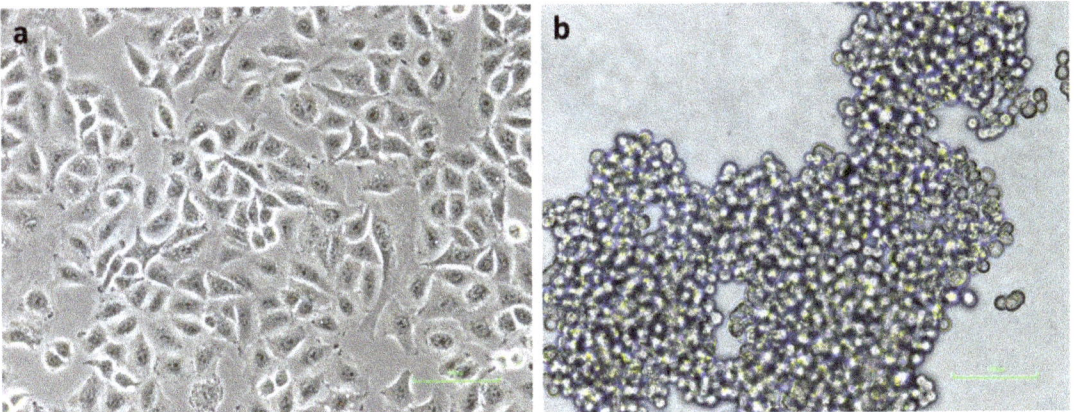

Figure 3. Morphology of (**a**) A549 attached cells cultured under adherent condition and (**b**) A549 floating cells cultured under non-adherent condition. Original magnification of ×200. Scale bar is 100 μm.

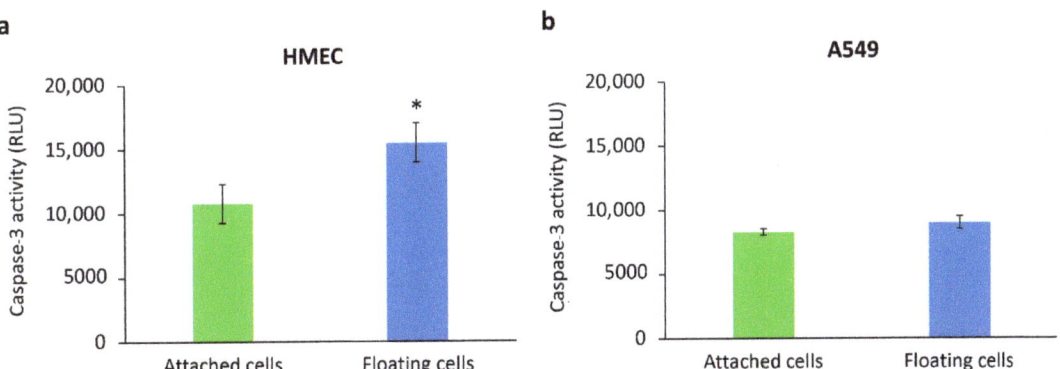

Figure 4. Caspase-3 activation in (**a**) human mammary epithelial cells/HMECs and (**b**) A549 cells after culturing under adherent condition (attached cells) and non-adherent condition (floating cells) for 24 h. Data are reported as mean ± SD from three independent experiments, * $p < 0.05$ compared to the attached cells.

2.3. Comparison of the Cytotoxic Effect of Paclitaxel-PLHNPs and Free Drug in Anoikis-Resistant A549 Cells

We applied the model of A549 floating cells to evaluate the potential of our formulated paclitaxel-PLHNPs against anoikis-resistant cancer. The A549 attached and floating cells were treated with free paclitaxel and paclitaxel-PLHNPs at concentrations equivalent to 1–1000 nM of paclitaxel for 72 h. As shown in Figure 5, both free paclitaxel and paclitaxel-PLHNPs displayed a similar pattern of dose-dependent cytotoxicity in attached and floating cells. The anoikis-resistant floating cells showed tolerance to free paclitaxel with 53.1-fold resistance, compared with the attached cells (Table 2). Treatment with paclitaxel-PLHNPs could reduce the paclitaxel tolerance of floating cells to a 10.3-fold resistance (Table 2). Our previous studies demonstrated that the acquisition of paclitaxel tolerance in floating lung cancer cells was not due to overexpression of drug transporters such as MDR1/P-gp, but was associated with upregulation of βIVa-tubulin gene, a paclitaxel-resistant β-tubulin isotype [15,34].

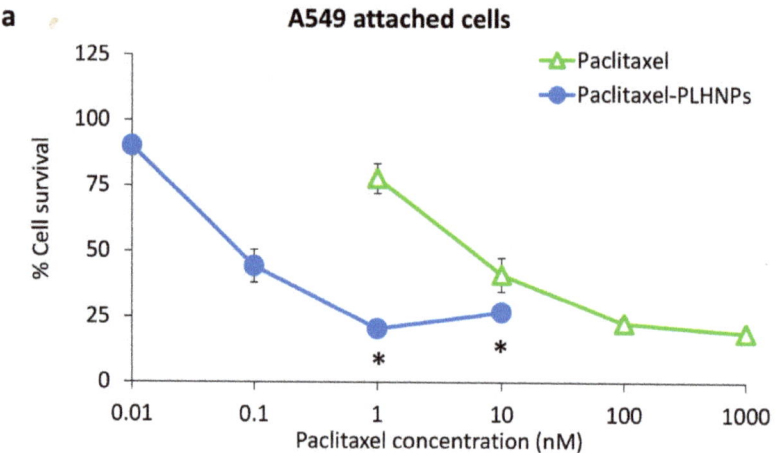

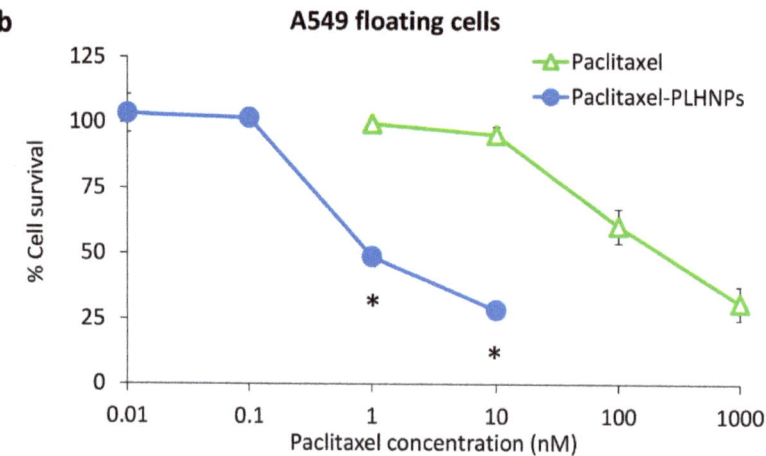

Figure 5. Relative survival rate (%) of (**a**) A549 attached cells and (**b**) A549 floating cells treated with paclitaxel and paclitaxel-PLHNPs. * $p < 0.05$ compared to free paclitaxel.

Table 2. IC_{50} values of paclitaxel in A549 attached and A549 floating cells. * $p < 0.05$ compared to free paclitaxel.

	IC_{50} at 72 h		Fold Resistance
	A549 Attached Cells	A549 Floating Cells	
Free paclitaxel	* [7.88 ± 1.38 nM	* [418.56 ± 194.17 nM	53.1
Paclitaxel-PLHNPs	0.11 ± 0.05 nM]	1.13 ± 0.57 nM]	10.3
Fold change in IC_{50}	71.6-fold decreased	370.4-fold decreased	

The remaining 10.3-fold resistance in A549 floating cells after paclitaxel-PLHNPs treatment indicated the contribution of other mechanisms to the paclitaxel tolerance of the floating cells. Several mechanisms can confer paclitaxel resistance in cancer, such as MDR1/P-gp overexpression, altered expression or mutation of β-tubulin isotypes, changes in expression of anti- or pro-apoptotic proteins, and equally important is that each of these mechanisms could separately contribute to the resistance [35]. Since paclitaxel acts by interfering with microtubule dynamics during mitosis in dividing cells, a reduction in cell growth capability may decrease the sensitivity of cells to paclitaxel cytotoxicity. Supporting evidence was reported from gene expression microarray analysis in paclitaxel-resistant MCF-7 breast cancer cells exhibiting slow growth rate, where the cell doubling time and expression of a cell cycle inhibitor gene CDKNA2/p16 were increased in the paclitaxel-resistant cells [36]. Similarly, we observed that our A549 floating cells exhibited a slower growth compared to the attached cells (unpublished data). Prolonging cell division reduces the chance of paclitaxel interfering with microtubule dynamics, making the A549 floating cells have less response to paclitaxel. This resistant mechanism could not be overcome by increasing intracellular drug concentration, and the mechanism might contribute to the remaining 10.3-fold resistance in A549 floating cells after paclitaxel-PLHNPs treatment.

In A549 attached cells, paclitaxel-PLHNPs showed significantly enhanced cytotoxicity of the drug with a 71.6-fold decrease in IC_{50} value compared to free paclitaxel (Table 2). Interestingly, treatment with paclitaxel-PLHNPs showed a greater cytotoxic effect in A549 floating cells compared with attached cells, as demonstrated by a 370.4-fold decrease in IC_{50} value. These results indicated that the PLHNP platform markedly improved the effectiveness of paclitaxel in anoikis-resistant lung cancer cells. This is in line with a report by Lee et al. [37] demonstrating five-fold higher cytotoxicity of doxorubicin encapsulated in human serum albumin nanoparticles (HSA + DOX NPs), compared with free doxorubicin, in anoikis-resistant MDA-MB-231 breast cancer cells.

The effectiveness of paclitaxel-PLHNPs to overcome paclitaxel tolerance in A549 floating cells might be due to the ability of PLHNPs to deliver the hydrophobic drug into the cells, leading to increased intracellular drug concentration, as proven in our previous study for the use of PLHNPs to deliver a hydrophobic photosensitizer into multidrug-resistant lung cancer cells [34].

3. Conclusions

This study highlights the importance of paclitaxel delivery using PLHNPs to maximize the efficacy of the drug for overcoming paclitaxel tolerance in metastasizing lung cancer cells. The anoikis resistance property of the A549 floating cells was confirmed by the absence of caspase-3 activation, in contrast to the anoikis induction observed in HMEC floating cells. The self-assembly of PLGA-core and lipid-shell hybrid nanoparticles were formulated through a modified nanoprecipitation technique leading to an average particle diameter of 103.0 nm and strongly negative surface charge with a zeta potential of −52.9 mV. The paclitaxel-PLHNPs showed the ability to increase the therapeutic effect of paclitaxel in A549 floating cells with lower IC_{50} values compared with free paclitaxel.

Our results suggest that the PLHNP platform has great potential for delivering hydrophobic drugs to treat floating cancer cells during metastasis in the circulatory systems of the body. Further studies in animal models are required for validating the capability of paclitaxel-PLHNPs to reduce metastasis and recurrence.

However, the development of PLHNPs encapsulating other drugs is also necessary, because anoikis-resistant cells of different cancer types might acquire tolerance to distinct drugs. We also suggest that a targeted delivery system or combination with other treatments is required to improve effectiveness for overcoming drug tolerance of anoikis-resistant cancer cells, in order to reduce the metastatic rate and prevent cancer recurrence.

4. Materials and Methods

4.1. Materials and Chemicals

Poly(D,L-lactide-co-glycolide) (PLGA) with a 50:50 monomer ratio, soybean lecithin consisting of 95% phosphatidylcholine, paclitaxel, 3-(4,5-Dimethyl-2-thiazolyl)-2,5-diphenyl-2H-tetrazolium bromide (MTT), and poly(2-hydroxyethyl methacrylate) or polyHEMA were purchased from Sigma-Aldrich (St. Louis, MO, USA). 1,2-Distearoyl-sn-glycero-3-phosphoethanolamine-N-[carboxy(polyethylene glycol)-2000] (DSPE-PEG-COOH) was obtained from Avanti Polar Lipids, Inc. (Alabaster, AL, USA). Roswell Park Memorial Institute (RPMI) 1640 medium, fetal bovine serum (FBS), and antibiotic–antimycotic solution were supplied by Gibco (Grand Island, NY, USA). All other reagents were of analytical grade and used as received without further purification. Ultrapure water purified by Milli-Q-plus system (Millipore, MA, USA) was used throughout the study.

4.2. Nanoparticle Preparation and Characterization

PLGA-lipid hybrid nanoparticles or PLHNPs were prepared through a previously reported nanoprecipitation method combined with self-assembly [38]. In brief, 5 mg of PLGA polymer was dissolved in 2 mL acetonitrile. Lecithin and DSPE-PEG-COOH (3:1, molar ratio) at 20% PLGA weight were dissolved in 10 mL of 4% ethanol. Polymer solution was then slowly dropped into preheated lipid aqueous solution (65 °C) under stirring. The mixture was subsequently stirred at room temperature for 1.5 h. Paclitaxel-PLHNPs were formulated with a similar method where 0.5 mg of paclitaxel was added to 2 mL of PLGA/acetonitrile solution. The resulting nanoparticles were collected and washed three times with water through an Amicon Ultra-15 centrifugal filter, 10 kDa MWCO (Millipore, MA, USA). The nanoparticles were filter-sterilized and stored at 4 °C for later use.

Hydrodynamic size and size distributions were analyzed by dynamic light scattering (DLS). The zeta potential was determined via electrophoretic mobility. Measurements were performed on samples appropriately diluted with using Zetasizer Nano ZS90 (Malvern Instruments, UK). Paclitaxel content in the nanoparticles was quantified by analyzing the absorbance of paclitaxel in nanoparticles and comparing it to the standard calibration curve of the drug. Nanoparticle stability tests were performed to investigate the effect of storage temperature and duration of storage on the nanoparticle stability in terms of particle size.

4.3. Cell Culture

A549 human lung adenocarcinoma cell line was obtained from the American Type Culture Collection (ATCC, Rockville, MD, USA). Normal human mammary epithelial cells (HMECs) were purchased from LONZA (Walkersville, MD, USA). A549 cells were grown as monolayer cultures in RPMI 1640 supplemented with 10% (v/v) FBS and 1% (v/v) antibiotic–antimycotic. HMEC cells were cultured in Mammary Epithelial Cell Growth Medium (MEGM). All cells were maintained at 37 °C in a humidified incubator with 5% CO_2.

Floating cells were obtained by cultivating the cells under non-adherent culture conditions using polyHEMA-coated culture plates which were prepared according to a previous report with some modifications [15]. Briefly, 96-well plates were coated with a solution of 30 mg/mL polyHEMA in 95% ethanol, followed by drying at 37 °C for 48 h in a non-CO_2 incubator for ethanol evaporation. The dried coated plates were sterilized by exposure to UV light for 20 min prior to beginning each test.

4.4. Caspase-3 Activity Assay

Anoikis cell death was assessed by determining caspase-3 activation using Caspase-Glo 3/7 Assay (Promega, Medison, WI, USA). Briefly, 1×10^4 cells (100 µL/well) were seeded into non-coated or polyHEMA-coated 96-well plates. After 24 h, Caspase-Glo 3/7 reagent (100 µL) was added to each well. Then, the plates were incubated at room temperature in the dark for 1 h. The resulting luminescence was measured with a luminescence microplate reader (Molecular Devices, Sunnyvale, CA, USA).

4.5. Cytotoxicity Assay

A549 cells were plated into normal 96-well plates at a density of 5×10^3 cells (100 µL/well) to obtain attached cells. The cells were plated into polyHEMA-coated 96-well plates at a density of 1×10^4 cells (100 µL/well) to obtain floating cells. After 24 h incubation, the cells were exposed to different concentrations of paclitaxel or paclitaxel-PLHNPs (25 µL/well) for 72 h. A modified MTT assay for non-adherent culture was employed to determine cell viability at the end of treatment [34]. A 25 µL of fresh culture medium containing MTT was added to each well to reach a final concentration of 0.5 mg/mL and incubated for 4 h, followed by adding 100 µL of lysis solution (20% SDS in 10 mM HCl) to solubilize formazan crystals produced by viable cells, then the plates were kept in the dark for 48 h. After that, absorbance was measured at 550 nm and subtracted from a reference wavelength at 650 nm, using a microplate reader. The cell survival rate was expressed as percentage compared with control. Fold resistance was calculated from ratio of the IC_{50} value in floating cells to the IC_{50} value in attached cells.

4.6. Statistical Analysis

Data are expressed as mean with standard deviations (SD) of three independent experiments. A software package PASW Statistics 18 for Windows (SPSS Inc., Chicago, USA) was employed for statistical analysis. A *p*-value less than 0.05 was considered statistically significant.

Author Contributions: Conceptualization, S.P. and K.L.; methodology, S.P., K.A., and K.L.; validation, S.P., K.A., and K.L.; formal analysis, S.P., K.A., and K.L.; investigation, S.P., K.A., and K.L.; resources, P.C. and N.N.; data curation, S.P., K.A., and K.L.; writing—original draft preparation, S.P.; writing—review and editing, K.L., N.N., and J.S.; visualization, S.P. and K.L.; supervision, K.L.; project administration, K.L.; funding acquisition, K.L. and J.S. All authors have read and agreed to the published version of the manuscript.

Funding: This work was supported by Thailand Science Research and Innovation (TSRI), Chulabhorn Research Institute (Grant No. 36821/4274353), and the grant from Center of Excellence on Environmental Health and Toxicology (EHT), OPS, Ministry of Higher Education, Science, Research and Innovation.

Institutional Review Board Statement: Not applicable.

Informed Consent Statement: Not applicable.

Data Availability Statement: The data presented in this study are available on request from the corresponding author.

Acknowledgments: The authors would like to thank Pantipa Subhasitanont, Thiwaree Sornprachum, Phichamon Phetchahwang, Narittira Onnok, Phreeranat Montatip, and Photsathorn Mutapat for their technical support. Special thanks should be given to Voraratt Champattanachai for kindly providing HMECs.

Conflicts of Interest: The authors declare no conflict of interest.

Sample Availability: Samples of the compounds are available from the authors.

References

1. Fares, J.; Fares, M.Y.; Khachfe, H.H.; Salhab, H.A.; Fares, Y. Molecular principles of metastasis: A hallmark of cancer revisited. *Signal Transduct. Target. Ther.* **2020**, *5*, 28. [CrossRef]
2. Adeshakin, F.O.; Adeshakin, A.O.; Afolabi, L.O.; Yan, D.; Zhang, G.; Wan, X. Mechanisms for modulating anoikis resistance in cancer and the relevance of metabolic reprogramming. *Front. Oncol.* **2021**, *11*, 626577. [CrossRef] [PubMed]
3. Taddei, M.; Giannoni, E.; Fiaschi, T.; Chiarugi, P. Anoikis: An emerging hallmark in health and diseases. *J. Pathol.* **2012**, *226*, 380–393. [CrossRef] [PubMed]
4. Chiarugi, P.; Giannoni, E. Anoikis: A necessary death program for anchorage-dependent cells. *Biochem. Pharmacol.* **2008**, *76*, 1352–1364. [CrossRef] [PubMed]
5. Wang, X.; Zhang, H.; Chen, X. Drug resistance and combating drug resistance in cancer. *Cancer Drug Resist.* **2019**, *2*, 141–160. [CrossRef] [PubMed]
6. Hawash, M.; Kahraman, D.C.; Ergun, S.G.; Cetin-Atalay, R.; Baytas, S.N. Synthesis of novel indole-isoxazole hybrids and evaluation of their cytotoxic activities on hepatocellular carcinoma cell lines. *BMC Chem.* **2021**, *15*, 66. [CrossRef]
7. Sung, H.; Ferlay, J.; Siegel, R.L.; Laversanne, M.; Soerjomataram, I.; Jemal, A.; Bray, F. Global cancer statistics 2020: Globocan estimates of incidence and mortality worldwide for 36 cancers in 185 countries. *CA Cancer J. Clin.* **2021**, *71*, 209–249. [CrossRef]
8. Arbour, K.C.; Riely, G.J. Systemic Therapy for locally advanced and metastatic non–small cell lung cancer: A review. *JAMA* **2019**, *322*, 764–774. [CrossRef]
9. Nooreldeen, R.; Bach, H. Current and future development in lung cancer diagnosis. *Int. J. Mol. Sci.* **2021**, *22*, 8661. [CrossRef]
10. Bodor, J.N.; Kasireddy, V.; Borghaei, H. First-line therapies for metastatic lung adenocarcinoma without a driver mutation. *J. Oncol. Pract.* **2018**, *14*, 529–535. [CrossRef] [PubMed]
11. Sharifi-Rad, J.; Quispe, C.; Patra, J.K.; Singh, Y.D.; Panda, M.K.; Das, G.; Adetunji, C.O.; Michael, O.S.; Sytar, O.; Polito, L.; et al. Paclitaxel: Application in modern oncology and nanomedicine-based cancer therapy. *Oxid. Med. Cell. Longev.* **2021**, *2021*, 3687700. [CrossRef] [PubMed]
12. Meena, A.S.; Sharma, A.; Kumari, R.; Mohammad, N.; Singh, S.V.; Bhat, M.K. Inherent and acquired resistance to paclitaxel in hepatocellular carcinoma: Molecular events involved. *PLoS ONE* **2013**, *8*, e61524. [CrossRef] [PubMed]
13. Murphy, W.K.; Fossella, F.V.; Winn, R.J.; Shin, D.M.; Hynes, H.E.; Gross, H.M.; Davilla, E.; Leimert, J.; Dhingra, H.; Raber, M.N.; et al. Phase II study of taxol in patients with untreated advanced non-small-cell lung cancer. *J. Natl. Cancer Inst.* **1993**, *85*, 384–388. [CrossRef] [PubMed]
14. Chang, A.Y.; Kim, K.; Glick, J.; Anderson, T.; Karp, D.; Johnson, D. Phase II Study of taxol, merbarone, and piroxantrone in stage IV non-small-cell lung cancer: The eastern cooperative oncology group results. *J. Natl. Cancer Inst.* **1993**, *85*, 388–394. [CrossRef]
15. Atjanasuppat, K.; Lirdprapamongkol, K.; Jantaree, P.; Svasti, J. Non-adherent culture induces paclitaxel resistance in H460 lung cancer cells via ERK-mediated up-regulation of βIVa-tubulin. *Biochem. Biophys. Res. Commun.* **2015**, *466*, 493–498. [CrossRef]
16. Ma, Y.; Yu, S.; Ni, S.; Zhang, B.; Kung, A.C.F.; Gao, J.; Lu, A.; Zhang, G. targeting strategies for enhancing paclitaxel specificity in chemotherapy. *Front. Cell Dev. Biol.* **2021**, *9*, 626910. [CrossRef]
17. Raza, F.; Zafar, H.; Khan, M.W.; Ullah, A.; Khan, A.U.; Baseer, A.; Fareed, R.; Sohail, M. Recent advances in the targeted delivery of paclitaxel nanomedicine for cancer therapy. *Mater. Adv.* **2022**, *3*, 2268–2290. [CrossRef]
18. Hawash, M.; Jaradat, N.; Eid, A.M.; Abubaker, A.; Mufleh, O.; Al-Hroub, Q.; Sobuh, S. Synthesis of novel isoxazole–carboxamide derivatives as promising agents for melanoma and targeted nano-emulgel conjugate for improved cellular permeability. *BMC Chem.* **2022**, *16*, 47. [CrossRef] [PubMed]
19. Eid, A.M.; Hawash, M. Biological evaluation of safrole oil and safrole oil nanoemulgel as antioxidant, antidiabetic, antibacterial, antifungal and anticancer. *BMC Complement. Med. Ther.* **2021**, *21*, 159. [CrossRef]
20. Matsumura, Y.; Maeda, H. A new concept for macromolecular therapeutics in cancer chemotherapy: Mechanism of tumoritropic accumulation of proteins and the antitumor agent smancs. *Cancer Res.* **1986**, *46 Pt 1*, 6387–6392.
21. Tanaka, M.; Hattori, Y.; Ishii, T.; Tohnai, R.; Itoh, S.; Kawa, Y.; Kono, Y.; Urata, Y.; Satouchi, M. The efficacy of carboplatin plus nanoparticle albumin-bound paclitaxel after cisplatin plus pemetrexed in non-squamous non-small-cell lung cancer patients. *Respir. Investig.* **2020**, *58*, 269–274. [CrossRef]
22. Nadaf, S.J.; Killedar, S.G.; Kumbar, V.M.; Bhagwat, D.A.; Gurav, S.S. Pazopanib-laden lipid based nanovesicular delivery with augmented oral bioavailability and therapeutic efficacy against non-small cell lung cancer. *Int. J. Pharm.* **2022**, *628*, 122287. [CrossRef] [PubMed]
23. Majumder, J.; Minko, T. Multifunctional lipid-based nanoparticles for codelivery of anticancer drugs and siRNA for treatment of non-small cell lung cancer with different level of resistance and EGFR mutations. *Pharmaceutics* **2021**, *13*, 1063. [CrossRef] [PubMed]
24. Shi, M.; Gu, A.; Tu, H.; Huang, C.; Wang, H.; Yu, Z.; Wang, X.; Cao, L.; Shu, Y.; Wang, H.; et al. Comparing nanoparticle polymeric micellar paclitaxel and solvent-based paclitaxel as first-line treatment of advanced non-small-cell lung cancer: An open-label, randomized, multicenter, phase III trial. *Ann. Oncol.* **2021**, *32*, 85–96. [CrossRef]
25. Fonseca, C.; Simões, S.; Gaspar, R. Paclitaxel-loaded PLGA nanoparticles: Preparation, physicochemical characterization and in vitro anti-tumoral activity. *J. Control. Release* **2002**, *83*, 273–286. [CrossRef] [PubMed]

26. Jiménez-López, J.; El-Hammadi, M.M.; Ortiz, R.; Cayero-Otero, M.D.; Cabeza, L.; Perazzoli, G.; Martin-Banderas, L.; Baeyens, J.M.; Prados, J.; Melguizo, C. A novel nanoformulation of PLGA with high non-ionic surfactant content improves in vitro and in vivo PTX activity against lung cancer. *Pharmacol. Res.* **2019**, *141*, 451–465. [CrossRef] [PubMed]
27. Yuan, X.; Ji, W.; Chen, S.; Bao, Y.; Tan, S.; Lu, S.; Wu, K.; Chu, Q. A novel paclitaxel-loaded poly(d,l-lactide-co-glycolide)-Tween 80 copolymer nanoparticle overcoming multidrug resistance for lung cancer treatment. *Int. J. Nanomed.* **2016**, *11*, 2119–2131. [CrossRef]
28. Sivadasan, D.; Sultan, M.H.; Madkhali, O.; Almoshari, Y.; Thangavel, N. Polymeric lipid hybrid nanoparticles (plns) as emerging drug delivery platform—A comprehensive review of their properties, preparation methods, and therapeutic applications. *Pharmaceutics* **2021**, *13*, 1291. [CrossRef] [PubMed]
29. Shah, S.; Famta, P.; Raghuvanshi, R.S.; Singh, S.B.; Srivastava, S. Lipid polymer hybrid nanocarriers: Insights into synthesis aspects, characterization, release mechanisms, surface functionalization and potential implications. *Colloids Interface Sci. Commun.* **2022**, *46*, 100570. [CrossRef]
30. Chan, J.M.; Zhang, L.; Yuet, K.P.; Liao, G.; Rhee, J.W.; Langer, R.; Farokhzad, O.C. PLGA-lecithin-PEG core-shell nanoparticles for controlled drug delivery. *Biomaterials* **2009**, *30*, 1627–1634. [CrossRef]
31. De, S.; Robinson, D.H. Particle size and temperature effect on the physical stability of PLGA nanospheres and microspheres containing Bodipy. *AAPS PharmSciTech* **2004**, *5*, 18–24. [CrossRef] [PubMed]
32. Sakuma, Y.; Takeuchi, T.; Nakamura, Y.; Yoshihara, M.; Matsukuma, S.; Nakayama, H.; Ohgane, N.; Yokose, T.; Kameda, Y.; Tsuchiya, E.; et al. Lung adenocarcinoma cells floating in lymphatic vessels resist anoikis by expressing phosphorylated Src. *J. Pathol.* **2010**, *220*, 574–585. [CrossRef]
33. Porter, A.G.; Jänicke, R.U. Emerging roles of caspase-3 in apoptosis. *Cell Death Differ.* **1999**, *6*, 99–104. [CrossRef]
34. Pramual, S.; Lirdprapamongkol, K.; Jouan-Hureaux, V.; Barberi-Heyob, M.; Frochot, C.; Svasti, J.; Niamsiri, N. Overcoming the diverse mechanisms of multidrug resistance in lung cancer cells by photodynamic therapy using pTHPP-loaded PLGA-lipid hybrid nanoparticles. *Eur. J. Pharm. Biopharm.* **2020**, *149*, 218–228. [CrossRef] [PubMed]
35. Yusuf, R.Z.; Duan, Z.; Lamendola, D.E.; Penson, R.T.; Seiden, M.V. Paclitaxel resistance: Molecular mechanisms and pharmacologic manipulation. *Curr. Cancer Drug Targets* **2003**, *3*, 1–19. [CrossRef] [PubMed]
36. Kars, M.D.; Işeri, Ö.D.; Gündüz, U. A microarray based expression profiling of paclitaxel and vincristine resistant MCF-7 cells. *Eur. J. Pharmacol.* **2011**, *657*, 4–9. [CrossRef] [PubMed]
37. Lee, H.; Park, S.; Kim, J.B.; Kim, J.; Kim, H. Entrapped doxorubicin nanoparticles for the treatment of metastatic anoikis-resistant cancer cells. *Cancer Lett.* **2013**, *332*, 110–119. [CrossRef] [PubMed]
38. Pramual, S.; Lirdprapamongkol, K.; Svasti, J.; Bergkvist, M.; Jouan-Hureaux, V.; Arnoux, P.; Frochot, C.; Barberi-Heyob, M.; Niamsiri, N. Polymer-lipid-PEG hybrid nanoparticles as photosensitizer carrier for photodynamic therapy. *J. Photochem. Photobiol. B Biol.* **2017**, *173*, 12–22. [CrossRef] [PubMed]

Article

PLLA Coating of Active Implants for Dual Drug Release

Katharina Wulf [1,*,†], Madeleine Goblet [2,†], Stefan Raggl [3], Michael Teske [1], Thomas Eickner [1], Thomas Lenarz [2], Niels Grabow [1] and Gerrit Paasche [2,*]

1. Institute for Biomedical Engineering, University Medical Center Rostock, 18119 Rostock, Germany; michael.teske@uni-rostock.de (M.T.); thomas.eickner@uni-rostock.de (T.E.); niels.grabow@uni-rostock.de (N.G.)
2. Department of Otorhinolaryngology, Head and Neck Surgery and Hearing4all Cluster of Excellence, Hannover Medical School, 30625 Hannover, Germany; goblet.madeleine@mh-hannover.de (M.G.); lenarz.thomas@mh-hannover.de (T.L.)
3. MED-EL Medical Electronics, Fuerstenweg 77a, 6020 Innsbruck, Austria; stefan.raggl@medel.com
* Correspondence: katharina.wulf@uni-rostock.de (K.W.); paasche.gerrit@mh-hannover.de (G.P.); Tel.: +49-381-54345520 (K.W.); +49-511-5323808 (G.P.)
† These authors contributed equally to this work.

Abstract: Cochlear implants, like other active implants, rely on precise and effective electrical stimulation of the target tissue but become encapsulated by different amounts of fibrous tissue. The current study aimed at the development of a dual drug release from a PLLA coating and from the bulk material to address short-term and long-lasting release of anti-inflammatory drugs. Inner-ear cytocompatibility of drugs was studied in vitro. A PLLA coating (containing diclofenac) of medical-grade silicone (containing 5% dexamethasone) was developed and release profiles were determined. The influence of different coating thicknesses (2.5, 5 and 10 µm) and loadings (10% and 20% diclofenac) on impedances of electrical contacts were measured with and without pulsatile electrical stimulation. Diclofenac can be applied to the inner ear at concentrations of or below 4×10^{-5} mol/L. Release of dexamethasone from the silicone is diminished by surface coating but not blocked. Addition of 20% diclofenac enhances the dexamethasone release again. All PLLA coatings serve as insulator. This can be overcome by using removable masking on the contacts during the coating process. Dual drug release with different kinetics can be realized by adding drug-loaded coatings to drug-loaded silicone arrays without compromising electrical stimulation.

Keywords: PLLA coating; dual drug delivery; spiral ganglion neuron; impedance measurements; cochlear implant; diclofenac

Citation: Wulf, K.; Goblet, M.; Raggl, S.; Teske, M.; Eickner, T.; Lenarz, T.; Grabow, N.; Paasche, G. PLLA Coating of Active Implants for Dual Drug Release. *Molecules* **2022**, *27*, 1417. https://doi.org/10.3390/molecules27041417

Academic Editors: Marek Brzeziński and Małgorzata Baśko

Received: 3 February 2022
Accepted: 17 February 2022
Published: 19 February 2022

Publisher's Note: MDPI stays neutral with regard to jurisdictional claims in published maps and institutional affiliations.

Copyright: © 2022 by the authors. Licensee MDPI, Basel, Switzerland. This article is an open access article distributed under the terms and conditions of the Creative Commons Attribution (CC BY) license (https://creativecommons.org/licenses/by/4.0/).

1. Introduction

Cochlear implants (CI) are currently the most effective treatment options for severe to profound hearing loss. During cochlear implantation, an electrode array consisting of different numbers of platinum contacts on a silicone carrier is inserted into the scala tympani of a cochlea. Cochlear nerve cells, the spiral ganglion neurons (SGN), can then be electrically stimulated by application of pulses of constant current. Clinical results with CI are typically good; for example, most patients can communicate via the telephone again [1]. Nevertheless, there are several known limitations. First, after hearing loss, SGN also start to degenerate [2]. Second, for insertion of a CI electrode, the cochlea has to be opened and the electrode array is positioned in the scala tympani. This causes some additional trauma which is considered to be a risk for surviving SGN [3]. As a reaction of the human body to this trauma, but also to the implanted foreign body, fibrous tissue is formed around the electrode array [4]. As shown in postmortem studies, the amount of tissue formation can be variable from a few cells to the formation of new bone [5]. The increase in electrical impedance at the stimulating contacts, as reported for the first two to three weeks after implantation [6], was shown to be correlated with the tissue response after implantation [7].

Furthermore, when the tissue formation is not uniform along the electrode array, it might also affect the specificity of the electrical stimulation.

Currently, there are several approaches under investigation to reduce trauma and the formation of fibrous tissue after cochlear implantation. Amongst them are surface patterning of the electrode array [8], application of drugs via pumps [9], cells [10], coatings [11], or from a reservoir, such as the silicone of the electrode array [12,13], as well as intraoperative deposition of steroids either directly [6] or by using a catheter [14]. To the best of our knowledge, besides one report on three patients receiving mononuclear cells obtained from bone marrow with the cochlear implant [10], only intraoperative deposition of steroids and steroid elution from the silicone of CI electrodes have been used clinically so far [12,14] and were shown to reduce or delay the impedance increase after implantation. As elution from the silicone results in a slow release [15], combination with a faster release from a surface coating might be a promising way to effectively address the tissue reaction right after implantation and in the long term.

The release of active substances can basically be divided into two types: diffusion-controlled drug release and chemically controlled drug release [16]. Diffusion-controlled release is further divided into membrane-associated and matrix-associated release. No matter which of these two is considered, both behave according to Fick's first law of diffusion [16]. Characteristically, the release depends on the concentration gradient. At the beginning, the drug-release system is fully loaded whereas the tissue environment does not contain any drug. This results in a so-called initial burst release, a strong increase in concentration of the active ingredient in the tissue. In the further course of time, the release continues to level off.

In contrast, chemically controlled drug release requires steps that occur before the actual release. In so-called "swelling-controlled systems", the active ingredient is distributed in a polymer matrix but cannot diffuse out of the material, e.g., due to a small pore size. After a solvent is added the polymer swells, causing the pores to enlarge in such a way that diffusion is no longer inhibited, and the active ingredient is released. In degradation-controlled systems, bonds have to be cleaved before the active ingredient is released. These bonds belong to the polymer in which the active ingredient is incorporated. Furthermore, there are systems in which the active ingredient is covalently bound to the polymer, e.g., as side chains. Hence, there is also a bond that has to be cleaved, before the active ingredient can diffuse out [17–19].

For dual drug release—which means the release of two active ingredients, each exhibiting a different release behavior—it is advantageous to use different mechanisms of release control. Therefore, drugs incorporated in polymer matrices can be used together with, e.g., drugs that are covalently immobilized at the surface of the dedicated polymer. Different polymers with different diffusion-controlled properties, e.g., different pore sizes, may also be used.

This principle was used to combine vascular endothelial growth factor (VEGF) and paclitaxel for application to the cardiovascular system [20,21]. There, paclitaxel was incorporated into a PDLLA polymer coating and VEGF was covalently attached to the surface of the polymer on either films or nanofiber non-woven. The resulting dual drug release improved endothelial cell viability in vitro, even in the presence of paclitaxel, which alone resulted in significantly decreased viability.

Furthermore, polymers with different diffusion properties can be used to obtain a dual drug release. For example, a silicone matrix can be loaded with an active ingredient that shows a relatively slow release. A faster-releasing coating on the silicone body then leads to either a faster release of the same or a different drug [15].

Dexamethasone (DMS) was already incorporated in coatings intended for CI [11,22]. Growth factors and other substances applied to the cochlea and being released from a coating were IGF1, HGF [23], BDNF [24], NT-3 [25] and Ara-C [22]. Most of these examples were intended to enhance the survival of SGN but not for reduction of fibrous tissue formation. A further search for additional substances that might reduce the inflammatory reaction and are already approved for other applications revealed diclofenac (DCF) [26,27] and enalapril [28,29] as possible candidates. DCF is a nonsteroidal anti-inflammatory drug. Its anti-inflammatory action can be explained by the inhibition of the cyclooxygenase in vitro and in vivo [30,31]. Enalapril is an angiotensin-converting-enzyme inhibitor and can reduce local inflammation after myocardial infarction [32].

Therefore, the aim of the current study was to investigate DCF and enalapril for their safety when applied to cells from the inner ear. In a second step, these substances should be included in a surface PLLA coating for a fast initial release and this PLLA coating shall be combined to DMS-loaded silicone of the electrode array for a slower long-term release of DMS. Release characteristics and the influence of the coatings on electrode contact impedances were investigated.

2. Results

2.1. Cell Culture

Diclofenac (DCF) and enalapril were tested regarding their effects on freshly isolated SGN in comparison to the known effects of dexamethasone (DMS). At concentrations of 2×10^{-4} mol/L, surviving SGN were barely found with all three substances (Figure 1a). Survival increased to about 100% at a concentration of 8×10^{-6} mol/L for DMS and DCF and remained stable for lower substance concentrations. After addition of enalapril, the highest survival of SGN with about 76.8% was achieved at a concentration of 8×10^{-6} mol/L. In contrast, neurite length was not affected for all three substances (Figure 1b). Here, only some reduction and fluctuations were observed at concentrations of 2×10^{-4} mol/L and 4×10^{-5} mol/L, where cell numbers were reduced. Based on the results it was decided to concentrate on DCF and DMS in further experiments.

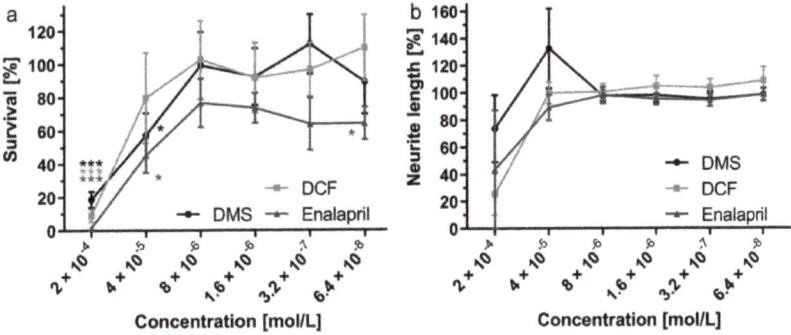

Figure 1. Survival (**a**) and neurite length (**b**) of freshly isolated SGN after addition of different concentrations of DCF, DMS or enalapril. * $p < 0.05$; *** $p < 0.001$ against controls. No differences between drugs were detected.

2.2. PLLA Coating

In order to achieve a stable PLLA coating on silicone, the silicone surface (Sil) (Figure 2, left) was activated with O_2 plasma and an intermediate layer of Silicone-(3-glycidyloxypropyl)trimethoxysilane (Sil-GOPS, Figure 2, left) was generated using the crosslinker GOPS and functionalized via PLLA-NH_2. A stable PLLA coating resulting in Silicone- PLLA (Sil-PLLA, Figure 2, left) could be deposited on this intermediate layer.

In order to illustrate the surface morphology of the different layers, SEM images were taken. The Sil surface (Figure 2a) morphology only slightly changed with addition of GOPS (Figure 2b). However, with addition of the PLLA systems the modified Sil surface became more structured (Figure 2c).

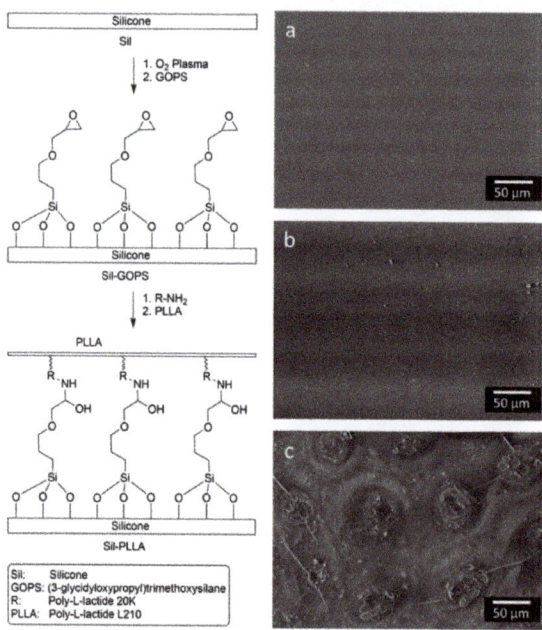

Figure 2. Left: General reaction scheme for the coating of silicone surfaces (Sil) with PLLA via the cross linker GOPS (Sil-GOPS) and the PLLA functionalized with amino groups (Sil-PLLA); **Right**: representative SEM micrographs of Sil (**a**); Sil-GOPS (**b**) and Sil-PLLA (**c**) surfaces.

2.3. Contact Angle Measurements

The hydrophilicity of the silicone surface changed with each reaction step as shown by the contact angles (Figure S2, Supplementary Materials). After addition of GOPS the contact angle decreased significantly, by about 50°. Due to the rather thin interlayer with PLLA-NH$_2$, the measured contact angle changed only slightly. Afterwards, the surface was coated with high molecular weight PLLA and the contact angle again did not change significantly.

2.4. ATR-FTIR

In order to characterize the chemical changes on the Sil surface, the samples were analyzed by ATR-FTIR. The changes in characteristic IR bands indicate a chemical change in the surface composition. Significant differences in the range of 3500–3000 cm^{-1} and 1800–1700 cm^{-1} can be seen comparing the IR spectra Sil, Sil-GOPS, Sil-PLLA-NH$_2$ and Sil-PLLA (Figure 3). Furthermore, the IR spectra for all PLLA coated silicone samples Sil-PLLA-NH$_2$ and Sil-PLLA reveal a prominent band around 1751 cm^{-1} (Figure 3), which denotes carbonyl (C=O) stretching vibration, characteristic for ester bonds found in the used coating polymers. Furthermore, all FTIR spectra for the investigated samples were compared with the specific material used, shown in the supporting materials (Figures S3–S5, Supplementary Materials).

2.5. Drug Release

As shown in Figure 4a, the DMS release is influenced by the addition of the PLLA coating. For the uncoated silicone samples, the released amounts of DMS are significantly higher compared to the coated samples. Around 24 µg DMS was released after 92 days, in contrast to the coated samples where 7 to 10 µg DMS was released in the same amount of time. Furthermore, the additionally incorporated DCF in the PLLA coating (Sil-DCF/PLLA) increases the DMS release significantly from day 1 compared to the other coated samples.

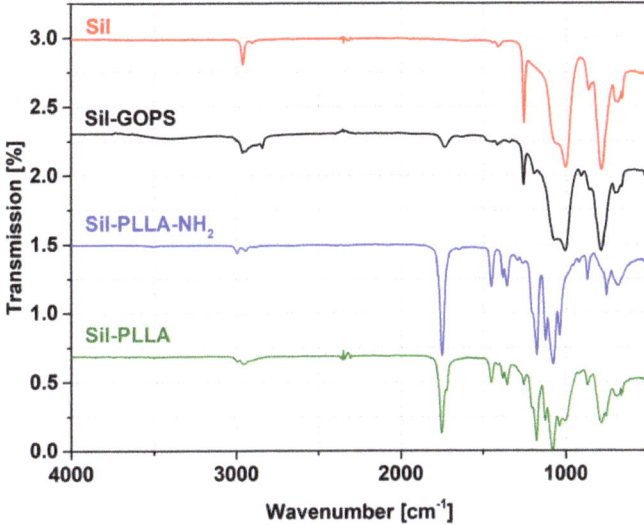

Figure 3. FTIR spectra of investigated Sil based samples Sil, Sil-GOPS, Sil-PLLA-NH$_2$, Sil-PLLA in the range of 4000–500 cm^{-1}. The prominent band between 1764–1712 cm^{-1} corresponds to PLLA signals.

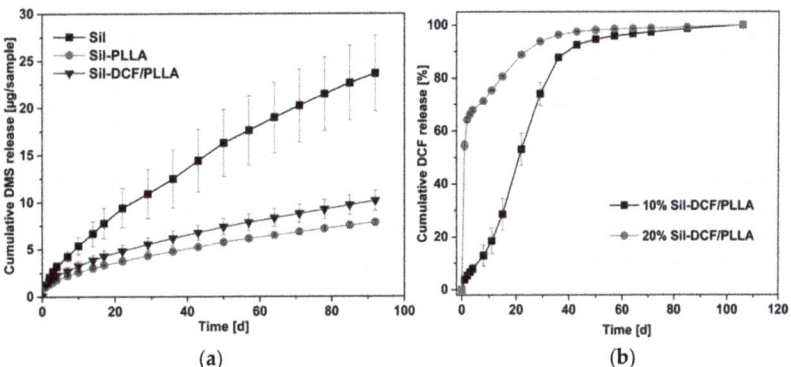

Figure 4. Cumulative in vitro release of incorporated substances from the samples with and without PLLA coating; (**a**): DMS release from uncoated and coated samples (Ø = 6 mm; DCF: PLLA 5: 95 wt%, $N = 3$) Release was with $p < 0.05$ significantly different between all investigated systems after 1 day. (**b**): DCF release from PLLA coated samples (DCF: PLLA 10: 90 wt% and 20: 80 wt%, $N = 3$). Release was with $p < 0.05$ significantly different between days 0 to 71.

Besides the influence of the PLLA coating on the DMS release, also the release of DCF from the PLLA coating was characterized. In vitro release studies with 10 and 20% DCF in the coating were performed. As shown in Figure 4b, a significantly higher burst release was detected for the samples with 20% DCF content compared to the 10% samples. After only one day of release, more than 50% of DCF was already released from the 20% DCF-containing samples in contrast to the 10% DCF-containing samples with only 4% released DCF. With 10% DCF in the coating, it took about 20 days to release 50% of the DCF.

2.6. Impedance Results

The influence of different PLLA-coatings with varying thicknesses (2.5; 5; 10 µm; $N = 5$ each) on electrical contacts was investigated using flat silicone samples, each having included three Pt-contacts comparable to cochlear implant electrode arrays (Figure 5a–c).

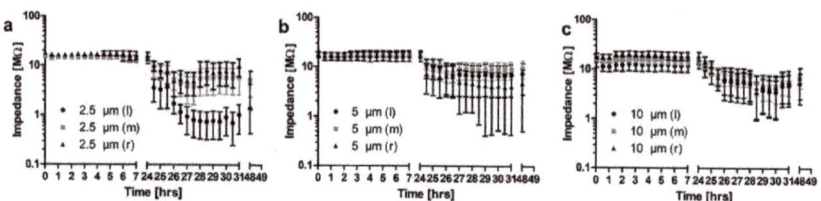

Figure 5. Mean (± SEM) impedance values as measured for contacts coated with PLLA at a thickness of 2.5 μm (**a**), 5 μm (**b**) or 10 μm (**c**). All measurements from $t = 0$ to 24 h were performed with the samples being immersed in 0.9% NaCl. Between all later measurements, left and right contacts were electrically stimulated. $N = 5$ each; l—left contact of the samples, m—middle, r—right contacts.

Initial impedances measured at 1 kHz were above 10 MΩ for 44 of the 45 contacts. The last contact showed impedances between 1 and 10 MΩ. Impedances for all contacts were stable during 24 h incubation in 0.9% NaCl. During the following 24 h of electrical stimulation of two of the three contacts (left and right contacts) on each sample, impedances of stimulated contacts were much more variable which is indicated by larger standard errors of the mean (SEM) in Figure 5a–c. Impedance could remain stable or drop to less than 10 kΩ (examples provided in Figure 6).

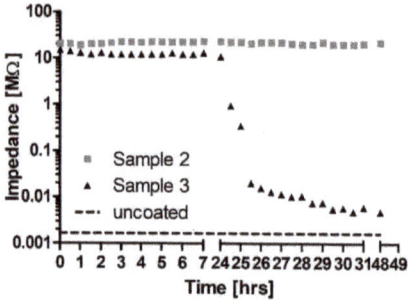

Figure 6. Impedance development over time on two contacts, both coated with 5 μm PLLA. Electrical stimulation started right after the 24 h measurement. The dashed horizontal line indicates mean impedance values of uncoated contacts ($N = 6$).

Average values for uncoated contacts ($N = 6$) were 1.65 kΩ and are indicated by a horizontal dashed line in Figure 6 for comparison. An overview on number of contacts and measured impedances after 24 h of electrical stimulation is provided in Table 1.

Table 1. Overview on the number of contacts with very low or very high (unchanged) impedances after 24 h of electrical stimulation ($N = 5$ samples per condition with 2 stimulated contacts on each sample).

2.5 μm		5 μm		10 μm	
<10 kΩ	>1 MΩ	<10 kΩ	>1 MΩ	<10 kΩ	>1 MΩ
2/10	4/10	6/10	4/10	1/10	4/10

Similar results were found when 10% or 20% DCF were incorporated into a 10 μm PLLA coating (Figure 7). With 10% DCF, initial impedances were >10 MΩ for 11 out of 12 contacts whereas with 20% DCF, initial impedances were between 1 MΩ and 10 MΩ for 11 out of 12 contacts. For both concentrations, impedances could drop under electrical stimulation to below 10 kΩ (2/8 with 10% DCF and 4/8 with 20%) or remain at >1 MΩ (1/8 for both concentrations).

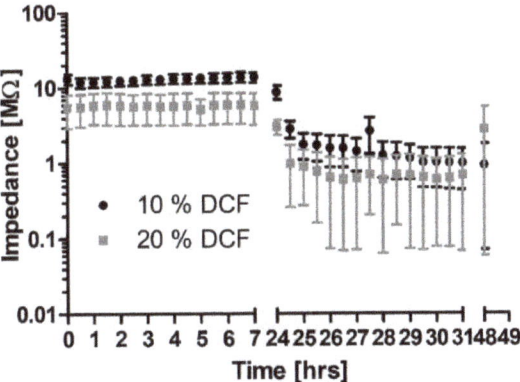

Figure 7. Mean (± SEM) impedance values as measured for contacts coated with PLLA and loaded with 10% or 20% DCF. All measurements from $t = 0$ to 24 h were performed with the samples being immersed in 0.9% NaCl. Between all later measurements, contacts were electrically stimulated ($N = 8$).

2.7. Effect of Electrical Stimulation on Coating

All samples were also examined for morphological changes after impedance measurements. As shown in Figure 8, for each coating thickness of 2.5, 5 and 10 µm, undamaged coating around the platinum contacts (Figure 8a–c) and platinum contacts with cracks and erosion of the coating (Figure 8a`–c`) were found.

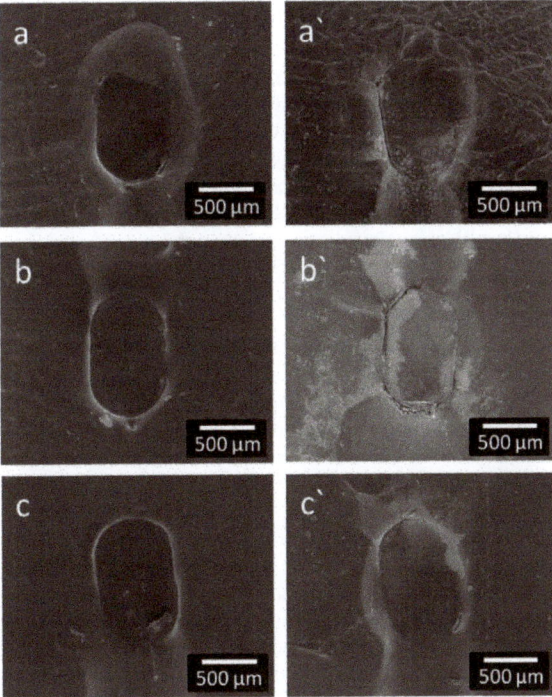

Figure 8. Representative SEM micrographs of platinum contacts after impedance measurements with intact (**a–c**) and damaged (**a`–c`**) PLLA coating of different thicknesses of 2.5 µm (**a,a`**); 5 µm (**b,b`**) and 10 µm (**c,c`**).

2.8. Prevention of Coating of Contacts

In order to prevent the electrical contacts from being coated, masking was attached to the surface of the research electrode contacts prior to the PLLA coating process. After clearing the masked areas, impedances as measured at 1 kHz were between 5 and 24 kΩ (mean 14.0 ± 5.5 kΩ). When measuring the same contacts with the clinical system, impedances were between 2.71 and 4.13 kΩ (mean 3.31 ± 0.51 kΩ).

3. Discussion

Formation of fibrous tissue around the electrode carrier after implantation remains one of the challenges in cochlear implantation. This increases electrical impedances at the stimulating contacts and can reduce the specificity of the stimulation and therefore potentially compromise the hearing outcome with a CI. Potentially induced trauma to the cochlea during electrode insertion and also a foreign body reaction are considered possible reasons for the tissue reaction. To address these, the current project aimed at the combination of a long-term release of DMS from the silicone body of the implant and a short-term release of other suitable substances from a coating on the surface of the electrode array. DCF and enalapril were identified as suitable substances as both are considered to reduce the inflammatory reaction and are already approved for other applications [23,25].

3.1. Cell Culture

No substance applied to the inner ear should evoke toxic effects on spiral ganglion neurons. Therefore, both substances and DMS were first tested with freshly isolated SGN. Nearly no SGN survival at a concentration of 2×10^{-4} mol/L and a slightly reduced survival at 4×10^{-5} mol/L seem to indicate toxic effects at these concentrations. This can most likely be explained by the amount of solvent in the culture wells. DCF and DMS were dissolved in ethanol. Cell growth of HepG2 cells was strongly affected by an ethanol concentration of 2.5% [33], but the cytotoxic concentration differed depending on the cell type. As we had about 2% of ethanol in the samples with a substance concentration of 2×10^{-4} mol/L, we speculate that the reduced cell survival can be attributed to the amount of ethanol in the wells. At lower concentrations, no differences compared to controls were detected for addition of DCF. Therefore, application of DCF is considered safe at least at concentrations of 4×10^{-5} mol/L and below.

Enalapril was dissolved in DMSO. This led to a DMSO concentration of 0.28 mol/L in wells with an enalapril concentration of 2×10^{-4} mol/L. A slightly reduced cell survival can be expected in the range of 0.4 mol/L DMSO in the wells for fibroblasts and the solvent has to be considered toxic at a concentration of 0.7 mol/L [34,35]. Therefore, we expect that neuronal survival at 2×10^{-4} mol/L is most likely influenced by effects of the solvent. The DMSO concentration at 4×10^{-5} mol/L was 0.056 mol/L. Therefore, and according to the published results, cell survival should nearly be unaffected [34,35]. Addition of enalapril never resulted in SGN survival above 80%. As survival of SGN with addition of enalapril was also different to controls at more concentrations, its application to the inner ear was considered not safe. Therefore, in all further experiments the focus was put on DCF.

3.2. PLLA Coating

In order to reach a local long-term release of DMS and DCF, the DMS should be incorporated in the silicone carrier and the DCF should be embedded in a coating for the initial release. Coating silicone carriers is still a challenge due to their inert surface properties [36]. A desired PLLA coating of DMS-containing silicone without crosslinking resulted in a nonsufficient adhesion (data not shown). By first contact with water, the coating immediately peeled off. Therefore, a coating using the covalent binding of GOPS to silicone via the silane group and intermolecular forces between the PLLA-NH$_2$-moiety and the PLLA-bulk was generated, with the PLLA-NH$_2$ providing additional stability of the coating. The contact angle measurements revealed the alteration of the surface hydrophilic properties after each step of the coating. In contrast to the literature, contact angles after GOPS treatment remained quite high (80° vs. 57°) [37]. This goes back to the modification step. While the processes used in the literature are wet chemical in nature, our modification is a plasma-chemical process. The surface density of the groups generated by the O_2 plasma for binding GOPS is thus lower than in wet-chemical processes. As a result, the silicone surface is not completely masked by the GOPS modification and still contributes to the contact angle. The slight increase in the contact angle after the binding of PLLA-NH$_2$ was expected, because of the resulting thin PLLA-layer that presents a more hydrophobic surface. The contact angle after PLLA-coating remains the same as in the NH$_2$-PLLA modification step as the material at the surface before and after the coating process remains PLLA. Chemical changes of the surface composition after each reaction step could also be confirmed by IR measurements. IR spectra of the polymer-coated samples exhibit several characteristic bands of the pure polymer as well as bands that correspond to the pure silicone.

3.3. Drug Release

Profiles of released DMS were detected for Sil, Sil-PLLA and Sil-DCF/PLLA. As DMS release decreases after PLLA-coating, the coating acts as diffusion barrier. Surprisingly, the incorporation of DCF in the PLLA coating led to a significant increase in the release of DMS from 7 to 10% after 13 weeks in vitro DMS release. This is likely due to the incorporation of DCF, which in turn leads to an altered spacing of the polymer chains of PLLA from each other, favoring the formation of enlarged pores compared to PLLA without DCF. To our knowledge, this is the first time that DCF sodium was incorporated in PLLA. Therefore, the behavior has also not been observed before, especially not in a dual drug-release system. The large increase in burst DCF release with 20% DCF in the layer compared to the 10% supports the hypothesis of enlarged pores by DCF in general and by the increased concentration in particular.

3.4. Impedance Measurements

As the function of cochlear or other active implants depends on precise electrical stimulation, and it is to be expected that any coatings of the electrode array will also be deposited on the stimulating contacts, the possible influence of the coating on electrical impedances was investigated. For this purpose, a setup was developed that allowed impedance measurements under reproducible and controlled conditions. The size of the stimulating contacts was chosen to be comparable to the clinically used Cl electrodes. As polymers, and especially the used PLLA, swell slightly by uptake of water after immersion in aqueous solutions [38], samples were just placed in physiological saline and impedances monitored for 24 h. Impedances were stable at very high values, indicating that the PLLA coating acts as insulator independently of the thickness of the coating. Parameters for electrical stimulation were adapted from Peter et al. [39]. The stimulating current was chosen to be safe for spiral ganglion neurons, at least in their in vitro setting. We can only speculate why impedances were reduced on some contacts with the electrical stimulation and on others, no change was detected. Furthermore, cracks in the coating after electrical stimulation were only found at contacts with reduced impedances and always at the transition from the contacts to the surrounding silicone. One possible reason could be the handling of the samples. Silicone is flexible whereas the Pt contacts are rigid. When samples were unintentionally bent during handling, tension within the polymer layer on the surface would be increased especially where the silicone meets the platinum. This could have introduced first cracks that facilitated water uptake and subsequently current flow. In addition, the influence of local voltage peaks at the edges of the Pt contacts cannot be excluded.

Interestingly, impedances of contacts with coatings containing 10% or 20% DCF were different by an order of magnitude right from the first measurement a few seconds after immersion in saline, but stable for at least 24 h. As shown, when there is 20% DCF in the polymer, more than 50% of it is released within 24 h. However, it is very unlikely that this amount of substance is released within a few seconds and the release remains halted for the rest of the first 24 h. Therefore, there must be different explanations for having lower impedances with 20% DCF. It might be that the high amount of the slightly hygroscopic DCF, which is incorporated in PLLA, leads to an increased and/or faster water uptake. This can result in altered spacing between the polymer chains of PLLA, favoring the formation of enlarged pores. Moreover, the increased concentration of the diclofenac sodium salt also increases the ion concentration within the coating, which may lead to increased conductivity and hence lower impedance.

As PLLA acts as insulator, strategies must be developed to avoid coating of contacts or to remove the coating from the stimulating contacts. For the current investigation, masking was additionally added on the contacts before coating. Removal of the masking after coating reliably reduced impedances at contacts as measured at 1 kHz. For a direct comparison with impedances of CI electrodes as measured with the clinical systems using rectangular pulses for the measurements, this approach was additionally taken for measurements of contacts where the coating was removed. Measured values were slightly increased compared to uncoated electrode contacts in the current setting and comparable to known impedance values of animal CI electrodes right before implantation [7] or commercial electrodes shortly after implantation [14].

4. Materials and Methods

4.1. Ethical Statement

The experiments with primary cells were conducted in accordance with the German "Law on Protecting Animals" (§4) and the European Directive 2010/63/EU for protection of animals used for experimental purpose, and registered (no. 2016/118) with the local authorities (Lower Saxony State Office for Consumer Protection and Food Safety (LAVES), Oldenburg, Germany).

4.2. Materials

The investigated silicone (Sil) and Pt-contact samples were provided by MED-EL (Innsbruck, Austria). All silicone samples used in this study contained 5 wt% DMS, which was added during the manufacturing process.

Poly-L-lactide (PLLA, L210) was purchased from Evonik (Schwerte, Germany). The crosslinker (3-glycidyloxypropyl)trimethoxysilane (GOPS) was purchased from Sigma-Aldrich (Taufkirchen, Germany) and PLLA-NH$_2$ was provided by VWR (Dresden, Germany).

4.3. Preparation of Substances

Dexamethasone (Sanofi, Paris, France) and diclofenac (Sigma-Aldrich) were dissolved in ethanol (Carl Roth, Karlsruhe, Germany) whereas enalapril (Selleckchem, Munich, Germany) was dissolved in DMSO (#A3672 AppliChem GmbH, Darmstadt, Germany) at concentrations of 10 mmol/L. The stock solutions were further diluted with complemented cell culture medium to reach a concentration twice as high as the intended test concentrations.

4.4. Spiral Ganglion Cell Culture

Freshly isolated spiral ganglion cells were prepared from neonatal Sprague-Dawley rats (p3–5) following the protocol published by Schulze et al. [40]. In brief, after rapid decapitation, the cochleae were prepared from both halves of the skull and the bony shell of the cochleae was removed. After separating the spiral ganglia from the cochleae, these were enzymatically dissociated with HBSS containing 0.1% trypsin (Biochrom, Berlin, Germany) and 0.01% DNase I (Roche, Basel, Switzerland) for about 20 min followed by gentle trituration in 1 mL serum-free Panserin 401 (PAN Biotech, Aidenbach, Germany), supplemented with HEPES (23.4 µmol/mL, Invitrogen, Carlsbad, CA, USA), glucose (0.15%; Braun AG, Melsungen, Germany), penicillin (30 U/mL; Biochrom, Berlin, Germany), PBS (0.172 mg/mL; Gibco, Thermo Fisher Scientific, Waltham, MA, USA), N2-supplement (0.1%, Invitrogen, Carlsbad, CA, USA), and insulin (8.7 µg/mL; Biochrom, Berlin, Germany) until a homogeneous solution was achieved. Viable cells were counted and seeded at a concentration of 1×10^4 cells per well in a 96-multiwell culture plate (TPP, Trasadingen, Switzerland), coated with 0.1 mg/mL poly-D/L-ornithine (Sigma-Aldrich) and 0.01 mg/mL laminin (natural from mouse, Life Technologies, Carlsbad, CA, USA). Cells were cultivated for 48 h in a mixture of complemented Panserin, supplemented with brain-derived neurotrophic factor (BDNF, Invitrogen, Carlsbad, CA, USA), and the different dilutions of the abovementioned substances. The final BDNF concentration was 50 ng/mL and for the test substances 2×10^{-4} to 6.4×10^{-8} mol/L. Each concentration was tested 6 times ($N = 6$) with three repetitions ($n = 3$) on each plate. After 48 h, the cells were fixed by addition of a 1:1 mixture of acetone (J. T. Baker, Deventer, Netherlands) and methanol (Carl Roth) for 10 min and washed three times with PBS.

4.5. Immunhistochemistry

Following fixation, cells were incubated with the monoclonal mouse 200 kD neurofilament antibody (Novocastra, Leica, Wetzlar, Germany) for 1 h at 37 °C, 5% CO_2. After rinsing with PBS, the secondary biotinylated anti-mouse antibody (Vector Laboratories Inc., Burlingame, CA, USA) was added for 30 min at room temperature before rinsing again with PBS. ABC complex solution (Vectastain® Elite® ABC-Kit, Vector Laboratories, Burlingame, CA, USA) was added to the cells using the protocol of the Vectastain® Elite® ABC Kit. Addition of diaminobenzidine (Peroxidase Substrate Kit DAB; Vector Laboratories, Burlingame, CA, USA) visualized the stained SGNs.

All surviving neurons of each well exhibiting a neurite length of at least three cell soma diameters were counted using a transmission light microscope (Olympus CKX41, Hamburg, Germany). For neurite length measurements, the five longest neurons in each field of view (one in the center and four around the perimeter of the well) were measured using the imaging software cellSens (Olympus, Hamburg, Germany) (compare Figure S6, Supplementary Materials). The survival rate was calculated by normalizing average cell numbers for each condition to average cell numbers in BDNF-treated control wells of the same 96-well plate before averaging across different plates. The same procedure was followed for evaluation of the neurite length. If not otherwise stated, values are presented as mean ± SD.

Statistical analysis of cell-culture results was performed by repeated measures ANOVA followed by Dunnett's posttest using GraphPad Prism version 5.02 (GraphPad, La Jolla, CA, USA). p values of less than 0.05 were considered to be statistically significant.

4.6. Coating of the Silicone Surface

The cleaned silicone surfaces were activated via O_2-plasma using 100 W power at 0.3 mbar for 1 min in a plasma chamber (Diener, Ebhausen, Germany). Then the samples were incubated in pure GOPS for 4 h at 90 °C. The activated samples were rinsed 3 times with ethanol and dried at 80 °C overnight under vacuum at 40 mbar.

The coating of the activated silicone samples was prepared via an established and characterized in-house manufactured spray-coating process. First, the activated silicone samples were spray coated with a thin polymer layer of PLLA-NH_2 using a chloroform PLLA-NH_2 (2 wt%) spray solution. Afterwards the samples were dried at 80 °C overnight and coated with pure polymer at thicknesses of 2.5, 5 or 10 µm (measured via microscopy (Olympus SZX16, Olympus, Hamburg, Germany)) or polymer/drug mixture, containing DCF to PLLA at ratios 10:90 wt% or 20:80 wt% in order to reach a layer thickness of about 10 µm. A chloroform PLLA (0.2 wt%) spray solution was used.

The surfaces were examined in a QUANTA FEG 250 (FEI Company, Germany) scanning electron microscope (SEM). A Contact Angle System (OCA 20, DataPhysics Instruments GmbH, Filderstadt, Germany) was used for analyzing surface modifications by contact angle measurements of ultra-pure water sessile drops. Presented mean values and standard deviations were calculated from $N = 5$ samples. Data were analyzed by Wilcoxon's test using SPSS 27.0.

Attenuated total reflection—Fourier-transform infrared spectroscopy (ATR-FTIR)—measurements of the investigated silicone samples were performed using a Bruker Vertex 70 IR-spectrometer (Bruker, Ettlingen, Germany) equipped with a DLaTGS-detector. Each spectrum has been recorded in the range of 4000–500 cm^{-1} at a spectral resolution of 4 cm^{-1} and with 32 scans on the average using a Graseby Golden Gate Diamond ATR-unit (Bruker, Ettlingen, Germany). All spectra were analyzed using OPUS software (Bruker, Ettlingen, Germany) and were subsequently atmosphere and baseline corrected.

4.7. In Vitro Drug Release

The in vitro drug release studies were carried out under quasi-stationary and sink conditions. Between defined withdrawal time-periods each polymer sample (Ø = 6 mm) was left in the dark at 37 °C immersed in 1 mL artificial perilymph (145 mM NaCl; 2.7 mM KCl; 2 mM MgSO$_4$; 1.2 mM CaCl$_2$; 5 mM HEPES at a pH of 7.3). At the specific time, the elution medium was completely removed, replaced by 1 mL fresh artificial perilymph, and the drug concentration was quantified by HPLC. The chromatography was performed under isocratic conditions with a mobile phase consisting of acetonitrile/ultrapure water (50.5/49.5) (v/v), 0.15 M acetic acid and 4.7 mM trimethylamine at a pH 4.35. The flow rate was set to 1.0 mL/min. UV detection was conducted at a wavelength of 275 nm. The retention times for DMS and DCF are 3.8 and 8.3 min, respectively. MV and SD were calculated from N = 3 samples. In order to compare the release, the released amounts were normalized and referred to as the total amount of DMS (100%) and DCF (100%) in the samples. Data were analyzed by Kruskal-Wallis test using SPSS 27.0.

4.8. Impedance Measurements of Coated Samples

For impedance measurements of electrical contacts, flat rectangular silicone samples (1 cm by 1 cm) were generated (Figure S1, Supplementary Materials) with three Pt-contacts (approximate size: 480 × 800 µm) being fixed to one side of the sample. Two of these samples were placed in a beaker filled with 0.9% NaCl such that the samples were positioned approximately parallel at a distance of 2 cm with the contacts facing each other. Measurements were performed between one contact of sample 1 and one contact of sample 2 at 1 kHz by using a 3522-50 LCR-HiTESTER (Hioki, Ueda, Japan). One of the samples in the setup remained uncoated and served as reference electrode whereas the other was the test subject being either uncoated or having received one of the different coatings. Impedances were measured at room temperature without additional electrical stimulation after placing the samples in the beaker (t = 0), then every 30 min during the first 7 h and again 24 h after start. Pulsatile electrical stimulation (biphasic, pulse width: 400 µs, interphase gap: 120 µs, repetition rate: 1 kHz; current: 0.44 mA) was applied for the next 24 h to two of the three contacts (left and right contacts) by using a pulse generator (TGP 110, AIM-TTI, Huntingdon, UK). Impedance measurements were continued on all three contacts of the test sample according to the measurement regime of the first day. After the impedance measurements, the investigated surfaces were examined by scanning electron microscopy (SEM).

In additional measurements with coated research electrodes (MED-EL), the clinical MAESTRO software together with an attached MAX-box (MED-EL) was also used.

Data were analyzed using repeated measures ANOVA (two-way) followed by Bonferroni posttest.

5. Conclusions

After proving cytocompatibility of DCF for application in the inner ear and developing a coating strategy for drug-loaded medical-grade silicone, an approach to realize a dual drug release from cochlear implants was presented that combines a short-term release from a polymeric coating with a long-lasting release from the silicone body of the electrode array. The effect of the coating on electrode contact impedances was characterized and a strategy to overcome the insulation of the contacts was presented. The developed dual drug release for cochlear implant electrode arrays can now be investigated in vivo.

Supplementary Materials: The following supporting information can be downloaded online. Figure S1: Silicone sample as used for impedance measurements; (**a**): uncoated sample with three Pt-contacts and the connector; (**b**): Enlargement of one contact—coated version. Figure S2: Water contact angle $\Theta W \pm$ standard deviation (SD) on silicone surfaces after each reaction step for sessile drop method ($N = 5$). *** $p < 0.001$. Figure S3: Fourier transform infrared spectra of investigated Sil-GOPS in comparison with pure GOPS in the range of 3500–500 cm^{-1}; (**1**) prominent band at 1254 cm^{-1} depicts Si-CH$_2$ bond from the GOPS structure (**2**) prominent band around 1254 cm^{-1} represents oxirane group from the GOPS structure. Figure S4: Fourier transform infrared spectra of investigated Sil-PLLA-NH$_2$ in comparison with pure PLLA-NH$_2$ in the range of 4000–500 cm^{-1}; prominent band at 1751 cm^{-1} corresponds to the C=O stretching vibration from the PLLA structure. Figure S5: Fourier transform infrared spectra of investigated Sil-PLLA in comparison with pure PLLA in the range of 4000–500 cm^{-1}; prominent band at 1751 cm^{-1} corresponds to the C=O stretching vibration from the PLLA structure. Figure S6: Microscopic image of stained spiral ganglion neurons (dark cell bodies) with 5 traced neurites (red). Treatment: 3.2×10^{-7} mol/L DCF. Scale bar: 50 µm.

Author Contributions: Conceptualization, T.L., N.G., methodology, S.R., M.T., T.E., G.P.; validation, T.E.; formal analysis, K.W., M.G., investigation, K.W., M.G.; writing—original draft preparation, K.W., M.G.; writing—review and editing, T.E., G.P.; supervision, G.P.; project administration, resources, funding acquisition, T.L., N.G. All authors have read and agreed to the published version of the manuscript.

Funding: This study was supported by German Ministry for Education and Research (BMBF) as part of RESPONSE-partnership for innovation in implant technology, FKZ: 03ZZ0914D (Hannover), 03ZZ0914A (Rostock), and 03ZZ0914K (MED-EL).

Institutional Review Board Statement: Not applicable.

Informed Consent Statement: Not applicable.

Data Availability Statement: All the data that support the findings of this study are available on request from the corresponding author.

Acknowledgments: The authors would like to thank Dalibor Bajer, Andrea Rohde, Caroline Dudda, Katja Hahn, Jasmin Bohlmann and Babette Hummel for the technical assistance.

Conflicts of Interest: The authors declare no conflict of interest. The funders had no role in the design of the study; in the collection, analyses, or interpretation of data; in the writing of the manuscript, or in the decision to publish the results.

Sample Availability: Samples of the compounds used during the current study are available from the corresponding author on reasonable request.

References

1. Anderson, I.; Baumgartner, W.-D.; Böheim, K.; Nahler, A.; Arnoldner, C.; Arnolder, C.; D'Haese, P. Telephone use: What benefit do cochlear implant users receive? *Int. J. Audiol.* **2006**, *45*, 446–453. [CrossRef] [PubMed]
2. Webster, M.; Webster, D.B. Spiral ganglion neuron loss following organ of corti loss: A quantitative study. *Brain Res.* **1981**, *212*, 17–30. [CrossRef]
3. Scheper, V.; Paasche, G.; Miller, J.M.; Warnecke, A.; Berkingali, N.; Lenarz, T.; Stöver, T. Effects of delayed treatment with combined GDNF and continuous electrical stimulation on spiral ganglion cell survival in deafened guinea pigs. *J. Neurosci. Res.* **2009**, *87*, 1389–1399. [CrossRef] [PubMed]
4. Xu, J.; Shepherd, R.K.; Millard, R.E.; Clark, G.M. Chronic electrical stimulation of the auditory nerve at high stimulus rates: A physiological and histopathological study. *Hear. Res.* **1997**, *105*, 1–29. [CrossRef]
5. Somdas, M.A.; Li, P.M.M.C.; Whiten, D.M.; Eddington, D.K.; Nadol, J.B. Quantitative evaluation of new bone and fibrous tissue in the cochlea following cochlear implantation in the human. *Audiol. Neurootol.* **2007**, *12*, 277–284. [CrossRef]
6. Paasche, G.; Bockel, F.; Tasche, C.; Lesinski-Schiedat, A.; Lenarz, T. Changes of postoperative impedances in cochlear implant patients: The short-term effects of modified electrode surfaces and intracochlear corticosteroids. *Otol. Neurotol.* **2006**, *27*, 639–647. [CrossRef]
7. Wilk, M.; Hessler, R.; Mugridge, K.; Jolly, C.; Fehr, M.; Lenarz, T.; Scheper, V. Impedance Changes and Fibrous Tissue Growth after Cochlear Implantation Are Correlated and Can Be Reduced Using a Dexamethasone Eluting Electrode. *PLoS ONE* **2016**, *11*, e0147552. [CrossRef]
8. Linke, I.; Fadeeva, E.; Scheper, V.; Esser, K.-H.; Koch, J.; Chichkov, B.N.; Lenarz, T.; Paasche, G. Nanostructuring of cochlear implant electrode contacts induces delayed impedance increase in vivo. *Phys. Status Solidi A* **2015**, *212*, 1210–1215. [CrossRef]

9. Borenstein, J.T. Intracochlear drug delivery systems. *Expert Opin. Drug Deliv.* **2011**, *8*, 1161–1174. [CrossRef]
10. Roemer, A.; Köhl, U.; Majdani, O.; Klöß, S.; Falk, C.; Haumann, S.; Lenarz, T.; Kral, A.; Warnecke, A. Biohybrid cochlear implants in human neurosensory restoration. *Stem Cell Res. Ther.* **2016**, *7*, 148. [CrossRef]
11. Xu, M.; Ma, D.; Chen, D.; Cai, J.; He, Q.; Shu, F.; Tang, J.; Zhang, H. Preparation, characterization and application research of a sustained dexamethasone releasing electrode coating for cochlear implantation. *Mater. Sci. Eng. C Mater. Biol. Appl.* **2018**, *90*, 16–26. [CrossRef] [PubMed]
12. Briggs, R.; O 'Leary, S.; Birman, C.; Plant, K.; English, R.; Dawson, P.; Risi, F.; Gavrilis, J.; Needham, K.; Cowan, R. Comparison of electrode impedance measures between a dexamethasone-eluting and standard Cochlear™ Contour Advance® electrode in adult cochlear implant recipients. *Hearing Research* **2020**, *390*, 107924. [CrossRef] [PubMed]
13. Bas, E.; Bohorquez, J.; Goncalves, S.; Perez, E.; Dinh, C.T.; Garnham, C.; Hessler, R.; Eshraghi, A.A.; van de Water, T.R. Electrode array-eluted dexamethasone protects against electrode insertion trauma induced hearing and hair cell losses, damage to neural elements, increases in impedance and fibrosis: A dose response study. *Hear. Res.* **2016**, *337*, 12–24. [CrossRef] [PubMed]
14. Prenzler, N.K.; Salcher, R.; Lenarz, T.; Gaertner, L.; Warnecke, A. Dose-Dependent Transient Decrease of Impedances by Deep Intracochlear Injection of Triamcinolone With a Cochlear Catheter Prior to Cochlear Implantation-1 Year Data. *Front. Neurol.* **2020**, *11*, 258. [CrossRef]
15. Bohl, A.; Rohm, H.W.; Ceschi, P.; Paasche, G.; Hahn, A.; Barcikowski, S.; Lenarz, T.; Stöver, T.; Pau, H.-W.; Schmitz, K.-P.; et al. Development of a specially tailored local drug delivery system for the prevention of fibrosis after insertion of cochlear implants into the inner ear. *J. Mater. Sci. Mater. Med.* **2012**, *23*, 2151–2162. [CrossRef]
16. Peppas, N.A.; Narasimhan, B. Mathematical models in drug delivery: How modeling has shaped the way we design new drug delivery systems. *J. Control. Release Off. J. Control. Release Soc.* **2014**, *190*, 75–81. [CrossRef]
17. Kamaly, N.; Yameen, B.; Wu, J.; Farokhzad, O.C. Degradable Controlled-Release Polymers and Polymeric Nanoparticles: Mechanisms of Controlling Drug Release. *Chem. Rev.* **2016**, *116*, 2602–2663. [CrossRef]
18. Gao, J.; Wang, H.; Zhuang, J.; Thayumanavan, S. Tunable enzyme responses in amphiphilic nanoassemblies through alterations in the unimer-aggregate equilibrium. *Chem. Sci.* **2019**, *10*, 3018–3024. [CrossRef]
19. Fredenberg, S.; Wahlgren, M.; Reslow, M.; Axelsson, A. The mechanisms of drug release in poly(lactic-co-glycolic acid)-based drug delivery systems—A review. *Int. J. Pharm.* **2011**, *415*, 34–52. [CrossRef]
20. Wulf, K.; Teske, M.; Matschegewski, C.; Arbeiter, D.; Bajer, D.; Eickner, T.; Schmitz, K.-P.; Grabow, N. Novel approach for a PTX/VEGF dual drug delivery system in cardiovascular applications-an innovative bulk and surface drug immobilization. *Drug Deliv. Transl. Res.* **2018**, *8*, 719–728. [CrossRef]
21. Wulf, K.; Arbeiter, D.; Matschegewski, C.; Teske, M.; Huling, J.; Schmitz, K.-P.; Grabow, N.; Kohse, S. Smart releasing electrospun nanofibers-poly: L.lactide fibers as dual drug delivery system for biomedical application. *Biomed. Mater.* **2020**, *16*, 15022. [CrossRef] [PubMed]
22. Yu, H.; Tan, H.; Huang, Y.; Pan, J.; Yao, J.; Liang, M.; Yang, J.; Jia, H. Development of a rapidly made, easily personalized drug-eluting polymer film on the electrode array of a cochlear implant during surgery. *Biochem. Biophys. Res. Commun.* **2020**, *526*, 328–333. [CrossRef] [PubMed]
23. Kikkawa, Y.S.; Nakagawa, T.; Ying, L.; Tabata, Y.; Tsubouchi, H.; Ido, A.; Ito, J. Growth factor-eluting cochlear implant electrode: Impact on residual auditory function, insertional trauma, and fibrosis. *J. Transl. Med.* **2014**, *12*, 280. [CrossRef] [PubMed]
24. Chikar, J.A.; Hendricks, J.L.; Richardson-Burns, S.M.; Raphael, Y.; Pfingst, B.E.; Martin, D.C. The use of a dual PEDOT and RGD-functionalized alginate hydrogel coating to provide sustained drug delivery and improved cochlear implant function. *Biomaterials* **2012**, *33*, 1982–1990. [CrossRef]
25. Richardson, R.T.; Wise, A.K.; Thompson, B.C.; Flynn, B.O.; Atkinson, P.J.; Fretwell, N.J.; Fallon, J.B.; Wallace, G.G.; Shepherd, R.K.; Clark, G.M.; et al. Polypyrrole-coated electrodes for the delivery of charge and neurotrophins to cochlear neurons. *Biomaterials* **2009**, *30*, 2614–2624. [CrossRef]
26. Al-Nimer, M.S.; Hameed, H.G.; Mahmood, M.M. Antiproliferative effects of aspirin and diclofenac against the growth of cancer and fibroblast cells: In vitro comparative study. *Saudi Pharm. J. SPJ Off. Publ. Saudi Pharm. Soc.* **2015**, *23*, 483–486. [CrossRef]
27. Jayant, R.D.; McShane, M.J.; Srivastava, R. In vitro and in vivo evaluation of anti-inflammatory agents using nanoengineered alginate carriers: Towards localized implant inflammation suppression. *Int. J. Pharm.* **2011**, *403*, 268–275. [CrossRef]
28. Botta, R.; Lisi, S.; Marcocci, C.; Sellari-Franceschini, S.; Rocchi, R.; Latrofa, F.; Menconi, F.; Altea, M.A.; Leo, M.; Sisti, E.; et al. Enalapril reduces proliferation and hyaluronic acid release in orbital fibroblasts. *Thyroid* **2013**, *23*, 92–96. [CrossRef]
29. Yu, M.; Zheng, Y.; Sun, H.-X.; Yu, D.-J. Inhibitory effects of enalaprilat on rat cardiac fibroblast proliferation via ROS/P38MAPK/TGF-β1 signaling pathway. *Molecules* **2012**, *17*, 2738–2751. [CrossRef]
30. Todd, P.A.; Sorkin, E.M. Diclofenac sodium. A reappraisal of its pharmacodynamic and pharmacokinetic properties, and therapeutic efficacy. *Drugs* **1988**, *35*, 244–285. [CrossRef]
31. Brogden, R.N.; Heel, R.C.; Pakes, G.E.; Speight, T.M.; Avery, G.S. Diclofenac sodium: A review of its pharmacological properties and therapeutic use in rheumatic diseases and pain of varying origin. *Drugs* **1980**, *20*, 24–48. [CrossRef] [PubMed]
32. Borchert, T.; Hess, A.; Lukačević, M.; Ross, T.L.; Bengel, F.M.; Thackeray, J.T. Angiotensin-converting enzyme inhibitor treatment early after myocardial infarction attenuates acute cardiac and neuroinflammation without effect on chronic neuroinflammation. *Eur. J. Nucl. Med. Mol. Imaging* **2020**, *47*, 1757–1768. [CrossRef] [PubMed]

33. Nguyen, S.T.; Nguyen, H.T.-L.; Truong, K.D. Comparative cytotoxic effects of methanol, ethanol and DMSO on human cancer cell lines. *Biomed. Res. Ther.* **2020**, *7*, 3855–3859. [CrossRef]
34. Miller, F.; Hinze, U.; Chichkov, B.; Leibold, W.; Lenarz, T.; Paasche, G. Validation of eGFP fluorescence intensity for testing in vitro cytotoxicity according to ISO 10993-5. *J. Biomed. Mater. Res. B Appl. Biomater.* **2017**, *105*, 715–722. [CrossRef]
35. Moskot, M.; Jakóbkiewicz-Banecka, J.; Kloska, A.; Piotrowska, E.; Narajczyk, M.; Gabig-Cimińska, M. The role of dimethyl sulfoxide (DMSO) in gene expression modulation and glycosaminoglycan metabolism in lysosomal storage disorders on an example of mucopolysaccharidosis. *Int. J. Mol. Sci.* **2019**, *20*, 304. [CrossRef]
36. Zhang, D.-X.; Yoshikawa, C.; Welch, N.G.; Pasic, P.; Thissen, H.; Voelcker, N.H. Spatially Controlled Surface Modification of Porous Silicon for Sustained Drug Delivery Applications. *Sci. Rep.* **2019**, *9*, 1367. [CrossRef]
37. Festag, G.; Steinbrück, A.; Wolff, A.; Csaki, A.; Möller, R.; Fritzsche, W. Optimization of gold nanoparticle-based DNA detection for microarrays. *J. Fluoresc.* **2005**, *15*, 161–170. [CrossRef]
38. Udayakumar, M.; Kollár, M.; Kristály, F.; Leskó, M.; Szabó, T.; Marossy, K.; Tasnádi, I.; Németh, Z. Temperature and time dependence of the solvent-induced crystallization of poly(l-lactide). *Polymers* **2020**, *12*, 1065. [CrossRef]
39. Peter, M.N.; Warnecke, A.; Reich, U.; Olze, H.; Szczepek, A.J.; Lenarz, T.; Paasche, G. Influence of in vitro electrical stimulation on survival of spiral ganglion neurons. *Neurotox. Res.* **2019**, *36*, 204–216. [CrossRef]
40. Schulze, J.; Kaiser, O.; Paasche, G.; Lamm, H.; Pich, A.; Hoffmann, A.; Lenarz, T.; Warnecke, A. Effect of hyperbaric oxygen on BDNF-release and neuroprotection: Investigations with human mesenchymal stem cells and genetically modified NIH3T3 fibroblasts as putative cell therapeutics. *PLoS ONE* **2017**, *12*, e0178182. [CrossRef]

Communication

Long-Term Evaluation of Poly(lactic acid) (PLA) Implants in a Horse: An Experimental Pilot Study

Júlia Ribeiro Garcia Carvalho [1], Gabriel Conde [1], Marina Lansarini Antonioli [1], Clarissa Helena Santana [2], Thayssa Oliveira Littiere [1], Paula Patrocínio Dias [3], Marcelo Aparecido Chinelatto [3], Paulo Aléscio Canola [1], Fernando José Zara [1] and Guilherme Camargo Ferraz [1,*]

1. School of Agricultural and Veterinarian Sciences—FCAV, São Paulo State University—UNESP, 14884-900 Jaboticabal, São Paulo, Brazil; juliargc@hotmail.com (J.R.G.C.); gabriel.conde@unesp.br (G.C.); m.antonioli@unesp.br (M.L.A.); t.littiere@unesp.br (T.O.L.); paulo.canola@unesp.br (P.A.C.); fjzara@fcav.unesp.br (F.J.Z.)
2. Veterinary School, Federal University of Minas Gerais—UFMG, 31270-901 Belo Horizonte, Minas Gerais, Brazil; santana.chs@gmail.com
3. São Carlos School of Engineering—EESC, University of São Paulo—USP, 13566-590 São Carlos, São Paulo, Brazil; pauladpdias@usp.br (P.P.D.); mchinelatto@sc.usp.br (M.A.C.)
* Correspondence: guilherme.c.ferraz@unesp.br

Abstract: In horses, there is an increasing interest in developing long-lasting drug formulations, with biopolymers as viable carrier alternatives in addition to their use as scaffolds, suture threads, screws, pins, and plates for orthopedic surgeries. This communication focuses on the prolonged biocompatibility and biodegradation of PLA, prepared by hot pressing at 180 °C. Six samples were implanted subcutaneously on the lateral surface of the neck of one horse. The polymers remained implanted for 24 to 57 weeks. Physical examination, plasma fibrinogen, and the mechanical nociceptive threshold (MNT) were performed. After 24, 28, 34, 38, and 57 weeks, the materials were removed for histochemical analysis using hematoxylin-eosin and scanning electron microscopy (SEM). There were no essential clinical changes. MNT decreased after the implantation procedure, returning to normal after 48 h. A foreign body response was observed by histopathologic evaluation up to 38 weeks. At 57 weeks, no polymer or fibrotic capsules were identified. SEM showed surface roughness suggesting a biodegradation process, with an increase in the median pore diameter. As in the histopathological evaluation, it was not possible to detect the polymer 57 weeks after implantation. PLA showed biocompatible degradation and these findings may contribute to future research in the biomedical area.

Keywords: biocompatibility; biodegradation; biomaterial; polylactide-based materials; polymer; scanning electron microscopy

1. Introduction

Biodegradable materials have been widely studied for medical applications during the last decades due to their numerous benefits over non-biodegradable materials. Importantly, since these materials disappear after degradation, the implants do not need to be removed. Furthermore, implants prepared using biodegradable polymers can prevent recurrences in the event of fractures, as they can be designed to degrade at a rate that will slowly transfer the load to the healing bone, thus allowing for adequate healing [1,2]. However, these materials also have some technical limitations that make them difficult to process and use as a final product, such as limited control over physicochemical properties, because of the intrinsic raw material variability and difficulties in adjusting degradation rates [1,3].

Poly(lactic acid) (PLA) is one of the most widely used biopolymers and has the potential to be used as an alternative to high-cost, non-biodegradable biocompatible materials. PLA holds promising applications in several areas, particularly in packaging, agricultural products, disposable materials, and the biomedical industry [4,5]. Among its advantages we list its biodegradability, renewability and low environmental impact when discarded. In addition, PLA has good biological safety, mechanical resistance, and processability profiles [6,7]. In the biomedical area, PLA has been used as a degradable suture, bone fixation device, material for surgical implants, drug delivery systems, and scaffolds for human and rodent tissue engineering applications [1,6,8,9].

Therefore, the medical application of PLA in horses can be an exciting opportunity for the experimental development of new biomaterials, as this species undergoes a fast and excessive healing process, which tends to follow abnormal repair reactions such as the formation of exuberant granulation tissue [10,11]. Moreover, with a global herd totaling about 61 million heads and a relevant world market, this animal species is part of the cattle production chain, utilized as a companion animal or in numerous sports, with emphasis on Olympic sports [12,13]. Thus, PLA may be an interesting alternative with potential application in equine medicine.

In the equine species, therapeutic polymer implants can mimic tissues, promote cell proliferation, and tissue reconstitution, forming a conjunctive capsule. Moreover, they can be combined with drugs for systemic or local action. However, polymer implants must be biocompatible and bioabsorbable [4,8]. There are few studies on the use of polymer implants in horses, with most of them involving their use to repair bone fractures [13–17]. Therefore, polymeric materials represent potential alternatives to the use of autologous and heterologous grafts, which have been limited to fracture correction in equine medicine [13,15]. Furthermore, biomaterials have been used for the treatment of joint damage [18–20] and drug delivery [21]. Among these studies, only one used PLA for internal fixation of fractures of horses' proximal sesamoid bone. PLA proved superior to metallic implants because the animals presented a lower degree of lameness and better-quality bone remodeling [14].

Our research group recently evaluated the biocompatibility and biodegradability of PLA and a polymer blend based on PLA and poly(ε-caprolactone) (PCL) compatibilized with a copolymer derived from ε-caprolactone and tetrahydrofuran, which was implanted subcutaneously in horses. However, the materials were only implanted for 24 weeks [22]. Herein, we expand these findings by evaluating the biocompatibility and biodegradability of six PLA implants in a horse up to 57 weeks. Such long-term evaluation of polymer implants has not been performed previously in horses. Our data shows that the PLA implants do not provoke toxic reactions and can be applied safely in vivo for extended times in the equine species.

2. Results

2.1. Clinical Evaluation

Clinical evaluation was carried out during the experiment to verify the health evolution of the horse and/or the possible presence of systemic changes induced by the implantation of the biomaterials. In general, there was no discomfort associated with the implantation of the materials, and no behavioral and/or appetite changes were observed. Intestinal motility, hydration status, apparent mucous color, and capillary filling time were within the normal range for the species. RR and RT values were within reference values for adult horses throughout the implantation period (8–16 mpm; 37–38.3 °C; [23]), (Figure S1b,c). HR values increased at times up to 144 h after implantation; however, they returned to normal values for the equine species after 168 h (28–44 bpm; [23]) (Figure S1a).

2.2. Plasma Fibrinogen

PF concentration was determined to check for systemic inflammatory processes due to the implantation of the biopolymers. PF values remained within the range of reference for the equine species (100–400 mg/dL; [24]) throughout the evaluation period (Figure S2).

2.3. Mechanical Nociceptive Threshold (MNT)

Compared to baseline, MNT reduced 12 and 24 h ($F = 9.431$; $p \leq 0.001$) after implantation. Forty-eight hours after implantation MNT was reversed (Figure 1).

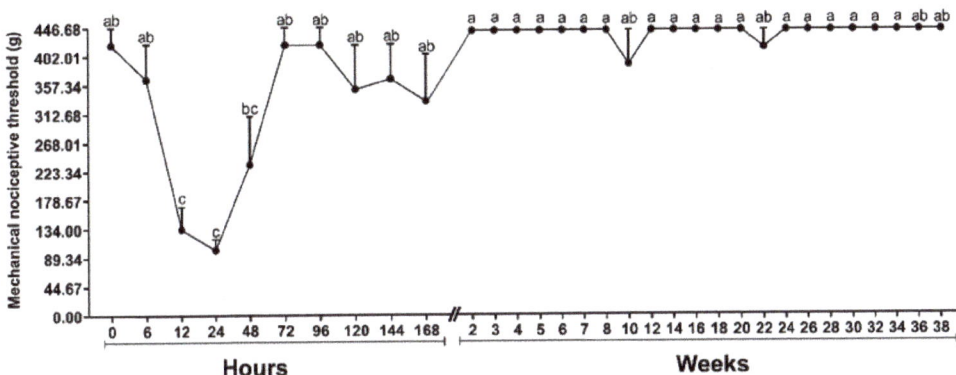

Figure 1. Graphic representation of the means ± standard error of the mechanical nociceptive threshold (MNT) of a horse submitted to the implantation of six polymers of poly(lactic acid) (PLA). Means followed by the same letter do not differ by the Tukey's test ($p < 0.05$) in each moment.

2.4. Histopathological Analysis

Histopathologic evaluation revealed fibrotic capsular formation involving the polymer at all implants surgical removal times (Figure 2), except at 57 weeks; however, over time, the capsules became more organized, as indicated by the progression in the scores of capsule characterizations (Table 1). We observed cellular growth, characterized as fragmentation of the biomaterial and the invasion of the fibrotic tissue, and the consequent surrounding of the evaluated fragments, which got smaller at late implant removal times (Figure 3 and Table 1). We also detected lymphoplasmacytic and histiocytic inflammatory infiltrates associated with the fibrotic capsule, including epithelioid macrophages and giant multinucleated cells with intracytoplasmic polymer fragments that increased with time (Figure 4 and Table 1). These features revealed phagocytosis of the polymer, which was more evident at late implant removal times (Table 1). Further, the fibrotic capsule of all implants showed angioplasia (Figure 2). The sample removed after 57 weeks of implantation presented hemorrhage, some neutrophils, moderate angioplasia, and an abundant presence of collagen fibers. However, no polymer material or fibrotic capsule were identified, hampering the scoring of the fragmented tissue (Figure 5).

Table 1. Score classification of the histopathological lesions in the implant samples at each surgical removal time of a horse implanted with six polymers of poly(lactic acid) (PLA). Intensity of lesions was classified as mild (1), moderate (2), marked (3), and severe (4).

Polymer	Capsule Characterization	Infiltrate/ Inflammation	Cellular Growth	Phagocytosis	Angioplasia
PLA24	1	1	1	2	1
PLA24F	2	1	3	1	2
PLA28	2	2	2	1	1
PLA34	2	2	3	1	2
PLA38	3	3	4	3	2
PLA57	Ns	Ns	Ns	Ns	Ns

PLA24: 24 weeks following implantation; PLA24F: 24 weeks following implantation, formalin-fixed; PLA28: 28 weeks following implantation PLA34: 34 weeks following implantation; PLA38: 38 weeks following implantation; PLA57: 57 weeks following implantation. Ns: not scored.

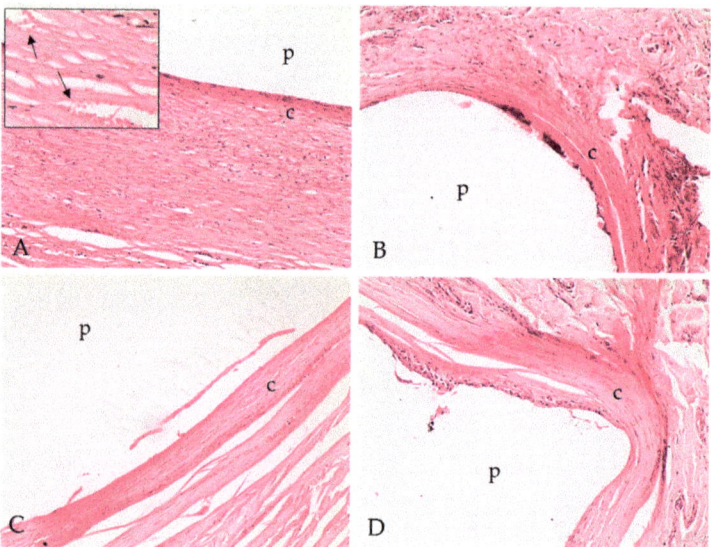

Figure 2. Photomicrographs of implant polymer sites with the polymer (p) and fibrotic capsular formation (c) involving the material. (**A**) Fibrotic capsular formation at 24 weeks; (**B**) 28 weeks; (**C**) 34 weeks; and (**D**) 38 weeks. Hematoxylin and eosin stain, 100×. (**A**) Inlet: fibrotic capsular formation at 24 weeks with small vases proliferation (arrows), characterizing angioplasia. Hematoxylin and eosin stain, 400×.

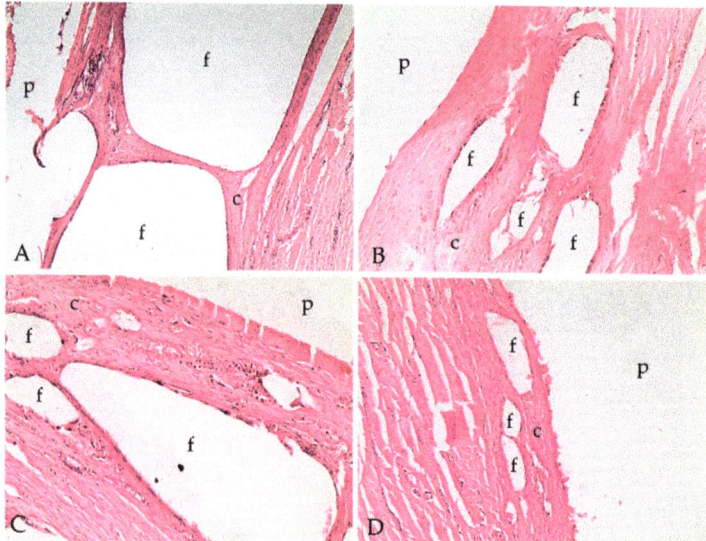

Figure 3. Photomicrographs of implant polymer sites with the capsular formation (c), polymer (p) and cellular growth with fragmentation of the polymer (f). (**A**) Cellular growth and polymer fragments at 24 weeks; (**B**) 28 weeks; (**C**) 34 weeks; and (**D**) 38 weeks. Hematoxylin and eosin stain; 40× (**A**) and 100× (**B–D**).

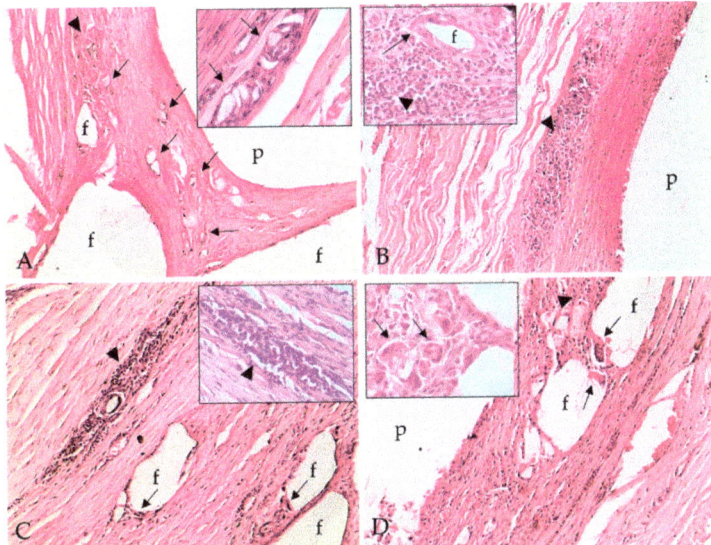

Figure 4. Photomicrographs of implant polymer sites with the polymer (p), lymphoplasmacytic inflammatory infiltrate (arrowhead), and epithelioid macrophages and multinucleated giant cells, with intracytoplasmic polymer material, thus characterizing polymer phagocytosis (arrows). (**A**) Inflammatory infiltrate and polymer phagocytosis at 24 weeks; (**B**) 28 weeks; (**C**) 34 weeks; and (**D**) 38 weeks. Hematoxylin and eosin stain; 100×. Inlet (**A**,**D**): multinucleated giant cells with intracytoplasmic fragments of polymer material, characterizing phagocytosis (arrows). Inlet (**B**,**C**): lymphoplasmacytic inflammatory infiltrate (arrowhead). Inlet (**C**): epithelioid macrophages (arrow) delimiting a polymer fragment (f). Hematoxylin and eosin stain; 400×.

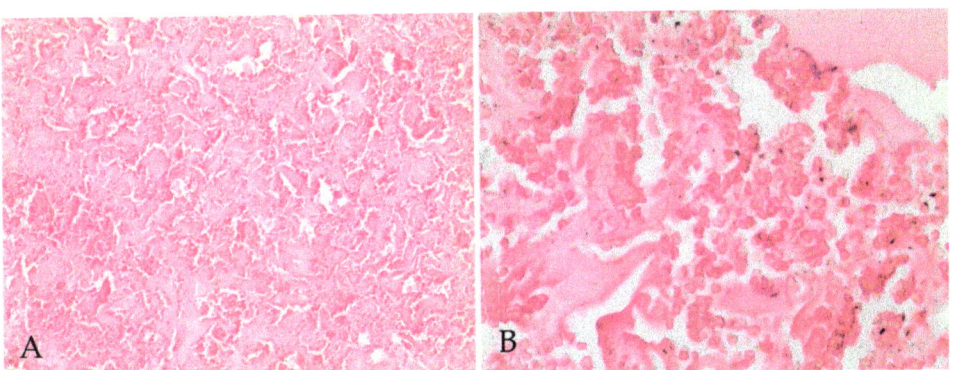

Figure 5. Photomicrographs of implant polymer site removed 57 weeks after implantation. No polymer was identified at this time (**A**) Site of implantation with abundant disorganized collagen fibers, no integral or fragmented polymer material or fibrotic capsule and intense diffuse hemorrhage. (**B**) Site of implantation evidencing the intense diffuse hemorrhage, characterized by many erythrocytes out of vessels. Hematoxylin and eosin; 100× (**A**) and 400× (**B**).

2.5. Scanning Electron Microscopy (SEM)

Figure 6 reveals SEM micrographs of PLA surfaces implanted and exposed to biodegradation for 24 (Figure 6a), 28 (Figure 6b), 34 (Figure 6c), and 38 weeks (Figure 6d) and their respective pore size distribution histograms. The distribution of pore diameters was not normal. The median pore size is shown in Table 2. The data revealed surface irregularity on the surface, porous morphological appearance, and the presence of cracks at all times, indicating a process of biodegradation on the material. Moreover, there was an increase in pore diameter at all times compared to the first moment (24 weeks) (Kruskal-Wallis, $H = 423.113$; $p \leq 0.001$) (Table 2 and Figure 7).

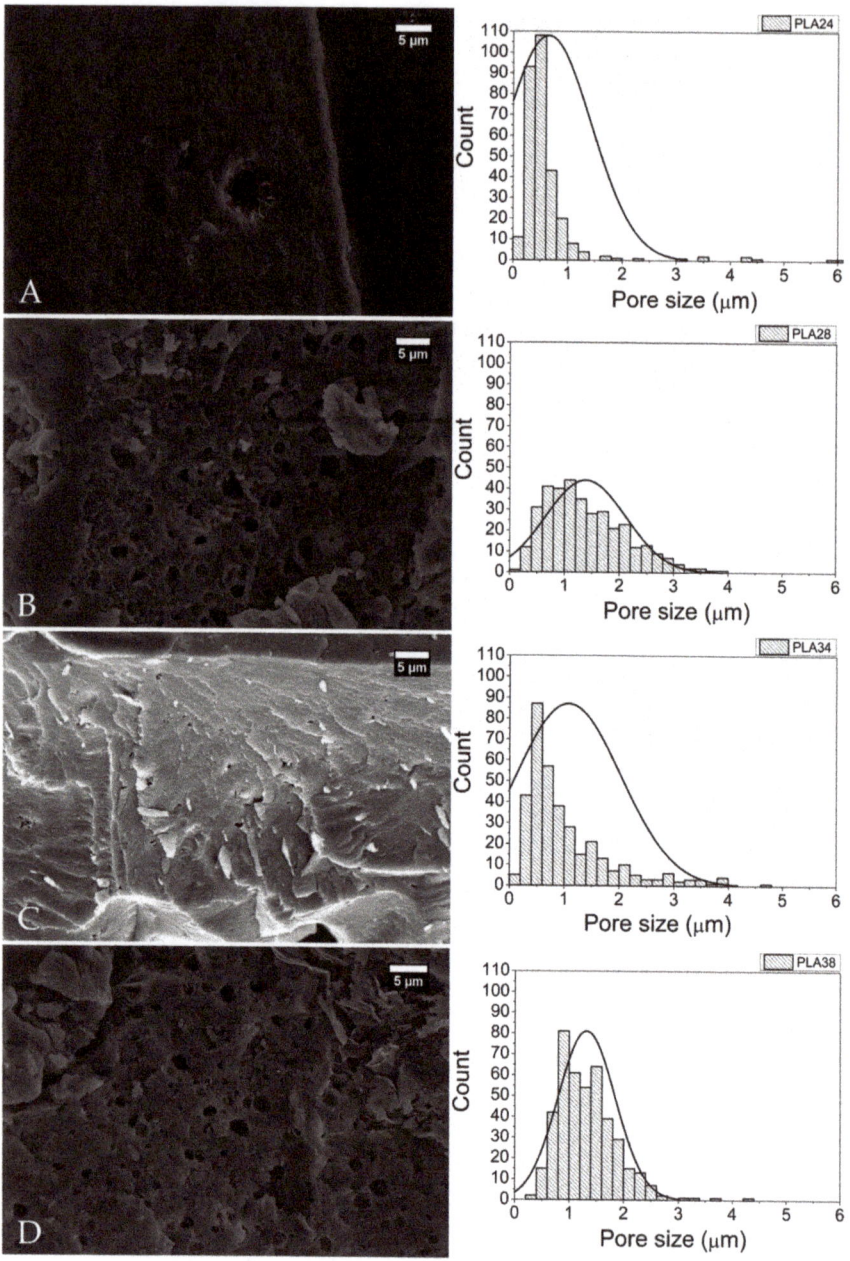

Figure 6. SEM micrographs and pore size distribution histograms of the surface of PLA implanted subcutaneously in one horse. (**A**) Twenty-fourweeks following implantation; (**B**) 28 weeks following implantation; (**C**) 34 weeks following implantation; and (**D**) 38 weeks following implantation.

Table 2. Pore median diameter of PLA implanted subcutaneously in one horse and removed at different times.

Polymer	Pore Median Diameter (µm) (IQR)
PLA24	0.477 (0.354–0.624)
PLA28	1.251 (0.815–1.846) *
PLA34	0.751 (0.487–1.353) *
PLA38	1.254 (0.914–1.618) *

PLA24: 24 weeks following implantation; PLA28: 28 weeks following implantation; PLA34: 34 weeks following implantation; PLA38: 38 weeks following implantation. * Indicates larger pore diameter. IQR, interquartile range.

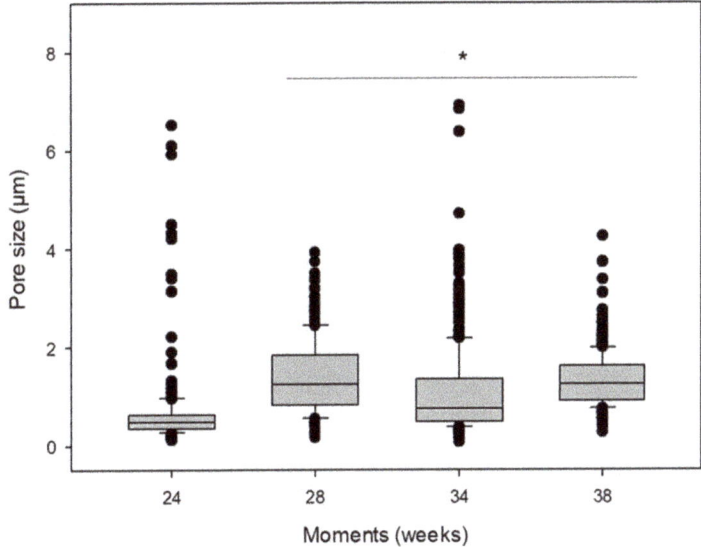

Figure 7. Medians and amplitude of pore size of PLA implanted subcutaneously in one horse. * Indicates increase in pore size when compared to pore size at 24 weeks following implantation (Kruskal-Wallis, $H = 423.113$; $p < 0.001$).

Furthermore, over time, the polymer underwent degradation and changes in its morphology up to 57 weeks. Afterward it was no longer possible to detect the polymer within the skin fragment, thus impeding the removal of polymer fragments for analysis without the skin (Figure 8). We must recognize a limitation of our study. Owning to logistic restrictions, we have not been able to produce aa SEM image from a non-implanted PLA sample. However, this limitation does not compromise the study findings and more information on this non-implanted sample can be found in Carvalho et al., 2020 [22].

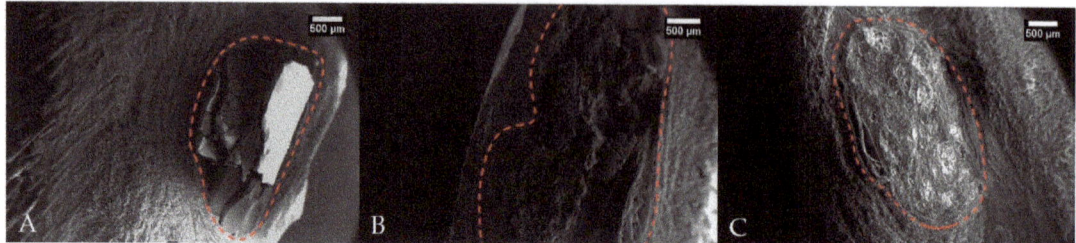

Figure 8. SEM micrographs of skin fragments with PLA implanted in one horse (**A**) 34 weeks following implantation; (**B**) 38 weeks following implantation; and (**C**) 57 weeks following implantation. Dotted red lines delimits the area of the implants.

3. Discussion

We previously studied the changes related to the safe biodegradation of the PLA polymer implanted subcutaneously in horses for 24 weeks [22]. Here we evaluated the changes incurred in a horse over 57 weeks.

We found that PLA implantation did not induce systemic inflammatory responses in the horse studied. HR values increased up to 144 h after implantation. This may have occurred because there was more frequent handling of the animal in this evaluation period, which probably triggered a mild anticipatory sympathetic response to general management. Indeed, HR decreased from the seventh day of implantation, remaining within normal values for equine species. Likewise, the concentrations of plasma fibrinogen, an acute-phase protein commonly used to diagnose and monitor various inflammatory conditions in equine medicine [25], revealed that the injury caused by the skin incisions and the implantation of the polymers was not able to elicit a systemic inflammatory response.

On the other hand, MNT evaluation revealed the presence of a local inflammatory response, characterized by a decrease in nociception. Von Frey Filaments, used to assess the skin MNT, measures cutaneous hyperalgesia and allodynia, and usefully mimics clinical conditions that present increased cutaneous sensitivity [26]. In a study that evaluated the preemptive analgesic effect of epidural ketamine before performing skin incisions with sutures in horses, VFF were also able to quantify skin sensitivity [27].

Histopathological analysis revealed the formation of a fibrotic capsule delimiting the biomaterial from the initial stage of the inflammatory response, remaining present until 38 weeks post-implantation. Encapsulation of biomaterials occurs due to a chronic foreign body reaction [28]. In comparison with PLA24, the PLA38 presented a fibrotic capsule with multiple layers of fibroblasts and essential changes. Among these changes, we detected severe cell growth of fibrous tissue and polymer fragmentation, marked lymphohistioplasmocytic inflammation, marked phagocytosis of the material, and moderate angioplasia, which are characteristic alterations of a chronic inflammatory response [29].

Inflammation around the material, characterized by the presence mainly of lymphocytes and macrophages, accompanied the cell growth of the fibrous tissue amidst the biopolymers, which represented the local chronic inflammation and organization of this tissue. The presence of macrophages in the capsule and at the material's interface was observed in other studies, being related to phagocytosis and implant clearance [8,30]. Interestingly, accentuated phagocytosis was observed in both PLA24 and PLA38 samples. The intense phagocytosis observed in PLA38 was possibly related to an intense fragmentation and biodegradation of the biomaterial, which incited macrophages to phagocytose the resulting fragments for tissue cleaning. Fragmentation of the material leads to cell growth, and the resulting capsule insulated the material while some fibrous tissue infiltrated the polymer, thus promoting further fragmentation.

The proliferation of fibrous tissue around polymer fragments may be associated with the presence of pores and cracks in the materials, as observed by SEM. These pores promote cell growth as they provide a greater surface area [31]. Fifty-seven weeks after implantation, the polymer was no longer detected within the skin fragment. The assessment interval between 38 and 57 weeks is significant and relatively large, and it is not possible to determine what occurred in this period. However, a progression on polymer fragmentation associated with tissue clearance possibly happened to the point when the material was no longer detected. Thus, the fibrotic capsule was replaced by fibrovascular tissue rich in organized and well-differentiated collagen, similar to the collagen observed deep in the dermis. We also speculate that the intense phagocytosis observed in one of the polymers at 24 weeks following implantation is probably related to an initial foreign body response.

In addition to macrophages and multinucleated giant cells, lymphocytes were often present in the infiltrates around the implants. These findings are probably associated with a chronic inflammatory response at the injury site, which is usually characterized by the presence of macrophages, multinucleated giant cells, monocytes, and lymphocytes, in addition to fibrosis and angioplasia [32]. Furthermore, angioplasia was observed at all evaluated times, with mild intensity in PLA24 and moderate intensity in the other materials. The process of well-differentiated vascular proliferation is essential for tissue nutrition, maintenance of cell proliferation, and migration of inflammatory cells [8].

SEM data, obtained from the sample removed 24 weeks following implantation, revealed signs of biodegradation of the PLA. However, few pores were present with an median diameter of 0.477 μm. As post-implantation time progressed, we observed a gradual increase in the number of surface pores and a large variation in pore size. In vivo biodegradation of PLA depends on the intrinsic characteristics of the material, such as the D-PLA content, crystallinity, and molar mass and the recipient's conditions such as temperature, pH, and presence of cellular infiltrates [33]. The PLA used herein has a low content of D-PLA, which hampers its biodegradation. Indeed, chains with a predominance of L-PLA can organize into crystalline structures that are less susceptible to the permeation of water and extracellular enzyme infiltrates [34]. The biodegradation time established for the low content D-PLA implant used herein can be considered high when compared to that reported by Tschakaloff et al. (1994) [35], which showed that 70% of the mass of a L,D-PLA implant in rabbits biodegraded within 14 days.

Another important factor is the size of the PLA chains since only small molecules can enter cells and be used for cell metabolism. The average molar mass of the PLA used in this study is high (around 80,000 g.mol^{-1}). The main in vivo biodegradation mechanism of PLA is the hydrolysis of ester groups, which leads to a slow reduction in chain size by chain scission to small oligomers or unitary lactic acid molecules. Once reduced to small lactic acid oligomers, they can be phagocytosed and take part in cellular metabolism [33].

The distribution of the histograms revealed that the biodegradation pattern of the polymer occurs with the emergence of small pores on the material's surface promoted by the hydrolysis and breakage of polymer chains by extracellular enzymes [33]. Small superficial pores allow the infiltration of water and extracellular enzymes into the polymeric bulk, thus promoting the appearance of more small pores inside the material. Eventually, the chains between the small pores are consumed, and the pores become larger. For this reason, the median pore size was very similar between samples with different post-implantation times, but the analysis of the micrographs revealed a higher number of pores as the post-implantation time progressed until the polymer was completely biodegraded.

The present in vivo study evaluating the long-term biocompatibility and biodegradability of PLA implants in a horse is innovative and suggest that this biopolymer can be safely used in equine medicine. However, PLA has low flexibility and resistance, high apolarity, and low degradation rate, which are limiting properties for its use in human and veterinary medicine. Several approaches could be employed to overcome these limitations or to adjust the physical properties of the currently available PLA, such as the production of polymeric blends. Interestingly, although the PLA used herein was not of medical grade, it proved to be biocompatible and provided encouraging results that could pave the way for new low-cost biomaterials. In any case, future studies addressing the effectiveness of

PLA for further uses, such as drug development, tissue engineering, equine surgery, and general medicine are warranted.

4. Materials and Methods

4.1. Ethics Statement

The study followed the Ethical Principles in Animal Experimentation adopted by the Brazilian College of Animal Experimentation and was approved by the Ethics Committee on Animal Use of the CEUA–FCAV/Universidade Estadual Paulista (UNESP) under protocol n° 006548/17.

4.2. Material Preparation

The polymer was prepared at the Department of Materials Engineering at the São Carlos School of Engineering, University of São Paulo, Brazil, according to Dias e Chinelatto (2019) [36]. The material used was Poly(acid lactic) (PLA grade: Ingeo 3251D), manufactured by NatureWorks Co. Ltd. with 80,000 g.mol^{-1}, 1.4% of D-PLA content, 48 MPa tensile strength, 2.5% strain at break, and 16 J.m^{-1} Izod impact resistance. Polymer implants (1 cm^2 and 1 mm thick) were produced via hot pressing at 180 °C. Sterilization was performed with ethylene oxide (Acecil Central de Esterilização Comércio e Indústria Ltda, Campinas, São Paulo, Brazil).

4.3. Animals

We used one adult horse, gelding, crossbreed, weighing 350 kg, 15 years old, from the didactic herd of the Laboratory of Equine Exercise Physiology and Pharmacology (LAFEQ/DMFA/FCAV), UNESP. The animal was kept in a paddock, fed with corn silage, water, and mineral salt ad libitum, in addition to 0.2% of body weight of a mash feed once a day. Before study commencement, the horse underwent a complete physical examination, in addition to hematological and biochemical tests, to determine its health status. The horse was previously treated with anthelmintic (repeated every four months) and vaccinated against rabies, tetanus toxoid, east and west equine encephalomyelitis, and equine influenza types A1 and A2.

4.4. Polymers Implantation

An overview of the implantation sites is depicted in Figure 9. The lateral surface (LS) of the neck was used as the implant site. The animal evaluated here was used in the pilot study that preceded our previous research [22]. The animal received the PLA implant in different locations between the right and left LS, aiming to investigate whether there would be interference from the inflammatory process between the implants, and whether the difference in thickness of the subcutaneous tissue observed via ultrasound would interfere with the inflammatory process.

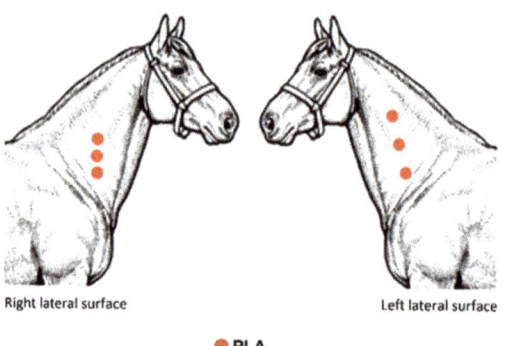

Figure 9. Schematic illustration showing the evaluation sites of a horse implanted with PLA.

Before implanting the biomaterial, six areas of approximately 8 cm2 on the LS (three on the left and three on the right) were shaved, and antisepsis was performed with 2% chlorhexidine digluconate and 70% ethyl alcohol. Subsequently, the animal was sedated with intravenous administration of xylazine hydrochloride (1 mg.kg^{-1}). Afterward, local anesthetic infiltration was performed around each incision site with 2.0 mL of 2% lidocaine hydrochloride. A 2 cm horizontal incision was made using a #15 scalpel blade in the determined area on the LS. A space between the skin and cutaneous muscle was obtained by blunt divulsion and the polymer implanted. It is important to highlight that the divulsion was performed ventrally to the incision so that the skin suture did not interfere in the evaluations. Then, skin suture was performed in a simple interrupted pattern with nylon 0. Postoperative procedures consisted of cleaning the area using gauze and 0.9% saline solution and fly repellent ointment around the surgical wound once a day for 10 days. No analgesic and/or anti-inflammatory medication was provided during the experimental period. The stitches were removed on the 7th postoperative day. The same surgeon performed all surgical procedures.

4.5. Evaluation Methods

4.5.1. Clinical Evaluation

Surgical wounds were evaluated regarding the integrity of the sutures and the presence of secretion. Physical examination quantified the heart rate (HR), respiratory rate (RR), rectal temperature (RT) as well as the intestinal motility, hydration status, apparent mucous color, and capillary filling time. Physical examination was performed before the commencement of the study and 6, 12, 24, 48, 72, 96, 120, 144, 168 h after implantation and then weekly up to 8 weeks, fortnightly up to 38 weeks, and 57 weeks. The same evaluator obtained all the measurements.

4.5.2. Plasma Fibrinogen (PF)

To determine PF concentration, the chronometric technique described by Clauss (1957) [37] was used. Blood samples (3.6 mL) were collected in tubes containing 3.2% sodium citrate. Plasma was separated and frozen at −80 °C. After thawing, 200 µL aliquots of the samples were diluted in a buffer solution containing sodium barbital at a 1:10 ratio. Subsequently, 100 µL of thrombin (Fibrinogênio hemostasis, Labtest diagnóstica, Brazil) was added. Clot formation time was determined at 37 °C using a coagulometer (COAG 1000; Wama Diagnóstica, Brazil), which automatically converted the time obtained into fibrinogen concentration (mg/dL^{-1}). PF was determined before the commencement of the study and 6, 12, 24, 48, 72, 96, 120, 144, 168 h after implantation and again at 2 and 3 weeks.

4.5.3. Mechanical Nociceptive Threshold (MNT)

A commercial set of Von Frey Filaments (VFF) were used to assess the skin MNT (Touch-TestTM Sensory Evaluators, Stoelting Company, Wood Dale, IL, USA). Six filaments of sizes 5.07 to 6.65 were used, which represent applied forces ranging from 11.8 to 446.7 g, respectively. The filaments were applied perpendicularly to the horse's skin until the nylon thread started to bend. Four applications, separated approximately 1 cm from each other, were performed around the implantation site, at intervals of 3 s. Initially, the thinnest filament was used and, when no aversive response was observed, the next filament was used until the animal showed an aversive reaction or the largest filament was used. An aversive reaction was defined as a movement of the tail, ears, or head, kicking, or stepping to the side. Simple motion reflexes upon first touching the filament on the skin were not accepted as an aversion response, and in these cases, the test was repeated after 10 s. Evaluations were performed before the commencement of the study and 6, 12, 24, 48, 72, 96, 120, 144, 168 h after implantation, then weekly until 8 weeks, and fortnightly until 38 weeks. The same operator performed all measurements, and these were made with the horse in a quadrupedal position in an area without movement restrictions. The values obtained were converted into force (g) according to the table provided by the manufacturer.

4.5.4. Collection of Biopsy Specimens

The implants were removed along with surrounding tissue fragments. Two implants were removed after 24 weeks of implantation (PLA24 and PLA24F), one implant was removed at 28 weeks (PLA28), one implant at 34 weeks (PLA34), one at 38 weeks (PLA38), and the final implant was removed at 57 weeks (PLA57). As in the implantation procedure, the animal was properly sedated, and local infiltrative anesthesia was performed before the surgical removal of the fragment. One of the fragments removed after 24 weeks of implantation (PLA24F) was fixed in 10% buffered formalin for 24 h. The other implants and surrounding tissue fragments were fixed in 3.5% glutaraldehyde solution with PBS (pH 7.2) for 24–48 h and subsequently subjected to three washes using 5% glucose in 0.1 m sodium cacodylate buffer solution for 5 min. As we did not know whether fixation in glutaraldehyde would interfere with the histological evaluation and because this horse was used in the pilot study by Carvalho et al., 2020 [22], at 24 weeks two polymers were removed, one of which was fixed in formalin, and the other one in glutaraldehyde. After the slides were evaluated by an experienced pathologist, who confirmed that fixation in glutaraldehyde did not interfere with the histological quality, we proceeded to prepare the samples fixed in glutaraldehyde for SEM analysis. All materials were cut into two fragments, one for histopathological analysis and one for scanning electron microscopy (SEM), except for the PLA24F material, which was submitted to histopathological analysis only. Materials were stored in 70% alcohol until further use.

4.5.5. Histopathological Analysis

For histopathologic evaluation, tissue samples were processed for paraffin embedding, and 5 µm sections were obtained and stained with hematoxylin and eosin. Tissue samples were evaluated in a light optic microscope by a pathologist with no knowledge about the implant's removal time, thus eliminating bias. Lesions were classified semiquantitatively using the scores described by De Jong et al. (2005) [30], with some modifications. Briefly, classification included the characterization of the capsule involving the implant, the presence of an inflammatory infiltrate, cellular growth within the implant, and polymer's phagocytosis. The intensity of each category was evaluated from mild to severe. The intensity of the inflammatory infiltrates, cellular growth within the implant, and polymer's phagocytosis was evaluated as (1) minimal; (2) moderate; (3) marked; and (4) severe. In addition to scoring as proposed by De Jong et al. (2005) [30], the angioplasia in the capsule involving the polymers was also classified as (1) mild, when few small vases were observed within the fibrotic capsule; (2) moderate, when there were some moderately hyperemic vessels grouped around the capsule; and (3) marked, when diffusely hyperemic vessels were observed within the capsule.

4.5.6. Scanning Electron Microscopy (SEM)

Morphological characterization of the materials was performed by SEM. Implant samples with surrounding tissue fragments underwent a dehydration process in an increasing alcohol solution and dried at a critical point. Samples that had the polymer removed from the obtained fragments were dried in an oven at 35 °C for 12 h. All samples were mounted on supports and sputtered with gold in a vacuum. The material surfaces were analyzed in a scanning electron microscope (Zeiss EVO 10, Zeiss, Oberkochen, Germany) operated at 10–15 kV. Images were examined with Fiji ImageJ 1.50i software, and the average pore diameter was determined manually from at least 300 pores. For non-circular pores, the greatest distance was considered.

4.6. Statistical Analysis

All variables were checked for normal distribution using the Shapiro-Wilk test. To evaluate the MNT, the six implants were considered as repetitions ($n = 6$) for the same animal. MNT was subjected to analysis of variance (ANOVA) for repeated measurements in unilateral time, followed by Tukey's post-hoc test. Pore diameter evaluated by SEM images were submitted to the Kruskal–Wallis H-test for one-way analysis of variance (ANOVA), followed by the Dunn's post-hoc test. All statistical analyzes were performed using Sigma-Plot software, version 12.0, with a significance level of $p < 0.05$.

5. Conclusions

Over time the polymer changed its morphology, indicating biocompatible degradation. The kinetics of PLA biodegradation revealed herein reinforce PLA's potential for clinical use and may contribute to future research efforts to enable its use in the biomedical industry, including equine medicine.

Supplementary Materials: The following are available online, Figure S1: Physiological variables of a horse submitted to the implantation of six poly(lactic acid) (PLA) samples. (A) Heart rate (HR), (B) respiratory rate (RR) and (C) rectal temperature (RT). Figure S2: Plasma fibrinogen concentration (mg/dL) of a horse submitted to the implantation of six polymers of poly(lactic acid) (PLA).

Author Contributions: Conceptualization, G.C.F., J.R.G.C., G.C., P.A.C., M.A.C.; methodology, G.C.F., J.R.G.C., G.C., M.L.A., C.H.S., P.P.D., M.A.C., P.A.C., F.J.Z.; formal analysis, J.R.G.C., G.C., M.L.A., C.H.S., T.O.L., P.P.D.; investigation, G.C.F., J.R.G.C., P.P.D.; data curation, G.C.F., J.R.G.C., G.C., M.L.A., C.H.S., P.P.D.; writing—original draft preparation, J.R.G.C., C.H.S., P.P.D.; writing—review and editing, G.C.F., G.C., T.O.L.; supervision, G.C.F., M.A.C., P.A.C., F.J.Z.; project administration, G.C.F., J.R.G.C.; funding acquisition, G.C.F., J.R.G.C. All authors have read and agreed to the published version of the manuscript.

Funding: This research was funded by The São Paulo Research Foundation–FAPESP, grant number 2017/10959-4, 2015/26738-1, 2018/13685-5, and 2019/16779-3 and the National Council for Scientific and Technological Development (CNPq, process no. 132700/2017-4). Additionally, this study was financed in part by the Coordenação de Aperfeiçoamento de Pessoal de Nível Superior—Brasil (CAPES)—Finance Code 001.

Institutional Review Board Statement: The study was conducted according to the the Ethical Principles in Animal Experimentation adopted by the Brazilian College of Animal Experimentation and was approved and supervised by the Ethics Committee on Animal Use of the CEUA–FCAV/Universidade Estadual Paulista (UNESP) under protocol n° 006548/17.

Informed Consent Statement: Not applicable.

Data Availability Statement: The raw/processed data required to reproduce these findings are available from the authors upon request.

Acknowledgments: We thank Áureo Evangelista Santana and the Clinical Pathology Laboratory of the Veterinary Hospital (FCAV/UNESP) for their excellent technical assistance. The authors are also grateful to Claudia Fiorillo and to The Electron Microscopy Laboratory (FCAV/UNESP) facility for their technical support. Also, we thank the Laboratory of Equine Exercise Physiology and Pharmacology (LAFEQ) for their technical assistance. We also thank the Graduate Course Program in Veterinary Medicine (FCAV/UNESP), the Graduate Course Program in Animal Science (FCAV/UNESP) and the Department of Animal Morphology and Physiology (FCAV/UNESP) for the academic support.

Conflicts of Interest: The authors declare no conflict of interest.

References

1. Lasprilla, A.J.R.; Martinez, G.A.R.; Lunelli, B.H.; Jardini, A.L.; Maciel-Filho, R. Poly-lactic acid synthesis for application in biomedical devices—A review. *Biotechnol. Adv.* **2012**, *30*, 321–328. [CrossRef] [PubMed]
2. Abdeljawad, M.B.; Carette, X.; Argentati, C.; Martino, S.; Gonon, M.-F.; Odent, J.; Morena, F.; Mincheva, R.; Raquez, J.-M. Interfacial compatibilization into PLA/Mg composites for improved in vitro bioactivity and stem cell adhesion. *Molecules* **2021**, *26*, 5944. [CrossRef]
3. Sharma, S.; Sudhakara, P.; Singh, J.; Ilyas, R.A.; Asyraf, M.R.M.; Razman, M.R. Critical review of biodegradable and bioactive polymer composites for bone tissue engineering and drug delivery applications. *Polymers* **2021**, *13*, 2623. [CrossRef] [PubMed]
4. Saini, P.; Arora, M.; Ravi Kumar, M.N.V. Poly(lactic acid) blends in biomedical applications. *Adv. Drug Deliv. Rev.* **2016**, *107*, 47–59. [CrossRef] [PubMed]
5. Zhao, X.; Hu, H.; Wang, X.; Yu, X.; Zhou, W.; Peng, S. Super tough poly(lactic acid) blends: A comprehensive review. *RSC Adv.* **2020**, *10*, 13316. [CrossRef]
6. Finotti, P.F.M.; Costa, L.C.; Capote, T.S.O.; Scarel-Caminaga, R.M.; Chinelatto, M.A. Immiscible poly(lactic acid)/poly(ε-caprolactone) for temporary implants: Compatibility and cytotoxicity. *J. Mech. Behav. Biomed. Mater.* **2017**, *68*, 155–162. [CrossRef] [PubMed]
7. Huang, S.; Xue, Y.; Yu, B.; Wang, L.; Zhou, C.; Ma, Y. A review of the recent developments in the bioproduction of polylactic acid and its precursors optically pure lactic acids. *Molecules* **2021**, *26*, 6446. [CrossRef] [PubMed]
8. Ciambelli, G.S.; Perez, M.O.; Siqueira, G.V.; Candella, M.A.; Motta, A.C.; Duarte, M.A.T.; Alberto-Rincon, M.C.; Duek, E.A.R. Characterization of poly (L-co-D, L Lactic Acid) and a study of polymer-tissue interaction in subcutaneous implants in wistar rats. *Mater. Res.* **2013**, *16*, 28–37. [CrossRef]
9. Conde, G.; Carvalho, J.R.G.; Dias, P.P.; Moranza, H.G.; Montanhim, G.L.; Ribeiro, J.O.; Chinelatto, M.A.; Moraes, P.C.; Taboga, S.R.; Bertolo, P.H.L.; et al. In vivo biocompatibility and biodegradability of poly(lactic acid)/poly(ε-caprolactone) blend compatibilized with poly(ε-caprolactone-b-tetrahydrofuran) in Wistar rats. *Biomed. Phys. Eng. Express* **2021**, *7*, 035005. [CrossRef]
10. Chvapil, M.; Pfister, T.; Escalada, S.; Ludwig, J.; Peacock, E.E. Dynamics of the healing of skin wounds in the horse as compared with the rat. *Exp. Mol. Pathol.* **1979**, *30*, 349–359. [CrossRef]
11. Theoret, C. Physiology of wound healing. In *Equine Wound Management*, 3rd ed.; Theoret, C., Schumacher, J., Eds.; John Wiley & Sons, Inc.: Hoboken, NJ, USA, 2017; pp. 1–13.
12. FAO—Food and Agriculture Organization of the United Nations. 2019. Available online: http://www.fao.org/faostat/en/#data/QCL (accessed on 23 August 2021).
13. Golafshan, N.; Vorndran, E.; Zaharievski, S.; Brommer, H.; Kadumudi, F.B.; Dolatshahi-Pirouz, A.; Gbureck, W.; van Weeren, R.; Castilho, M.; Malda, J. Tough magnesium phosphate-based 3D-printed implants induce bone regeneration in an equine defect model. *Biomaterials* **2020**, *261*, 120302. [CrossRef]
14. Pyles, M.D.; Alves, A.L.G.; Hussni, C.A.; Thomassian, A.; Nicoletti, J.L.M.; Watanabe, M.J. Parafusos biaobsorvíveis na reparação de fraturas experimentais de sesamóides proximais em equinos. *Cienc. Rural* **2007**, *37*, 1367–1373. [CrossRef]
15. Moreira, R.C.; Graaf, G.M.M.V.D.; Pereira, C.A.; Zoppa, A.L.D.V.D. Mechanical evaluation of bone gap filled with rigid formulations castor oil polyurethane and chitosan in horses. *Cienc. Rural* **2016**, *46*, 2182–2188. [CrossRef]
16. Nóbrega, F.S.; Selim, M.B.; Arana-Chavez, V.E.; Correa, L.; Ferreira, M.P.; Zoppa, A.L.V. Histologic and immunohistochemical evaluation of biocompatibility of castor oil polyurethane polymer with calcium carbonate in equine bone tissue. *Am. J. Vet. Res.* **2017**, *78*, 1210–1214. [CrossRef]
17. Selim, M.B.; Nóbrega, F.S.; Facó, L.L.; Hagen, S.C.F.; Zoppa, A.L.V.; Arana-Chavez, V.E.; Corrêa, L. Histological and radiographic evaluation of equine bone structure after implantation of castor oil polymer. *Vet. Comp. Orthop. Traumatol.* **2018**, *31*, 405–412. [CrossRef] [PubMed]
18. Rumbaugh, M.L.; Burba, D.J.; Tetens, J.; Oliver, J.L.; Williams, J.; Hosgood, G.; LeBlanc, C.J. Effects of intra-articular injection of liquid silicone polymer in the equine middle carpal joint. In Proceedings of the 50th Annual Convention of the American Association of Equine Practitioners, Denver, CO, USA, 4–8 December 2004; American Association of Equine Practitioners: Lexington, KY, USA, 2004.
19. Barnewitz, D.; Endres, M.; Krüger, I.; Becker, A.; Zimmermann, J.; Wilke, I.; Ringe, J.; Sittinger, M.; Kaps, C. Treatment of articular cartilage defects in horses with polymer-based cartilage tissue engineering grafts. *Biomaterials* **2006**, *27*, 2882–2889. [CrossRef] [PubMed]
20. Albert, R.; Vásárhelyi, G.; Bodó, G.; Kenyeres, A.; Wolf, E.; Papp, T.; Terdik, T.; Módis, L.; Felszeghy, S. A computer-assisted microscopic analysis of bone tissue developed inside a polyactive polymer implanted into an equine articular surface. *Histol. Histopathol.* **2012**, *27*, 1203–1209. [CrossRef] [PubMed]
21. Petit, A.; Redout, E.M.; Van de Lest, C.H.; Grauw, J.C.; Müller, B.; Meyboom, R.; van Midwoud, P.; Vermonden, T.; Hennink, W.E.; van Weeren, P.R. Sustained intra-articular release of celecoxib from in situ forming gels made of acetyl-capped PCLA-PGE-PCLA triblock copolymers in horses. *Biomaterials* **2015**, *53*, 426–436. [CrossRef] [PubMed]
22. Carvalho, J.R.G.; Conde, G.; Antonioli, M.L.; Dias, P.P.; Vasconcelos, R.O.; Taboga, S.R.; Canola, P.A.; Chinelatto, M.A.; Pereira, G.T.; Ferraz, G.C. Biocompatibility and biodegradation of poly (lactic acid) (PLA) and an immiscible PLA/poly (ε-caprolactone) (PCL) blend compatibilized by poly (ε-caprolactone-b-tetrahydrofuran) implanted in horses. *Polym. J.* **2020**, *52*, 629–643. [CrossRef]

23. Byars, T.D.; Gonda, K.C. Equine history, physical examination, records, and recognizing abuse or neglect in patients. In *Large Animal Internal Medicine*, 5th ed.; Smith, B.P., Ed.; Elsevier: Riverport Lane, MO, USA, 2015; pp. 13–20.
24. Kaneko, J.J.; Harvey, J.W.; Bruss, M.L. *Clinical Biochemistry of Domestic Animals*, 6th ed.; Elsevier: Oxford, UK, 2008; p. 884.
25. Johns, J.L. Alterations in blood proteins. In *Large Animal Internal Medicine*, 5th ed.; Smith, B.P., Ed.; Elsevier: Riverport Lane, MO, USA, 2015; pp. 386–392.
26. Gregory, N.S.; Harris, A.L.; Robinson, C.R.; Dougherty, P.M.; Fuchs, P.N.; Sluka, K.A. An overview of animal models of pain: Disease models and outcome measures. *J. Pain* **2013**, *14*, 1255–1269. [CrossRef]
27. Rédua, M.A.; Valadão, C.A.A.; Duque, J.C.; Balestrero, L.T. The pre-emptive effect of epidural ketamine on wound sensitivity in horses tested by using Von Frey filaments. *Vet. Anaesth. Analg.* **2002**, *29*, 200–206. [CrossRef] [PubMed]
28. Kastellorizios, M.; Papadimitrakopoulos, F.; Burgess, D.J. Prevention of foreign body reaction in a pre-clinical large animal model. *J. Control. Release* **2015**, *202*, 101–107. [CrossRef] [PubMed]
29. Ringler, D.J. Inflamação e reparo. In *Patologia Veterinária*, 6th ed.; Jones, T.C., Hunt, R.D., King, N.W., Eds.; Editora Manole Ltda: Barueri, Brasil, 2000; pp. 119–166.
30. De Jong, W.H.; Bergsma, J.E.; Robinson, J.E.; Bos, R.R.M. Tissue response to partially in vitro predegraded poly-L-lactide implants. *Biomaterials* **2005**, *26*, 1781–1791. [CrossRef]
31. Dhandayuthapani, B.; Yoshida, Y.; Maekawa, T.; Kumar, D.S. Polymeric scaffolds in tissue engineering application: A review. *Int. J. Polym. Sci.* **2011**, *2011*, 290602. [CrossRef]
32. Morais, J.M.; Papadimitrakopoulos, F.; Burgess, D.J. Biomaterials/tissue interactions: Possible solutions to overcome foreign body response. *AAPS J.* **2010**, *12*, 188–196. [CrossRef]
33. Silva, D.; Kaduri, M.; Poley, M.; Adir, O.; Krinsky, N.; Shainsky-Roitman, J.; Schroeder, A. Biocompatibility, biodegradation and excretion of polylactic acid (PLA) in medical implants and theranostic systems. *Chem. Eng. Sci.* **2018**, *340*, 9–14. [CrossRef] [PubMed]
34. Zin, M.R.M.; Mahendrasingam, A.; Konkel, C.; Narayanan, T. Effect of D-isomer content on strain-induced crystallization behaviour of Poly(lactic acid) polymer under high speed uniaxial drawing. *Polymer* **2021**, *216*, 123422. [CrossRef]
35. Tschakaloff, A.; Losken, H.W.; von Oepen, R.; Michaeli, W.; Moritz, O.; Mooney, M.P.; Losken, A. Degradation kinetics of biodegradable DL-polylactic acid biodegradable implants depending on the site of implantation. *Int. J. Oral. Maxillofac. Surg.* **1994**, *23*, 443–445. [CrossRef]
36. Dias, P.P.; Chinelatto, M.A. Effect of poly(ε-caprolactone-b-tetrahydrofuran) triblock copolymer concentration on morphological, termal and mechanical properties of immiscible PLA/PCL blends. *J. Renew. Mater.* **2019**, *7*, 129–138. [CrossRef]
37. Clauss, A. Gerinnugsphysiologische schnellmethode zur bestimmung des fibrinogens. *Acta. Haematol.* **1957**, *17*, 237–246. [CrossRef]

Article

Creation of a Stable Nanofibrillar Scaffold Composed of Star-Shaped PLA Network Using Sol-Gel Process during Electrospinning

Karima Belabbes [1], Coline Pinese [1,*], Christopher Yusef Leon-Valdivieso [1], Audrey Bethry [1] and Xavier Garric [1,2]

[1] Polymers for Health and Biomaterials, IBMM, CNRS, ENSCM, University of Montpellier, 34090 Montpellier, France; belabbeskarimapharm@gmail.com (K.B.); christopher-yusef.leon-valdivieso@umontpellier.fr (C.Y.L.-V.); audrey.bethry@umontpellier.fr (A.B.); xavier.garric@umontpellier.fr (X.G.)
[2] Department of Pharmacy, Nîmes University Hospital, 30900 Nîmes, France
* Correspondence: coline.pinese@umontpellier.fr

Citation: Belabbes, K.; Pinese, C.; Leon-Valdivieso, C.Y.; Bethry, A.; Garric, X. Creation of a Stable Nanofibrillar Scaffold Composed of Star-Shaped PLA Network Using Sol-Gel Process during Electrospinning. *Molecules* **2022**, *27*, 4154. https://doi.org/10.3390/molecules27134154

Academic Editors: Marek Brzeziński, Małgorzata Baśko and Domenico Lombardo

Received: 18 May 2022
Accepted: 27 June 2022
Published: 28 June 2022

Publisher's Note: MDPI stays neutral with regard to jurisdictional claims in published maps and institutional affiliations.

Copyright: © 2022 by the authors. Licensee MDPI, Basel, Switzerland. This article is an open access article distributed under the terms and conditions of the Creative Commons Attribution (CC BY) license (https://creativecommons.org/licenses/by/4.0/).

Abstract: PLA nanofibers are of great interest in tissue engineering due to their biocompatibility and morphology; moreover, their physical properties can be tailored for long-lasting applications. One of the common and efficient methods to improve polymer properties and slow down their degradation is sol-gel covalent crosslinking. However, this method usually results in the formation of gels or films, which undervalues the advantages of nanofibers. Here, we describe a dual process sol-gel/electrospinning to improve the mechanical properties and stabilize the degradation of PLA scaffolds. For this purpose, we synthesized star-shaped PLAs and functionalized them with tri-ethoxysilylpropyl groups (StarPLA-PTES) to covalently react during nanofibers formation. To achieve this, we evaluated the use of (1) a polymer diluent and (2) different molecular weights of StarPLA on electrospinnability, StarPLA-PTES condensation time and crosslinking efficiency. Our results show that the diluent allowed the fiber formation and reduced the condensation time, while the addition of low-molecular-weight StarPLA-PTES improved the crosslinking degree, resulting in stable matrices even after 6 months of degradation. Additionally, these materials showed biocompatibility and allowed the proliferation of fibroblasts. Overall, these results open the door to the fabrication of scaffolds with enhanced stability and prospective long-term applications.

Keywords: functionalized polymers; silylated PLA; crosslinking in situ; hybrid network; soft tissues regeneration

1. Introduction

Polylactide (or poly(lactic acid) (PLA)) is a promising material for biomedical applications due to their biocompatibility, degradability and mechanical properties [1–3]. This aliphatic polyester has been used in a wide range of medical devices (implants, surgical sutures, stents, etc.), scaffolds for tissue regeneration and drug delivery systems [4–7]. Conventional PLA processing for such applications includes film casting, injection molding, blow molding and foaming; these techniques, however, quite often present some important drawbacks. For instance, casted PLA films tend to lack porosity, a critical issue since nutrient and oxygen transport to cells cannot be assured. Similarly, the creation of pores through solidification is challenged with injection molding [8]. Lastly, even if foams exhibit desirable porosity and morphology for tissue engineering [9], their formulation increases the hydrophilicity of PLA, which in turn can speed up its degradation in a predominantly hydrophilic medium [10]. Alternatively, different research groups have shown a growing interest in the use of electrospinning to manufacture PLA scaffolds [11–13]: the nanofibrous scaffolds produced by this technique have morphological and architectural characteristics similar to those of the natural extracellular matrix [14,15], e.g., a very high surface/volume ratio and high porosity with appropriate pore size [13].

In addition to choosing an efficient and rather facile method to process PLA, modulation of the final properties can widen its versatility and applicability. In order to tune the mechanical properties and the degradation rate of PLA materials, many strategies have been used at the molecular level, including copolymerization, blending or crosslinking [16–18]. Among these, only crosslinking allows the formation of covalent bonds of PLA homopolymers, which would result in a strong and stable 3D network while keeping their inherent physicochemical properties. This approach can be classified into two groups: chemical crosslinking and crosslinking by exposure to low energy light or ionizing radiation [19,20]. Chemical crosslinking allows better control over the structure of the resulting materials due to the specificity of the reactive groups (crosslinkers), while irradiation, even if it is a solvent-free method (thus performed in softer conditions), normally leads to random crosslinking distributions [16,17]. There are two main paths to chemically crosslink PLA: reaction of peroxides on non-functional PLA of high-molecular-weight [21–23] and addition (or condensation) reactions of functionalized PLA oligomers [24,25], which results in stronger bonding (with no modification of PLA chains) compared to the use of peroxide.

While chemical crosslinking is regarded as a useful tool to enhance the physical features of PLA, cytotoxicity remains as a legitimate concern when the final material has prospective applications for health. In order to preserve the cytocompatibility of crosslinked PLA, two conditions have to be met: (1) crosslinkers grafted on PLA oligomers must react ideally without the use of catalysts and (2) no toxic by-products should be formed as result of this process. To this end, the crosslinking of siloxane functionalized PLA via sol-gel has been proposed as it offers chemoselectivity and mild reaction conditions [26–28]. We previously demonstrated that hydrolysis and condensation of alkoxysilane-bearing polymers could render, in soft conditions, modulable hydrogel or film networks linked through Si-O-Si bonds [29–32]. This method is based on two main steps: hydrolysis of the alkoxysilanes attached to the polymer to form silanol groups and condensation of the latter to form siloxane bonds. The combination of electrospinning and sol-gel process is rarely used but has been described, especially for the fabrication of ceramic or titanium oxide hybrid nanofibers [33–36]. These nanofibers are then made from TEOS "TetraEthyl Ortho Silicate" which reacts in a sol-gel process by trapping the elements without integrating them in the network. Other works also describe the use of water-soluble polymers such as PVA in such processes to facilitate electrospinning while allowing the sol-gel reaction of TEOS [37–39]. Very few studies use functionalized polymers to react via sol-gel. While such polymers are described, most studies in the literature focus on gels or films [40–42], which limit the advantages and potential of PLA as a biomaterial.

In this study, we took up the challenge for the first time to induce the hydrolysis of silanol-bearing PLA, followed by the condensation of reactive groups during (in situ) electrospinning to obtain crosslinked and stable nanofibrous scaffolds for medical applications.

In our research, we chose to work with 4-arm functionalized PLA in order to increase the number of reactive extremities, and therefore to promote the formation of siloxane bonds during electrospinning, facilitating the creation of a 3D network. We also sought to optimize the condensation of siloxane bonds by (i) introducing a diluent polymer that promotes the formation of electrospun fibers and (ii) varying the molecular weight (hence the arm size) of the star-shaped PLA in the electrospinning solution. The efficiency of the condensation was evaluated by quantifying the gel fraction and the degradation profile. Finally, the cytocompatibility and cell proliferation of the crosslinked nanofibers were also tested.

2. Results and Discussion

The main challenge in our study was to initiate the sol-gel reaction during the electrospinning process while allowing the formation of PLA nanofibers for the creation of a stable 3D polymeric network in a single step which, to the best of our knowledge, has never been reported in the literature. The creation of the network between the StarPLA-PTES is based on a hydrolysis of the triethoxysilane (SiOEt$_3$) groups of PTES in an acidic environment which create silanol groups in the electrospinning syringe. Then, the silanol groups condensed with each other mainly during the travel time between the charged syringe and collector. For this purpose, StarPLAs were selected to increase the number of reactive chain ends, and promote their condensation, polymer entanglement and network formation as already demonstrated [15].

2.1. Synthesis and Functionalization of StarPLA

The synthesis of the tetrafunctional StarPLAs is presented in Figure 1A. StarPLAs were obtained by polymerizing D,L-lactide onto 4-arm pentaerythritol by ring opening polymerization with Sn(Oct)$_2$ as catalyst. We produced three different StarPLAs (StarPLA5k, StarPLA12k and StarPLA25k) with different lengths of PLA arms (Figure 1B). ^{1}H-NMR analysis showed good agreement between theoretical and experimental molecular weights (Mn), suggesting that no transesterification reactions occurred. Mn values obtained by SEC were higher than those calculated by ^{1}H-NMR, especially for StarPLA25k; this difference is however, explained by the SEC conditions (calibration, solvent) which render results that are rather relative for a particular polymer. All the synthesized star polymers presented (1) a narrow molecular weight distribution, with dispersity (Đ) values ranging from 1.15 to 1.63 and (2) a monodisperse profile, which is consistent with the synthesis of this type of polymer structure [43,44].

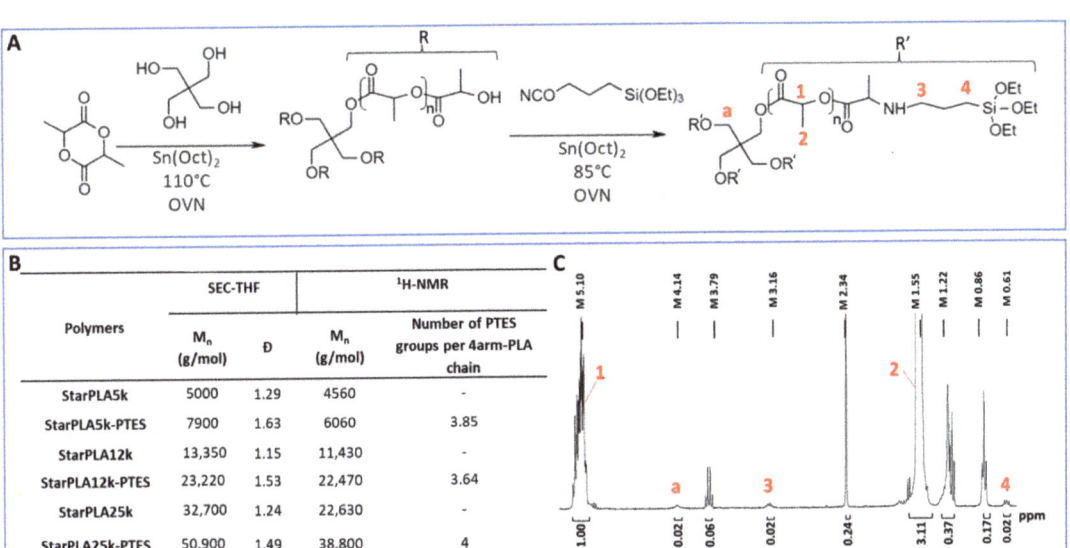

Figure 1. Synthesis and characterization of StarPLA and tetra(triethoxysilyl) StarPLA: (**A**) General synthesis and functionalization reactions; (**B**) ^{1}H-NMR and SEC characterization of StarPLA and tetrafunctionalized StarPLA; (**C**) ^{1}H-NMR spectrum of StarPLA25k-PTES.

Then, the tetrafunctionalized StarPLA were obtained by reacting the isocyanate group of PTES with the hydroxyl groups at the chain end of the StarPLAs, resulting in the formation of a urethane bond in the presence of $Sn(Oct)_2$. The degree of functionalization of PLA was calculated from ^{1}H-NMR spectroscopy using the ratio between the methylene proton of the lactic units at 5.1 ppm and the methylene protons of triethoxysilane at 0.6 ppm (Figure 1C). The number of grafted PTES functions per 4-arm polymers was 3.85, 3.64 and 4 for StarPLA5k-PTES, StarPLA12k-PTES and StarPLA25k-PTES, respectively, which corresponds to a degree of functionalization of 94.5%, 91% and 100%, respectively.

2.2. Creation of StarPLA-PTES Nanofibers by Sol-Gel Process during Electrospinning

In a previous study, we showed the feasibility of producing two-dimensional polymer networks based on StarPLA-PTES [31]. Briefly, we triggered the hydrolysis of the siloxane functions of the polymer in solution and then poured it into a mold; the subsequent evaporation of the solvent caused the silanol groups to condense, leading to the formation of a crosslinked network. We aim here to fabricate nanofibrillar scaffolds made of StarPLA-PTES using electrospinning by triggering (i) the hydrolysis of PTES groups and (ii) the condensation of silanol groups (enhanced by solvent evaporation during polymer fiber ejection and stretching) within the same nanofiber-forming process.

Among the different parameters used to evaluate the suitability of this double process, we considered the electrospinnability window as a crucial one. This parameter is the time required for the condensation of the silanol functions to alter the fluidity of the polymer solution such that it is no longer possible to electrospin it. We first started by studying the electrospinnability of StarPLA25k-PTES, due to its relevant molecular weight that would promote the polymer chain entanglements and formation of electrospun nanofibers. The hydrolysis of ethoxy groups was activated prior to electrospinning by introducing 10 µL of HCL (0.1 M in ethanol) per mL of polymers solution without any diluent (Figure 2A, Sample A). We observed that the condensation of the PTES functions into Si-O-Si bonds was very fast and generated immediately a gel in the syringe demonstrating that the macromolecular network was formed even before the solution could be electrospun. This very short gelation time is not generally found for nanofibers made from TEOS. TEOS, being a very short chain molecule, requires a long ageing time (between 1 to 4 h) [37,38] for the hydrolysis/condensation to generate a solution of sufficient viscosity to be electrospun. In order to slow down the condensation time and allow the electrospinning process to occur, we used a linear PLA of 200,000 g·mol^{-1} as a diluent polymer to serve as a spacer between the reactive species and we evaluated its impact on the formation of fibers, the condensation time and the gel fraction. For this purpose, we varied the diluent/StarPLA25k-PTES mass ratio by 0.25, 0.5, 1 and 3 (Figure 2A, samples B, C, D and E respectively). In order to compare the samples with each other, we kept the electrospinning parameters unvaried during their fabrication. In contrast to the mixture without diluent polymer (sample A), the four samples containing the diluent increased the window of electrospinnability and thus permitted the generation of nanofibers that were free of defects (Figure 2B). As expected, the hydrolyzed functions were distanced from each other and this changed according to the diluent/star polymer ratio used. For instance, the total condensation time of mixture B, which contained 1/4 diluent, was 16 min and it increased to 120 min when the mass of diluent was three times bigger than that of the star polymer (mixture E). In terms of nanostructure, all the electrospun fibers were in the range of 360 to 560 nm, except for the 0.25 diluent/star polymer ratio (1.27 µm). In all the cases, the fiber size distribution was found to be homogeneous, which was confirmed by their small standard deviation (Figure 2A).

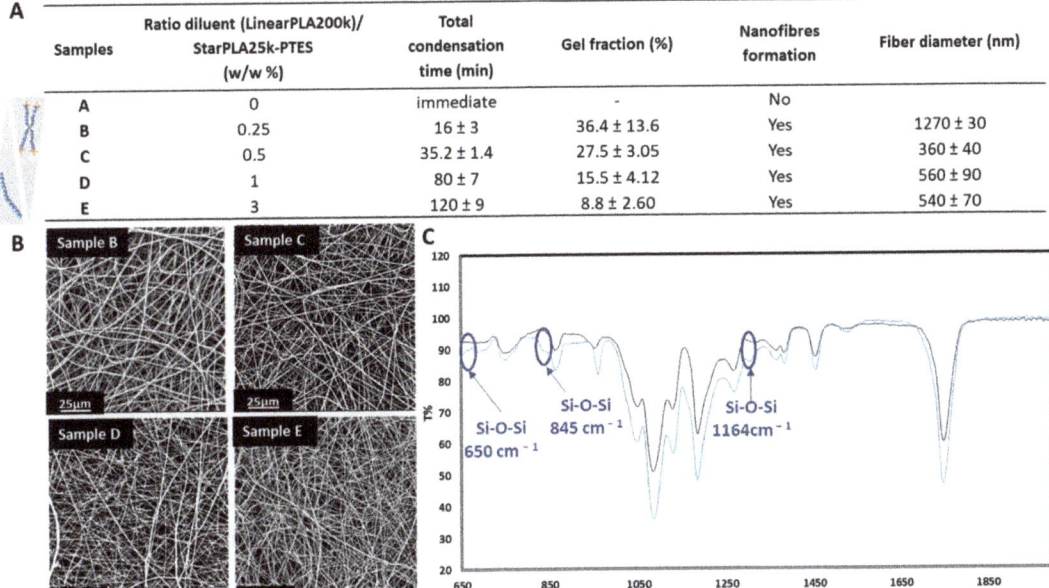

Figure 2. Crosslinking of nanofibers during electrospinning: (**A**) Polymer solution total condensation time, nanofibers gel fraction and fiber diameter; (**B**) SEM images of samples B, C, D and E; (**C**) FT-IR analysis of sample C without and with activation of the sol-gel reaction.

In order to confirm the condensation of PTES groups during electrospinning, we analyzed the obtained nanofibers by FT-IR. The results show the condensation of Si-OH groups into Si-O-Si in crosslinked samples (when hydrolysis was initiated) compared to non-crosslinked ones (no hydrolysis), by the appearance of signals at 1164 cm^{-1}, 845 cm^{-1} and 650 cm^{-1} (Figure 2C). It is well known that a crosslinked material is insoluble in a solvent that solubilizes it when not crosslinked. Thus, we qualitatively tested the nanofibers solubility in THF (good solvent for StarPLAs) and we noticed that the fibers of samples B and C were insoluble; sample D was partially soluble while sample E was totally dissolved in such solvent. Next, we evaluated the degree of crosslinking: we measured the gel fraction, which corresponds to the percentage of material actually crosslinked and therefore not solubilized by an organic solvent. This confirmed that the increase in the diluent/StarPLA25k-PTES ratio results in a decrease of network formation: we obtained gel fractions of 36.4, 27.5, 15.5 and 8.8% for diluent/StarPLA25k-PTES ratios of 0.25, 0.5, 1 and 3 (samples B to E, respectively). In summary, as the amount of polymer diluent in the solution increases (thus provoking more steric hindrance), the time required for the condensation of the PTES functions increases (i.e., longer electrospinnability window) but the efficiency of the crosslinking is, importantly, reduced.

In order to maximize the network formation during electrospinning, we increased the density of PTES functions in sample C by introducing low-molecular-weight tetrafunctionalized polymers (StarPLA12k-PTES and StarPLA5k-PTES). We chose sample C because it represents a good compromise between the time needed to obtain a nanofiber layer (driven by the condensation time) and the degree of crosslinking of the nanofibers (given by the gel fraction). First, we added StarPLA12k-PTES to an equivalent amount of StarPLA25k-PTES to form sample F (Figure 3A). This increase of the number of PTES moles from 10.1 to 18.4 × 10^{-5} mol·g^{-1} slightly enhanced the gel fraction from 27.5 to 35.7%, although a large variability was also obtained (Figure 3B).

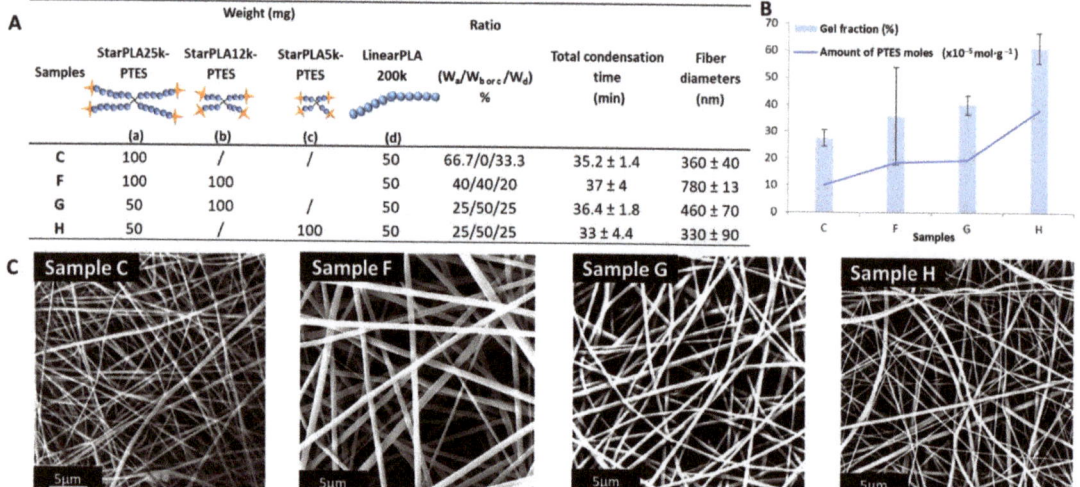

Figure 3. Effect of the density of PTES groups on the crosslinking efficiency of the nanofiber network: (**A**) Influence of molecular weight of the added polymer on the total condensation time and fiber diameter; (**B**) Evolution of the gel fraction as a function of the density of PTES groups; (**C**) SEM images of samples C, F, G and H.

To evaluate the influence of the length of StarPLA arms without significantly modifying the amount of PTES moles and in view of reducing the standard deviation between the gel fractions obtained, we promoted the predominance of StarPLA12k-PTES in the mixture by halving the mass quantity of StarPLA25k-PTES contained in the sample F (which resulted in sample G). The gel fraction of the nanofibers obtained from sample G was increased by 4% compared to the mixture F with no change in the condensation time.

Finally, to further improve the crosslinking rate, we replaced StarPLA12k-PTES with the low-molecular-weight StarPLA5k-PTES (sample H), keeping the same weight ratio as in sample G. This rendered a significant increase in the gel fraction from 40.0 to 61.1%. This enhancement in the condensation of silanol groups by the addition of low-molecular-weight polymers was directly correlated to the increase in the density of the crosslinking groups in solution (i.e., the polymer/functional groups ratio decreases with shorter chains). Moreover, the improvement of the gel fraction and thus the creation of a polymer network in samples G and H (which is also confirmed with the insolubility of the fibers in THF) did not affect the electrospinnability window (~35 min), which is the time needed to obtain scaffolds that are sufficiently stable for handling in our case. We believe this conservation of the electrospinnability window is related to the low viscosity of the solutions due to the use of short chain length polymers [45], which creates space between the reactive functions in comparison with polymers of longer chains, which induce a longer time to generate the condensation.

The resulting nanofibers had diameters between 360 and 780 nm (Figure 3A,C). This is a significant reduction of the fiber size by adding low-molecular-weight polymers. Indeed, the shorter the chain length of the polymer arm, the finer the fiber diameter. This phenomenon is explained by the decrease in the solution viscosity due to low-molecular-weight polymers (given by the Mark–Houwink equation), which in turn has a direct impact on the fiber size [46,47].

2.3. Nanofibers Behavior in a Physiological Environment-Like over Time

Hydrolysis is the primary mechanism for degradation of aliphatic polyesters [48,49]. In this study, we investigated the influence of the sol-gel process and the formation of siloxane bonds on the rate of hydrolytic degradation of nanofibers. In this context, we tracked the evolution of degradation over 6 months of samples G and H (the most promising ones in terms of resulting gel fraction), and compared them to the reference sample C, which is composed of only one type of functionalized StarPLA, in order to assess the influence of the crosslinking group density on the hydrolysis of the scaffolds (Table 1).

Table 1. Scaffold properties during 6 months of degradation in vitro.

Samples	Degradation Time (Months)	Gel Fraction (%)	Mass Loss (%)	Young Modulus E (Mpa)	Glass Transition (°C)
C	0	27.5 ± 3	/	19.1 ± 5	60.7
	1	76.2 ± 2	5.5 ± 4	27.8 ± 7	49.5
	3	80 ± 3	0	23 ± 1	52.6
	6	84.7 ± 5	7.9 ± 5	19.8 ± 8	54.2
G	0	40 ± 4	/	37.9 ± 19	54.8
	1	85.2 ± 9	5.8 ± 4	31.5 ± 9	48.3
	3	81.8 ± 1	1.2 ± 2.1	82.3 ± 7	48.4
	6	81.9 ± 7	1.7 ± 0.6	35.2 ± 3	48
H	0	61 ± 6	/	20.8 ± 4	52.1
	1	65.3 ± 14	8.7 ± 7	20.3 ± 3	48.8
	3	65.7 ± 3	1.5 ± 2.2	32.8 ± 11	50
	6	72.3 ± 5	1.0 ± 1	37.1 ± 11	48

For both C and G scaffolds, the gel fraction increased in the first month of degradation from 27.5 to 76.2% and from 40 to 85.2%, respectively, and stabilized for the remaining months. For scaffold H, the gel fraction did not change during the whole degradation period. The increase in the gel fraction for C and G can be attributed to an increase in hydrolysis reactions of unreacted PTES groups during electrospinning, which subsequently condensed. The following stabilization is then related to a total consumption of the functions (i.e., a saturation of crosslinked groups). In the case of scaffold H, this saturation is highly likely to have been reached during the same electrospinning process, hence the unchanged values of gel fraction from the beginning of the degradation period. The reconstituted bonds during degradation, together with the bonds already present, led to the formation of a stable covalent network in the nanofibers, which explains the low mass loss during the 6 months. Arsenie et al. also reported that the degradation of a StarPLA-PTES film crosslinked with siloxane bonds was stable during their degradation assay period (8 weeks) [15].

In contrast, non-crosslinked StarPDLAs degrade rapidly: it has been shown that 6-arm StarPLAs films can lose up to 18% of their mass in 45 days [50] which might be disadvantageous for prolonged use. The stiffness of the scaffolds was also preserved during the whole degradation period. Indeed, after 6 months the Young's modulus only (and slightly) evolved for H nanofibers (from 20.8 ± 4 to 37.1 ± 11 Mpa), while for C and G nanofibers it remained practically unchanged (from 19.1 ± 5 to 19.8 ± 8 Mpa and from 37.9 ± 19 to 35.2 ± 3 Mpa, respectively).

Although the properties of the nanofibrillar scaffolds were stable overall, SEM images showed (i) the appearance of some cracks or fissures on the G and H nanofibers and (ii) the beginning of loss of shape of the C nanofibers, suggesting that degradation was starting after 6 months (Figure 4).

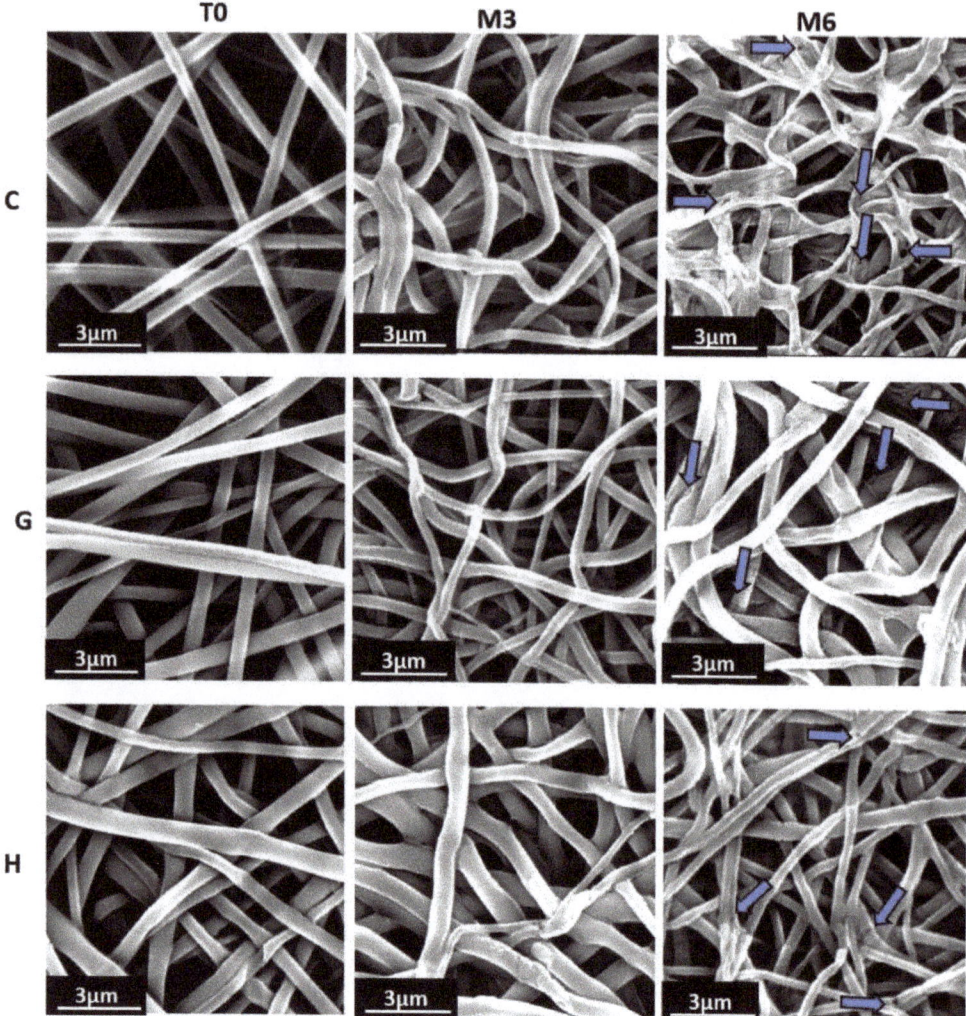

Figure 4. Morphology of samples C, G and H before (T0) and after 3 and 6 months of degradation. The blue arrows show degradation cracks on the surface of nanofibers at month six.

2.4. Biological Evaluation In Vitro

While the biocompatibility of PLA is widely mentioned in the literature [3,51], we wanted to evaluate whether the fabrication of StarPLA-PTES nanofibers as well as the presence of covalent crosslinking in the final network impart any cytotoxic effects in the final material, for potential use in tissue engineering (hence the use of L929 cell line as target). To this end, we selected mixtures G and H, corresponding to a PTES density of 19.3 and 38×10^{-5} mol·g^{-1}. Compared to the viability on TCPS control, the results show a L929 viability of 70% and 74% for nanofibers G and H, respectively (Figure 5A). These are acceptable cell survival levels according to the ISO 10993-5 norm.

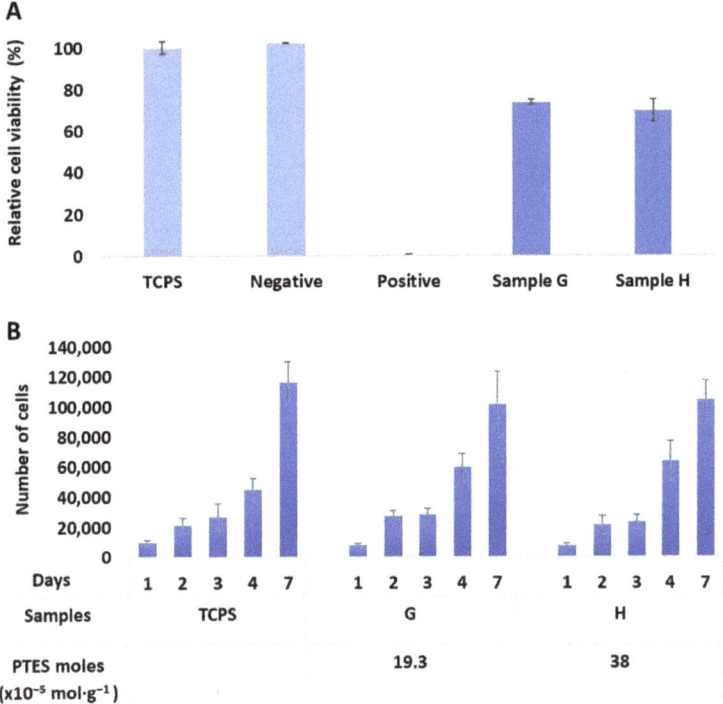

Figure 5. Biological evaluation in vitro: (**A**) Samples G and H were not cytotoxic after 24 h of contact with L929 cells compared to TCPS (tissue culture polystyrene); negative and positive controls were also tested (high density polyethylene film and polyurethane film + 0.1% zinc diethyldithiocarbamate, respectively). (**B**) The proliferation of L929 cells on G and H nanofibers was similar to that on TCPS over 7 days ($p > 0.05$).

We next evaluated the L929 cell proliferation on the same samples over 7 days (Figure 5B). The results show that there is no significant difference between the proliferation on the two nanofibers and on TCPS control, with a calculated p value between 0.07 and 0.9 for any given time point. Thus, we can conclude that the presence of Si-O-Si bonds (at the levels used in this study) did not negatively influence the proliferation of L929. Overall, these studies show that the nanofibers obtained were non-cytotoxic, permitted good cell proliferation and therefore are good candidates for future biomedical applications.

3. Materials and Methods

3.1. Materials

D,L-lactide was provided by Corbion (Gorinchem, The Netherlands). Tin (II) 2-ethylhexanoate (Sn(Oct)$_2$, 95%), pentaerythritol, diethylene glycol (DEG), (3-isocyanatopropyl) triethoxysilane (IPTES), dichloromethane, heptane, toluene, tetrahydrofuran, hydrochloric acid (37%) and hydrochloric acid (37%) (1 M in methanol), dichloromethane (DCM), trifluoroethanol, trifluoroacetic acid and phosphate buffer solution (PBS) were purchased from Sigma-Aldrich (St Quentin Fallavier, France).

CellTiter Glo assay was provided by Promega G7571 (Charbonnières-les-Bains, France). PrestoBlue® assay (A13262) and Clariostar plate reader (A13626) were acquired from Invitrogen (Illkirch, France). Negative RM-C and positive RM-A were supplied by Hatano Research Intitue, Food and Drug Safety Center, Hadano, Japan).

3.2. Synthesis of Linear and 4-Arm Star Poly(Lactide)

We synthesized LinearPLA of 200 kg·mol^{-1} and 4-arms StarPLA of 25 kg·mol^{-1}, 12 kg·mol^{-1} and 5 kg·mol^{-1}, i.e., LinearPLA200k, StarPLA25k, StarPLA12k and StarPLA5k, respectively. Polymers were synthesized by ring opening polymerization, using a procedure previously described [31]. Typically, for the synthesis of StarPLA-25k: D,L-lactide (30 g, 208 mmol, 14 eq), pentaerythritol as multifunctional initiator (0.163 g, 1.19 mmol, 1 eq) and SnOct$_2$ as catalyst (0.194 g, 4.79 mmol, 0.4 eq) were introduced in a flask. After 2 h under vacuum, the flask was sealed and maintained at 120 °C for five days. The obtained polymers were solubilized in DCM and then precipitated in cold heptane (4 °C). Finally, the recovered polymer was vacuum dried overnight. The average reaction yields were 88 ± 9%. StarPLA12k and StarPLA5k were obtained by varying the molar ratio monomers/initiator: 20.8 and 8.67 respectively. The linearPLA200k was synthesized with the same protocol of polymerization of the StarPLA, but this time using diethylene glycol as initiator in a monomer/initiator molar ratio of 694.4. The molecular weight of the StarPLAs was determined by ^{1}H-NMR from the signal ratio between methylene protons of pentaerythritol and the methyl proton of the lactic unit.

^{1}H-NMR (400 MHz, CDCl$_3$) δ (ppm) = 5.15 (m, CHCH$_3$); 4.34 (m, CHCH$_3$OH); 4.14 (s, CCH$_3$O); 3.5 (s, CCH$_3$OH); 1.54 (m, CHCH$_3$)

3.3. Functionalization of 4-Arm Starpoly(Lactide) with Triethoxysilane

Polymers were functionalized using a procedure previously described [31]. Typically, the StarPLA25k (5 g, 2.5 mmol) was put in a flask under vacuum for 4 h. The polymer was solubilized in anhydrous toluene (100 mL) under inert gas during 1 h. Then, the IPTES (3.71 g, 15 mmol, 6 eq) was added in the solution and SnOct$_2$ as catalyst (0.162 g, 0.4 mmol, 0.16 eq). The reaction was carried out at 75 °C, with constant stirring and under inert atmosphere for 24 h. The obtained StarPLA25k-PTES was purified in cold heptane and vacuum dried in order to eliminate traces of solvent. StarPLA12k and StarPLA5k were functionalized using the same protocol as for StarPLA25k.

The degree of functionalization was determined by ^{1}H-NMR from the signal ratio between the methyl lactic proton (δ = 5.1 ppm) and the methylene proton linked to triethoxysilane (δ = 0.6 ppm).

^{1}H-NMR (400 MHz, CDCl$_3$) δ(ppm) = 5.15(m, CHCH$_3$); 4.34(m, CHCH$_3$OH); 4.14 (s, CCH$_3$O); 3.83 (m, OCH$_2$CH$_3$); 3.5 (s, CCH$_3$OH); 3.17 (m, NHCH$_2$); 1.54 (m, CHCH$_3$); 1.2 (m, OCH$_2$CH$_3$); 0.64–0.54 (m, CH$_2$CH$_2$Si).

3.4. Electrospinning

StarPLA-PTES were dissolved in TFE at concentrations ranging from 15 to 25 wt%. In order to activate the sol gel process, 10 μL of HCL (0.1 M in ethanol) was added per mL of polymer solution. After one minute of agitation using a vortex, the mixture was loaded into a syringe with a 21-G needle and fixed on a syringe pump (KDS100, KD Scientific,). To start the electrospinning process, a voltage of 15 kV was applied to the polymer solution, which was dispensed at a flow rate of 0.5 mL·h^{-1}. The nanofibers were collected on a flat collector that was placed 15 cm from the syringe needle. The experiments were performed at 25 ± 1 °C with a relative humidity of 48 ± 3%. All the tests were carried out under the same conditions.

3.5. Characterization Methods

Size exclusion chromatography (SEC) was conducted using a Shimadzu LC-200AD Prominence system (Shimadzu, Marne-la-Vallée, France) equipped with a PLgel MIXED-C guard column (Agilent, 5 μm, 50 mm length × 7.5 mm diameter), two PLgel MIXED-C columns (Agilent, 5 μm, 300 mm length × 7.5 mm diameter) and a RID-20A refractive index signal, using poly (ethylene glycol) as calibration standard with a flow rate = 1 mL·min^{-1}. The polymer was solubilized in THF at 10 mg·mL^{-1}; 200 μL of this solution was injected in the system. ^{1}H-NMR measurements were carried out at 300 MHz with an AMX300 Bruker spectrophotometer (Cambridge Scientific Products, Watertown, USA) at room temperature using deuterated chloroform as solvent. Fourier transform infrared analysis was performed using a Perkin Elmer Spectrum 100 (PerkinElmer France, Villebon-sur-Yvette, France); four scans were performed at room temperature in the range of 4000–650 cm^{-1}.

For scanning electron microscopy (SEM), the samples were sputter coated with a 10 nm (2 min) thick gold film and imaged under a scanning electron microscope (Phenom-world ProX) using an accelerating voltage of 15 kV. Micrographs were analyzed with ImageJ (https://imagej.nih.gov/ij/download.html (accessed on 12 April 2021)). The diameter of the fibers was measured with the same software by drawing a perpendicular line on both edges of the fiber (number of fibers measured for each condition = 50).

The differential scanning calorimetry (DSC) analysis was conducted under argon with a Mettler Toledo 3 Star DSC system. A total of 3 mg of each nanofiber sample was weighed into a standard aluminum dish. The thermal cycle consisted of a heating sweep from 0 °C to 200 °C (10°C·min^{-1}) followed by cooling to −10 °C (10°C·min^{-1}) and a second heating sweep to 300 °C (10°C·min^{-1}). The glass transition temperature (Tg) was measured on the first ramp of heating.

3.6. Gel Fraction

For each condition, gel fraction was measured by weighing 3 samples (w_i) followed by immersion in 3 mL of DCM under agitation for 3 h. Next, the DCM solution was centrifugated (3000 rpm for 10 min) to separate the insoluble and soluble fractions. After drying under vacuum for 24h, the insoluble fractions were weighed ($w_{insoluble}$). Gel fraction was calculated with Equation (1) after verifying that the addition of soluble and insoluble phases was 100% of the initial sample weight:

$$\text{Gel fraction (\%)} = (w_{insoluble} / w_i) \times 100 \tag{1}$$

3.7. Tensile Tests

Mechanical properties were measured on C, G, H samples (3 cm wide × 1 cm length, n = 4) after different degradation time points, using uniaxial tensile testing on an Instron 3344 testing system at a speed of 5 mm·min^{-1} and at 37 ± 1 °C. The Young's modulus (E, MPa) was calculated based on the initial linear section of the stress–strain curve and was reported as the mean value of the measurements.

3.8. Nanofibers Behavior in a Physiological Environment-Like over Time

Electrospun scaffolds (10 mm length × 30 mm wide, 7.5 ± 3 mg) were weighed (m_i = initial mass) and then immersed in 6 mL of PBS (pH = 7.4) at 37 °C under constant stirring. Samples were removed from PBS at months one, three and six and their mass loss, gel fraction and mechanical properties were evaluated using quadruplicates.

The water uptake and mass loss were calculated from Equations (2) and (3), respectively:

$$\text{Water uptake (\%)} = ((m_w - m_i)/m_i) \times 100 \tag{2}$$

$$\text{Mass loss (\%)} = ((m_i - m_d)/m_i) \times 100 \tag{3}$$

where m_w is the hydrated mass of the scaffolds and m_d is mass after vacuum drying overnight.

The gel fraction and mechanical properties of degraded samples were evaluated following protocols described in Sections 3.6 and 3.7, respectively.

3.9. Biological Evaluation In Vitro

NCTC-Clone 929 cells (mouse fibroblast cell line (ECACC 85011425)), passage 32, were cultured in 500 mL MEM with 5 mL glutamax (1% stabilized glutamine), 50 mL horse serum, and 100 U/mL penicillin and streptomycin 100 µg/mL.

3.9.1. Cytotoxicity Assay

The in vitro cytocompatibility of scaffolds was tested following EN ISO 10993-5 standard protocols (n = 3). Samples were irradiated with UV-C (λ = 254 nm, 5 min, 80 W) for decontamination. Then, 1.5 mL of complete cell culture medium (DMEM 4.5 g/L D-glucose supplemented with 5% FBS, 1 mM L-glutamine, 100 U/mL penicillin and 100 µg/mL streptomycin) was added to the scaffolds and kept at 37 °C under constant stirring for 72 h. Next, 0.1 mL of this extracting medium was added to murine fibroblasts L929 P32 previously seeded at 1.10^4 cells per well in a 96-well plate the night before to allow their adhesion. After 24 h of incubation at 37 °C in a humid environment, cell viability was assessed based on ATP quantification using CellTiter Glo (Promega G7571) according to the supplier instructions. Briefly, 50 µL of media was removed from each well, then 50 µL of CellTiter Glo reagent was immediately added to each well containing the cells and incubated for 10 min in the dark. Finally, luminescence was read and the results for each sample were compared to TCPS. Negative (high-density polyethylene film) and positive (polyurethane containing 0.1% zinc diethyldithiocarbamate) controls were also tested.

3.9.2. Proliferation Assay

Samples (n = 4) were cut into disks of 2 cm diameter, sterilized with UV-C irradiation (λ = 254 nm, 2.5 min on each side, 80 W) and placed in non-treated 24-well plates. Scaffolds were kept in a fixed position with the use of O-rings. L929 cells were seeded on top of the scaffolds at 8.10^4 cells per well and were placed at 37 °C and 5% CO_2. The number of cells was assessed after 1, 2, 3, 4 and 7 days of contact with the scaffolds using a PrestoBue assay that evaluates the transformation of weakly fluorescent blue resazurin into highly fluorescent red resorufin through the mitochondrial activity of the cell. Briefly, this reagent was placed at a volume ratio of 1:10 in each well and incubated in the dark for 40 min at 37 °C. Fluorescence intensity was then read in a CLARIOstar® microplate reader (wavelength: excitation 558 nm, emission 590 nm). After each measurement, the supernatant was replaced with fresh medium to continue the cell culture until day 7.

3.9.3. Statistical Analysis

R software version 3.5.2 (R Foundation, Vienna, Austria) was used to perform statistical analysis. Significance was assessed by non-parametric Kruskall–Wallis test with repeated measures followed by Dunn's post-test. Values of $p > 0.05$ were considered as not statistically significant.

4. Conclusions

Electrospun PLA scaffolds are of great interest in biomedical applications, not only because of their biocompatibility and nanostructure (which resembles the extracellular matrix) but also because they display appropriate degradation times for tissue engineering. In this study, we created a crosslinked PLA network to further extend the degradation time of such scaffolds. We demonstrated that it is possible to create a stable and slow degrading 3D network by activating the sol-gel process during electrospinning. The presence of a polymer diluent was needed to allow the formation of fibers; the crosslinking degree of the network (as well as the final properties) could be modulated by adjusting the proportion of diluent and the molecular weight of the star polymers in the mixture. In addition, these nanofibrillar materials were not cytotoxic and allowed the proliferation of L929 cell line. Future studies will include the incorporation of bioactive peptides (functionalized with IPTES) into the StarPLA network, using our condensation method to fabricate nanofibrillar structures that are not only stable over long periods, but that also trigger a biological response in cells.

Author Contributions: Conceptualization, C.P.; validation, C.P. and X.G.; formal analysis, A.B.; investigation, K.B.; writing—original draft preparation, K.B.; writing—review and editing, C.P., X.G. and C.Y.L.-V.; supervision, C.P.; All authors have read and agreed to the published version of the manuscript.

Funding: This research was partially funded by the ministry of higher education of Algeria (N°004Bis/PG./France/2019/2020).

Data Availability Statement: The data presented in this study are available on request from the corresponding author.

Acknowledgments: We thank Cartigen platform for the access to the SEM equipment.

Conflicts of Interest: The authors declare no conflict of interest.

Sample Availability: All samples are available from the authors.

References

1. Garlotta, D. A Literature Review of Poly(Lactic Acid). *J. Polym. Environ.* **2001**, *9*, 63–84. [CrossRef]
2. Ramot, Y.; Haim-Zada, M.; Domb, A.J.; Nyska, A. Biocompatibility and safety of PLA and its copolymers. *Adv. Drug Deliv. Rev.* **2016**, *107*, 153–162. [CrossRef] [PubMed]
3. Da Silva, D.; Kaduri, M.; Poley, M.; Adir, O.; Krinsky, N.; Shainsky-Roitman, J.; Schroeder, A. Biocompatibility, biodegradation and excretion of polylactic acid (PLA) in medical implants and theranostic systems. *Chem. Eng. J.* **2018**, *340*, 9–14. [CrossRef] [PubMed]
4. Lasprilla, A.J.R.; Martinez, G.A.R.; Lunelli, B.H.; Jardini, A.L.; Filho, R.M. Poly-lactic acid synthesis for application in biomedical devices—A review. *Biotechnol. Adv.* **2012**, *30*, 321–328. [CrossRef] [PubMed]
5. Dürselen, L.; Dauner, M.; Hierlemann, H.; Planck, H.; Claes, L.E.; Ignatius, A. Resorbable polymer fibers for ligament augmentation: Resorbable Polymer Fibers. *J. Biomed. Mater. Res.* **2001**, *58*, 666–672. [CrossRef]
6. Coutu, D.L.; Yousefi, A.-M.; Galipeau, J. Three-dimensional porous scaffolds at the crossroads of tissue engineering and cell-based gene therapy. *J. Cell. Biochem.* **2009**, *108*, 537–546. [CrossRef]
7. Kellomäki, M.; Niiranen, H.; Puumanen, K.; Ashammakhi, N.; Waris, T.; Törmälä, P. Bioabsorbable scaffolds for guided bone regeneration and generation. *Biomaterials* **2000**, *21*, 2495–2505. [CrossRef]
8. Choi, E.J.; Son, B.; Hwang, T.S.; Hwang, E.-H. Increase of degradation and water uptake rate using electrospun star-shaped poly(d,l-lactide) nanofiber. *J. Ind. Eng. Chem.* **2011**, *17*, 691–695. [CrossRef]
9. Mikos, A.G.; Thorsen, A.J.; Czerwonka, L.A.; Bao, Y.; Langer, R.; Winslow, D.N.; Vacanti, J.P. Preparation and characterization of poly(l-lactic acid) foams. *Polymer* **1994**, *35*, 1068–1077. [CrossRef]
10. Martin, I.; Shastri, V.P.; Padera, R.F.; Yang, J.; Mackay, A.J.; Langer, R.; Vunjak-Novakovic, G.; Freed, L.E. Selective differentiation of mammalian bone marrow stromal cells cultured on three-dimensional polymer foams. *J. Biomed. Mater. Res.* **2001**, *55*, 229–235. [CrossRef]
11. Yang, F.; Murugan, R.; Ramakrishna, S.; Wang, X.; Ma, Y.-X.; Wang, S. Fabrication of nano-structured porous PLLA scaffold intended for nerve tissue engineering. *Biomaterials* **2004**, *25*, 1891–1900. [CrossRef] [PubMed]
12. Kumbar, S.G.; Nukavarapu, S.P.; James, R.; Nair, L.S.; Laurencin, C.T. Electrospun poly(lactic acid-co-glycolic acid) scaffolds for skin tissue engineering. *Biomaterials* **2008**, *29*, 4100–4107. [CrossRef] [PubMed]
13. Kumbar, S.G.; James, R.; Nukavarapu, S.P.; Laurencin, C.T. Electrospun nanofiber scaffolds: Engineering soft tissues. *Biomed. Mater.* **2008**, *3*, 34002. [CrossRef] [PubMed]
14. Geng, X.; Kwon, O.; Jang, J. Electrospinning of chitosan dissolved in concentrated acetic acid solution. *Biomaterials* **2005**, *26*, 5427–5432. [CrossRef] [PubMed]
15. Chong, E.; Phan, T.; Lim, I.; Zhang, Y.; Bay, B.; Ramakrishna, S.; Lim, C. Evaluation of electrospun PCL/gelatin nanofibrous scaffold for wound healing and layered dermal reconstitution. *Acta Biomater.* **2007**, *3*, 321–330. [CrossRef]
16. Li, X.; Kang, H.; Shen, J.; Zhang, L.; Nishi, T.; Ito, K.; Zhao, C.; Coates, P. Highly toughened polylactide with novel sliding graft copolymer by in situ reactive compatibilization, crosslinking and chain extension. *Polymer* **2014**, *55*, 4313–4323. [CrossRef]
17. Chalermpanaphan, V.; Choochottiros, C. Synthesis of unsaturated aliphatic polyester-based copolymer: Effect on the ductility of PLA blend and crosslink. *Polym. Bull.* **2022**, *79*, 2003–2017. [CrossRef]
18. Liu, H.; Zhang, J. Research progress in toughening modification of poly(lactic acid). *J. Polym. Sci. Part B Polym. Phys.* **2011**, *49*, 1051–1083. [CrossRef]
19. Bednarek, M.; Borska, K.; Kubisa, P. New Polylactide-Based Materials by Chemical Crosslinking of PLA. *Polym. Rev.* **2021**, *61*, 493–519. [CrossRef]
20. Bednarek, M.; Borska, K.; Kubisa, P. Crosslinking of Polylactide by High Energy Irradiation and Photo-Curing. *Molecules* **2020**, *25*, 4919. [CrossRef]

21. Takamura, M.; Sugimoto, M.; Kawaguchi, S.; Takahashi, T.; Koyama, K. Influence of extrusion temperature on molecular architecture and crystallization behavior of peroxide-induced slightly crosslinked poly(L-lactide) by reactive extrusion. *J. Appl. Polym. Sci.* **2012**, *123*, 1468–1478. [CrossRef]
22. Yang, S.; Wu, Z.-H.; Yang, W.; Yang, M.-B. Thermal and mechanical properties of chemical crosslinked polylactide (PLA). *Polym. Test.* **2008**, *27*, 957–963. [CrossRef]
23. Yang, S.-L.; Wu, Z.-H.; Meng, B.; Yang, W. The effects of dioctyl phthalate plasticization on the morphology and thermal, mechanical, and rheological properties of chemical crosslinked polylactide: Effects of Dioctyl Phthalate Plasticization. *J. Polym. Sci. Part B Polym. Phys.* **2009**, *47*, 1136–1145. [CrossRef]
24. Storey, R.F.; Warren, S.C.; Allison, C.J.; Wiggins, J.S.; Puckett, A.D. Synthesis of bioabsorbable networks from methacrylate-endcapped polyesters. *Polymer* **1993**, *34*, 4365–4372. [CrossRef]
25. Helminen, A.O.; Korhonen, H.; Seppälä, J.V. Structure modification and crosslinking of methacrylated polylactide oligomers: Methacrylated Polylactide Oligomers. *J. Appl. Polym. Sci.* **2002**, *86*, 3616–3624. [CrossRef]
26. Pooyan, S.; Box, P.O. Sol-gel process and its application in Nanotechnology. *J. Polym. Eng. Technol.* **2005**, *5*, 38–41.
27. Owens, G.J.; Singh, R.K.; Foroutan, F.; Alqaysi, M.; Han, C.-M.; Mahapatra, C.; Kim, H.-W.; Knowles, J.C. Sol–gel based materials for biomedical applications. *Prog. Mater. Sci.* **2016**, *77*, 1–79. [CrossRef]
28. Livage, J.; Sanchez, C.; Henry, M.; Doeuff, S. The chemistry of the sol-gel process. *Solid State Ion.* **1989**, *32–33*, 633–638. [CrossRef]
29. Echalier, C.; Pinese, C.; Garric, X.; Van Den Berghe, H.; Jumas Bilak, E.; Martinez, J.; Mehdi, A.; Subra, G. Easy Synthesis of Tunable Hybrid Bioactive Hydrogels. *Chem. Mater.* **2016**, *28*, 1261–1265. [CrossRef]
30. Echalier, C.; Levato, R.; Mateos-Timoneda, M.A.; Castaño, O.; Déjean, S.; Garric, X.; Pinese, C.; Noël, D.; Engel, E.; Martinez, J.; et al. Modular bioink for 3D printing of biocompatible hydrogels: Sol–gel polymerization of hybrid peptides and polymers. *RSC Adv.* **2017**, *7*, 12231–12235. [CrossRef]
31. Arsenie, L.V.; Pinese, C.; Bethry, A.; Valot, L.; Verdie, P.; Nottelet, B.; Subra, G.; Darcos, V.; Garric, X. Star-poly(lactide)-peptide hybrid networks as bioactive materials. *Eur. Polym. J.* **2020**, *139*, 109990. [CrossRef]
32. Pinese, C.; Jebors, S.; Echalier, C.; Licznar-Fajardo, P.; Garric, X.; Humblot, V.; Calers, C.; Martinez, J.; Mehdi, A.; Subra, G. Simple and Specific Grafting of Antibacterial Peptides on Silicone Catheters. *Adv. Heal. Mater.* **2016**, *5*, 3067–3073. [CrossRef] [PubMed]
33. Li, D.; McCann, J.T.; Xia, Y.; Marquez, M. Electrospinning: A Simple and Versatile Technique for Producing Ceramic Nanofibers and Nanotubes. *J. Am. Ceram. Soc.* **2006**, *89*, 1861–1869. [CrossRef]
34. Choi, S.-S.; Lee, S.G.; Im, S.S.; Kim, S.H.; Joo, Y.L. Silica nanofibers from electrospinning/sol-gel process. *J. Mater. Sci. Lett.* **2003**, *22*, 891–893. [CrossRef]
35. Liu, J.; Fu, B.; Dai, J.; Zhu, X.; Wang, Q. Study the Mechanism of Enhanced Li Storage Capacity through Decreasing Internal Resistance by High Electronical Conductivity via Sol-gel Electrospinning of Co_3O_4 Carbon Nanofibers. *ChemistrySelect* **2019**, *4*, 3542–3546. [CrossRef]
36. Mirmohammad Sadeghi, S.; Vaezi, M.; Kazemzadeh, A.; Jamjah, R. 3D networks of TiO_2 nanofibers fabricated by sol-gel/electrospinning/calcination combined method: Valuation of morphology and surface roughness parameters. *Mater. Sci. Eng. B* **2021**, *271*, 115254. [CrossRef]
37. Pirzada, T.; Arvidson, S.A.; Saquing, C.D.; Shah, S.S.; Khan, S.A. Hybrid Carbon Silica Nanofibers through Sol–Gel Electrospinning. *Langmuir* **2014**, *30*, 15504–15513. [CrossRef]
38. Pirzada, T.; Arvidson, S.A.; Saquing, C.D.; Shah, S.S.; Khan, S.A. Hybrid Silica–PVA Nanofibers via Sol–Gel Electrospinning. *Langmuir* **2012**, *28*, 5834–5844. [CrossRef]
39. Dejob, L.; Toury, B.; Tadier, S.; Grémillard, L.; Gaillard, C.; Salles, V. Electrospinning of in situ synthesized silica-based and calcium phosphate bioceramics for applications in bone tissue engineering: A review. *Acta Biomater.* **2021**, *123*, 123–153. [CrossRef]
40. Storey, R.F.; Wiggins, J.S.; Mauritz, K.A.; Puckett, A.D. Bioabsorbable composites. I: Fundamental design considerations using free radically crosslinkable matrices. *Polym. Compos.* **1993**, *14*, 7–16. [CrossRef]
41. Helminen, A.; Korhonen, H.; Seppälä, J.V. Biodegradable crosslinked polymers based on triethoxysilane terminated polylactide oligomers. *Polymer* **2001**, *42*, 3345–3353. [CrossRef]
42. Storey, R.F.; Hickey, T.P. Degradable polyurethane networks based on d,l-lactide, glycolide, ε-caprolactone, and trimethylene carbonate homopolyester and copolyester triols. *Polymer* **1994**, *35*, 830–838. [CrossRef]
43. Michalski, A.; Brzezinski, M.; Lapienis, G.; Biela, T. Star-shaped and branched polylactides: Synthesis, characterization, and properties. *Prog. Polym. Sci.* **2019**, *89*, 159–212. [CrossRef]
44. Kim, S.H.; Han, Y.-K.; Kim, Y.H.; Hong, S.I. Multifunctional initiation of lactide polymerization by stannous octoate/pentaerythritol. *Makromol. Chem.* **1992**, *193*, 1623–1631. [CrossRef]
45. Shenoy, S.L.; Bates, W.D.; Frisch, H.L.; Wnek, G.E. Role of chain entanglements on fiber formation during electrospinning of polymer solutions: Good solvent, non-specific polymer–polymer interaction limit. *Polymer* **2005**, *46*, 3372–3384. [CrossRef]
46. Zong, X.; Kim, K.; Fang, D.; Ran, S.; Hsiao, B.S.; Chu, B. Structure and process relationship of electrospun bioabsorbable nanofiber membranes. *Polymer* **2002**, *43*, 4403–4412. [CrossRef]
47. Jacobs, V.; Anandjiwala, R.D.; Maaza, M. The influence of electrospinning parameters on the structural morphology and diameter of electrospun nanofibers. *J. Appl. Polym. Sci.* **2010**, *115*, 3130–3136. [CrossRef]
48. Zong, X.; Ran, S.; Kim, K.-S.; Fang, D.; Hsiao, B.S.; Chu, B. Structure and Morphology Changes during in Vitro Degradation of Electrospun Poly(glycolide-*co*-lactide) Nanofiber Membrane. *Biomacromolecules* **2003**, *4*, 416–423. [CrossRef]

49. Percec, V. (Ed.) *Hierarchical Macromolecular Structures: 60 Years after the Staudinger Nobel Prize II*; Advances in Polymer Science; Springer International Publishing: Cham, Switzerland, 2013; Volume 262, ISBN 978-3-319-03718-9.
50. Yuan, W.; Zhu, L.; Huang, X.; Zheng, S.; Tang, X. Synthesis, characterization and degradation of hexa-armed star-shaped poly(l-lactide)s and poly(d,l-lactide)s initiated with hydroxyl-terminated cyclotriphosphazene. *Polym. Degrad. Stab.* **2005**, *87*, 503–509. [CrossRef]
51. Grizzi, I.; Garreau, H.; Li, S.; Vert, M. Hydrolytic degradation of devices based on poly(dl-lactic acid) size-dependence. *Biomaterials* **1995**, *16*, 305–311. [CrossRef]

Article

Composite Fiber Spun Mat Synthesis and In Vitro Biocompatibility for Guide Tissue Engineering

Rodrigo Osorio-Arciniega [1], Manuel García-Hipólito [2], Octavio Alvarez-Fregoso [2] and Marco Antonio Alvarez-Perez [1,*]

[1] Laboratorio de Bioingeniería de Tejidos, DEPeI, Facultad de Odontología, Universidad Nacional Autónoma de México, Circuito Exterior s/n. Cd. Universitaria, Coyoacán 04510, Mexico; rosorioarci@gmail.com
[2] Instituto de Investigaciones en Materiales, Circuito Exterior s/n. Cd. Universitaria, Coyoacán 04510, Mexico; maga@unam.mx (M.G.-H.); oaf@unam.mx (O.A.-F.)
* Correspondence: marcoalv@unam.mx

Abstract: Composite scaffolds are commonly used strategies and materials employed to achieve similar analogs of bone tissue. This study aims to fabricate 10% wt polylactic acid (PLA) composite fiber scaffolds by the air-jet spinning technique (AJS) doped with 0.5 or 0.1 g of zirconium oxide nanoparticles (ZrO_2) for guide bone tissue engineering. ZrO_2 nanoparticles were obtained by the hydrothermal method and characterized by X-ray diffraction (XRD) and scanning electron microscopy (SEM). SEM and fourier-transform infrared spectroscopy (FTIR) analyzed the synthesized PLA/ZrO_2 fiber scaffolds. The in vitro biocompatibility and bioactivity of the PLA/ZrO_2 were studied using human fetal osteoblast cells. Our results showed that the hydrothermal technique allowed ZrO_2 nanoparticles to be obtained. SEM analysis showed that PLA/ZrO_2 composite has a fiber diameter of 395 nm, and the FITR spectra confirmed that the scaffolds' chemical characteristics are not affected by the synthesized technique. In vitro studies demonstrated that PLA/ZrO_2 scaffolds increased cell adhesion, cellular proliferation, and biomineralization of osteoblasts. In conclusion, the PLA/ZrO_2 scaffolds are bioactive, improve osteoblasts behavior, and can be used in tissue bone engineering applications.

Keywords: composite scaffold; guide tissue engineering; polylactic acid (PLA) nanofibers; zirconium ceramic; biocompatibility

1. Introduction

Periodontal disease is a chronic inflammatory condition caused by a highly pathogenic biofilm. If left untreated, it may result in irreversible destruction of the supporting periodontal tissues, which consists of the hard and soft tissues surrounding and supporting the teeth. In severe cases, reduced periodontal support can lead to tooth loss, requiring tissue bone augmentation and regeneration, followed by surgical procedures to restore the reconstitution of the complex structure of the tissue [1].

In clinical applications, various attempts have been made to repair the bone/periodontal apparatus over the last four decades. These include root surface conditioning, graft materials, non-resorbable membranes of polytetrafluoroethylene (e-PTFE), and bioceramic scaffolds [2]. However, the regeneration of defective or damaged bone/periodontal tissue has been challenging in reconstructive surgery. Furthermore, it is still considered a common cause of permanent functional loss and post-traumatic morbidity [3].

Hence, guiding the regeneration processes of the bone/periodontal apparatus's complex architecture still represents one of the most significant challenges in modern dentistry. From an anatomical and physiological perspective, the functional integration of composite scaffolds and/or polymeric matrices doped with nanoparticles that synchronously guide the tissue's regeneration is still demanding [4].

In recent years, scaffolds that combine a biodegradable polymers matrix, such as PCL, PLGA, or PLA doped with bioceramics (i.e., hydroxyapatite, TiO_2, magnesium-calcium silicate, β-Tricalcium phosphate, mineral trioxide aggregate, calcium phosphate) have received considerable attention as promising biomaterials with potential for next-generation bone tissue engineering scaffolds, since they provide peculiar bioactive signals to improve osteoconductivity and to enhance mesenchymal stem cells adhesion, migration, and differentiation [5–10]. In addition, among the different calcium ceramics used in the application in bone and periodontal surgery, there is also the use of zirconium oxide (ZrO_2) due to its excellent physicochemical properties, its high chemical and dimensional stability, its low ionic and thermal conductivity, its excellent mechanical resistance, fracture toughness, its low cytotoxicity, and its biocompatibility response [11,12]. Moreover, the use of the thermoplastic aliphatic polyester as polylactic acid (PLA) in bone tissue engineering is because it has numerous interesting properties, including excellent mechanical properties, thermal stability, processability, and low environmental impact. However, its most important characteristic is that it is Food and Drug Administration (FDA) approved for the low toxicity and degradation products (H_2O and CO_2) that do not interfere with tissue healing. Indeed, the hydrolyzes of the constituent of PLA could be incorporated into the tricarboxylic acid cycle and excreted, making it a natural choice for biomedical applications [13,14].

Recently, the synthesis of polymeric nano- or microfibers with bioceramics via electrospinning has been widely applied in bone tissue engineering; due to its large surface/volume ratio, the process allows modification of their physical or chemical properties and enables precise control over the shapes and structures of the fibers, which is often reported in the literature [8,15–19]. Likewise, another versatile, low-cost, safe (does not employ high voltage), scalable, and versatile method that produces micro/nanoscale fibers from different synthetic and natural polymers is the air-jet spinning (AJS) method. This method utilizes a specialized spinning system nozzle, such as a commercial airbrush, a surface for collecting polymer fibers, and compressed gas through which the polymer solution and a pressurized gas are simultaneously ejected to form the fiber morphology, allowing the design of sheet-like platforms that could be applied in bone and tissue engineering [20–23]. Furthermore, poly(lactic acid) fiber scaffolds were successfully fabricated, and our in vitro biological response of mesenchymal stem cells showed that the polymer concentration and fiber size influenced the biocompatibility response. Furthermore, our results indicated that nanofibrous topography possesses the potential to enhance cell adhesion and proliferation and improve the cues to guide the fiber orientation by the cells; additionally, our in vivo studies showed that PLA fiber spun scaffolds are not cytotoxic in a Wistar rat model [24–26]. Moreover, our previous results, where we reported the synthesis of polymer composites composed of PLA/ZrO_2 to attempt optimization of the fabrication by AJS, showed that composites have a fibrous morphology with a random distribution, with a diameter of fibers and mechanical properties depending on polymer and zirconia concentrations, suggesting that the PLA/ZrO_2 composite may be used as a biomaterial [27]. Thus, we propose the preparation of a composite by using a polymer solution of 10% wt of PLA with 0.1 and 0.5 g of ZrO_2 nanoparticles by the AJS method, with the final goal of finding a biomaterial for bone/periodontal tissue engineering, with the presence of bioceramic as an environmental cues signal needed to guide osteoblast cell responses.

2. Results

2.1. Characterization of ZrO_2 Nanoceramic

A hydrothermal technique was used to synthesize zirconium oxide nanoparticles. This allows a nanoceramic with homogeneous semispherical morphology, with dimensions of around 25 to 80 nm, to be obtained; this could be due to the presence of agglomerates constituted by several small polycrystallites in the range of 10–20 nm in the dimension (Figure 1a). Moreover, the X-ray diffraction (XRD) pattern showed a considerable broadening of the peaks due to the nanostructure of the crystalline grains (Figure 1b). The XRD of the ZrO_2 showed a crystalline behavior with two phases. Based on the XRD spectrum of the JCPDS card No. 03-065-0461, the monoclinic phase showed the peaks (1 0 1) at 30.26°, (1 1 0) at 35.31°, (1 1 2) at 50.28° and (−2 0 2) at 62.93°. A cubic phase was also identified, with peaks (0 1 0) at 24°, (1 1 1) at 28°, (−1 1 1) at 31.5°, (0 2 1) at 38.5°, (1 2 1) at 41°, (2 0 2) at 45.25°, (−2 0 2) at 55°, and (3 1 1) at 60.28°, according to the JCPDS card No. 03-065-0461. Using the Debye-Scherrer formula for the line broadening fitting curve XRD program, the particle size was evaluated. The average particle diameter for the monoclinic phase was around 25–35 nm, considering that the grains were spheres. For the cubic phase, the average size was estimated at around 6–14 nm.

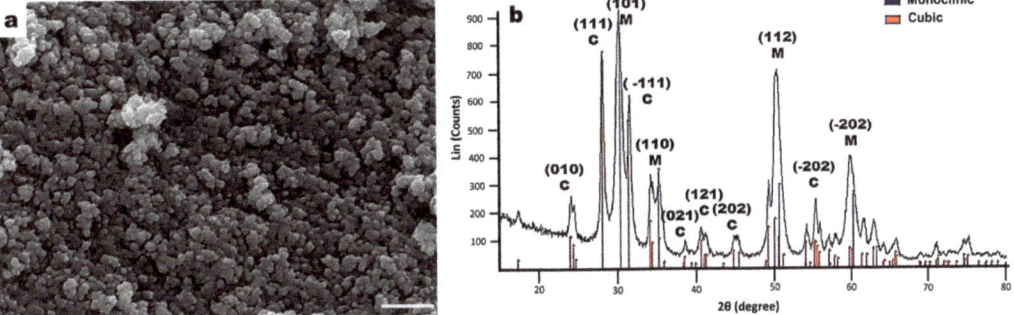

Figure 1. Characterization of zirconium oxide (ZrO$_2$) nanoceramic. (**a**) Scanning electron microscopy images of zirconium oxide nanoparticles. ZrO$_2$ nanoparticles with homogeneous morphology and semispherical surface; the nanoparticles tend to agglomerate; scale bar = 200 nm. (**b**) X-ray diffraction patterns of ZrO$_2$ nanoparticles obtained with incident angles at 30.5°, 31.7°, 35.2°, 50.2°, 60.1°, and 63.2°.

2.2. Characterization of PLA/ZrO$_2$ Nanocomposite Fiber Scaffold

Figure 2 shows the analysis of the fiber membrane scaffolds of 10% wt of PLA and PLA/ZrO$_2$ nanocomposite fiber scaffold by scanning electron microscopy (SEM). The PLA fiber scaffold is composed of smooth and uniform fibers with minimal bead formation and a diameter of around 400 nm with random orientation (Figure 2). The morphological analysis demonstrated that PLA/ZrO$_2$ scaffolds have a rough surface due to the zirconium nanoceramic. The particles are observed through the fiber and sometimes as agglomerates around the interconnected strands with random orientation (Figure 2b,c). The analysis of the diameter sizes of the PLA/ZrO$_2$ nanocomposite fiber scaffold showed that incorporating the nanoceramic (0.1 and 0.5 g) increased the fiber diameter. Moreover, the fibers were in the range of 100 to 800 nm, with an average diameter of 395 nm.

The chemical structures of PLA fiber membranes synthesized with different concentrations of the ZrO$_2$ nanoceramic (0.1 and 0.5 g) were obtained using fourier-transform infrared spectroscopy (FTIR) spectroscopy. The results were compared to identify structural changes by incorporating the nanoceramic onto the PLA polymer matrix (Figure 3). The infrared absorbance spectra showed the typical characteristic of PLA; i.e., the absorption bands around 1750 cm^{-1} corresponding to the (C=O) ester carbonyl group; at 1445 and 1380 cm^{-1}, corresponding to the absorbance bands of the C-H bending vibration of CH$_3$; at 1350 cm^{-1} corresponding to the bending vibration of carbonyl CH; at 960 to 830 cm^{-1}, corresponding to the backbone stretching and CH$_3$ rocking; at the region of 3200 to 2800 cm^{-1}, corresponding to the symmetric and asymmetric stretch of CH; and at 1260 cm^{-1} and 1100 cm^{-1}, corresponding to the lactide C-O stretch. However, the typical spectral of PLA was accompanied by the absorption band peaks of the ZrO$_2$ nanoceramic; especially, a band at 758 cm^{-1} without structural change on the PLA fiber after doping with the nanoceramic.

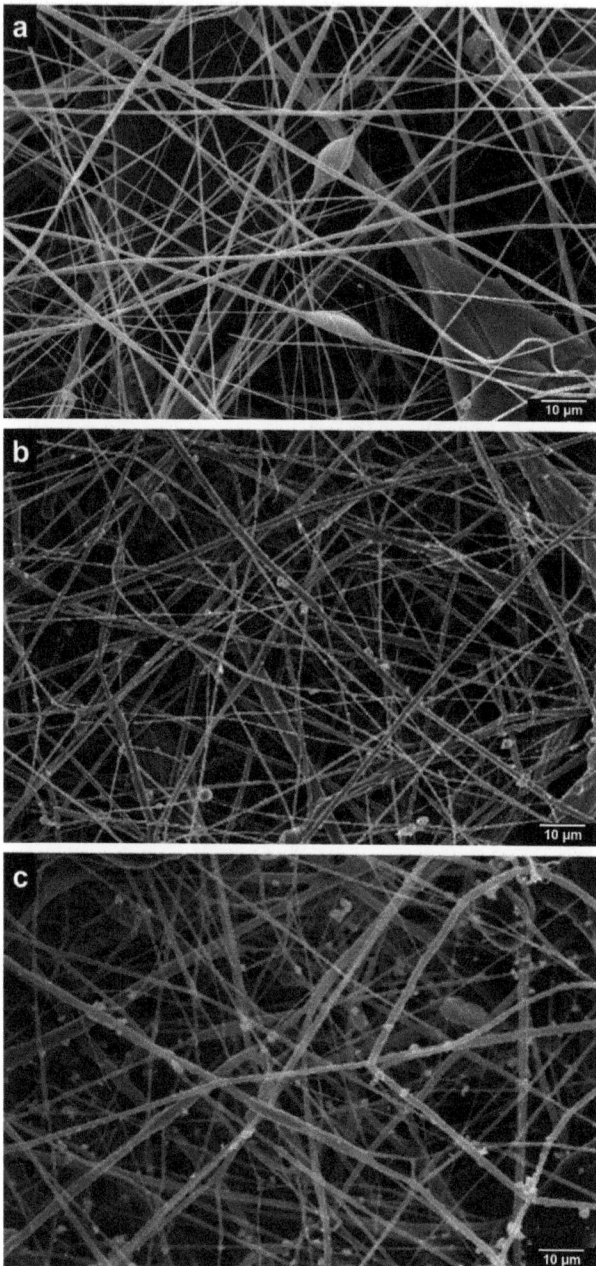

Figure 2. Scanning electron microscopy assessment of the microstructure of the polylactic acid (PLA) scaffolds loaded with different concentrations of zirconium oxide nanoparticles (ZrO_2). (**a**) PLA scaffolds synthesized with the air-jet spinning technique. (**b**) PLA scaffold loaded with 0.1 g of ZrO_2. (**c**) PLA scaffold loaded with 0.5 g of ZrO_2. The PLA/ZrO_2 scaffolds show dispersed and unsaturated ZrO_2 nanoparticles along the PLA fibers, scale bar = 10 μm.

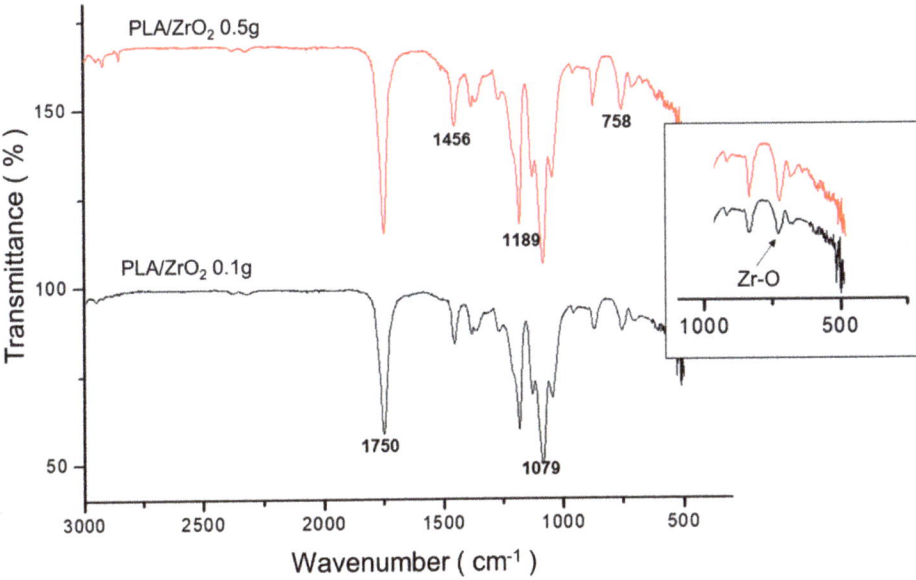

Figure 3. FTIR-IR spectrum of PLA/ZrO$_2$ fiber membrane scaffolds. The absorption bands typical characteristic of PLA: around 1750 cm^{-1}, the (C=O) ester carbonyl group; at 1445 and 1380 cm^{-1}, the absorbance bands of C-H bending vibration of CH$_3$; at 1350 cm^{-1}, the bending vibration of carbonyl CH; at 960 to 830 cm^{-1}, corresponding to the backbone stretching and CH$_3$ rocking; at the region of 3200 to 2800 cm^{-1}, to the symmetric and asymmetric stretch of CH; and at 1260 cm^{-1} and 1100 cm^{-1}, corresponding to the lactide C-O stretch, with 1090–1189 cm^{-1} (C-O=C) and 758 cm^{-1} attributed to (Zr–O). The insert image of the absorption band of Zr-O at 758 cm^{-1} was magnified for show the differences between the PLA/ZrO$_2$ with 0.1 g and 0.5 g composite scaffold.

2.3. Biocompatibility Assay

The biocompatibility of the PLA/ZrO$_2$ nanocomposite fiber scaffold was analyzed in in vitro cell culture to investigate the cell adhesion and cell viability of hFOB 1.19 cells. The cellular adhesion response of hFOB 1.19 cells at 4 and 24 h over the surface of the PLA/ZrO$_2$ nanocomposite fiber scaffold are presented as the percentage of attached cells in relation to control tissue culture plates (Figure 4a).

The cell adhesion of hFOB 1.19 cells was favorable. It exceeded 100% at 4 h and ≥150% at 24 h of attachment onto the PLA/ZrO$_2$ doped with 0.5 g of the nanoceramic with statistical differences compared to the PLA/ZrO$_2$ doped with 0.1 g of the nanoceramic and the PLA fiber spun mat at $p < 0.05$. However, there were no statistical differences between the adhesion of hFOB 1.19 cells onto the PLA/ZrO$_2$ doped with 0.1 g of the nanoceramic and the PLA fiber spun mat. The cell attachment for the former conditions was around 80% at 4 h and ≥120% at 24 h.

Concerning the cell viability, we performed the MTT assay to confirm that the PLA/ZrO$_2$ nanocomposite fiber scaffold is not toxic to the cells (Figure 4b). The results are presented as the optical absorbance at 570 nm. The histogram in Figure 4b suggests that, in all scaffolds, high levels of MTT oxidation are present. However, the higher conversion rate of MTT was found in PLA/ZrO$_2$ doped with 0.5 g of the nanoceramic from day 3, and with a constant increment until 21 days of cell culture. Furthermore, the MTT conversion rate of the PLA/0.5 g ZrO$_2$ scaffold was followed by PLA/ZrO$_2$ doped with 0.1 g of the nanoceramic and by the PLA fiber scaffold. Furthermore, statistical differences were found between the viability of hFOB cell culture onto the PLA/ZrO$_2$ nanocomposite fiber scaffold and hFOB cell culture in the PLA fiber scaffold at $p < 0.05$.

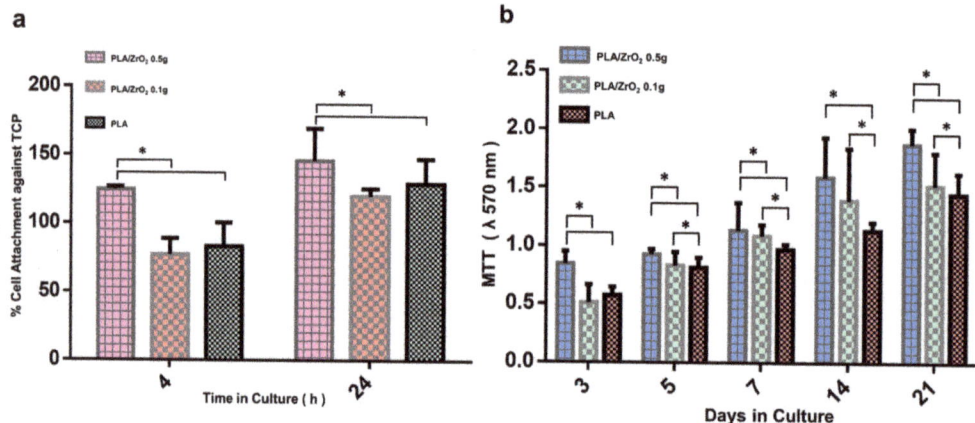

Figure 4. The hFOB biological response, after culture onto the PLA/ZrO$_2$ scaffolds. (**a**) The cellular adhesion response of hFOB at 4 and 24 h, evaluated with violet crystal. The cell adhesion onto the PLA/ZrO$_2$ nanocomposite fiber scaffold surface is presented as the percentage of attached cells in relation to control tissue culture plates. (**b**) Metabolic activity was evaluated with MTT assay at 3, 5, 7, 14, and 21 days to confirm that the PLA/ZrO$_2$ nanocomposite fiber scaffold is not toxic to the cells. The viability of the hFOB cell increased time-dependently, with the best results observed for the PLA/0.5 g of ZrO$_2$ after 21 days of culture. Asterisk (*) mean that scaffolds showed a statistical significance ($p < 0.05$).

2.4. Cell-Material Interaction

Figure 5 showed the cell morphology and the cell spreading pattern interaction between hFOB cells onto the PLA fiber scaffold and the PLA/ZrO$_2$ composite fiber scaffold doped with 0.5 g of the nanoceramic. The fluorescence images of the morphology showed that human osteoblasts cultured onto PLA/ZrO$_2$ had a well-attached cell with elongated morphology and filopodia extensions (Figure 5c,d), in comparison with the less spreading and less elongated morphology shown by osteoblasts cultured onto the control PLA fiber scaffold (Figure 5a,b). Additionally, SEM micrographs showed that cells onto PLA presented a rounded cytoplasm, with few spreading cells that exhibited a flat shape (Figure 5b), compared with preferential spread on the entire surface of the PLA/ZrO$_2$ composite fiber scaffold, wherein some points present an elongated morphology in direct contact with the nanofiber morphologies of the scaffolds due to the presence of cues exerted by zirconia nanoparticles (Figure 5d).

2.5. Biomineralization Assay

For the analysis of the biomineralization assay, we selected only the PLA/ZrO$_2$ of 0.5 g scaffold because it demonstrated the best results for cell adhesion and cell viability. The bioactivity of hFOB cells onto the composite was analyzed by Alizarin Red S staining (ARS) after 3 and 21 days (Figure 6). The images of light microscopy showed that after three days of culture, on control media, there was almost a very weak staining onto the PLA scaffold and the PLA/ZrO$_2$ composite fiber scaffold, in comparison with a very light signal of the red staining in both the PLA spun mat and the PLA/ZrO$_2$ composite fiber scaffold cultured on osteogenic media. However, after 21 days of culture, on control media, redder staining onto the PLA scaffold and the PLA/ZrO$_2$ composite fiber scaffold could be seen, in comparison with a darker red precipitate throughout the surface on both the PLA fiber scaffold and the PLA/ZrO$_2$ composite fiber scaffold, which indicates calcium deposits (Figure 6a). Likewise, the semi-quantitative analysis of the red alizarin staining reveals that the PLA/ZrO$_2$ composite scaffold presents a higher concentration related to the calcium precipitates at 3 and 21 days of culture than the PLA fibers in both culture media (Figure 6b).

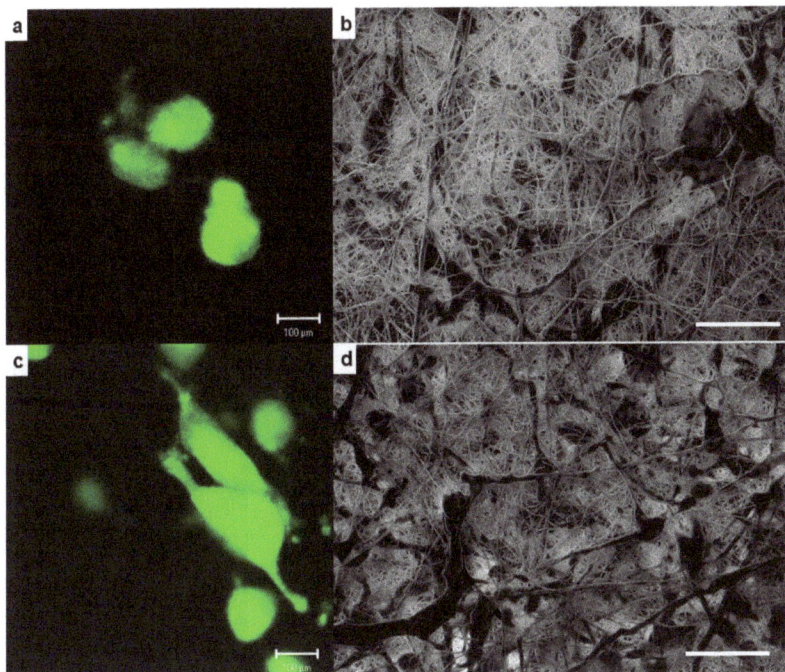

Figure 5. Representative fluorescence images of the cellular morphology and SEM micrographs of the spreading pattern interaction of hFOB cells cultured onto PLA fiber scaffold (**a**,**b**) and PLA/0.5 g ZrO$_2$ scaffolds (**c**,**d**) after 24 h of culture. Scale bar on SEM images = 100 μm.

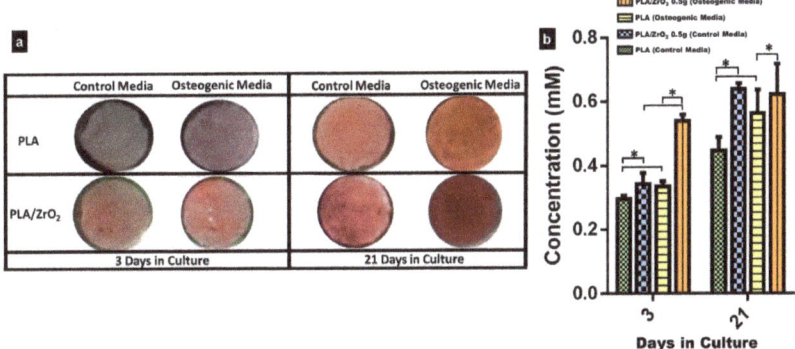

Figure 6. Alizarin red staining. The hFOB cells were cultured onto the PLA and PLA/ZrO$_2$ with the presence or absence of an osteogenic medium. (**a**) The calcified extracellular matrix deposits produced by hFOB cells onto the PLA/0.5 g ZrO$_2$ composite scaffolds are red. The area of the calcified deposits onto the scaffolds was time-dependent and increased with the culture time. However, the calcium deposit onto the PLA/0.5 g ZrO$_2$ scaffolds in the osteogenic medium, compared with the standard medium, was more visible, and the concentration of the alizarin was higher than in the PLA scaffolds. (**b**) The calcium (mM) concentration in PLA/0.5 g ZrO$_2$ scaffolds after 3 and 21 days of culture in osteogenic medium and standard medium. Asterisk (*) mean that the scaffolds showed a statistical significance ($p < 0.05$).

Moreover, the PLA scaffold and PLA/ZrO$_2$ composite fiber scaffold were analyzed by FTIR after 3 and 21 days of culture in osteoinductive culture media and compared to the control culture. From the spectra, the peaks that are characteristic of the amide I and amide II groups could be observed at 1650 cm^{-1} and 1533 cm^{-1}, corresponding to the extracellular collagen matrix (Figure 7).

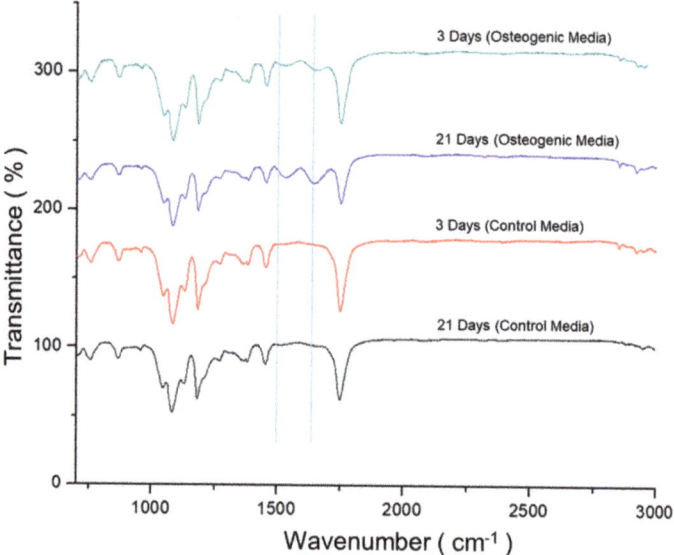

Figure 7. FTIR-IR spectrum of PLA/ZrO$_2$ scaffolds after 3 and 21 days of culture with hFOB. The PLA/ZrO$_2$ scaffolds seeded with hFOB cells were cultured with and without the presence of the osteogenic medium. The PLA/0.5 g ZrO$_2$ scaffold spectrum after 3 and 21 days of culture with osteogenic medium showed new peaks that are characteristic of the amide I and amide II groups at 1650 cm^{-1} and 1533 cm^{-1}; the bands that correspond to the PLA are also present.

3. Discussion

Tissue engineering (TE) is a multidisciplinary area of research aimed at repairing, replacing, or regenerating tissues or organs to restore impaired function due to defects and aging. Nanotechnology for TE application focuses on imitating the size and role of the extracellular matrix (ECM) with a biomimetic characteristic for improving the migration, proliferation, and differentiation of cells [28,29]. This ECM biomimetic related to bone tissue could be obtained by synthesizing a scaffold with a combination of different biomaterials as polymers and bioceramic, via several manufacturing methods that allow the combining of its properties for translation to the new composite phase, to improve its functionality [5–10,30]. Thus, in this study, we used hydrothermal synthesis and air-jet spinning (AJS) technology to engineer a multifunctional composite scaffold as a platform for bone tissue engineering applications.

First, we used hydrothermal synthesis, defined as a homogeneous reaction in the presence of aqueous solvents under temperature and pressure conditions to dissolve and recrystallize insoluble materials under normal conditions, to obtain the fiber's ceramic phase composite scaffold. Our XRD and SEM images show that hydrothermal synthesis allows a homogeneous thermally stable crystalline nanoceramic of ZrO$_2$ to be obtained. Furthermore, the nanoceramic showed agglomerates of crystallites, of nearly uniform size, between 10–80 nm on average for the particle diameter, with nanometric dimensions considering that the grains are spheres, and the results showed that nanocrystals coincide with the monoclinic and cubic structural phases specific to ZrO$_2$; the results are consistent with previous studies reported [31–33].

For bone application, nanocomposite fiber biomaterials are a relatively new class of materials that combine a homogeneous size of the crystalline structure with biopolymeric matrix structure, that enables the improvement of the fibrillar design nanocomposite [34,35]. In this regard, we explore the use of AJS to design a polymeric fibrillar matrix of 10% wt of PLA, doped with two different concentrations (0.1 and 0.5 g) of the nanoceramic of ZrO_2. As demonstrated by our results, the SEM images of the PLA/ZrO_2 nanocomposite had a random orientation of the fiber morphology with a submicron/nanometer range between 100 nm up to 1 µm in diameter, with an average size of 395 nm; by FTIR, it was possible to detect differences between the intensity peak of 0.5 and 0.1 g of ZrO_2 that could be related to the mass fraction of the nanoceramic for 0.1 g (~0.0099 g) and for 0.5 g (~0.047 g) in the composite. This difference in the peak intensity of samples showed that, with increased concentrations of nanoceramic, the corresponding interaction of the Zr-O bonds with the characteristic bands of the functional groups of PLA (carbonyl group, esters CH and CH_3) at 758 cm^{-1} could be detected; furthermore, this different concentration improves the anisotropic properties of the scaffold that could be related to the adhesion between the interface of the fiber matrix and the nanoceramic particles, in concordance with previous studies reported [25–27,36,37].

In bone tissue engineering, the goal of a nanocomposite is intended to provide a suitable surface for supporting the tissue regeneration process, and therefore requires osteoconduction or osteoinduction properties [35]. Our study found that the doped polymer matrix with the nanoceramic of ZrO_2 could be considered adequate to the bone because the result showed appropriate surface property cues and physicochemical stability that have been defined as important parameters for influencing cell cytocompatibility and osteogenesis [37–40]. In our study, we analyzed the biocompatibility of the PLA/ZrO_2 composite fiber scaffold by cell adhesion and viability response of human osteoblasts cells hFOB 1.19. In the adhesion assay and viability evaluation, we found that the composite scaffold of PLA doped with 0.5 g of the nanoceramic had higher cell attachment and proliferation of cells. This cellular behavior could be influenced by the optimal surface topography, the physicochemical properties of the material, the nanometric surface features, and the critical size of the doped nanophase that plays a role in mediating the cell behavior response [14,41].

Moreover, the physical, chemical, and surface topography property cues are critical parameters that induce cellular recognition signals. Our results showed that hFOB cells attached well to the fiber composite, covering long area extensions with very spreading, elongated, and lamellae morphology. Thus, the morphology of the topographic cues (spherical nanostructure) of the nanocomposite of PLA related to the presence of ZrO_2 demonstrated an excellent surface that stimulated cell attachment and proliferation and supported the extracellular matrix deposit by the osteoblast cells, as shown by alizarin red staining. This is an important effect of the nanoscale because AJS allows the synthesis of a more suitable scaffold with the physical integration of nanoceramic, conferring a homogenous distribution for cell binding. Furthermore, these could be improved by the adsorption of biomolecules by the media that can activate signaling pathways that change the cell behavior, recreating a microenvironment similar to the native tissue, indicating the recognition or sense of the surface by the osteoblast cells, favorably inducing a good cell-material interaction and differentiation [13].

In our study, we evaluated the potential of the 0.5 g of PLA/ZrO_2 fiber nanocomposite in promoting the mineral deposit of the ECM of the hFOB by a differentiation assay. Our analysis by alizarin red staining (ARS) and FTIR showed that osteoblasts could deposit a mineral-like tissue over the nanocomposite surface after 21 days of culture. This indicates the importance of the chemical components at the nanoscale, and the submicron-sized topographical cues that could modulate the cell osteoblasts behavior. Moreover, the chemical component related to the insertion of the ZrO_2 nanoceramic onto the fiber polymer matrix may confer a possible cue role on the bioactivity related to the osteogenesis potential that influences the proliferation and differentiation. The mineral deposit evaluated by ARS and FTIR results indicate that topographical properties of the nanocomposite were related to the size and morphology of the fiber, the porosity, but also to the osteogenic stimuli (inductive factors) presence on the cell culture media, together performing a specific role supporting osteoblast cell differentiation. Thus, our results are in agreement with different studies that reported that zirconia nanofibers or ZrO_2 composites may have pro-osteogenic properties, have good mechanical biocompatibility, and improve the metabolism of bone precursor cells. Furthermore, they could also be considered as a self-regulation material that possesses an excellent osteoinductive capacity that could modify the deposition of ECM by bone cells during differentiation or in response to the surface material, because the osteoblast cell functions could be substrate-sensitive, either with respect to stiffness, roughness, or to the size of the composition of the material [16,17,42–46].

4. Materials and Methods

4.1. Synthesis of Nanoceramic of Zirconium Oxide

Zirconium oxide nanoceramic was synthesized via the hydrothermal technique. The volume ratio of zirconium tetrachloride ($ZrCl_4$) and sodium hydroxide (NaOH) was 3:1. The obtained solution was mixed with pH 14 and let to settle for 20 h, until a substitution reaction happened between $ZrCl_4$ and NaOH, obtaining sodium chloride (NaCl) and zirconium oxide (ZrO_2) as products. After that, the reaction product was removed, washed five times with deionized water, then the clean ZrO_2 was loaded into a Teflon vessel for autoclaving. Afterward, the autoclave was placed in a furnace (Barnstead model 1500), heated, and maintained at 200 °C for 2 h, then cooled to room temperature. The powder obtained was crushed on an agate mortar. The characterization of the ceramic was made by X-ray diffraction analysis and scanning electron microscopy (SEM).

The crystal size was calculated by the equation of Scherrer, where d is the mean size of the particle, K is the Scherrer constant, λ is the wavelength, β is the peak at half the maximum intensity (FWHM), and θ is the Bragg angle (in degrees). The following equation is applied only for crystals with sizes in the range of 1–100 nm.

$$d = \frac{K\lambda}{\beta \cos\theta}$$

4.2. Polymer Fiber Composite Fabrication of PLA/ZrO_2

Fibrous composite spun scaffolds were fabricated via the AJS process from PLA polymeric solutions of 10% wt. First, polymeric solutions of 10% wt of PLA were prepared: PLA pellets ($C_3H_6O_3$; MW 192,000 from Nature Works, Minnetonka, MN, USA) were dissolved in chloroform ($CHCl_3$ from J. T. Baker) and stirred for 20 h. After that, acetone was added, and the solution was stirred for 30 min to obtain a homogeneous solution. The volume ratio of chloroform/acetone was 3:1. Then, the polymeric solution was prepared with two different amounts of nanoceramic of ZrO_2 (0.1 and 0.5 g) for composite scaffolds. During their preparation, solutions were incubated in an ultrasonic bath for 1 h to guarantee the homogenization of nanoparticles onto the polymeric solution. In all cases, the polymeric solution was placed in a commercially available airbrush ADIR model 699 with a 0.3 mm nozzle diameter and with a gravitational feed of the solution to synthesize the fiber membrane scaffolds. The airbrush was connected to a pressurized argon tank (CAS number 7740-37, concentration > 99%, PRAXAIR Mexico, Nuevo Leon, Mexico). For deposition of the fibers, a pressure of 30 psi with 11 cm of distance from the nozzle to the target was held constant. The fiber scaffold characterization was analyzed by a Fourier transform infrared (FTIR) spectrophotometer, employing an IRAffinity-1S (Shimadzu, Kyoto, Japan) within the 400–4000 cm^{-1} range. The morphology and structure of the fibers were observed with a field emission scanning electron microscope (FE-SEM, JSM-7800F, JEOL). Fiber diameter was measured from SEM micrographs employing image analysis software (ImageJ, National Institutes of Health, Gaithersburg, MD, USA).

4.3. Cell Culture

The human fetal osteoblast cell line (hFOB 1.19, ATCC CRL-11372) was used to evaluate the cell biocompatibility response of the PLA fibers spun mat, and the PLA/ZrO_2 spun composite scaffolds. Briefly, hFOB 1.19 cells were cultured in 75 cm^2 culture flasks, in Ham's F12/Dulbecco's modified Eagle's Media (1:1, Ham's F12-DMEM, Sigma-Aldrich, St. Louis, MO, USA), supplemented with 10% fetal bovine serum (FBS, Biosciences, Collierville, TN, USA), 2.5 mM L-glutamine, and an antibiotic solution (streptomycin 100 µg/mL and penicillin 100 U/mL, Sigma-Aldrich, St. Louis, MO, USA). The cells were incubated in a 100% humidified environment at 37 °C in 95% air and 5% CO_2. The hFOB 1.19 cells were used between 2–6 passages of culture for all the experimental procedures.

4.4. Cell Adhesion

As the first attempt to investigate how hFOB cells would interact with the PLA/ZrO_2 nanocomposite fiber material, adhesion assays were performed. For this purpose, hFOB 1.19 cells were seeded at 1×10^4 cells/mL onto the composite scaffold in triplicate and were allowed to adhere under standard cell culture conditions for 4 and 24 h. After each time, nanocomposite fiber scaffolds were rinsed three times using phosphate-buffered saline (PBS) to remove non-adherent cells. Next, the adherent cells were fixed with 3.5% paraformaldehyde, and the evaluation of cell attachment was performed according to the crystal violet assay. Briefly, cells were incubated with 0.1% crystal violet solution for 15 min and then washed carefully with deionized water until clear water was obtained. Then, the dye was extracted with 0.1% of sodium dodecyl sulfate (SDS), and the optical absorption was quantified by spectrophotometry at 550 nm with an ELISA plate reader (ChroMate, Awareness Technology). Cells attached in triplicate onto a conventional tissue culture plate (TCP) were considered as 100% cell attachment of hFBO 1.19 cells, and cells seeded in triplicate onto PLA fiber membrane scaffolds were used as the control.

4.5. Viability Cells (MTT Assay)

The cell viability response of hFOB 1.19 cultured onto PLA/ZrO_2 composite fiber membrane scaffolds in triplicate was analyzed with the MTT assay. The assay is based on the ability of mitochondrial dehydrogenases to oxidize thiazolyl blue (MTT), a tetrazolium salt (3-[4,5-dimethylthiazolyl-2-y]-2,5-diphenyltetrazolium bromide), into an insoluble blue formazan product. Cells were seeded at 1×10^4 cell/mL density in triplicate and incubated for 3, 5, 7, 14, and 21 days of culture. After each time, PLA/ZrO_2 nanocomposite fiber scaffolds were washed with PBS and incubated with fresh cultured medium containing 0.5 mg/mL of MTT for 4 h at 37 °C in the dark. Then, the medium was removed, and 500 µL of dimethyl sulfoxide (DMSO) was added to each well. After 60 min of slow shaking, the absorbance was read at 570 nm. Cells seeded in triplicate onto PLA fiber membrane scaffold were used as the control. During the experimental time, the culture medium was changed every two days.

4.6. Cell-Material Interaction

We examine the cell morphology, the spreading patterns, and cell-material interaction of hFOB 1.19 cells cultured onto the PLA fiber scaffold and PLA/ZrO_2 of 0.5 g nanocomposite fiber scaffold, because it demonstrated the best results for adhesion and viability. Briefly, hFOB 1.19 cells were cultured at a density of 1×10^4 cells/mL onto the scaffolds for 24 h in triplicate. Then, the samples were analyzed using SEM and fluorescence microscopy (AMSCOPE). For SEM analysis, after 24 h of incubation, scaffolds were washed three times with PBS, fixed with 4% formaldehyde, then dehydrated with a graded series of ethanol (25–100%) and air-dried. Next, the samples were sputter-coated with a thin layer of gold and examined by SEM. For fluorescence observation, before seeding the hFOB 1.19 cells onto the scaffolds, cells were incubated with CellTracker™ Green (CMFDA, 5-chloromethylfluorescein diacetate) in phenol red-free medium at 37 °C for 30 min. Subsequently, the cells were washed with PBS and incubated for 1 h in a complete medium. Then, hFOB 1.19 cells were trypsinized and counted to the desired cell concentration (1×10^4 cells/mL), incubated for 24 h onto the scaffolds, and examined under the microscope.

4.7. Biomineralization Assay

As previously mentioned, the PLA/ZrO$_2$ of 0.5 g nanocomposite fiber membrane scaffold demonstrated the best results in the biocompatibility assays. Therefore, it was used for biomineralization assays. Briefly, hFOB 1.19 cells were seeded onto the PLA fiber scaffold and PLA/ZrO$_2$ of 0.5 g scaffold at a density of 1×10^4 cell/mL and left to adhere overnight. Then, a group of scaffolds (PLA and PLA/ZrO$_2$) were incubated with mineralizing media (complete medium, 50 μg/mL of ascorbic acid, 10 mM glycerol-2-phosphate, and 10^{-7} M of dexamethasone) and, together with the control group (PLA and PLA/ZrO$_2$ scaffolds), were maintained in standard media without any osteogenic factors for 3 and 21 days of culture. After a time, PLA and PLA/ZrO$_2$ composite fiber membrane scaffolds were washed three times with PBS. Then, cells were fixed with 4% formaldehyde for 1 h, washed five times carefully with distilled water, and then stained with a saturated solution (40 mM) of Alizarin Red S pH 4.2 for 30 min at room temperature. After several washes with distilled water to remove the dye excess, scaffolds were examined under an optical microscope.

For the quantitative analysis of ARS staining, the fiber membrane scaffolds were transferred to a 1.5 mL microcentrifuge tube, and ARS was extracted with 10% acetic acid, incubated for 30 min, and then heated at 85 °C for 10 min. Samples were then cooled for 5 min on ice and centrifuged at 20,000 rpm for 15 min; the reaction was stopped with 10% ammonium hydroxide. Finally, the dye solution sample was read at 405 nm in the spectrophotometer. The concentration of ARS was determined by correlating the absorbance of the experimental samples with a standard known curve of ARS dye concentrations. The experiments were repeated twice, and three fiber scaffolds were used in each experiment. Moreover, the mineral-like tissue deposited onto the PLA and PLA/ZrO$_2$ composite fiber membrane scaffold at 3 and 21 days of culture was analyzed by FTIR.

4.8. Statistical Analysis

Statistical analysis was performed using a two-way ANOVA (GraphPad Prism 7 software) followed by Tukey's post-hoc test to evaluate the differences among groups. All the results are presented as means ± standard deviation. A *p*-value of less than 0.05 was considered statistically significant.

5. Conclusions

In summary, our results indicate that thermally stable ZrO$_2$ nanoceramic was successfully synthesized via the hydrothermal method with an average size of 10 to 35 nm. The air-jet spinning technique allowed the design of a nanocomposite scaffold of PLA/ZrO$_2$ composed of fiber morphology, with an average diameter of 395 nm, with random distribution. The biocompatibility of osteoblasts analyzed in the function of cellular adhesion and cellular viability onto the PLA/ZrO$_2$ fiber nanocomposite did not show any cytotoxic effects. Moreover, 0.5 g of the PLA/ZrO$_2$ nanocomposite influenced a good cell functionality, providing a microenvironment correlated to the nanometric and interconnectivity of the fiber scaffold that increased the cell bioactivity over the topography, shown by the mineral-like deposit. However, further studies may be focused on the structural properties of the 0.5 g PLA/ZrO$_2$ fiber nanocomposite in critical-size bone models as a future potential to be applied in bone tissue engineering.

Author Contributions: Conceptualization, R.O.-A., O.A.-F. and M.A.A.-P.; investigation, R.O.-A. and M.G.-H.; methodology, R.O.-A. and M.G.-H.; validation, R.O.-A., O.A.-F. and M.A.A.-P.; data curation and formal analysis, R.O.-A. and M.G.-H.; writing—original draft preparation, R.O.-A., M.G.-H. and M.A.A.-P.; writing—review and editing, O.A.-F. and M.A.A.-P.; funding acquisition, M.A.A.-P. All authors have read and agreed to the published version of the manuscript.

Funding: The authors are grateful for the financial support of the DGAPA-UNAM: PAPIIT IN213821 project and the CONACYT by the particular program of Fondo Sectorial de Investigación para la Educación A1-S-9178 project.

Institutional Review Board Statement: Not applicable.

Informed Consent Statement: Not applicable.

Data Availability Statement: All data generated for this study are included in the manuscript.

Acknowledgments: We appreciate the support provided by the CONACYT through its scholarship (number: 613415 and CVU: 745379) to Rodrigo Osorio Arciniega during his Doctoral studies in the Programa de Maestría y Doctorado en Ciencias Médicas, Odontológicas y de la Salud at UNAM. The authors want to thank Raul Reyes Ortíz and Adriana Tejeda Cruz from IIM-UNAM for their technical assistance and support.

Conflicts of Interest: The authors declare that the research was conducted without any commercial or financial relationships that could be construed as a potential conflict of interest.

References

1. Könönen, E.; Gursoy, M.; Gursoy, U.K. Periodontitis: A multifaceted disease of tooth-supporting tissues. *J. Clin. Med.* **2019**, *8*, 1135. [CrossRef]
2. Ausenda, F.; Rasperini, G.; Acunzo, R.; Gorbunkova, A.; Pagni, G. New Perspectives in the use of biomaterials for periodontal regeneration. *Materials* **2019**, *12*, 2197. [CrossRef]
3. Amghar-Maach, S.; Gay-Escoda, C.; Sánchez-Garcés, M.Á. Regeneration of periodontal bone defects with dental pulp stem cells grafting: Systematic Review. *J. Clin. Exp. Dent.* **2019**, *11*, e373–e381. [CrossRef]
4. Gloria, A.; De Santis, R.; Ambrosio, L. Polymer-based composite scaffolds for tissue engineering. *J. Appl. Biomater. Biomech.* **2010**, *8*, 57–67.
5. Kao, C.T.; Chen, Y.J.; Ng, H.Y.; Lee, A.K.; Huang, T.H.; Lin, T.F.; Hsu, T.T. Surface modification of calcium silicate via mussel-inspired polydopamine and effective adsorption of extracellular matrix to promote osteogenesis differentiation for bone tissue engineering. *Materials* **2018**, *11*, 1664. [CrossRef]
6. Tsai, K.Y.; Lin, H.Y.; Chen, Y.W.; Lin, C.Y.; Hsu, T.T.; Kao, C.T. Laser sintered magnesium-calcium silicate/poly-ε-caprolactone scaffold for bone tissue engineering. *Materials* **2017**, *10*, 65. [CrossRef] [PubMed]
7. Fujikura, K.; Obata, A.; Kasuga, T. Cellular migration to electrospun poly (lactic acid) fiber mats. *J. Biomater. Sci. Polym. Ed.* **2012**, *23*, 1939–1950. [CrossRef]
8. Su, C.J.; Tu, M.G.; Wei, L.J.; Hsu, T.T.; Kao, C.T.; Chen, T.H.; Huang, T.H. Calcium silicate/chitosan-coated electrospun poly (lactic acid) fibers for bone tissue engineering. *Materials* **2017**, *10*, 501. [CrossRef] [PubMed]
9. Chiu, Y.C.; Fang, H.Y.; Hsu, T.T.; Lin, C.Y.; Shie, M.Y. The characteristics of mineral trioxide aggregate/polycaprolactone 3-dimensional scaffold with osteogenesis properties for tissue regeneration. *J. Endod.* **2017**, *43*, 923–929. [CrossRef] [PubMed]
10. Huang, X.; Qi, Y.; Li, W.; Shi, Z.; Weng, W.; Chen, K.; He, R. Enhanced integrin-mediated human osteoblastic adhesion to porous amorphous calcium phos-phate/poly (L-lactic acid) composit. *Chin. Med. J.* **2014**, *127*, 3443–3448. [CrossRef]
11. Piconi, C.; Maccauro, G. Zirconia as a ceramic biomaterial. *Biomaterials* **1999**, *20*, 1–25. [CrossRef]
12. Herath, H.M.T.U.; Di Silvio, L.; Evans, J.R.G. Response to zirconia surfaces with different topographies. *Mater. Sci. Eng. C* **2015**, *57*, 363–370. [CrossRef]
13. Razak, S.I.A.; Sharif, N.F.A.; Rahman, W.A.W.A. Biodegradable polymers and their bone applications: A Review. *Int. J. Basic Appl. Sci.* **2012**, *12*, 31–49.
14. Julien, J.M.; Bénézet, J.C.; Lafranche, E.; Quantin, J.C.; Bergeret, A.; Lacrampe, M.F.; Krawczak, P. Development of poly (lactic acid) cellular materials: Physical and morphological characterizations. *Polymer* **2012**, *53*, 5885–5895. [CrossRef]
15. Gadad, A.P.; Vannuruswamy, G.; Vijay Kumar, S.V.; Dandagi, P.M.; Mastiholimath, V.S. Nanofibers: Fabrication, characterization and their biomedical applications. *Indian Drugs* **2013**, *50*, 5–19. [CrossRef]
16. Rodaev, V.V.; Razlivalova, S.S.; Zhigachev, A.O.; Vasyukov, V.M.; Golovin, Y.I. Preparation of zirconia nanofibers by electrospinning and calcination with zirconium acetylacetonate as precursor. *Polymers* **2019**, *11*, 1067. [CrossRef] [PubMed]
17. Tyurin, A.I.; Rodaev, V.V.; Razlivalova, S.S.; Korenkov, V.V.; Zhigachev, A.O.; Vasyukov, V.M.; Golovin, Y.I. Morphology and mechanical properties of 3y-tzp nanofiber mats. *Nanomaterials* **2020**, *10*, 2097. [CrossRef]
18. Dejob, L.; Toury, B.; Tadier, S.; Grémillard, L.; Gaillard, C.; Salles, V. Electrospinning of in situ synthesized silica-based and calcium phosphate bioceramics for applications in bone tissue engineering: A review. *Acta Biomater.* **2021**, *123*, 123–153. [CrossRef] [PubMed]
19. Lim, J.W.; Jang, K.J.; Son, H.; Park, S.; Kim, J.E.; Kim, H.B.; Seonwoo, H.; Choung, Y.H.; Lee, M.C.; Chung, J.H. Aligned nanofiber-guided bone regeneration barrier incorporated with equine bone-derived hydroxyapatite for alveolar bone regeneration. *Polymers* **2020**, *13*, 60. [CrossRef]

20. DeFrates, K.; Markiewicz, T.; Xue, Y.; Callaway, K.; Gough, C.; Moore, R.; Bessette, K.; Mou, X.; Hu, X. Air-jet spinning corn zein protein nanofibers for drug delivery: Effect of biomaterial structure and shape on release properties. *Mater. Sci. Eng. C* **2021**, *118*, 111419. [CrossRef] [PubMed]
21. Dadol, G.C.; Lim, K.J.A.; Cabatingan, L.K.; Tan, N.P.B. Solution blow spinning-polyacrylonitrile-assisted cellulose acetate nanofiber membrane. *Nanotechnology* **2020**, *31*, 345602. [CrossRef]
22. Bienek, D.R.; Hoffman, K.M.; Tutak, W. Blow-spun chitosan/PEG/PLGA nanofibers as a novel tissue engineering scaffold with antibacterial properties. *J. Mater. Sci. Mater. Med.* **2016**, *27*, 146. [CrossRef]
23. Abdal-hay, A.; Sheikh, F.A.; Lim, J.K. Air jet spinning of hydroxyapatite/poly (lactic acid) hybrid nanocomposite membrane mats for bone tissue engineering. *Colloids Surf. B Biointerfaces* **2013**, *102*, 635–643. [CrossRef]
24. Suarez-Franco, J.L.; Vázquez-Vázquez, F.C.; Pozos-Guillen, A.; Montesinos, J.J.; Alvarez-Fregoso, O.; Alvarez-Perez, M.A. Influence of diameter of fiber membrane scaffolds on the biocompatibility of hPDL mesenchymal stromal cells. *Dent. Mater. J.* **2018**, *37*, 465–473. [CrossRef]
25. Granados-Hernández, M.V.; Serrano-Bello, J.; Montesinos, J.J.; Alvarez-Gayosso, C.; Medina-Velázquez, L.A.; Alvarez-Fregoso, O.; Alvarez-Perez, M.A. In Vitro and In Vivo biological characterization of poly (lactic acid) fiber scaffolds synthesized by air jet spinning. *J. Biomed. Mater. Res. Part B Appl. Biomater.* **2018**, *106*, 2435–2446. [CrossRef]
26. Mendieta-Barrañon, I.; Chanes-Cuevas, O.A.; Alvarez-Perez, M.A.; González-Alva, P.; Medina, L.A.; Aguilar-Franco, M.; Serrano-Bello, J. Physicochemical and tissue response of PLA nanofiber scaffolds sterilized by different techniques. *Odovtos-Int. J. Dent. Sci.* **2019**, *21*, 77–88. [CrossRef]
27. Albanés-Ojeda, E.A.; Calderón-Olvera, R.M.; García-Hipólito, M.; Chavarría-Bolaños, D.; Vega-Baudrit, R.; Álvarez-Perez, M.A.; Alvarez-Fregoso, O. Physical and Chemical Characterization of PLA/ZrO2 Nanofibers Composites Synthesized by Air-jet Spinning. *Indian J. Fibre Text. Res. (IJFTR)* **2020**, *45*, 57–64.
28. Sahay, R.; Kumar, P.S.; Sridhar, R.; Sundaramurthy, J.; Venugopal, J.; Mhaisalkarc, S.G.; Ramakrishna, S. Electrospun composite nanofibers and their multifaceted applications. *J. Mater. Chem.* **2012**, *22*, 12953–12971. [CrossRef]
29. Jiang, T.; Carbone, J.E.; Lo, K.W.H.; Laurencin, C.T. Electrospinning of polymer nanofibers for tissue regeneration. *Prog. Polym. Sci.* **2015**, *46*, 1–24. [CrossRef]
30. Rajak, D.K.; Pagar, D.D.; Menezes, P.L.; Linul, E. Fiber-reinforced polymer composites: Manufacturing, properties, and applications. *Polymers* **2019**, *11*, 1667. [CrossRef]
31. Byrappa, K.; Yomishura, M. *Handbook of Hydrothermal Technology*, 2nd ed.; William Andrew Publishing: New York, NY, USA, 2013; pp. i–iii. [CrossRef]
32. Denry, I.; Kelly, J.R. State of the art of zirconia for dental applications. *Dent. Mater.* **2008**, *24*, 299–307. [CrossRef]
33. Assal, P.A. The osseointegration of zirconia dental implants. *Schweiz Mon. Zahnmed* **2013**, *123*, 644–654.
34. Teimouri, A.; Ebrahimi, R.; Emadi, R.; Beni, B.H.; Chermahini, A.N. Nanocomposite of silk fibroin/chitosan/Nano ZrO$_2$ for tissue engineering applications: Fabrication and morphology. *Int. J. Biol. Macromol.* **2015**, *76*, 292–302. [CrossRef]
35. Bharadwaz, A.; Jayasuriya, A.C. Recent trends in the application of widely used natural and synthetic polymer nanocomposites in bone tissue regeneration. *Mater. Sci. Eng. C* **2020**, *110*, 110698. [CrossRef]
36. Gómez-Pachón, E.Y.; Sánchez-Arévalo, F.M.; Sabina, F.J.; Maciel-Cerda, A.; Mon-tiel Campos, R.; Batina, N.; Morales-Reyes, I.; Vera-Graziano, R. Characterization and modelling of the elastic properties of poly (lactic acid) nanofibre scaffolds. *J. Mater. Sci.* **2013**, *48*, 8308–8319. [CrossRef]
37. Carrasco, F.; Pagès, P.; Gámez-Pérez, J.; Santana, O.O.; Maspoch, M.L. Processing of poly (lactic acid): Characterization of chemical structure, thermal stability and me-chanical properties. *Polym. Degrad. Stab.* **2010**, *95*, 116–125. [CrossRef]
38. Karpiński, R.; Jaworski, L.; Czubacka, P. The structural and mechanical properties of the bone. *J. Technol. Exploit. Mech. Eng.* **2017**, *3*, 43–50. [CrossRef]
39. Morgan, E.F.; Unnikrisnan, G.U.; Hussein, A.I. Bone mechanical properties in healthy and diseased states. *Annu. Rev. Biomed. Eng.* **2018**, *20*, 119–143. [CrossRef]
40. Hoveizi, E.; Nabiuni, M.; Parivar, K.; Rajabi-Zeleti, S.; Tavakol, S. Functionaliza-tion and surface modification of electrospun polylactic acid scaffold for tissue engi-neering. *Cell Biol. Int.* **2014**, *38*, 41–49. [CrossRef]
41. Ito, H.; Sasaki, H.; Saito, K.; Honma, S.; Yajima, Y.; Yoshinari, M. Response of os-teoblast-like cells to zirconia with different surface topography. *Dent. Mater. J.* **2013**, *32*, 122–129. [CrossRef] [PubMed]
42. Delgado-Ruíz, R.A.; Gomez Moreno, M.; Aguilar-Salvatierra, A.; Markovic, A.; Mate-Sánchez, J.E.; Calvo-Guirado, J.L. Human fetal osteoblast behavior on zirconia dental implants and zirconia disks with microstructured surfaces. An experimental In Vitro study. *Clin. Oral Implant. Res.* **2016**, *27*, e144–e153. [CrossRef]
43. Cadafalch Gazquez, G.; Chen, H.; Veldhuis, S.A.; Solmaz, A.; Mota, C.; Boukamp, B.A.; van Blitterswijk, C.A.; Ten Elshof, J.E.; Moroni, L. Flexible Yttrium-Stabilized Zirconia Nanofibers Offer Bioactive Cues for Osteogenic Differentiation of Human Mesenchymal Stromal Cells. *ACS Nano.* **2016**, *10*, 5789–5799. [CrossRef]
44. Seweryn, A.; Pielok, A.; Lawniczak-Jablonska, K.; Pietruszka, R.; Marcinkowska, K.; Sikora, M.; Witkowski, B.S.; Godlewski, M.; Marycz, K.; Smieszek, A. Zirconium Oxide Thin Films Obtained by Atomic Layer Deposition Technology Abolish the Anti-Osteogenic Effect Resulting from miR-21 Inhibition in the Pre-Osteoblastic MC3T3 Cell Line. *Int. J. Nanomed.* **2020**, *15*, 1595–1610. [CrossRef]

45. Ding, S.J.; Chu, Y.H.; Chen, P.T. Mechanical Biocompatibility, Osteogenic Activity, and Antibacterial Efficacy of Calcium Silicate-Zirconia Biocomposites. *ACS Omega* **2021**, *6*, 7106–7118. [CrossRef] [PubMed]
46. Prymak, O.; Vagiaki, L.E.; Buyakov, A.; Kulkov, S.; Epple, M.; Chatzinikolaidou, M. Porous Zirconia/Magnesia Ceramics Support Osteogenic Potential In Vitro. *Materials* **2021**, *14*, 1049. [CrossRef]

Article

Influence of PCL and PHBV on PLLA Thermal and Mechanical Properties in Binary and Ternary Polymer Blends

Raasti Naseem [1], Giorgia Montalbano [2], Matthew J. German [3], Ana M. Ferreira [1], Piergiorgio Gentile [1] and Kenneth Dalgarno [1,*]

[1] School of Engineering, Newcastle University, Newcastle upon Tyne NE1 7RU, UK
[2] Department of Applied Science and Technology, Politecnico di Torino, 10129 Torino, Italy
[3] School of Dental Sciences, Translational and Clinical Research Institute, Faculty of Medical Sciences, Newcastle University, Newcastle upon Tyne NE1 7RU, UK
* Correspondence: kenny.dalgarno@ncl.ac.uk

Abstract: PLLA, PCL and PHBV are aliphatic polyesters which have been researched and used in a wide range of medical devices, and all three have advantages and disadvantages for specific applications. Blending of these materials is an attractive way to make a material which overcomes the limitations of the individual polymers. Both PCL and PHBV have been evaluated in polymer blends with PLLA in order to provide enhanced properties for specific applications. This paper explores the use of PCL and PHBV together with PLLA in ternary blends with assessment of the thermal, mechanical and processing properties of the resultant polymer blends, with the aim of producing new biomaterials for orthopaedic applications. DSC characterisation is used to demonstrate that the materials can be effectively blended. Blending PCL and PHBV in concentrations of 5–10% with PLLA produces materials with average modulus improved by up to 25%, average strength improved by up to 50% and average elongation at break improved by 4000%, depending on the concentrations of each polymer used. PHBV impacts most on the modulus and strength of the blends, whilst PCL has a greater impact on creep behaviour and viscosity. Blending PCL and PHBV with PLLA offers an effective approach to the development of new polyester-based biomaterials with combinations of mechanical properties which cannot be provided by any of the materials individually.

Keywords: polymer blends; polyester biomaterials; PLLA; PCL; PHBV

1. Introduction

Orthopaedic procedures are seemingly ever increasing due to both an increase in the aging population and a rise in obesity rates, contributing to the increase in bone fractures. Permanent implants are expected to serve for the whole term of a patient's life, in contrast to temporary implants, which are required for a shorter time period to allow for the healing of broken bones [1]. Current orthopaedic implants of both categories are produced from metallic materials, which display adequate mechanical properties and fatigue and corrosion resistance; however, the stiffness of metallic materials is greater than that of bone, and this can cause stress shielding [2,3]. As a less stiff alternative to metal implants, aliphatic polyesters are a group of biocompatible and bioresorbable polymers used in a wide range of biomedical applications. Some of the most commonly studied polymers include polylactide (PLA), polycaprolactone (PCL) and polyhydroxyalkanoates (PHA) [4–6].

Each of these polymers has advantages and disadvantages for biomedical applications [7]. Poly l-lactide (PLLA) has been used in a range of medical devices [8], including various orthopaedic clinical applications [9], where its favourable mechanical properties when compared to other biopolymers offers a performance advantage. PLLA has a high tensile strength, low ductility and is a semi-crystalline polymer. The crystalline structure can make PLLA mechanically superior when compared to other polyesters in load-bearing applications [4,9–11]. PLLA has glass transition and melting temperatures of 55–80 °C and 170–180 °C, respectively. In contrast, polycaprolactone (PCL) is a soft, semi-crystalline polymer with a low melting and glass transition temperatures (55–60 °C and −54 °C, respectively) [12,13]. Although it possesses high ductility (with a tensile elongation at break of over 700%) and a high impact strength, it has a low tensile strength (~23 MPa) and Young's modulus [3,5,14–16]. Poly(3-hydroxybutyrate-co-3-hydroxyvalerate) (PHBV) is an aliphatic biodegradable polyester with a melting temperature between 80 and 160 °C and a glass transition temperature in the range of −5 to 20 °C depending on the HV (hydroxyvalerate) content of the copolymer, which can be adjusted to control the mechanical properties and degradation of the polymer [17]. PHBV has been shown to produce consistent favourable bone tissue adaptation response in addition to the elimination of any undesirable chronic inflammatory responses (up to 12 months after implantation) [18].

Both PLLA and PHBV are hard polymers with poor impact performance [4,5,9,19], which limits their use. The brittle nature of the polymers can be improved through blending with soft ductile polymers [12]. Blending of polymers is a simple yet effective method to obtain new materials with enhanced properties, as the limitations of the dominant component in the blend can be mitigated. Tuning of the physical and mechanical properties of a blend can be achieved with the selection of appropriate materials, adjustment of the blend compositions and appropriate preparation conditions [14]. The blending of PLA/PCL and PLA/PHBV as two co-polymer blends for use in biomedical applications has been previously investigated [6,20–23]. These studies have shown that it is possible to increase the fracture toughness or elongation at break of PLLA by blending with PCL or PHBV, but with reduced modulus or tensile strength [12,24–26]. Ternary blends of the three materials have not been previously assessed; however, the three materials together offer an attractive combination: PLLA providing strength, PCL giving ductility and PHBV enhancing biocompatibility of the material whilst also contributing to the mechanical properties [27].

Ternary blends with polymers which are not established biomaterials have previously been considered in terms of their ability to modify the behaviour of PLLA, PCL and/or PHBV using (i) polypropylene carbonate (PPC) in PLLA/PHBV/PPC blends [28], which showed reduced strength and modulus compared to PLLA alone; (ii) poly(butylene succinate) (PBS) in PLLA/PHBV/PBS blends [29], which showed reduced strength and modulus but increased elongation at break compared to PLLA alone; and (iii) montmorillonite (MMT) nanoclay in PLA/PCL/MMT-nanoclay blends [30], which increased modulus and strength compared to the PLLA/PCL blend, but reduced toughness and impact strength. These studies reinforce the observation from previous studies with PLLA/PCL and PLLA/PHBV blends that a formulation which provides increased ductility without sacrificing modulus and strength has yet to be found.

The aim of this study was, therefore, to characterise binary and ternary blends of PLLA, PCL and PHBV produced through twin screw extrusion for their suitability for use in orthopaedic applications in terms of mechanical performance and processability. Changes in microstructure are assessed via FTIR and DSC, with mechanical properties characterised in tension using modulus, strength, creep rate and viscosity as the key indicators of processability and mechanical behaviour over extended time periods. As orthopaedic devices are commonly applied for extended time periods, the creep behaviour of the blended materials was considered an important property to consider, and one which had not been studied in previous work on blended polyester formulations.

2. Materials and Methods
2.1. Materials

The base polymers were PLLA (PURASORB PL38, Corbion Purac, Amsterdam, The Netherlands), PHBV with 8% HV content (Sigma-Aldrich, St. Louis, MO, USA) and PCL (average Mn 80,000; Sigma-Aldrich). A range of two-component and three-component blends were evaluated using the three polymers detailed, using at least 70% PLLA, and at least 5% of either PHBV or PCL or both. Four specific blends were selected for in-depth study: PLLA/PHBV 85:15; PLLA/PCL 70:30; PLLA/PCL/PHBV 90:5:5 and 80:10:10.

2.2. Extrusion

Filament was extruded in batches of 20 g of polymer. A Microlab 10 mm twin screw melt extruder connected to a twin belt haul-off (Rondol, France) was used for filament production. Laser measurement at the end of the haul off machine allows for a live reading of filament diameter, which was nominally 1.75 mm. Figure 1 illustrates the locations of the 5 temperature zones along the extruder barrel, and Table 1 shows the extrusion temperatures in the zones for each material. Attempts to extrude PHBV alone were unsuccessful in producing usable filament due the high temperature sensitivity of the polymer and associated thermal degradation. Melt processing of PHBV is known to be difficult to achieve [31,32]. The extrusion temperatures used for extrusion of the blends were the same as for PLLA.

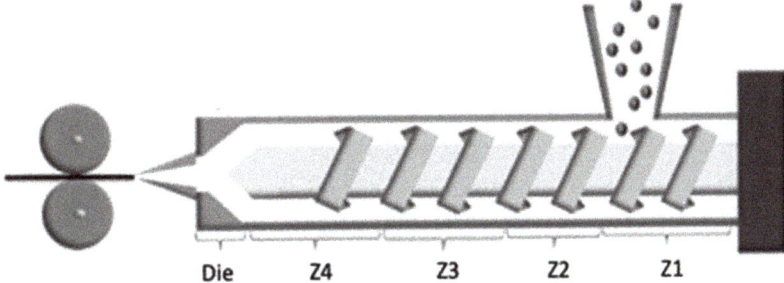

Figure 1. Schematic of extruder with temperature zones.

Table 1. Extrusion temperatures for each of the 5 temperature zones.

Polymer	Temperature (°C)				
	Z1	Z2	Z3	Z4	Die
PLLA and blended materials	125	240	240	220	180
PCL	95	120	100	100	90

2.3. Mechanical Properties

A universal tester (AGS-X autograph, Shimadzu, Kyoto, Japan) was used to conduct mechanical tests on polymer filaments using a 1 kN load cell. All tests were conducted at 22 °C. Filaments were subjected to a 5 N pre-load, the gauge length was 300 mm and materials were tested in triplicate. Additionally, 30 mm/min was used for the displacement-based test and termination was at the point of material failure or 300 mm displacement.

Creep tests were also carried out on filaments using the universal tester and a 1 kN load cell. A displacement loading rate of 10 mm/min was applied to the specimens until a load of 40 N was reached, then held for a duration of 3 h or until the sample reached a maximum displacement of 0.5 m. The initial gauge length was again 300 mm, with tests carried out in triplicate. The change in displacement for each polymer sample for the test duration was evaluated by calculating a steady state creep rate ($\dot{\varepsilon}$):

$$\dot{\varepsilon} = \frac{\varepsilon_{8000}}{\varepsilon_{50}} \tag{1}$$

where ε_{8000} and ε_{50} were the strains measured after 8000 and 50 s, respectively.

2.4. Attenuated Total Reflectance—Fourier Transform Infrared (ATR-FTIR)

An FT-IR Spectrometer (Spectrum Two, Perkin Elmer) was used for chemical analysis. Spectra were recorded from 4000 to 500 cm^{-1} and 16 scans per specimen (spectral resolution 4 cm^{-1}). All output data were baseline corrected and normalised using Spectrum Quant software (Perkin Elmer).

2.5. Differential Scanning Calorimetry (DSC)

Differential scanning calorimetry (DSC823e, Mettler Toledo) was used to assess the thermal profile of the materials. A heating rate of 10 °C/min was applied in the range of 25–240 °C. Aluminium crucibles were used as the reference and sample holder which contained a known weight of polymer prior to analysis. Material crystallinity was calculated using Equation (2) [33]:

$$Crystallinity\ (\%) = \left(\frac{\Delta H_m - \Delta H_{cc}}{\Delta H_{m100\%}} \right) \qquad (2)$$

where ΔH_m is the melting enthalpy, ΔH_{cc} is enthalpy of cold crystallisation and $\Delta H_{m100\%}$ is the enthalpy of melting for 100% crystalline polymer being investigated. For PLLA, PCL and PHBV, $\Delta H_{m100\%}$ is reported in the literature as 93.7 J/g, 139 J/g and 146.6 J/g, respectively [6,34,35]. For polymer blends, the crystallinity for each polymer in the blend was calculated according to the relative weight of that polymer in the blend and these quantities added together to give the total polymer crystallinity.

2.6. Rheological Assessment

Rheological analyses were conducted on three samples for each material using a DHR-2 controlled stress rotational rheometer (TA Instruments, New Castle, DE, USA) equipped with an environmental test chamber for temperature control during testing. Analysis was conducted using a 25 mm parallel plate geometry, keeping a constant temperature of 210 °C. Frequency sweep tests were performed to observe the variation in the storage modulus (G') and loss modulus (G") with increasing angular frequency (from 1 to 600 rad/s). The values obtained from the frequency sweep analysis were used to obtain the material flow ramps by applying the Cox–Merz equation, where the variation in the material viscosity was obtained for increasing shear rates (1–600 s^{-1}).

3. Results

3.1. Thermal and Chemical Analysis

DSC curves illustrating the thermal characteristics of the base polymers and blends are shown in Figure 2 and quantified in Table 2. For the base materials investigated pre- and post-extrusion (Figure 2a,b), there is a notable difference for PLLA and a subtle change for PCL post-extrusion in comparison to pre-extrusion. After extrusion, PLLA presents a crystallisation peak, along with lower T_g and T_m peaks. In contrast, the melting peak of PCL does not appear to be affected by the extrusion process. PHBV pre-extrusion shows two melt peaks (Figure 2c) which relate to the two constituents of the polymer, PHB and HV, with PHB representative of the higher melting peak. T_g and T_m remain close to that of PLLA in all blends, but the location of the peak is modified by parts of the blends becoming more mobile at lower temperatures.

The two-polymer blend profiles are illustrated in Figure 2d,e. Incorporation of 30% PCL in a blend with PLLA eliminates the crystallisation process for PLLA, whereas it remains present with PHBV incorporation. The step change glass transition for PLLA is accentuated with the incorporation of PCL (Figure 2d). Here, we see the fusion of the T_g of PLLA together with the T_m of PCL at 68 °C. For both two-polymer blends, the T_g and T_m are similar to those for PLLA post-extrusion. The three-polymer blend profiles are shown in Figure 2f,g, and there is a clear change in the glass transition and crystallisation peaks when more PCL/PHBV is incorporated into PLLA. The melting temperatures of the three-component blends remain in a 4 °C range of the PLLA post-extrusion T_m.

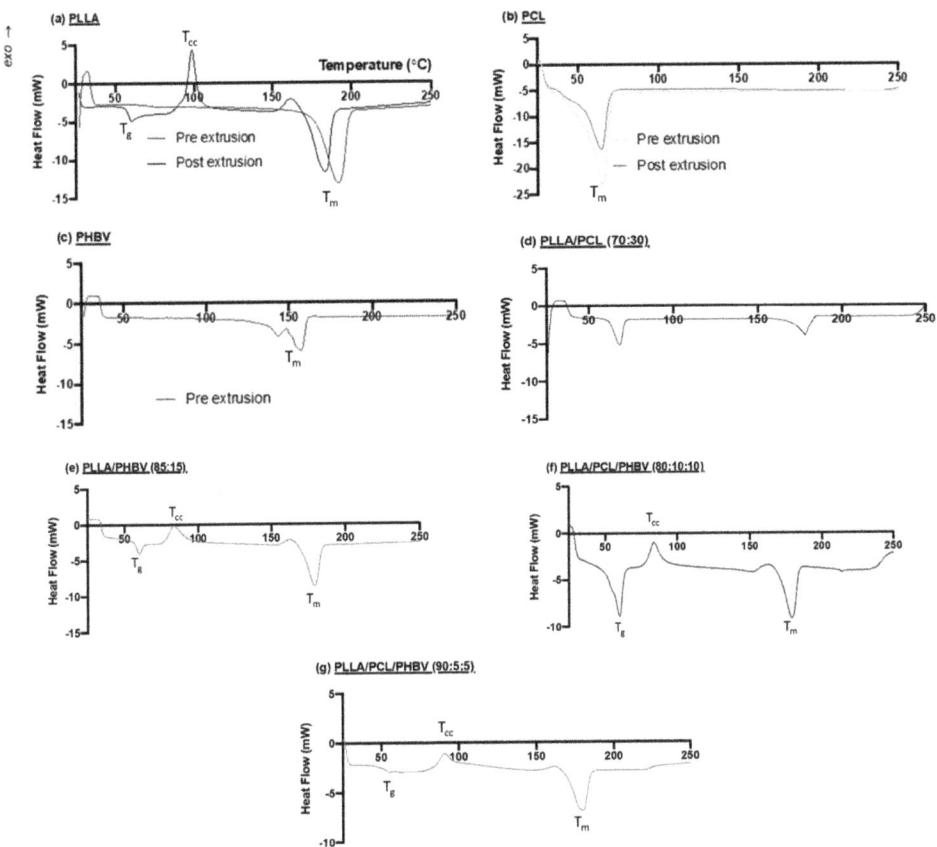

Figure 2. DSC curves (**a**) PLLA, (**b**) PCL, (**c**) PHBV, (**d**) PLLA/PCL (70:30), (**e**) PLLA/PHBV (85:15), (**f**) PLLA/PCL/PHBV (80:10:10) and (**g**) PLLA/PCL/PHBV (90:5:5).

Table 2. Thermal characteristics and crystallinity. *omr* indicates effect outside the measurement range; *nm* indicates effect not measurable. Results for blended materials are all post-extrusion. ΔH_{cc} for samples which did not show a cold crystallinity peak assumed to be 0. PLLA and PHBV cold crystallisation peaks or melt peaks closely align, so a single value is reported. PCL melting peaks were distinct, and so the blends containing PCL values are reported separately for PCL and the other components of the blend.

Polymer	T_g (°C)	T_{cc} (°C)	T_m (°C)	ΔH_m (J/g)	ΔH_{cc} (J/g)	X_c (%)
PLLA (Pre-extrusion)	76	*nm*	190	74.4	~0	79.4
PLLA (Post-extrusion)	60	98	182	57.7	29.2	30.4
PCL (Pre-extrusion)	*omr*	*nm*	64	86.7	~0	62.4
PCL (Post-extrusion)	*omr*	*nm*	64	73.4	~0	53.0
PHBV (Pre-extrusion)	*omr*	*nm*	143/157	44.0	~0	30.0
PLLA/PCL (70:30)	68	*nm*	178	31.3 (PLLA); 5.9 (PCL)	~0	24.7
PLLA/PHBV (85:15)	59	83	178	53.6	20.2	42.5
PLLA/PCL/PHBV (80:10:10)	59	83	178	34.5 (PLLA and PHBV); 27.4 (PCL)	14.1 (PLLA and PHBV)	21.7
PLLA/PCL/PHBV (90:5:5)	54	90	180	37.5 (PLLA and PHBV); 2.3 (PCL)	9.0 (PLLA and PHBV)	26.5

Table 2 also details the melt enthalpy (ΔH_m) values for the polymers and the cold crystallisation enthalpy (ΔH_{cc}), which allows the degree of crystallisation to be estimated. The effect of thermal processing on PLLA and PCL can be seen from the decline in material crystallinity from pre to post-extrusion, with decreases of 49.4% and 8.9% for PLLA and PCL, respectively. PHBV has the lowest melt enthalpy of all the polymers. For PLLA/PCL (70:30), the melting peak at ~178 °C is reduced by PCL incorporation into PLLA, lowering the melt enthalpy by 26.39 J/g. The melt enthalpy for PLLA/PHBV (85:15) is not as significantly affected by PHBV incorporation (4.12 J/g enthalpy reduction). The material crystallinity of PLLA/PCL is impacted when the two polymers are combined; however, the crystallinity of the PLLA/PHBV (85:15) blend is improved when combining these two polymers.

Figure 3 shows the ATR-FTIR spectra, which have been normalised and baseline corrected. All the spectra display similar profiles. Peaks at ~2920 and ~1720 cm^{-1} are characteristic peaks of C–H bonds and C=O carbonyl bonds, respectively. The peak located at ~1270 cm^{-1} represents the C–O saturated ester bonds [36]. PLLA shows an increase in the intensity of the peaks at ~1200 (C–C), ~1094 (C–O), ~1750 (C=O), ~1079–1189 (CH3) and ~1184/1088 (C–O–C) after extrusion [16,37–39]. With PCL, there is an increase in intensity of the fingerprint region (500–1500 cm^{-1}) post-extrusion. PLLA, PCL and PHBV all have ATR-FTIR profiles that are indicative of them being aliphatic polyesters.

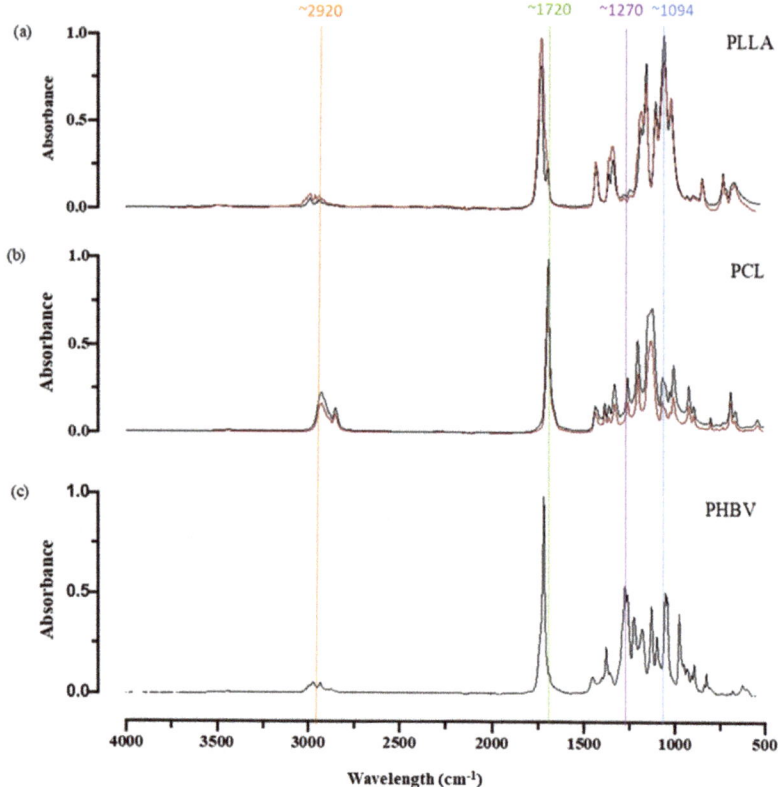

Figure 3. ATR-FTIR spectra for (**a**) PLLA, (**b**) PCL and (**c**) PHBV; the black lines are indicative of pre-extrusion and red lines of post-extrusion. Vertical lines indicate key wavelengths for ease of interpretation.

3.2. Mechanical Properties

The results of the tensile tests on extruded filaments are summarised in Figure 4 and Table 3. PCL, when tested as a single polymer, displays the anticipated low modulus and high elongation behaviour. In the binary blend with PLLA, the inclusion of PCL reduces the average modulus and yield stress compared to PLLA alone but increases elongation at break. The addition of PHBV to PLLA slightly increases the average modulus and yield stress compared to PLLA alone, but also increases the elongation at break.

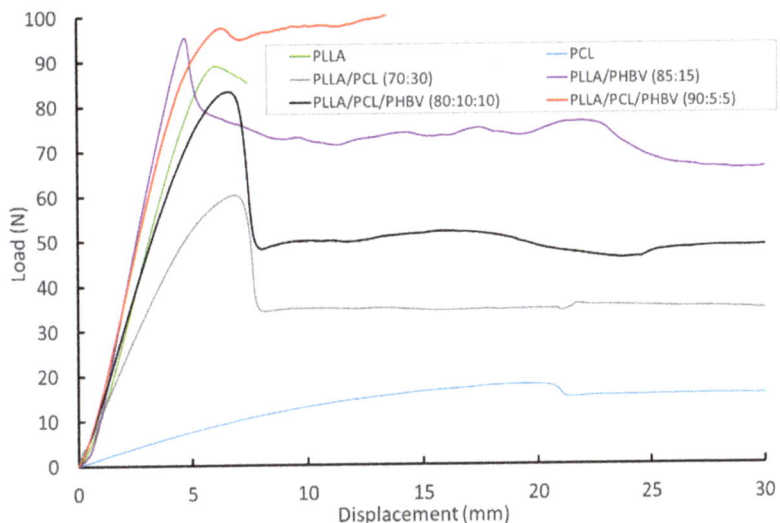

Figure 4. Representative load–displacement responses from filament tensile testing.

Table 3. Mechanical properties for polymer filaments. Mean ± s.d. $n = 3$.

Material	Ratio (wt %)	E (GPa)	Yield Stress (MPa)	Elongation at Break (%)
PLLA	100	2.23 ± 0.14	34.4 ± 1.90	2.69 ± 0.31
PCL	100	0.27 ± 0.12	10.30 ± 3.10	No break
PLLA/PCL	70/30	1.40 ± 0.07	21.40 ± 2.94	No break
PLLA/PHBV	85/15	2.80 ± 0.49	42.94 ± 5.90	66 ± 17
PLLA/PCL/PHBV	90/5/5	2.65 ± 0.17	53.29 ± 4.62	2.94 ± 0.21
PLLA/PCL/PHBV	80/10/10	1.94 ± 0.11	36.80 ± 5.0	133 ± 10

For the ternary blends, small amounts of PHBV increase the average modulus and yield stress, but as the amount of PCL increases, the effect of PCL on lowering the average modulus and yield stress also increases. For the ternary blends, the effect on the elongation at break is variable, but the average elongation at break achieved is always greater than that shown by PLLA alone.

Figure 5 illustrates curves for material strain under constant load with respect to time for each of the polymers and blends at room temperature (22 °C), and Table 4 shows the steady state creep rate for the materials. For PCL, the strain rate is very high (the maximum test displacement of 500 mm was reached in 3000 s), so no creep rate was calculated for this material. For all other materials, the absolute strain remains below 3% over the course of the test. For the PLLA/PCL/PHBV 80:10:10 and PLLA/PHBV 85:15 blends, there is a sharper increase in strain over time in comparison to the other materials. PLLA is the polymer with the lowest deformation rate. The deformation rate increases when PCL and

PHBV are added to the PLLA matrix, and the PLLA/PCL/PHBV (90/5/5) blend has the most comparable strain relationship with time to PLLA.

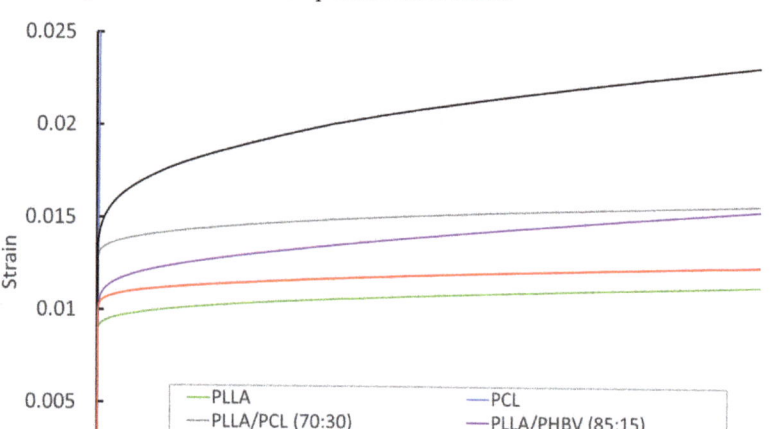

Figure 5. Representative time–strain curves for polymers and blend filaments under a 40 N constant load.

Table 4. Average creep rate at room temperature under a 40 N constant load, and measured viscosity at 210 °C.

Material	Average Creep Rate (s^{-1}) Mean ± s.d. $n = 3$	Viscosity at 1 Hz (Pa.s)	Viscosity at 600 Hz (Pa.s)
PLLA	1.31 ± 0.20	3148.3	941.2
PLLA/PCL (70/30)	1.35 ± 0.004	1245.9	391.4
PLLA/PHBV (85/15)	1.35 ± 0.034	2120.4	470.2
PLLA/PCL/PHBV (90/5/5)	1.33 ± 0.11	3482.3	960.7
PLLA/PCL/PHBV (80/10/10)	1.60 ± 0.10	2436.7	500.3

For the binary blends, the viscosity of the materials (Table 4) broadly follows the trends indicated by the modulus and strength: blending PHBV with PLLA increases the viscosity, while blending with PCL reduces the viscosity. In the ternary blends, the effect of the PCL dominates, giving a reduced viscosity compared to PLLA alone. All of the materials show a reduction in viscosity with increasing frequency.

4. Discussion

It is useful to note that initially, across all of the test types, the single polymers show properties in line with those reported in the literature for those polymers. The main aim in creating the polymer blends was to enhance the properties of PLLA, and we demonstrate that it is possible to use relatively small amounts of PCL and PHBV to achieve this aim. Considering first the two-component blends, the addition of both PCL and PHBV produced polymer blends which were less brittle than PLLA alone processed in the same way. Previous studies have shown that PLLA can be made less brittle through the addition of PCL [12], as long as the PCL is well dispersed within the PLLA. The extrusion process used here seems to have achieved this dispersion. A number of studies have blended either PHB or PHBV with PLLA and observed improved toughness [24–26]. Blending PLLA with PCL reduces the crystallinity compared to PLLA processed alone using the same processing parameters, with this reflected in a reduced modulus and tensile strength. Blending PLLA with PHBV, however, increases the crystallinity compared to PLLA processed alone using the same processing parameters, with this reflected in an increased modulus and tensile strength, giving an almost ideal outcome of improved modulus, strength and elongation at break from blending. PHBV is considered to act as a nucleation agent for the PLLA [25], giving more crystallisation but smaller crystals, thus enhancing the mechanical properties across the board.

Interestingly, in the three-component PLLA/PCL/PHBV blends, the addition of PCL causes the crystallinity to fall to below that observed for PLLA alone, whilst maintaining the modulus and elongation at break at levels similar to those for PLLA alone, suggesting that whilst the PCL reduces the total amount of crystallinity, the refined crystal structure stimulated by the PHBV still enhances the mechanical properties. Overall, the mechanical properties of the PLLA/PHBV blend and the two PLLA/PCL/PHBV blends offer similar modulus and strength to that of PLLA alone, whilst improving the elongation at break, although for the 90/5/5 PLLA/PCL/PHBV blend, the improvement in elongation at break was limited. This combination of improved properties has not been attained in previously reported work on binary or ternary blends with PLLA.

When considering the creep behaviour of the polymers, the proportion of PCL in the ternary blends seemed to have a greater influence than it had on the initial mechanical properties. That the creep behaviour of the blends cannot be inferred from the trends seen in the initial mechanical properties is important to bear in mind when developing blended polymers for applications which involve the polymer being under load for an extended time period. The 80/10/10 blend had a higher initial strain under load, and a greater rate of strain accumulation than the PLLA, PLLA/PHBV or 90/5/5 ternary blend. We consider that the reduced crystallinity, reduced modulus and increased volume of PCL in the 80/10/10 blend made it easier for the polymer material to creep through increased localised yielding [40], which then led to an increase in the steady state creep rate.

The addition of PCL into the blends also has a greater impact on viscosity than on the static properties. Testing of molten material means that any benefits which are dependent on crystallinity are no longer applicable, but it is interesting to note that in the ternary blends, it is the influence of PCL which dominates over the PHBV, which in the PLLA/PHBV blend causes an increase in viscosity. In essence, it appears that the low-viscosity PCL acts effectively as a plasticiser for the PLLA and PHBV. This makes the blends which contain PCL easier to melt to create a component shape using conventional polymer processing techniques or, as we have filament raw material, the additive manufacturing technique of fused deposition modelling.

In evaluating materials for potential use in orthopaedic applications on the basis of their mechanical properties, we should first note that the materials described here are appropriate for cancellous bone applications (modulus in the range 10 MPa to 3 GPa, strength 0.1 to 30 MPa), as the mechanical properties do not approach those of cortical bone (modulus ~18 GPa; strength ~70 MPa) [41]. With that constraint, the combination of initial mechanical properties and creep behaviour mean that the PLLA/PHBV blend and the 90:5:5 PLLA/PCL/PHBV blend would be the most promising materials. The enhanced elongation at break of the PLLA/PHBV blend offers further value where greater elasticity is required. Future work will consider the biological properties of the materials, but the track record of the constituent materials in biomedical applications gives confidence that the materials will have biological properties appropriate for orthopaedic applications.

5. Conclusions

Blends of PLLA, PHBV and PCL were produced using twin screw extrusion and characterised using FTIR, DSC, tensile testing (including tensile creep) and rheometry. FTIR and DSC showed that all three materials were effectively processed and incorporated into the blends. The tensile, creep and viscosity data show that in small concentrations, PHBV alone (up to 25%) and PCL/PHBV together (up to 10% of each polymer) can modify the behaviour of PLLA in a positive way, offering improved modulus, strength and elongation at break depending on the concentrations of each polymer used. Within the constraints of the range of concentrations of the three polymers used here, the influence of PHBV outweighs that of PCL in terms of the effect on modulus and strength, whereas for creep behaviour and viscosity, the influence of the PCL outweighs that of PHBV. At a concentration of 5% PCL in the ternary blends, the creep rate exhibited by the polymer blend was identical to that of PLLA alone. Overall, we consider that blending PCL and PHBV through melt processing with PLLA offers an effective approach to the development of new polyester-based biomaterials with combinations of properties which cannot be provided by any of the materials individually, and which have not been demonstrated through previous approaches to blending.

Author Contributions: Conceptualisation, K.D., A.M.F., P.G. and G.M.; methodology, R.N., G.M., M.J.G., A.M.F., P.G. and K.D.; investigation, R.N. and G.M.; resources, K.D., A.M.F., P.G. and M.J.G.; data curation, G.M. and R.N.; writing—original draft preparation, R.N. and G.M.; writing—review and editing, K.D., A.M.F., P.G. and M.J.G.; supervision, K.D., A.M.F., P.G. and M.J.G.; project administration, K.D.; funding acquisition, K.D., A.M.F., P.G. and M.J.G. All authors have read and agreed to the published version of the manuscript.

Funding: This project received funding from the European Union's Horizon 2020 research and innovation programme under grant agreement no. 814410.

Institutional Review Board Statement: Not applicable.

Informed Consent Statement: Not applicable.

Data Availability Statement: Data supporting this publication are openly available at the following dois:10.5281/zenodo.7235496; 10.5281/zenodo.7235502; 10.5281/zenodo.7235530; 10.5281/zenodo.7235574.

Conflicts of Interest: The authors declare no conflict of interest.

References

1. Jin, W.; Chu, P.K. Orthopedic implants. In *Encyclopedia of Biomedical Engineering 1–3*; Elsevier: Amsterdam, The Netherlands, 2018; pp. 425–439.
2. Moghaddam, N.S.; Andani, M.T.; Amerinatanzi, A.; Haberland, C.; Huff, S.; Miller, M.; Elahinia, M.; Dean, D. Metals for bone implants: Safety, design, and efficacy. *Biomanuf. Rev.* **2016**, *1*, 1. [CrossRef]
3. Hofmann, G. Biodegradable implants in orthopaedic surgery—A review on the state-of-the-art. *Clin. Mater.* **1992**, *10*, 75–80. [CrossRef]
4. Narayanan, G.; Vernekar, V.N.; Kuyinu, E.L.; Laurencin, C.T. Poly (lactic acid)-based biomaterials for orthopaedic regenerative engineering. *Adv. Drug Deliv. Rev.* **2016**, *107*, 247–276. [CrossRef] [PubMed]
5. Urquijo, J.; Guerrica-Echevarria, G.; Eguiazábal, J.I. Melt processed PLA/PCL blends: Effect of processing method on phase structure, morphology, and mechanical properties. *J. Appl. Polym. Sci.* **2015**, *132*. [CrossRef]
6. Patrício, T.; Bártolo, P. Thermal stability of PCL/PLA blends produced by physical blending process. In *Procedia Engineering*; Elsevier Ltd.: Amsterdam, The Netherlands, 2013; Volume 59, pp. 292–297.
7. Mofokeng, J.P.; Luyt, A.S. Dynamic mechanical properties of PLA/PHBV, PLA/PCL, PHBV/PCL blends and their nanocomposites with TiO2 as nanofiller. *Thermochim. Acta* **2015**, *613*, 41–53. [CrossRef]
8. da Silva, D.; Kaduri, M.; Poley, M.; Adir, O.; Krinsky, N.; Shainsky-Roitman, J.; Schroeder, A. Biocompatibility, biodegradation and excretion of polylactic acid (PLA) in medical implants and theranostic systems. *Chem. Eng. J.* **2018**, *340*, 9–14. [CrossRef]
9. Farah, S.; Anderson, D.G.; Langer, R. Physical and mechanical properties of PLA, and their functions in widespread applications—A comprehensive review. *Adv. Drug Deliv. Rev.* **2016**, *107*, 367–392. [CrossRef]
10. Balani, K.; Verma, V.; Agarwal, A.; Narayan, R. Physical, Thermal, and Mechanical Properties of Polymers. In *Biosurfaces*; John Wiley & Sons, Inc.: Hoboken, NJ, USA, 2015; pp. 329–344. [CrossRef]
11. Wagner, J.R.; Mount, E.M.; Giles, H.F. Polymer Structure. In *Extrusion*; Elsevier: Amsterdam, The Netherlands, 2014; pp. 225–232. [CrossRef]
12. Fortelny, I.; Ujcic, A.; Fambri, L.; Slouf, M. Phase Structure, Compatibility, and Toughness of PLA/PCL Blends: A Review. *Front. Mater.* **2019**, *6*, 206. [CrossRef]

13. Ulery, B.D.; Nair, L.S.; Laurencin, C.T. Biomedical applications of biodegradable polymers. *J. Polym. Sci. Part B Polym. Phys.* **2011**, *49*, 832–864. [CrossRef]
14. Gunatillake, P.; Mayadunne, R.; Adhikari, R. Recent developments in biodegradable synthetic polymers. *Biotechnol. Annu. Rev.* **2006**, *12*, 301–347.
15. Munajad, A.; Subroto, C. Fourier Transform Infrared (FTIR) Spectroscopy Analysis of Transformer Paper in Mineral Oil-Paper Composite Insulation under Accelerated Thermal Aging. *Energies* **2018**, *11*, 364. [CrossRef]
16. Mohammadi, M.S.; Ahmed, I.; Muja, N.; Rudd, C.; Bureau, M.N.; Nazhat, S.N. Effect of phosphate-based glass fibre surface properties on thermally produced poly(lactic acid) matrix composites. *J. Mater. Sci. Mater. Med.* **2011**, *22*, 2659–2672. [CrossRef] [PubMed]
17. Nair, L.S.; Laurencin, C.T. Biodegradable polymers as biomaterials. *Prog. Polym. Sci.* **2007**, *32*, 762–798. [CrossRef]
18. Pouton, C.W.; Akhtar, S. Biosynthetic polyhydroxyalkanoates and their potential in drug delivery. *Adv. Drug Deliv. Rev.* **1996**, *18*, 133–162. [CrossRef]
19. Doyle, C.; Tanner, E.; Bonfield, W. In vitro and in vivo evaluation of polyhydroxybutyrate and of polyhydroxybutyrate reinforced with hydroxyapatite. *Biomaterials* **1991**, *12*, 841–847. [CrossRef]
20. Gerard, T.; Budtova, T. Morphology and molten-state rheology of polylactide and polyhydroxyalkanoate blends. *Eur. Polym. J.* **2012**, *48*, 1110–1117. [CrossRef]
21. Ostafinska, A.; Fortelny, I.; Nevoralova, M.; Hodan, J.; Kredatusova, J.; Slouf, M. Synergistic effects in mechanical properties of PLA/PCL blends with optimized composition, processing, and morphology. *RSC Adv.* **2015**, *5*, 98971–98982. [CrossRef]
22. Todo, M.; Park, S.-D.; Takayama, T.; Arakawa, K. Fracture micromechanisms of bioabsorbable PLLA/PCL polymer blends. *Eng. Fract. Mech.* **2007**, *74*, 1872–1883. [CrossRef]
23. Sultana, N.; Wang, M. PHBV/PLLA-based composite scaffolds containing nano-sized hydroxyapatite particles for bone tissue engineering. *J. Exp. Nanosci.* **2008**, *3*, 121–132. [CrossRef]
24. Krishnan, S.; Pandey, P.; Mohanty, S.; Nayak, S.K. Toughening of Polylactic Acid: An Overview of Research Progress. *Polym. Technol. Eng.* **2015**, *55*, 1623–1652. [CrossRef]
25. El-Hadi, A.M. Increase the elongation at break of poly (lactic acid) composites for use in food packaging films. *Sci. Rep.* **2017**, *7*, 46767. [CrossRef] [PubMed]
26. Ferreira, B.M.P.; Zavaglia, C.A.C.; Duek, E.A.R. Films of PLLA/PHBV: Thermal, morphological, and mechanical characterization. *J. Appl. Polym. Sci.* **2002**, *86*, 2898–2906. [CrossRef]
27. Zhang, M.; Thomas, N.L. Blending polylactic acid with polyhydroxybutyrate: The effect on thermal, mechanical, and biodegradation properties. *Adv. Polym. Technol.* **2011**, *30*, 67–79. [CrossRef]
28. Hedrick, M.M.; Wu, F.; Mohanty, A.K.; Misra, M. Morphology and performance relationship studies on biodegradable ternary blends of poly(3-hydroxybutyrate-co-3-hydroxyvalerate), polylactic acid, and polypropylene carbonate. *RSC Adv.* **2020**, *10*, 44624–44632. [CrossRef]
29. Zhang, K.; Mohanty, A.K.; Misra, M. Fully Biodegradable and Biorenewable Ternary Blends from Polylactide, Poly(3-hydroxybutyrate-co-hydroxyvalerate) and Poly(butylene succinate) with Balanced Properties. *ACS Appl. Mater. Interfaces* **2012**, *4*, 3091–3101. [CrossRef]
30. Rao, R.U.; Venkatanarayana, B.; Suman, K.N.S. Enhancement of Mechanical Properties of PLA/PCL (80/20) Blend by Reinforcing with MMT Nanoclay. *Mater. Today Proc.* **2019**, *18*, 85–97.
31. Zhao, H.; Cui, Z.; Sun, X.; Turng, L.-S.; Peng, X. Morphology and Properties of Injection Molded Solid and Microcellular Polylactic Acid/Polyhydroxybutyrate-Valerate (PLA/PHBV) Blends. *Ind. Eng. Chem. Res.* **2013**, *52*, 2569–2581. [CrossRef]
32. Javadi, A.; Pilla, S.; Gong, S.; Turng, L.S. Biobased and Biodegradable PHBV-Based Polymer Blends and Biocomposites: Properties and Applications. In *Handbook of Bioplastics and Biocomposites Engineering Applications*; John Wiley and Sons: Hoboken, NJ, USA, 2011; pp. 372–396. [CrossRef]
33. Lajewski, S.; Mauch, A.; Geiger, K.; Bonten, C. Rheological Characterization and Modeling of Thermally Unstable Poly(3-hydroxybutyrate-co-3-hydroxyvalerate) (PHBV). *Polymers* **2021**, *13*, 2294. [CrossRef]
34. Matta, A.; Rao, R.U.; Suman, K.; Rambabu, V. Preparation and Characterization of Biodegradable PLA/PCL Polymeric Blends. *Procedia Mater. Sci.* **2014**, *6*, 1266–1270. [CrossRef]
35. Jia, S.; Yu, D.; Zhu, Y.; Wang, Z.; Chen, L.; Fu, L. Morphology, Crystallization and Thermal Behaviors of PLA-Based Composites: Wonderful Effects of Hybrid GO/PEG via Dynamic Impregnating. *Polymers* **2017**, *9*, 528. [CrossRef]
36. Lagazzo, A.; Moliner, C.; Bosio, B.; Botter, R.; Arato, E. Evaluation of the Mechanical and Thermal Properties Decay of PHBV/Sisal and PLA/Sisal Biocomposites at Different Recycle Steps. *Polymers* **2019**, *11*, 1477. [CrossRef] [PubMed]
37. Chieng, B.W.; Ibrahim, N.A.; Yunus, W.M.Z.W.; Hussein, M.Z. Poly(lactic acid)/poly(ethylene glycol) polymer nanocomposites: Effects of graphene nanoplatelets. *Polymers* **2014**, *6*, 93–104. [CrossRef]
38. Jing, N.; Jiang, X.; Wang, Q.; Tang, Y.; Zhang, P. Attenuated total reflectance/Fourier transform infrared (ATR/FTIR) mapping coupled with principal component analysis for the study of in vitro degradation of porous polylactide/hydroxyapatite composite material. *Anal. Methods* **2014**, *6*, 5590–5595. [CrossRef]
39. Prasad, B.; Borgohain, R.; Mandal, B. Advances in Bio-based Polymer Membranes for CO_2 Separation. In *Advances in Sustainable Polymers*; Springer: Singapore, 2019; pp. 277–307. [CrossRef]

40. Bergström, J.S.; Hayman, D. An Overview of Mechanical Properties and Material Modeling of Polylactide (PLA) for Medical Applications. *Ann. Biomed. Eng.* **2016**, *44*, 330–340. [CrossRef] [PubMed]
41. Morgan, E.F.; Unnikrisnan, G.U.; Hussein, A.I. Bone Mechanical Properties in Healthy and Diseased States. *Annu. Rev. Biomed. Eng.* **2018**, *20*, 119–143. [CrossRef]

Article

High-Toughness Poly(lactic Acid)/Starch Blends Prepared through Reactive Blending Plasticization and Compatibilization

Huan Hu, Ang Xu, Dianfeng Zhang, Weiyi Zhou, Shaoxian Peng and Xipo Zhao *

Hubei Provincial Key Laboratory of Green Materials for Light Industry, Collaborative Innovation Center of Green Light-Weight Materials and Processing, Hubei University of Technology, Wuhan 430068, China; huhuan13579@163.com (H.H.); xuang126912@163.com (A.X.); wodiao1066@163.com (D.Z.); zhouweiyi@hbut.edu.cn (W.Z.); psxbb@126.com (S.P.)
* Correspondence: xpzhao123@163.com

Academic Editors: Marek Brzeziński and Małgorzata Baśko
Received: 11 November 2020; Accepted: 7 December 2020; Published: 16 December 2020

Abstract: In this study, poly(lactic acid) (PLA)/starch blends were prepared through reactive melt blending by using PLA and starch as raw materials and vegetable oil polyols, polyethylene glycol (PEG), and citric acid (CA) as additives. The effects of CA and PEG on the toughness of PLA/starch blends were analyzed using a mechanical performance test, scanning electron microscope analysis, differential scanning calorimetry, Fourier-transform infrared spectroscopy, X-ray diffraction, rheological analysis, and hydrophilicity test. Results showed that the elongation at break and impact strength of the PLA/premixed starch (PSt)/PEG/CA blend were 140.51% and 3.56 kJ·m^{-2}, which were 13.4 and 1.8 times higher than those of pure PLA, respectively. The essence of the improvement in the toughness of the PLA/PSt/PEG/CA blend was the esterification reaction among CA, PEG, and starch. During the melt-blending process, the CA with abundant carboxyl groups reacted in the amorphous region of the starch. The shape and crystal form of the starch did not change, but the surface activity of the starch improved and consequently increased the adhesion between starch and PLA. As a plasticizer for PLA and starch, PEG effectively enhanced the mobility of the molecular chains. After PEG was dispersed, it participated in the esterification reaction of CA and starch at the interface and formed a branched/crosslinked copolymer that was embedded in the interface of PLA and starch. This copolymer further improved the compatibility of the PLA/starch blends. PEGs with small molecules and CA were used as compatibilizers to reduce the effect on PLA biodegradability. The esterification reaction on the starch surface improved the compatibilization and toughness of the PLA/starch blend materials and broadens their application prospects in the fields of medicine and high-fill packaging.

Keywords: poly(lactic acid); starch; reactive blending; plasticization; compatibilization

1. Introduction

With the increasing tension in petroleum resources and environmental issues, several bio-based degradable resources have been drawing wide attention in recent years. Poly(lactic acid) (PLA), which is the most promising bio-based degradable material, can be obtained through the polymerization [1] of lactic acid by fermenting plant starch. PLA has broad application prospects in daily lives, packaging industry, and the biomedical field [2–4] due to its good mechanical properties and biocompatibility. However, the high brittleness, low impact strength [5], and high cost of this material limit its development and applications. An effective way to overcome these drawbacks is to add other biodegradable materials to PLA [6]. Blending PLA with starch [7], which is a widely derived biodegradable material, yields high-performance and low-cost PLA/starch blends and expands the application range of renewable resources.

Starch contains numerous hydroxyl groups and has high polarity. However, when starch and PLA are blended, using the former as a filler cannot improve and even reduces the toughness of the PLA/starch blends. The layered microstructure along the flow direction is formed in the PLA/starch blends through the pressure-induced flow (PIF) process proposed by Zhang et al. [8]. During the compression process, PLA is oriented, whereas starch is deformed. This process increases the uniformity of the thickness of the PLA crystal layer. The layered structure allows the PLA/starch blend to absorb high impact energy, thereby improving the toughness of the material. However, the use of PIF to improve the toughness of the material is limited, and high pressure causes material damage. Small molecular plasticizers [9,10], such as polyethylene glycol (PEG) and citrate esters, can improve the ductility of PLA/starch blend materials but cannot change the immiscibility of PLA and starch. Plasticizers migrate with time and do not stabilize in the blends. Starch can be chemically modified through esterification, oxidation, crosslinking, and grafting [11] to promote the compatibility of the PLA/starch blend materials. To weaken the hydrogen bonding force among the starch molecular chains, the starch particle size or the polarity of the starch surface can be reduced to enhance the interaction force between PLA and starch. Wang et al. [12] prepared core–shell starch nanoparticles (CSS NPs). By esterifying the starch and subsequently polymerizing the emulsion, the CSS NPs utilized starch as the core to ensure that the material achieves a strong rigidity and polyethyl acrylate (PEA), which has a good affinity with PLA, as shell. Given the synergistic effect of starch and PEA, the PLA/CSS NP blends exhibited excellent toughness. Thielemans et al. [13] modified nanosized single crystal starch particles by using stearoyl chloride and PEG methyl ether. The crystal structure of the modified starch did not change, but the graft crystallized on the surface of the modified starch did. Appropriate grafts can weaken the hydrogen bonding force among starches and reduce the polarity of the starch surface. These changes help improve the compatibility of PLA/starch blends. In addition to the two-step method to modify PLA/starch composite materials, the chemical modification of starch can also be achieved through one-step reactive blending, and at the same time the compatibility of the PLA/starch composite system can be enhanced. Xiong et al. [14] synthesized a bio-based monofunctional epoxy compound Epicard and used it for the reactive compatibilization of PLA/starch composite materials. The epoxy groups on the Epicard can react with starch and form a hydrophobic layer on the surface of the latter, which will enhance the adhesion of the PLA/starch blends and improve the compatibility between PLA and starch. Jariyasakoolroj et al. [15] used chloropropyl trimethoxysilane (CPMS) modified starch (CP-starch) to melt blend with PLA to prepare PLA/starch composites. During the melt-blending process, CP-starch reacted with PLA to form a block copolymer, which acts as a nucleating agent and can significantly increase the crystallinity of the blend and improve the compatibility between PLA and starch. Modifying starch can effectively improve its surface polarity and increase the interaction force between starch and PLA. Reactive blending, which is a simple and practical method, can directly establish the compatibility of PLA/starch composite materials.

In this study, PLA/premixed starch (PSt)/PEG/citric acid (CA) blends were prepared through reactive blending to examine the effects of CA and PEG on the toughness of PLA/PSt (wt%/wt%, 70/30) blends. CA with high carboxyl underwent an esterification reaction with starch and consequently enhanced the interface adhesion between starch and PLA. As a plasticizer of PLA and starch, PEG effectively dispersed in the matrix and enhanced the mobility of the molecular chains. When the PEG that entered the matrix participated in the esterification reaction between CA and starch, a branched/crosslinked polyester was easily formed at the interface. This polyester further improved the compatibility of the PLA/starch blend materials. This PLA/starch composite material uses biodegradable raw materials, and at the same time, a highly filled PLA/starch composite material with excellent performance is obtained through reactive blending. Therefore, this composite material has broad application prospects in the medical field and packaging field.

2. Materials and Methods

2.1. Materials

The PLA (2003D; density and melting temperature were 1.24 g/cm^3 and 160 °C, respectively) was purchased from NatureWorks LLC (Minnetonka, MI, USA). The corn starch (St) was provided by Wuhan Huali Environmental Protection Technology Co., Ltd (Wuhan, Chian). The citric acid (CA) monohydrate and PEG 1000 (analytical purity) were purchased from Sinopharm Chemical Reagent Co., Ltd (Wuhan, Chian). The vegetable oil polyol (VOP) was provided by Dongguan Aoda Environmental New Materials Co., Ltd (Dongguan, Chian).

2.2. Preparation of the PLA/PSt/PEG/CA Sample

PLA and starch were placed in a 60 °C vacuum drying oven and dried for 24 h to reduce the moisture content of the latter to approximately 3%. The PSt was prepared by mixing dry starch with VOP for 15 min at a mass ratio of 7:3. The mixtures of PLA, PSt, PEG, and CA were melt-blended at 170 °C with a rotation speed of 60 rpm for 10 min. A blend of different components was obtained, and the corresponding formulations are presented in Table 1. The mixtures were then pressed into plates with thicknesses of 1, 2, and 4 mm by using a plate vulcanizer. These plates were cut and polished into standard splines for testing. The samples were completely dissolved in dichloromethane, centrifuged three times, washed with ethanol to separate the precipitate, and then dried in a vacuum oven at 60 °C for 48 h.

Table 1. Material formulations.

Sample	PLA (wt%)	PSt (wt%)	PEG (wt%)	CA (wt%)	St (wt%)	VOP (wt%)
1	100	-	-	-	-	-
2	100	-	1.19	0.17	-	1.19
3	90	10	1.19	0.17	10	1.19
4	80	20	1.19	0.17	20	1.19
5	70	30	1.19	0.17	30	1.19
6	60	40	1.19	0.17	40	1.19
7	50	50	1.19	0.17	50	1.19
8	70	30	-	-	30	-
9	70	30	1.19	-	30	1.19
10	70	30	-	0.17	-	-
11	100	-	1.19	-	-	-
12	100	-	-	0.17	-	-
13	70	-	-	-	30	-
14	100	-	-	-	-	5
15	70	-	1.19	-	30	-
16	70	-	-	0.17	30	-
17	100	-	-	0.17	-	5
18	70	-	1.19	0.17	30	-

2.3. Characterization

2.3.1. Mechanical Property Measurements

The hot pressed 2 mm-thick plate-shaped samples were cut and polished into dumbbell-shaped samples in accordance with the international standard ISO 527-2. The CMT-4204 electric tensile tester (SANS, Shenzhen, Chian) was used to test the samples at a tensile speed of 5 mm/min to obtain the stress–strain curve, tensile strength, and elongation at break. The 4 mm-thick samples were cut into a standard spline in accordance with ISO 179-1, and the impact energy was measured through the

Charpy impact test (SJJ-50, Chengde, Chian) with a range of 0.5 J. The impact strength of the sample was obtained using data calculation. The data of the samples in every test were measured on at least five splines.

2.3.2. Rheological Behavior Analysis

A rotating plate rheometer (AR2000EX, TA, Shanghai, Chian) was used to obtain the linear rheological behavior of the mixture with plate–plate geometry. The sample is a 1 mm-thick disc with a diameter of 25 mm. The set temperature was 170 °C, the frequency range was 0.1–100 Hz, and the strain was 0.1%.

2.3.3. Differential Scanning Calorimeter (DSC)

A DSC (Perkin Elmer, DSC 8000, Kunming, Chian) was used for crystallization performance analysis of the PLA/PSt/PEG/CA blend. The sample was heated from 30 to 200 °C at a rate of 10 °C/min and maintained at 200 °C for 3 min to eliminate the previous thermal history and then cooled to 30 °C at a rate of 10 °C/min. The sample was heated again from 30 to 200 °C at the same increase rate. The second heating curve was recorded, and the crystallinity of PLA was calculated as [16]

$$X_c = \frac{\Delta H_m - \Delta H_c}{\Delta H_f \omega_{PLA}} \times 100\% \tag{1}$$

where X_c represents the crystallinity of PLA, ΔH_m and ΔH_c are the melting enthalpy and cold crystallization enthalpies of the sample, respectively, and ΔH_f and ω_{PLA} are the melting enthalpy and PLA mass fraction of 100% crystallized PLA, respectively.

2.3.4. Scanning Electron Microscopy (SEM)

The morphology of the impact fracture surface was examined using a SEM (JSM-6390L V, JEOR, Tokyo, Japan) at an acceleration voltage of 10 kV. The 4 mm-thick sample was broken at room temperature, and the impact fracture surface was washed with ethanol.

2.3.5. X-Ray Diffraction (XRD)

The XRD analyzer from Panalytical B.V. (Empyream, Almelo, NL, USA) was used to test the pretreated powder samples with the target of Cu, test conditions: voltage = 40 kV, current = 30 mA, start angle = 5°, end angle = 40°, and scan step width = 0.02°.

2.3.6. Fourier-Transform Infrared (FTIR) Spectroscopy

The blend was dissolved in dichloromethane (CH_2Cl_2), washed with ethanol, and dried to obtain powder samples. The powders were mixed with potassium bromide at a weight ratio of 1:100 and subsequently compressed. Measurements were performed using an FTIR spectrometer (Iso10, Nicolet, Beijing, Chian) with a resolution of 4 cm^{-1} and scan number of 32.

2.3.7. Hydrophilicity Test

The water absorption rate of the test sample was tested in accordance with the plastic water absorption test method (GB1034-98). The blends were cut into small pieces with sizes of $10 \times 10 \times 2$ mm and placed in a vacuum oven at 50 °C for 48 h. The initial weight m_1 of a piece was measured using an analytical balance. Subsequently, the sample was soaked in deionized water for 24 h, wiped with filter paper, and weighed to determine the final weight m_2. The water absorption rate of the blend materials was determined as [17].

$$\text{Water absorption} = \frac{m_2 - m_1}{m_1} \times 100\% \tag{2}$$

3. Results and Discussions

3.1. Mechanical Properties of PLA/PSt/PEG/CA Blends

The mechanical properties of pure PLA and PLA/PSt/PEG/CA blends with different PSt mass ratios are shown in Figure 1. The elongation at break and the notched impact strength of the PLA are 10.45% and 1.58 kJ·m^{-2}, respectively, which reflects the brittleness of PLA. Compared with pure PLA, the elongation at break of PLA/PEG/CA blends increased slightly, while the tensile strength decreased, which may be due to the ability of low-molecular-weight PEG to plasticize PLA [18]. There is also an explanation that under the action of high temperature shear, a small amount of water and CA in PEG can promote the depolymerization of PLA [19], and the low-molecular-weight PLA produced by depolymerization can also have a plasticizing effect on PLA [20]. When the PSt content is less than 10wt%, the tensile strength of the blend decreases with the increase in PSt content, which is mainly due to the plasticization of PLA. The change in elongation at break is not obvious, which indicates that a new interaction force is generated between the blended components and a new structure is formed. This structure works together with the plasticized PLA to offset the defects caused by the incompatibility of PLA and starch, and this structure can improve the impact strength of the blend. The elongation at break and the impact strength of the blends show a significant upward trend as the PSt content increases and reaches the maximum at 30 wt%. This phenomenon can be attributed to the increased interaction between the components of the blend. Compared with the plasticizing effect of PEG and CA on PLA, the interaction force between the blending components at this time has a more significant effect on improving the toughness of the PLA/PSt/PEG/CA blend. When the PSt content exceeds 30 wt%, the elongation at break slowly decreases. The excess PSt forms an agglomeration, and concentration of stress occurs during the stretching process of the blend and results in increased defects. Conversely, the tensile strength of the blend monotonically decreases with the increase in the PSt content, which indicates that some components in the blend enhance the mobility of the PLA molecular chain. The results of the impact test are consistent with the trend of the elongation at break. The impact strength of the blend first increased and then decreased with the increase in PSt content, and reached the maximum when the PSt content was 30 wt%, indicating that the adhesion between PLA and starch was the largest at this time. When the PSt content exceeds 30 wt%, the possibility of PSt agglomeration in the blend increases, resulting in a decrease in the mechanical properties of the material. Therefore, a blend with a PSt content of 30 wt% was selected for follow-up research.

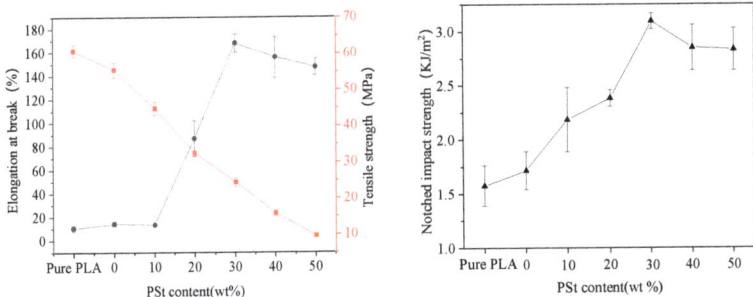

Figure 1. Mechanical properties of pure poly(lactic acid) (PLA) and PLA/premixed starch (PSt)/polyethylene glycol (PEG)/citric acid (CA) blends with different PSt contents.

To explore the contribution of each component of the PLA/PSt/PEG/CA blend to the mechanical properties of the blend, several blends with different components were fabricated. The test results are summarized in Table 2. Most raw materials blended with PLA alone caused the toughness of the blend to decrease, and only the PLA/PSt/CA blend and PLA/PSt/PEG/CA blend had a sudden change in toughness. This phenomenon indicates that the reason for the improved toughness of PLA/starch composites may be that the PEG and CA in the system improve the interface interaction between PLA and starch. Compared with pure PLA, the tensile strength and elongation at break of the PLA/PSt blends significantly declined. This phenomenon may be due to the weak interaction between the PLA and PSt components, which results in weak tangles among molecular chains and poor compatibility among components. After introducing a small amount of PEG into the PLA/PSt blend, the elongation at break of the new blend slightly increased and the tensile strength decreased. This result can be attributed to the plasticizing effect of PEG. PEG can enhance the mobility of the PLA molecular chain and increase the tensile toughness of the PLA/PSt/PEG blends. However, the PLA/PSt/CA blends showed obvious strain hardening, and their elongation at break reached 54.47%. After introducing PEG, the elongation at break of the PLA/PSt/PEG/CA blend significantly increased to 140.51%. The addition of CA increases the interaction force among the blend components, and at this time the negative effects of CA and water on PLA can be ignored. However, when CA and PEG are present in the blend at the same time, PEG acts more than just a plasticizer in the PLA/PSt/PEG/CA blend. Compared with that of the PLA/PSt/CA blend, the tensile toughness of the PLA/PSt/PEG/CA blend greatly improves. PEG may chemically react with other components during the blending process. This reaction can promote the movement of the PLA molecular chain and improve the compatibility among the blend components.

Table 2. Mechanical parameters of different blends.

Materials	Massratio	Tensile Strength/MPa	Elongation at Break/%	Notched Impact Strength/kJ·m^{-2}
Pure PLA	100	70.64 ± 2.5	10.45 ± 2.44	2.02 ± 0.15
PLA/PSt	70/30	45.65 ± 3.27	5.04 ± 0.12	2.08 ± 0.12
PLA/PSt/PEG	70/30/1.19	43.5 ± 1.22	7.15 ± 1.28	2.20 ± 0.05
PLA/PSt/CA	70/30/0.17	28.86 ± 2	54.47 ± 9.87	3.18 ± 0.01
PLA/PSt/PEG/CA	70/30/1.19/0.17	28.58 ± 1.38	140.51 ± 10.41	3.56 ± 0.06
PLA/PEG	100/1.19	65.6 ± 3.7	7.6 ± 0.67	-
PLA/CA	100/0.17	65.41 ± 2.76	8.88 ± 0.76	-
PLA/St	70/30	55.08 ± 4.9	5.55 ± 0.17	-
PLA/VOP	100/5	66.52 ± 2.39	10.41 ± 0.95	-
PLA/St/PEG	70/30/1.19	49.84 ± 2.33	4.41 ± 0.31	-
PLA/St/CA	70/30/0.17	47.66 ± 1.69	4.20 ± 0.19	-
PLA/CA/VOP	100/0.17/5	66.78 ± 0.64	7.5 ± 0.45	-
PLA/St/PEG/CA	70/30/1.19/0.17	33.48 ± 5.14	3.43 ± 0.67	-

3.2. Rheological Behavior of PLA/PSt/PEG/CA Blends

By testing the rheological properties of the blend, it is helpful to analyze the changes in the compatibility of the blend components. Generally, after PLA is blended with poorly compatible materials such as polyamide11 [21], starch [22], and cellulose [23], the viscosity of the composite material decreases. The introduction of compatibilizers, coupling agents or reactive compatibilizers can increase the interaction force effectively between the interfaces of the components of the composite material, thereby increasing the viscosity of the composite material. The rheological properties of the blends, such as storage modulus, complex viscosity, and loss modulus, were analyzed to explore the effects of PSt, PEG, and CA on the melt strength of the blend. The presence of PSt can change the viscosity and degree of shear-thinning [24] as compared with neat PLA. The dependence of the complex viscosity of different blends on frequency is illustrated in Figure 2a. The figure shows two types of the complex viscosity frequency dependence of the blends. In the low-frequency region, the viscosity of the blends after PSt addition decreases in varying degrees compared with that of pure PLA. Among the blends, the decrease in the viscosity of the PLA/PSt/PEG blends is the most obvious. This result can be attributed to the immiscibility of PLA and starch and the plasticizing effect of PEG on PLA [25]. The poorly compatible blend components are prone to phase separation, forming a defect structure with weak interaction force [26]. When external forces act on the blend, this defect structure cannot absorb energy, resulting in a decrease in the viscosity of the blend. PEG is dispersed among the PLA molecular chains during the blending process, and such dispersion enhances the mobility of the PLA molecular chains and reduces the viscosity [27] of the blend. The viscosity of the blend without CA does not change with the change in the low-frequency region, thereby showing the properties of a Newtonian fluid, as well as an obvious shear thinning in the high-frequency region. The platform of the curve of the viscosity of the CA-containing blend in the low-frequency region disappears, suggesting that CA can increase the shear-thinning strength of the blend. The degree of entanglement among the components of the blend hinders the movement of the PLA molecular chain and therefore results in increased viscosity in the low-frequency region.

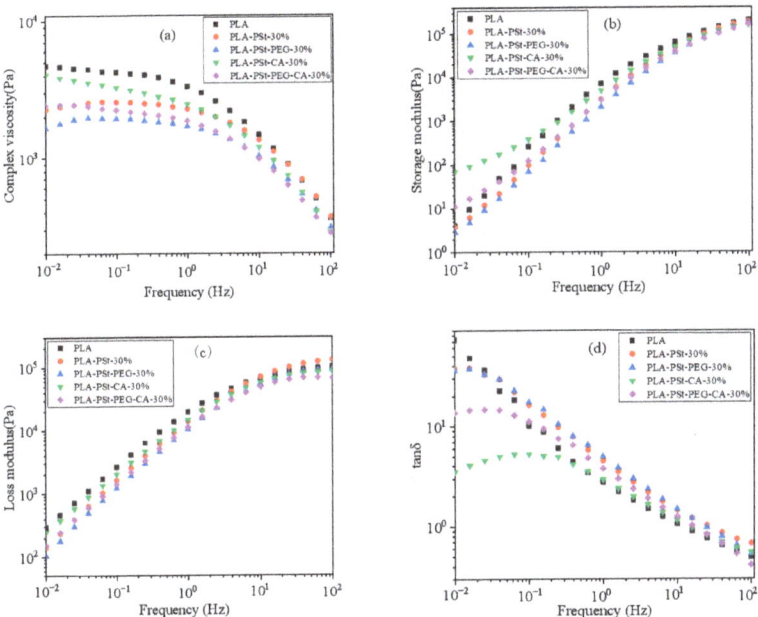

Figure 2. (**a**) Complex viscosity, (**b**) storage modulus, (**c**) loss modulus, and (**d**) loss tangent (tan δ) of PLA and different blends with PSt ratios of 30 wt%.

The behavior of the storage and loss moduli as a function of frequency is shown in Figure 2b,c, respectively. The storage modulus (G′) of PLA in the low-frequency region is lower than its loss modulus (G″); such a discrepancy implies that PLA shows typical viscous characteristics. This relationship slightly changes after adding PSt and PEG to the blend. When the CA component is added, a significant increase in the storage modulus of the blend is observed in the low-frequency region. This increment indicates that the existence of CA can promote the transformation of PLA from a liquid-like state to a solid-like one [28]. The change in tan δ during frequency scanning is used to estimate the dispersion of the filler or the dispersed phase in the blended system. The tan δ of liquid-like materials linearly decreases with the increase in frequency. The changes in the tan δ of several samples with frequencies at 170 °C are displayed in Figure 2d. The finding shows that PLA has a typical liquid-like characteristic [29] which slightly changes after adding PSt and PEG. After adding CA, the tan δ value of PLA/PSt/CA blend and PLA/PSt/PEG/CA blend in the low-frequency region significantly decreases, which means that a gel-like substance that can help improve the compatibility of PLA and PSt may appear in the system. The frequency dependence of tan δ weakens, indicating that the blend tends to change from a liquid-like state to a solid-like one.

3.3. Crystallization Behavior of PLA/PSt/PEG/CA Blends

PLA is a typical semi-crystalline polymer [30] whose mechanical properties are closely related to the thermal properties and crystallinity of a solid. The heat flow curves of several samples during the second temperature rise are shown in Figure 3 and Table 3. Compared with pure PLA, PLA/PSt blends have lower T_g and T_c, which may be due to the VOP entering the PLA matrix during the blending process destroying the crystallization of PLA. Secondly, the VOP attached to the surface of the starch blocks the interface between PLA and starch, inhibiting the heterogeneous nucleation [31] of PLA by starch. The results of analysis of mechanical properties and rheological behavior have explained the immiscibility of PLA and PSt. The crystallinity of PLA in the PLA/PSt blend calculated by formula (1) is reduced, which also confirms the reliability of the analysis. The T_g, T_c, and T_m of PLA/PSt/PEG blends further decreased, which was mainly due to the plasticization of PLA by PEG. PEG can enter between PLA macromolecules during melt mixing and establish physical interactions such as hydrogen bonding or dipole-dipole interactions in between atoms. As a result, some of the rigid homogeneous PLA-PLA interactions were replaced by heterogeneous PLA-PEG interactions. This phenomenon enhanced the mobility of PLA and decreased the energy consumption during glass transition (decreased T_g) [32]. Compared with the PLA/PSt blend, the T_c and T_m of the PLA/PSt/CA blend increased. This may be because the surface polarity of the starch changed after CA was introduced into the blend. The reason for the change may be that CA participated in the esterification reaction and formed a branched/crosslinked copolymer on the surface of the starch, which enhanced the interaction between the components of the blend [33]. Usually, T_g will increase as the interaction force between the components of the blend increases. However, the T_g of PLA/PSt/CA blends only increased slightly, which may be because CA tends to change the surface polarity of starch rather than acting on PLA. Moreover, the depolymerization of PLA by CA will destroy the crystallization of PLA and reduce X_c. Comparing PLA/PSt blends with PLA/PSt/CA blends, the result is that CA can slightly increase the T_g of the blends. However, when the PLA/PSt/PEG blend is compared with the PLA/PSt/PEG/CA blend, the result is that CA slightly reduces the T_g of the blend. This shows that when CA and PEG are contained in the blend, PEG may also participate in the esterification reaction in the melt-blending process. Some interesting facts have also been discovered from the perspective of crystallinity changes in Figure 4. After introducing PEG into the PLA/PSt blend, its crystallinity value increased by 4.32%; however, after introducing PEG into PLA/PSt/CA blend, its crystallinity value only increased by 2.05%. This may be because when the CA component is present in the system, CA restricts the dispersion of PEG in the PLA matrix, thereby attracting PEG to the CA side. Since PEG can plasticize PLA, the branched/crosslinked polymer containing PEG formed after the esterification reaction can more easily penetrate into the PLA matrix and form an amphiphilic bridging structure [34]. This structure is considered to be the "bridge" between PLA and starch, which can enhance the compatibility between PLA and starch. However, a part of the free branched/crosslinked polyester participates in the process of plasticizing [35] PLA and causes the T_g of the blend to decrease. The change of the interface structure between PLA and starch indicates that the compatibility of the two is increased. At this time, PSt can be regarded as a heterogeneous nucleating agent of PLA and promotes the crystallization of the PLA matrix. The increase in effective nucleation sites in the blend enhances the interfacial adhesion between PLA and starch and allows the blend exhibit excellent toughness in the macroscopic view, and this has been confirmed from the mechanical performance test results.

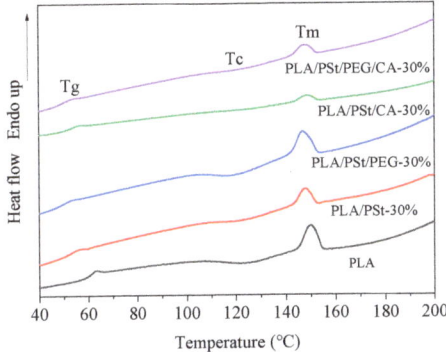

Figure 3. Differential Scanning Calorimeter (DSC) heat flow curves of pure PLA and PLA/PSt, PLA/PSt/PEG, PLA/PSt/CA, and PLA/PSt/PEG/CA blends.

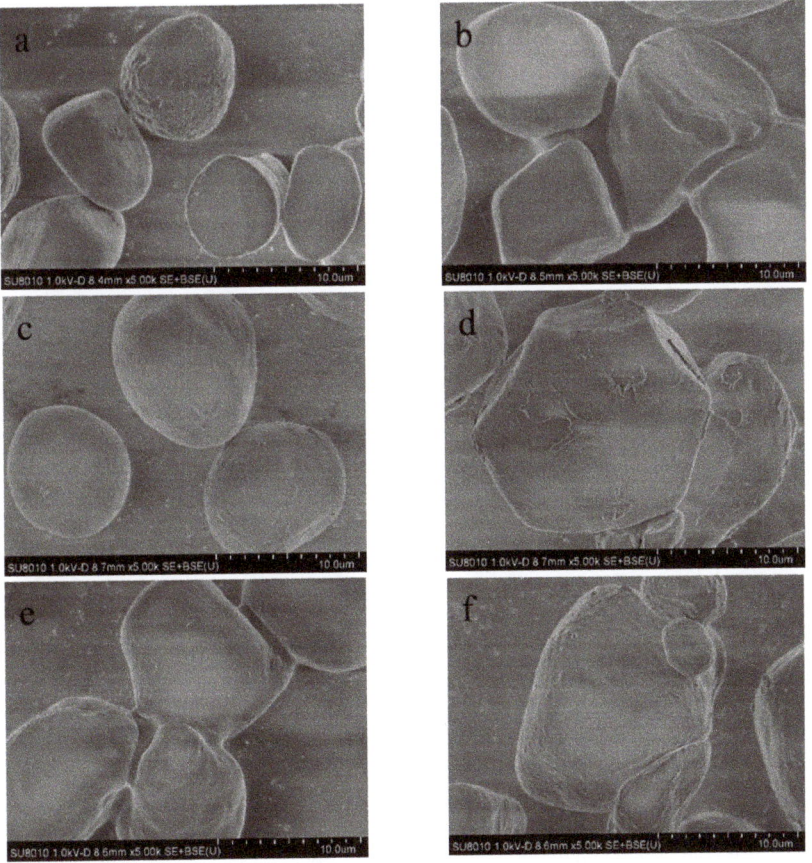

Figure 4. Surface morphology of several pretreated blend powder samples observed using SEM: (**a**) pure starch, (**b**) PSt, (**c**) PLA/PSt, (**d**) PLA/PSt/PEG, (**e**) PLA/PSt/CA, and (**f**) PLA/PSt/PEG/CA.

Table 3. Thermal properties of the materials.

Sample Code	T_g (°C)	T_c (°C)	ΔH_c (J/g)	T_m (°C)	ΔH_m (J/g)	X_c (%)
PLA	62.68	123.63	−9.07	150.05	13.09	6.14
PLA-PSt-30%	55.92	121.95	−5.69	147.66	7.85	3.30
PLA-PSt-PEG-30%	52.97	119.05	−9.94	146.54	15.04	7.78
PLA-PSt-CA-30%	55.98	126.14	−2.11	148.23	3.14	1.57
PLA-PSt-PEG-CA-30%	52.80	123.24	−3.64	147.01	6.01	3.62

3.4. Morphology of the Dispersed Phase

On the basis of the above analysis, CA is an important factor in toughening PLA/PSt/PEG/CA blends. To observe the effect of CA and PEG on the blend, a separate dispersed phase of PLA/PSt/PEG/CA blends was obtained by dissolving and washing. The surface morphology of the dried dispersed phase was observed using SEM. The morphologies of several samples examined under 5k× magnification are shown in Figure 4. Pure starch (Figure 4a) appears as an irregular sphere with a diameter of approximately 10 µm and has many small pits on the surface. As shown in Figure 4b, the surface of PSt is smooth. The starch in the pretreated PLA/PSt blends (Figure 4c) blend has similar size and surface morphology to pure starch. This similarity indicates that the compatibility between PLA and starch is still poor. Figure 4d shows that the size of starch particles in the PLA/PSt/PEG blend after the pretreatment increases to some extent, which may be due to the plasticizing effect of PEG on starch. The presence of numerous fine pits on the surface of starch (Figure 4d) indicates that PLA and starch are not effectively entangled, and starch only acts as a large particle filler in the PLA/PSt and PLA/PSt/PEG blends to reduce their macroscopic mechanical properties. However, although the particle size of starch in the PLA/PSt/CA blend does not significantly change after the addition of CA, the corresponding surface morphology greatly varies and several tiny protrusions form on the surface of the starch granules (Figure 4e). This phenomenon causes the agglomeration of starch, which is also evident in Figure 4f. In addition, the starch particles in Figure 4f seem to be affected by the plasticization of PEG, and the size of the starch particles increases to some extent. The addition of PEG promotes the plasticization of the PLA and starch phases, whereas the addition of CA enhances the compatibility among the components of the blend. The change of the starch surface structure improves the interface compatibility between PLA and starch, thereby increasing the effective nucleation sites in the blend. At this time, PSt can be regarded as a heterogeneous nucleating agent of PLA, which enhances the interfacial adhesion between PLA and starch. The blends show excellent toughness in the macroscopic view, which has been confirmed from the mechanical performance test results.

3.5. Crystallization of the Dispersed Phase

The XRD patterns of pure starch and pretreated PLA/PSt/CA and PLA/PSt/PEG/CA powder samples are shown in Figure 5. The shape of the XRD diffraction peaks of the three samples does not significantly change, which means that the crystalline type [36] of starch in the samples does not vary. The crystallinity of the three samples was obtained by fitting the XRD data. The crystallinity of the samples containing CA decreases compared with that of pure starch because the former destroys the amorphous area of starch to a certain extent, thereby enhancing the connection between PLA and the starch molecular chains and the compatibility between PLA and starch. In addition, the crystallinity of the blend further decreases after adding PEG, which suggests that PEG not only acts on PLA, but also plasticizes starch [37] to a certain extent. During the blending process, the crystalline area of starch is partly destroyed by the PEG and CA. This phenomenon facilitates the plasticization of starch and enhances starch dispersibility in PLA, and thus improves the compatibility between PLA and starch.

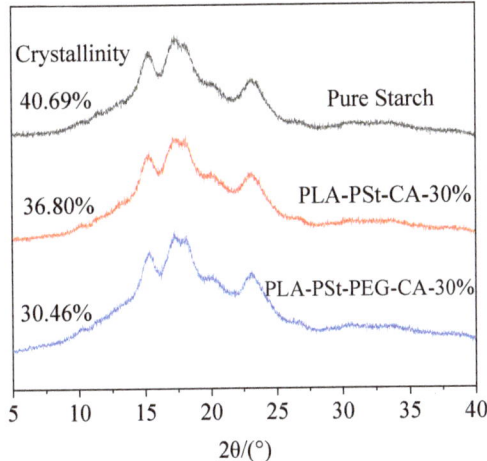

Figure 5. XRD patterns of pure starch and pretreated PLA/PSt/CA and PLA/PSt/PEG/CA powder samples.

3.6. FTIR Analysis of the Dispersed Phase

Chemical reactions may affect the surface morphology of starch, and this effect will influence the compatibility of the components of the blend. The dispersed phase was extracted from several blends, and FTIR spectroscopy was performed to analyze the changes in the compatibility among the components. The FTIR test results of several samples are presented in Figure 6. The PLA/PSt, PLA/PSt/PEG, PLA/PSt/CA, and PLA/PSt/PEG/CA blends are powder samples obtained through pretreatment. As shown in the figure, the C=O absorption peak near 1743 cm^{-1} can only be detected in VOP and PSt. No C=O absorption peak is observed in the FTIR spectra of the PLA/PSt and PLA/PSt/PEG powder samples after pretreatment. This result indicates that the PLA and VOP in the blend are completely removed, and PEG may display a plasticizing effect. However, C=O absorption peaks are detected in the PLA/PSt/CA and PLA/PSt/PEG/CA powder samples after pretreatment, which means that ester groups are still present in the powder samples at this time. CA participates in the esterification reaction during the blending process, and a branched/crosslinked copolymer forms. This branched/crosslinked copolymer is coated on the surface of the starch to prevent the removal of the ester group during the blending process. Another obvious change in peak position occurs near 2863 cm^{-1}, which can be attributed to the C–H symmetric stretching vibration absorption peak in VOP and PSt. The respective absorption peaks of pretreated PLA/PSt and PLA/PSt/PEG powder samples at 1743 and 2863 cm^{-1} disappear, implying that no chemical changes occur during the blending process. Starch only acts as a large-sized filler, and PEG acts as a plasticizer. The immiscibility between PLA and starch decreases the macroscopic tensile toughness of the blend. The esterification reaction might occur in the PLA/PSt/CA and PLA/PSt/PEG/CA powder samples after pretreatment, which might improve the surface activity of starch and increase the compatibility between PLA and starch. Furthermore, the blend shows a significant increase in tensile and impact toughness.

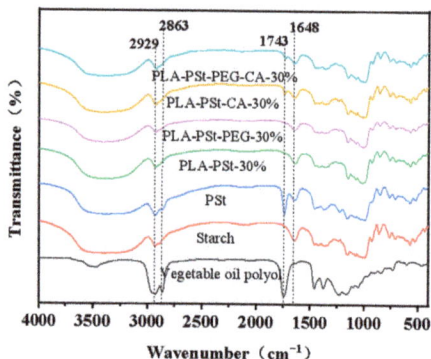

Figure 6. FTIR spectra of vegetable oil polyols, starch, PSt, and pretreated PLA/PSt, PLA/PSt/PEG, PLA/PSt/CA, and PLA/PSt/PEG/CA blend powder samples.

3.7. Hydrophilicity Test of PLA/PSt/PEG/CA Blends

The water absorption rates of several blends are shown in Figure 7. The addition of hydrophilic starch to PLA improves the water absorption rate of the blend materials. On this basis, this rate will be further improved by adding PEG or CA. These materials contain a large amount of hydrophilic -OH, which helps the blend to combine with H_2O molecules to form hydrogen bonds. From the results shown in Figure 7, PEG in the blend system seems to have other functions besides acting as a plasticizer. The water absorption rates of the PLA/PSt/CA and PLA/PSt/PEG/CA blends suggest that the hydrophilicity of the blend material decreases after the addition of hydrophilic PEG. This finding reveals that PEG may also participate in the esterification reaction along with CA during the reactive blending process. The polarity of PEG decreases and the content of hydrophilic groups decreases after the esterification reaction after the esterification reaction, resulting in a certain decrease in the water absorption rate of the blend. Not only that, under the combined action of PEG and CA, the interface between PLA and starch becomes more stable, and the interfacial gap becomes smaller, which results in starch not being able to be swollen fully. Secondly, while the crystalline and amorphous regions of starch are destroyed, PEG and CA will be dispersed near the active hydroxyl groups [38] of the starch, rendering these locations unable to accommodate more water molecules and reducing the water absorption of the blend.

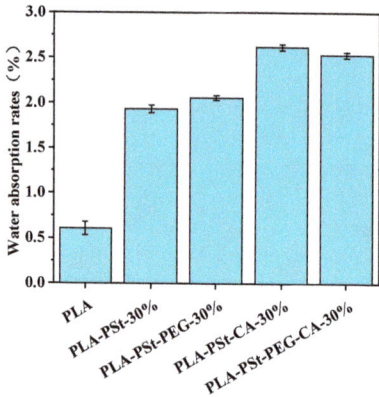

Figure 7. Water absorption rates of pure PLA and PLA/PSt, PLA/PSt/PEG, PLA/PSt/CA, and PLA/PSt/PEG/CA blends.

In view of the change of the hydrophilicity of the blends, the dynamic contact angles of several blends were analyzed. As shown in Figure 8, the contact angle of pure PLA is the largest in the initial state, indicating that PLA is relatively hydrophobic, which is determined by the structural properties of the macromolecular PLA itself. After introducing PSt, PEG and CA into PLA, the contact angles of the blends all decreased, which indicates that the blends are more hydrophilic. However, the contact angle of the PLA/PSt/PEG/CA blend increased compared with the PLA/PSt/CA blend, which is consistent with the conclusion of water absorption. This change further indicates that PEG may participate in the esterification reaction in the PLA/PSt/PEG/CA blend system.

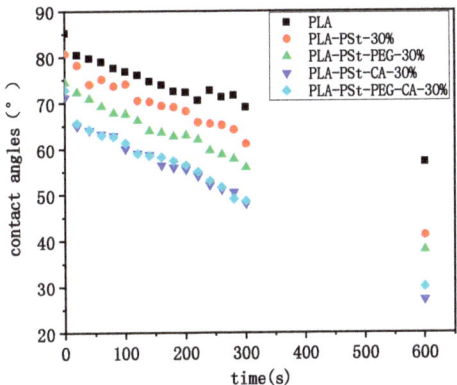

Figure 8. Dynamic contact angles of water in the first 10 min on the surface of pure PLA and PLA/PSt, PLA/PSt/PEG, PLA/PSt/CA, and PLA/PSt/PEG/CA blends.

In addition, low-molecular-weight PEG acts as a plasticizer to PLA and exerts a certain plasticizing effect on starch. This feature of PEG allows the branched/crosslinked copolymer with PEG chains that formed after the esterification reaction to penetrate the interface between PLA and starch and act as a bridge between the two to increase their interaction force. Meanwhile, the branched/crosslinked polyester forms a highly stable coating on the starch surface, which facilitates the compatibilization between PLA and starch.

4. Conclusions

In this study, the compatibilization between PLA and starch was improved through reactive blending, and the toughness of the PLA/PSt/PEG/CA blend was improved to a certain extent. The elongation at break of the blend with a PSt content of 30 wt% reached a maximum value of 140.51%, and the impact strength increased to 3.56 kJ/m^2. The addition of a small amount of CA did not change the crystal form of starch, but CA participated in the esterification reaction during the blending process. The generated crosslinked/branched polyester adhered to the surface of the starch granules and improved the surface activity of starch and the compatibility between starch and PLA. The particle size and shape of the starch in the blend did not significantly change, which indicated that the esterification reaction only occurred in the amorphous region of starch. The decrease in the hydrophilicity and the increase in the tensile toughness of the PLA/PSt/PEG/CA blend compared with that of PLA/PSt/CA suggested that PEG may also participate in the esterification reaction during melt blending.

The findings suggest that PEG, which acts as a plasticizer for PLA and starch, can lubricate the system. Moreover, when PEG is dispersed in the PLA and starch, the branched/crosslinked polyester containing PEG chains, which is formed by the esterification reaction, is easily embedded in the interface between PLA and starch. This polyester enhances the interaction between PLA and starch.

These results indicate that PEG can form a highly stable crosslinked/branched polyester on the starch surface after participating in the esterification reaction to further improve the compatibility between PLA and starch. Furthermore, the decrease in the water absorption can help enhance the stability of the blend material in a natural environment.

Author Contributions: Conceptualization, H.H. and X.Z.; Data curation, H.H.; Investigation, A.X. and D.Z.; Project administration, S.P.; Resources, W.Z.; Writing—original draft, H.H.; Writing—review & editing, H.H. All authors have read and agreed to the published version of the manuscript.

Funding: The work is funded by National Natural Science Foundation of China (51273060), the Open Fund of Hubei Provincial Key Laboratory of Green Materials for Light Industry.

Acknowledgments: The work is funded by National Natural Science Foundation of China (51273060), the Open Fund of Hubei Provincial Key Laboratory of Green Materials for Light Industry.

Conflicts of Interest: The authors declare no conflict of interest.

References

1. Sakdaronnarong, C.; Srimarut, N.; Lucknakhul, N.; Na-songkla, N.; Jonglertjunya, W. Two-step acid and alkaline ethanolysis/alkaline peroxide fractionation of sugarcane bagasse and rice straw for production of polylactic acid precursor. *Biochem. Eng. J.* **2014**, *85*, 49–62. [CrossRef]
2. Tajitsu, Y.; Kawase, Y.; Katsuya, K.; Tamura, M.; Sakamoto, K.; Kawahara, K.; Harada, Y.; Kondo, T.; Imada, Y. New wearable sensor in the shape of a braided cord (Kumihimo). *IEEE Trans. Dielectr. Electr. Insul.* **2018**, *25*, 772–777. [CrossRef]
3. Asmawi, N.N.M.; Sapuan, S.M.; Yusoff, M.Z.M.; Ilyas, R.A.; Sherwani, s.f. Nanocellulose Reinforced Thermoplastic Starch (TPS), Polylactic Acid (PLA), and Polybutylene Succinate (PBS) for Food Packaging Applications. *Front. Chem.* **2020**, *8*, 1–12.
4. Davachi, S.M.; Kaffashi, B. Polylactic acid in medicine. *Polym. Plast. Technol. Eng.* **2015**, *54*, 944–967. [CrossRef]
5. Xu, Z.; Yang, L.; Ni, Q.; Ruan, F.; Wang, H. Fabrication of high-performance green hemp/polylactic acid fibre composites. *J. Eng. Fibers Fabr.* **2019**, *14*, 1558925019834497. [CrossRef]
6. Zhao, X.P.; Hu, H.; Wang, X.; Yu, X.L.; Zhou, W.Y.; Peng, S.X. Super tough poly(lactic acid) blends: A comprehensive review. *RSC Adv.* **2020**, *10*, 13316–13368. [CrossRef]
7. Zaaba, N.F.; Ismail, H. A review on tensile and morphological properties of poly (lactic acid) (PLA)/thermoplastic starch (TPS) blends. *Polym. Plast. Technol. Mater.* **2019**, *58*, 1945–1964. [CrossRef]
8. Zhang, S.; Feng, X.; Zhu, S.; Huan, Q.; Han, K.; Ma, Y.; Yu, M. Novel toughening mechanism for polylactic acid (PLA)/starch blends with layer-like microstructure via pressure-induced flow (PIF) processing. *Mater. Lett.* **2013**, *98*, 238–241. [CrossRef]
9. Zhang, K.; Cheng, X.; Cheng, F.; Lin, Y.; Zhou, M.; Zhu, P. Poly(citrate glyceride): A hyperbranched polyester for starch plasticization. *Polym. Int.* **2018**, *67*, 399–404. [CrossRef]
10. Volpe, V.; De Feo, G.; De Marco, I.; Pantani, R. Use of sunflower seed fried oil as an ecofriendly plasticizer for starch and application of this thermoplastic starch as a filler for PLA. *Ind. Crop. Prod.* **2018**, *122*, 545–552. [CrossRef]
11. Koh, J.J.; Zhang, X.; He, C. Fully biodegradable Poly(lactic acid)/Starch blends: A review of toughening strategies. *Int. J. Biol. Macromol.* **2018**, *109*, 99–113. [CrossRef] [PubMed]
12. Wang, Y.; Hu, Q.; Li, T.; Ma, P.; Zhang, S.; Du, M.; Chen, M.; Zhang, H.; Dong, W. Core–Shell Starch Nanoparticles and Their Toughening of Polylactide. *Ind. Eng. Chem. Res.* **2018**, *57*, 13048–13054. [CrossRef]
13. Thielemans, W.; Belgacem, M.N.; Dufresne, A. Starch nanocrystals with large chain surface modifications. *Langmuir* **2006**, *22*, 4804–4810. [CrossRef] [PubMed]
14. Xiong, Z.; Dai, X.; Zhang, R.; Tang, Z.; Na, H. Preparation of Biobased Monofunctional Compatibilizer from Cardanol to Fabricate Polylactide/Starch Blends with Superior Tensile Properties. *Ind. Eng. Chem. Res.* **2014**, *53*, 10653–10659. [CrossRef]
15. Jariyasakoolroj, P.; Chirachanchai, S. Silane modified starch for compatible reactive blend with poly(lactic acid). *Carbohydr. Polym.* **2014**, *106*, 255–263. [CrossRef]

16. Meyvazeybek, Y.; Kaynak, C. Loss of thermoplastic elastomer toughening in polylactide after weathering. *J. Appl. Polym. Sci.* **2019**, *136*, 47177. [CrossRef]
17. 17. Halimatul, M.J.; Sapuan, S.M.; Jawaid, M.; Ishak, M.R.; Ilyas, R.A. Water absorption and water solubility properties of sago starch biopolymer composite films filled with sugar palm particles. *Polimery* **2019**, *64*, 596–604. [CrossRef]
18. Ferrarezi, M.M.; Taipina, M.D.; Silva, L.C.; Gonçalves, M.D. Poly(Ethylene Glycol) as a Compatibilizer for Poly(Lactic Acid)/Thermoplastic Starch Blends. *J. Polym. Environ.* **2013**, *21*, 151–159. [CrossRef]
19. Wang, N.; Zhang, X.X.; Han, N.; Fang, J.M. Effects of Water on the Properties of Thermoplastic Starch Poly(lactic acid) Blend Containing Citric Acid. *J. Thermoplast. Compos. Mater.* **2010**, *23*, 19–34. [CrossRef]
20. Avolio, R.; Castaldo, R.; Gentile, G.; Ambrogi, V.; Fiori, S.; Avella, M.; Cocca, M.; Errico, E. Plasticization of poly(lactic acid) through blending with oligomers of lactic acid: Effect of the physical aging on properties. *Eur. Polym. J.* **2015**, *66*, 533–542. [CrossRef]
21. Yu, X.; Wang, X.; Zhang, Z.; Jiangming, F. High-performance fully bio-based poly (lactic acid)/polyamide11 (PLA/PA11) blends by reactive blending with multi-functionalized epoxy. *Polym. Test.* **2019**, *78*, 105980. [CrossRef]
22. Noivoil, N.; Yoksan, R. Oligo (lactic acid)-grafted starch: A compatibilizer for poly (lactic acid)/thermoplastic starch blend. *Int. J. Biol. Macromol.* **2020**, *160*, 506–517. [CrossRef] [PubMed]
23. Aouat, T.; Kaci, M.; Devaux, E.; Campagne, C.; Lopez-Cuesta, J. Morphological, Mechanical, and Thermal Characterization of Poly (Lactic Acid)/Cellulose Multifilament Fibers Prepared by Melt Spinning. *Adv. Polym. Technol.* **2016**, *37*, 1193–1205. [CrossRef]
24. Chotiprayon, P.; Chaisawad, B.; Yoksan, R. Thermoplastic cassava starch/poly(lactic acid) blend reinforced with coir fibres. *Int. J. Biol. Macromol.* **2020**, *156*, 960–968. [CrossRef] [PubMed]
25. Li, F.; Liang, J.; Zhang, S.; Zhu, B. Tensile Properties of Polylactide/Poly(ethylene glycol) Blends. *J. Polym. Environ.* **2015**, *23*, 407–415. [CrossRef]
26. Wu, X.S.; Wu, X.S. Effect of Glycerin and Starch Crosslinking on Molecular Compatibility of Biodegradable Poly (lactic acid)-Starch Composites. *J. Polym. Environ.* **2011**, *19*, 912–917. [CrossRef]
27. Li, F.J.; Tan, L.C.; Zhang, S.D.; Zhu, B. Compatibility, steady and dynamic rheological behaviors of polylactide/poly (ethylene glycol) blends. *J. Appl. Polym. Sci.* **2015**, *133*. [CrossRef]
28. Fang, H.; Jiang, F.; Wu, Q.; Ding, Y.; Wang, Z. Supertough Polylactide Materials Prepared through In Situ Reactive Blending with PEG-Based Diacrylate Monomer. *ACS Appl. Mater. Interfaces* **2014**, *6*, 13552–13563. [CrossRef]
29. Nouri, S.; Dubois, C.; Lafleur, P.G. Effect of chemical and physical branching on rheological behavior of polylactide. *J. Rheol.* **2015**, *59*, 1045–1063. [CrossRef]
30. Liu, W.; He, S.; Yang, Y. Effect of stereocomplex crystal on foaming behavior and sintering of poly(lactic acid) bead foams. *Polym. Int.* **2019**, *68*, 516–526. [CrossRef]
31. Cai, J.; Liu, M.; Wang, L.; Yao, K.; Xiong, H. Isothermal crystallization kinetics of thermoplastic starch/poly(lactic acid) composites. *Carbohydr. Polym.* **2011**, *86*, 941–947. [CrossRef]
32. Chen, W.; Chen, H.; Yuan, Y.; Peng, S.; Zhao, X. Synergistic effects of polyethylene glycol and organic montmorillonite on the plasticization and enhancement of poly(lactic acid). *J. Appl. Polym. Sci.* **2019**, *136*, 47576. [CrossRef]
33. Zuo, Y.; Gu, J.; Yang, L.; Qiao, Z.; Tan, H.; Zhang, Y. Preparation and characterization of dry method esterified starch/polylactic acid composite materials. *Int. J. Biol. Macromol.* **2014**, *64*, 174–180. [CrossRef] [PubMed]
34. Wootthikanokkhan, J.; Kasemwananimit, P.; Sombatsompop, N.; Kositchaiyong, A.; Isarankura na Ayutthaya, S.; Kaabbuathong, N. Preparation of modified starch-grafted poly(lactic acid) and a study on compatibilizing efficacy of the copolymers in poly(lactic acid)/thermoplastic starch blends. *J. Appl. Polym. Sci.* **2012**, *126*, E389–E396. [CrossRef]
35. Shirai, M.A.; Grossmann, M.V.E.; Mali, S.; Yamashita, F.; Garcia, P.S.; Müller, C.M.O. Development of biodegradable flexible films of starch and poly(lactic acid) plasticized with adipate or citrate esters. *Carbohydr. Polym.* **2013**, *92*, 19–22. [CrossRef] [PubMed]
36. Xia, T.; Gou, M.; Zhang, G.; Li, W.; Jiang, H. Physical and structural properties of potato starch modified by dielectric treatment with different moisture content. *Int. J. Biol. Macromol.* **2018**, *118*, 1455–1462. [CrossRef] [PubMed]

37. Yu, Y.; Cheng, Y.; Ren, J.; Cao, E.; Fu, X.; Guo, W. Plasticizing effect of poly(ethylene glycol)s with different molecular weights in poly(lactic acid)/starch blends. *J. Appl. Polym. Sci.* **2015**, *132*. [CrossRef]
38. Sanyang, M.L.; Sapuan, S.M.; Jawaid, M.; Ishak, M.R.; Sahari, J. Effect of plasticizer type and concentration on physical properties of biodegradable films based on sugar palm (arenga pinnata) starch for food packaging. *J. Food Sci. Technol. Mysore* **2016**, *53*, 326–336. [CrossRef]

Sample Availability: Samples of the compounds are not available from the authors.

Publisher's Note: MDPI stays neutral with regard to jurisdictional claims in published maps and institutional affiliations.

© 2020 by the authors. Licensee MDPI, Basel, Switzerland. This article is an open access article distributed under the terms and conditions of the Creative Commons Attribution (CC BY) license (http://creativecommons.org/licenses/by/4.0/).

Article

Shear-Induced Crystallization of Star and Linear Poly(L-lactide)s

Joanna Bojda *, Ewa Piorkowska, Grzegorz Lapienis and Adam Michalski

Centre of Molecular and Macromolecular Studies, Polish Academy of Sciences, Sienkiewicza 112, 90-363 Lodz, Poland; epiorkow@cbmm.lodz.pl (E.P.); lapienis@cbmm.lodz.pl (G.L.); michadam@cbmm.lodz.pl (A.M.)
* Correspondence: jbojda@cbmm.lodz.pl

Abstract: The influence of macromolecular architecture on shear-induced crystallization of poly(L-lactide) (PLLA) was studied. To this aim, three star PLLAs, 6-arm with M_w of 120 and 245 kg/mol, 4-arm with M_w of 123 kg/mol, and three linear PLLAs with M_w of 121, 240 and 339 kg/mol, were synthesized and examined. The PLLAs were sheared at 170 and 150 °C, at 5/s, 10/s and 20/s for 20 s, 10 s and 5 s, respectively, and then cooled at 10 or 30 °C/min. Shear-induced crystallization during cooling was followed by a light depolarization method, whereas the crystallized specimens were examined by DSC, 2D-WAXS, 2D-SAXS and SEM. The effect of shear depended on the shearing conditions, cooling rate and polymer molar mass but it was also affected by the macromolecular architecture. The shear-induced crystallization of linear PLLA with M_w of 240 kg/mol was more intense than that of the 6-arm polymer with similar M_w, most possibly due to its higher M_z. However, the influence of shear on the crystallization of the star polymers with M_w close to 120 kg/mol was stronger than on that of their linear analog. This was reflected in higher crystallization temperature, as well as crystallinity achieved during cooling.

Keywords: poly(L-lactide); star poly(L-lactide); shear-induced crystallization; crystallinity; fibrillar nuclei

Citation: Bojda, J.; Piorkowska, E.; Lapienis, G.; Michalski, A. Shear-Induced Crystallization of Star and Linear Poly(L-lactide)s. *Molecules* **2021**, *26*, 6601. https://doi.org/10.3390/molecules26216601

Academic Editor: Ivan Gitsov

Received: 21 September 2021
Accepted: 22 October 2021
Published: 31 October 2021

Publisher's Note: MDPI stays neutral with regard to jurisdictional claims in published maps and institutional affiliations.

Copyright: © 2021 by the authors. Licensee MDPI, Basel, Switzerland. This article is an open access article distributed under the terms and conditions of the Creative Commons Attribution (CC BY) license (https://creativecommons.org/licenses/by/4.0/).

1. Introduction

In recent decades, bio-based polymers derived from annually renewable resources have drawn increasing attention [1], as biomass is the only source of available renewable carbon. Among them, polylactide (PLA) is the most promising polymer for the replacement of conventional thermoplastics, especially because it is also biodegradable (compostable) and can be used for a wide range of applications, including biomedical products, textiles, daily appliances, packaging, items used in agriculture and engineering [1–5].

Amorphous PLA with the glass transition temperature, T_g, in the range of 55–60 °C, is stiff and brittle at room temperature. The crystallizability of PLA strongly depends on its enantiomeric composition and worsens with increasing content of repeating units of different chirality in the chain [6,7]. However, even optically pure poly(L-lactide) (PLLA), if cooled sufficiently fast, remains amorphous and vitrifies. Then, upon heating from the glassy state, such amorphous PLLA can cold-crystallize. However, not only the enantiomeric composition but also other factors, including molar mass and macromolecular architecture, for instance, branching or star structure, are also important and influence the crystallization of PLA. Depending on crystallization temperature, PLA chains crystallize from melt in the ordered alpha or disordered alpha' orthorhombic forms [8], which can be identified not only by wide angle X-ray scattering but also by Raman spectroscopy or nuclear magnetic resonance spectroscopy [9,10].

Star polymers are especially interesting for many applications, for example in biomedicine or engineering [11,12]. Their rheological, thermal and mechanical properties can differ from those of their linear counterparts [11,13]. In addition, processing of star polymers can be carried out at lower temperatures than their linear analogs, which could be beneficial, especially in the case of polymers prone to thermal degradation like PLAs.

Star PLLAs were most often obtained by the bulk polymerization of cyclic lactide conducted at a temperature above the melting point of the monomer (~99 °C), in the presence of initiators with hydroxyl end groups, with stannous octoate (Sn(Oct)$_2$) as a catalyst [14–19]. Crystallization of star PLLAs was found to be affected by their molar masses and numbers of arms. Usually, PLLAs with number average molar masses, M_n, below 100 kg/mol were analyzed. The consequences of the relatively small molar mass of star PLLA are short arm length resulting in a small number or even absence of entanglements and a large number of chain ends as well as branching points. The presence of branching points, initiator moieties, especially bulky ones, in the middle of macromolecules, hydroxyl chain end groups enhancing hydrogen bonding, and chain directional change at the branching points, are the factors disturbing the segmental mobility [11,20]. Moreover, the branching points have to be excluded from the crystalline regions [21]. For example, for PLLAs with M_n about 35 kg/mol, nonisothermal crystallization peak temperature, T_c, and overall isothermal crystallization rate decreased upon the increasing number of arms up to six [22]. Moreover, an increase of T_g, cold crystallization peak temperature, T_{cc}, accompanied by a decrease of melting peak temperature, T_m, and crystal growth rate of 3-arm PLLAs, with M_n in the range of 13 to 63 kg/mol, were observed, as compared with those of linear PLLAs [21]. It is worth noting that crystallization of 6-arm PLLAs was rarely examined, and the studies were limited to polymers with M_n per one arm below 10 kg/mol. For instance, it was found [23] that for 1-, 2-, 4- and 6-arm PLLAs, T_m, T_{cc}, and crystallinity, χ_c, decreased with increasing number of arms at a fixed M_n. Recently, Bojda et al. [24] synthesized three star PLLAs, two 6-arm with weight average molar masses, M_w, of 120 and 245 kg/mol and one 4-arm with M_w of 123 kg/mol, and compared their crystallization with that of linear ones with M_w of 121, 291 and 339 kg/mol. At M_w close to 120 kg/mol the star architecture decreased the crystal growth rate in the temperature range of 120–145 °C. Crystal growth of PLLAs with M_w > 200 kg/mol was the slowest and unaffected by the macromolecular architecture. The slow crystal growth in these PLLAs resulted in their weak crystallization during cooling.

In turn, it is long known that the crystallization of polymers, including PLAs, is strongly influenced by flow, which plays a vital role during industrial processing. The flow-induced macromolecular orientation can strongly affect the crystallization kinetics and the resulting structure, which are controlled by the interplay between crystallization and chain relaxation. The fundamental processes governing the flow-induced crystallization of polymer melts were discussed by many authors [25–27]. The shear-induced crystallization strongly depends on the temperature of shearing, T_s, shear rate, $\dot{\gamma}$, and total strain. It is believed [28] that to induce the point-like nuclei and fibrillar nuclei, the shear rate has to exceed the inverse reptation time and the inverse Rouse relaxation time of the high molar mass tail, respectively, although when the flow is strong enough, but too short, intermediate regimes were also defined. It is worth noting that others postulated that mechanical work is a controlling parameter [26,29,30]. A very important factor is the polymer molecular characteristic, especially the high molar mass tail is crucial due to the vital role of the longest macromolecules in the flow-induced crystallization. [26,31,32]. Similarly to other polymers, the enhancement of point-like nucleation, formation of oriented nuclei and shish-kebab structures induced by shear were observed in PLAs [33–35]. The effect of shear on both isothermal [34–37] and nonisothermal crystallization of PLAs was studied. Bojda et al. [36] demonstrated that smaller content of D-lactide enhanced the effect of shear on nonisothermal crystallization on PLA and that higher crystallinity degree developed during slower post-shearing cooling. In turn, Kim et al. [38] compared shear-induced crystallization of linear and 4-arm PLLAs with M_w of about 2 kg/mol and found that that of the latter was slightly faster.

To the best of our knowledge, the effect of shear flow on the crystallization of star PLLAs with higher molar masses was not investigated. Only shear-induced isothermal crystallization of PLAs with long chain branching (LCB), prepared by γ irradiation, was studied in [39,40]. It was demonstrated that the shear-induced nucleation density in LCB PLA was strongly enhanced in comparison to linear PLA and increased with increasing LCB degree. Moreover, the transformation from spherulitic to oriented crystalline morphologies was observed. It was concluded that the shear-induced formation of the oriented crystalline morphology of LCB PLAs was related to the hindering of relaxation of the stretched LCB macromolecular chain network.

In the present study, shear-induced nonisothermal crystallization of star PLLAs with M_w close to 120 and 240 kg/mol was studied and compared with that of their linear analogs. In addition, the crystallization of linear PLLA with M_w of 339 kg/mol was also examined. The polymers were sheared at 170 and 150 °C and cooled at 10 or 30 °C/min. The crystallization was followed by a light depolarization technique, whereas the crystallized specimens were ex-situ examined with scanning electron microscopy (SEM), differential scanning calorimetry (DSC), small- and wide-angle X-ray scattering (2D-SAXS and 2D-WAXS).

2. Materials and Methods

Linear PLLAs having at one chain end benzyl alcohol and star-shaped PLLAs having as a core: di(trimethylolpropane) (4-arm PLLA-OH) or dipentaerythritol (6-arm PLA-OH), were synthesized in bulk at 130 °C by coordination polymerization using hydroxyl compound as an initiator and stannous octoate as a catalyst, as described previously [24]. The resulting PLLAs were dissolved in dichloromethane and precipitated into methanol, separated by filtration, and washed several times with methanol. The polymers were characterized with size exclusion chromatography (SEC) and ^{1}H NMR. After purification, the polymers were stabilized with 0.2 wt.% of Irganox 1010 and 0.2 wt.% of Irganox 1024, both from BASF (Ludwigshafen, Germany). The details of polymerization, purification, characterization and stabilization were previously described [19,24,41,42].

The average molar masses, number, M_n, weight, M_w, and z-average, M_z, and dispersity of linear and star-shaped PLLAs measured by SEC in dichloromethane are given in Table 1.

Table 1. The average molar masses, number, M_n, weight, M_w, and z-average, M_z, and dispersity M_w/M_n of star and linear PLLAs.

Sample Code	M_w (kg/mol)	M_n (kg/mol)	M_z (kg/mol)	M_w/M_n
L121	121	81	194	1.5
L240	240	157	414	1.3
L339	339	257	495	1.3
4S123	123	97	152	1.3
6S120	120	80	162	1.5
6S245	245	183	294	1.3

For studies of crystallization, 200 μm thick films were compression moulded at 200 °C for 3 min in a hydraulic press and quenched to room temperature, RT, between metal blocks.

Flow-induced crystallization was carried out in the Linkam CSS-450 optical shearing system (Linkam, Waterfield, UK) mounted in a polarizing light microscope (PLM) Nikon Eclipse 80i equipped with Nikon DS Fi1 video camera. The films were: heated to 210 °C at 30 °C/min and held at 210 °C for 3 min to erase the thermal history. Next the films were cooled at 30 °C/min to T_s of 170 or 150 °C and sheared at a rate, $\dot{\gamma}$, of 5, 10 and 20/s for 20, 10 and 5 s, respectively to reach the same strain of 100. After shearing, they were cooled to RT at a rate, v, of 10 or 30 °C/min. The shearing conditions were selected based on preliminary studies. Control specimens were subjected to a similar thermal treatment in quiescent conditions; they were held at T_s for 20 s without shearing.

The conversion of melt into the crystalline phase was followed using the light depolarization method. The intensity of transmitted depolarized light was measured during cooling and the relative volume conversion degree, $\alpha_{vr}(T)$, was calculated according to the expression:

$$\alpha_{vr}(T) = [I(T) - I(T_o)]/[I(T_e) - I(T_o)] \qquad (1)$$

where: I(T) denotes the intensity of transmitted depolarized light at temperature T, whereas T_o and T_e are the initial temperature and the final temperature of the measurement.

In the plate-plate geometry, shear rate varies along a radius, hence for ex-situ examination, specimens were cut from the films at proper distances from the centers, at which shear rates were equal to the selected values.

Crystallinity and thermal properties of the specimens were analyzed with differential scanning calorimetry (DSC) using TA Instrument DSC TA Q20 (New Castle, USA) during heating at 5 °C/min from RT, under nitrogen flow.

Crystal orientation in the films was examined with 2D-WAXS in the transmission mode, using a WAXS camera coupled to an X-ray generator (sealed-tube, fine point CuKα source, Ni filtered, operating at 50 kV and 35 mA) from Philips (Eindhoven, Netherlands). The incident beam was normal to the film plane. The lamellar structure was probed with 2D small angle X-ray scattering (2D-SAXS). Kiessig-type camera with the sample-detector distance of 1.2 m was coupled to GeniX Cu-LD X-ray system from Xenocs (Grenoble, France), with CuKα source operating at 50 kV and 1 mA. The incident beam was normal to the film plane. The scattering patterns were recorded with Pilatus 100K solid-state detector from Dectris (Baden, Switzerland).

To reveal their internal structure, selected sheared PLLA specimens were cut across their thickness parallel to the shearing direction, and the exposed surfaces were analyzed with scanning electron microscopy (SEM) using Jeol JSM-5500 LV (Tokyo, Japan). Before the examination, the specimens were etched according to the known method [36,43], at 37 °C, in a solution of 61 mg of Trizma base, 2 mg of sodium azide and 4 mg of Proteinase K (all from Sigma-Aldrich, St. Louis, MO, USA) in 5 mL of distilled water. After appropriate washing and drying, the specimens were sputtered with gold.

3. Results and Discussion

3.1. Crystallization

Exemplary DSC heating thermograms of star-shaped and linear PLLAs sheared at 170 °C and then cooled to RT at 30 °C/min, collected in Figure 1, exhibit glass transition with T_g at approx. 60–61 °C, cold crystallization exotherms and melting endotherms, with peaks at T_{cc} close to 100 °C and T_m above 170 °C, respectively. In addition, on the thermograms, small pre-melting exotherms are visible, with maxima close to 160 °C, which originated from the recrystallization of the disordered alpha' to the ordered alpha orthorhombic form [8]. During heating at 5 °C/min, the cold-crystallization occurred in a relatively low temperature range and the crystallization exotherms and the melting endotherms did not overlap, which facilitated integration of the peaks and calculation of enthalpies of the processes. The melting enthalpy, ΔH_{mc}, of crystals formed during cooling, before the heating in DSC, was calculated by subtracting the enthalpies of exothermic effects of crystallization and recrystallization, ΔH_{cc} and ΔH_{rc}, respectively, from the melting enthalpy, ΔH_m.

For the control specimens, ΔH_{mc} was small, or even close to zero, indicating that they were amorphous or with low crystallinity. However, ΔH_{mc} of the sheared specimens was markedly larger, proving that crystallinity developed in PLLAs during the post-shearing cooling. The differences between the thermograms in Figure 1 evidence that the effect of shear depended on the molar mass, but also on the macromolecular architecture of PLLAs studied.

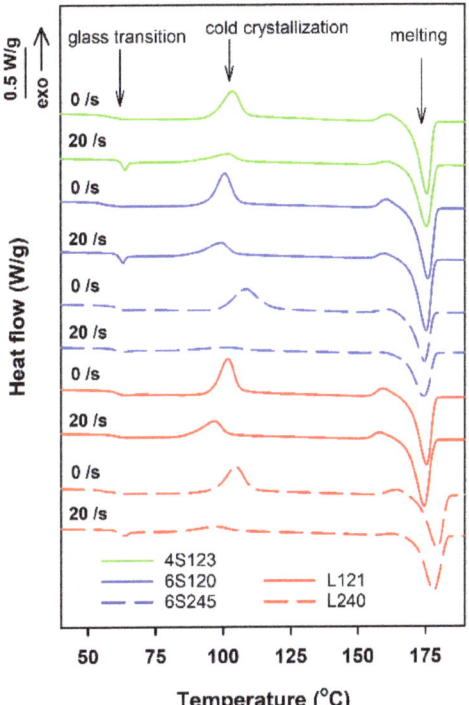

Figure 1. DSC heating thermograms of PLLAs sheared at 20/s for 5 s at 150 °C and next cooled at 30 °C/min, and thermograms of control specimens cooled at 30 °C/min. The curves shifted vertically for clarity.

It must be noted that such approach does not take into account the temperature dependence of heat of fusion, due to which ΔH_{cc} of the cold-crystallized crystals can be lower than their melting enthalpy. However, although $T_m - T_{cc}$ was up to about 70 °C, in most cases ΔH_{cc} was significantly smaller than ΔH_{mc}, therefore reducing the overestimation of the latter. Another effect that should be considered is the difference in the heat of fusion of the ordered alpha form and the disordered alpha' form of PLLA. It is known that between 100 and 120 °C PLLA crystallizes not only in the ordered alpha modification but also in the disordered alpha' form. Below 100 °C, PLLA crystallizes from the quiescent melt only in the alpha' form [8], although the alpha form was found after shear-induced crystallization at 96 °C [44]. Although the heat of fusion of the alpha' crystals is significantly lower than that of the alpha modification [45], the influence of that on ΔH_{mc} can be neglected, because of the alpha' to alpha recrystallization prior to melting. In addition, the alpha' to alpha recrystallization occurring near 160 °C and also reorganization occurring in the alpha phase prior to melting, further reduce the possible overestimation of ΔH_{mc}.

Figure 2 illustrates the effect of shearing conditions and post-shearing cooling on ΔH_{mc} and mass crystallinity, χ_c of linear and star PLLAs. The ΔH_{mc} values are averages, based on at least two or three measurements. The mass crystallinity, χ_c, was calculated from ΔH_{mc}, assuming that the heat of fusion of 100% crystalline PLLA is 106 J/g [46].

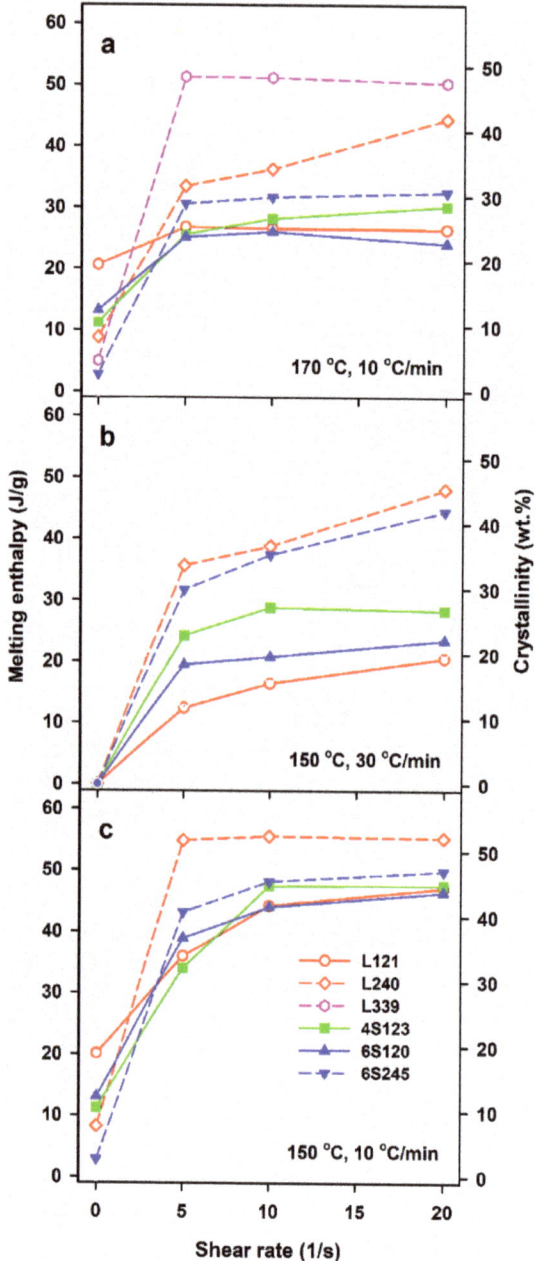

Figure 2. Melting enthalpy, ΔH_{mc}, of crystalline phase formed in PLLAs during cooling at 10 and 30 °C/min after shearing at 170°C (**a**) and 150 °C (**b**,**c**) versus shear rate, $\dot{\gamma}$.

The control specimens cooled at 30 °C/min were practically amorphous, whereas in those cooled at 10 °C/min crystallization occurred, although ΔH_{mc} and the corresponding χ_c were small, being the largest for L121, close to 20 J/g and 19%, respectively, as previously found for the same polymer [24]. Shearing at 150 and 170 °C enhanced crystallization in all PLLAs. In general, ΔH_{mc} increased with $\dot{\gamma}$, although weakly in most cases, or was even independent of $\dot{\gamma}$, most possibly due to the same final shear strain achieved during all experiments. Moreover, ΔH_{mc} of L339 sheared at 170 °C and L240 sheared at 150 °C, and next cooled at 10 °C/min, reached very high values even for $\dot{\gamma}$ of only 5/s.

Shearing at 170 °C followed by cooling at 30 °C/min resulted in rather low ΔH_{mc} of all PLLAs, up to about 15 J/g, except L240 and L339, for which ΔH_{mc} values were higher. A decrease of v to 10 °C/min enhanced the post-shearing crystallization in all PLLAs. As seen in Figure 2a, ΔH_{mc} values of PLLAs with M_w close to 120 kg/mol ranging from 24 to 30 J/g were similar. Slightly higher ΔH_{mc} of 31–32 J/g was found for 6S245, and even higher up to 44 J/g for L240. The effect of shear on the crystallization of L339 was the strongest, which was reflected in ΔH_{mc} close to 50 J/g. A decrease of T_s to 150 °C intensified the post-shearing crystallization at both cooling rates. L339 crystallized during shearing at 150 °C; hence, studies of its post-shearing nonisothermal crystallization were impossible. As seen in Figure 2b, ΔH_{mc} values of PLLAs with M_w close to 120 kg/mol, cooled at 30 °C/min, increased up to 21–29 J/g, and those of 6S245 and L240 to 44 and 48 J/g, respectively. Slower cooling, at 10 °C/min, resulted in higher ΔH_{mc}, in the range of 34–47 J/g for PLLAs with M_w close to 120 J/g, whereas in the range of 43–49 J/g and 55 J/g for 6S245 and L240, respectively, as s shown in Figure 2c.

DSC measurements allow us to determine only the final χ_c developed during post-shearing cooling, whereas the light depolarization method enables us to follow the increase of α_{vr} during crystallization. To compare the crystallinity increase in specimens with different final crystallinity, volume crystallinity $\alpha_v(T)$ equal to $\alpha_{vr}(T) \chi_v$, was plotted in Figure 3, where χ_v is the final volume crystallinity calculated based on χ_c and the densities of the amorphous and crystalline phases of PLA [47]. As it is explained above, χ_c was calculated based on the melting enthalpy of crystals formed during post-shearing cooling, ΔH_{mc}. It should be mentioned that the lower melting enthalpy of the alpha' phase was not accounted for because ΔH_{mc} was determined from the melting endotherm preceded by the pre-melting recrystallization of alpha' to alpha form. Differentiation of $\alpha_v(T)$ with respect to temperature permitted to obtain the temperature dependencies of crystallization rate. It appears that the lower T_s, slower cooling and higher M_w of PLLA resulted in the higher temperature range of crystallization. The effect of $\dot{\gamma}$, T_s and v, as well as of M_w of PLLA and its macromolecular architecture, on T_c was similar to that on ΔH_{mc}, as it is shown in Figure 4. T_c correlated with ΔH_{mc} and crystallinity, the higher the former the larger the latter. T_c increased with decreasing T_s and v, and with increasing $\dot{\gamma}$. The highest T_c values were found for L339, lower for L240 and 6S245, and even lower for 4S123, 6S120 and L121, showing the influence of M_w, but also of the macromolecular architecture.

The results show a crucial role of T_s and v. The lower T_s increased the relaxation times of macromolecules and lowered the energy barrier for nucleation, thus enhancing the shear-induced crystallization. In turn, the slower cooling enabled a longer time for crystallization before too low temperature was reached, increasing therefore T_c, ΔH_{mc} and χ_c, the latter determined based on ΔH_{mc}. However, not only the shearing conditions and v determined the post-shearing crystallization. T_c and ΔH_{mc} were strongly influenced by molar masses of PLLAs, as can be expected, but they were also affected by macromolecular architecture. Figure 2b shows that in the case of cooling at 30 °C/min, the shearing at 150 °C had the weakest effect on L121, stronger on 6S120, and the strongest on 4S123 crystallization. Figure 2a,c show that during cooling at 10 °C/min T_c and ΔH_{mc} values of all PLLAs with M_w near 120 kg/mol were similar. However, it must be reminded that in the temperature range of 120–145 °C the crystal growth rate of L121 was higher than that of the other PLLAs studied [24]. The crystallization kinetics is governed by both the crystal growth rate and the nucleation rate [48], hence, similar T_c and ΔH_{mc} values of L121, 6S120 and 4S123 are suggestive of much stronger nucleation in the two latter. The control specimens cooled at 30 °C/min were practically amorphous but crystallized during cooling at 10 °C/min. The shear-induced increase of ΔH_{mc} and T_c of specimens cooled at 10 °C/min is plotted in Figure 5 and Figure S1 in Supplementary Information (SI), respectively. The plots clearly show that the shear enhanced more the crystallization of 4S123 and 6S120 than that of L121, despite the higher M_z of the latter.

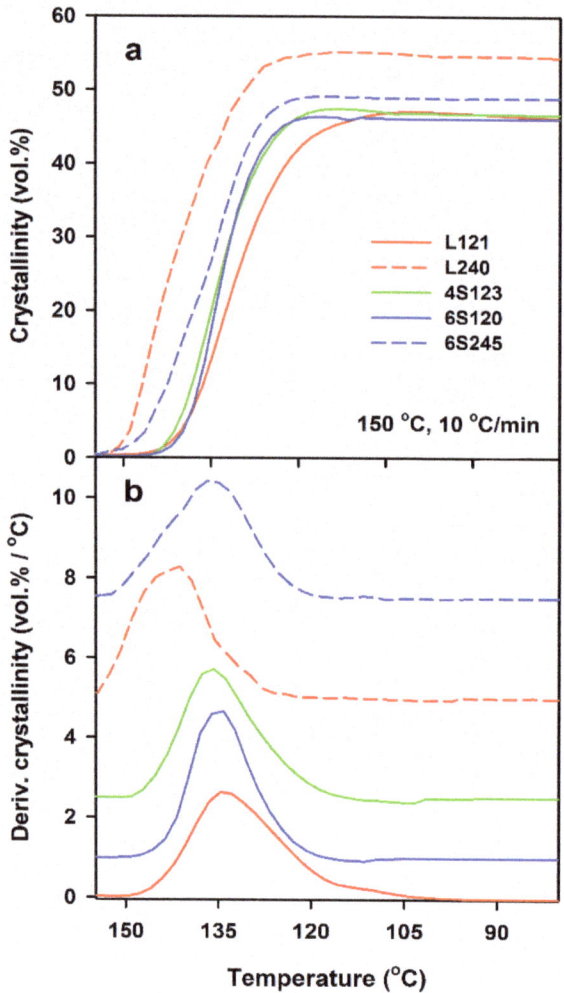

Figure 3. Development of crystallinity determined by light depolarization method (**a**) and derivative of crystallinity with respect to temperature (**b**) in PLLAs during cooling at 10 °C/min after shearing at 20/s, 150 °C. The curves in Figure 3b shifted vertically for clarity.

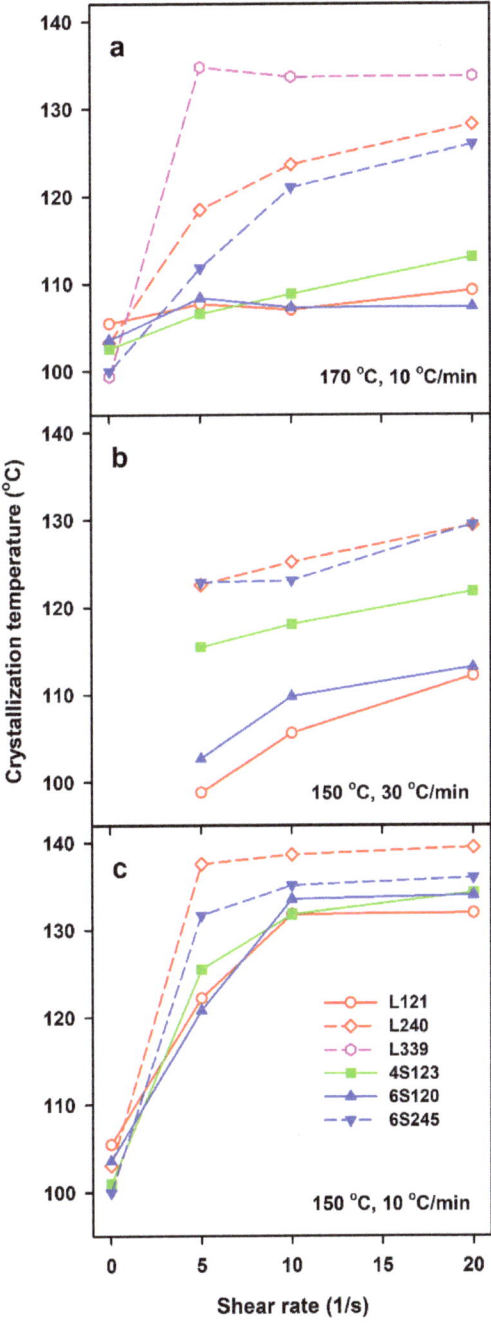

Figure 4. Crystallization peak temperature, T_c, during cooling of PLLAs at 10 and 30 °C/min after shearing at 170 °C (**a**) and 150 °C (**b**,**c**) versus shear rate, $\dot{\gamma}$.

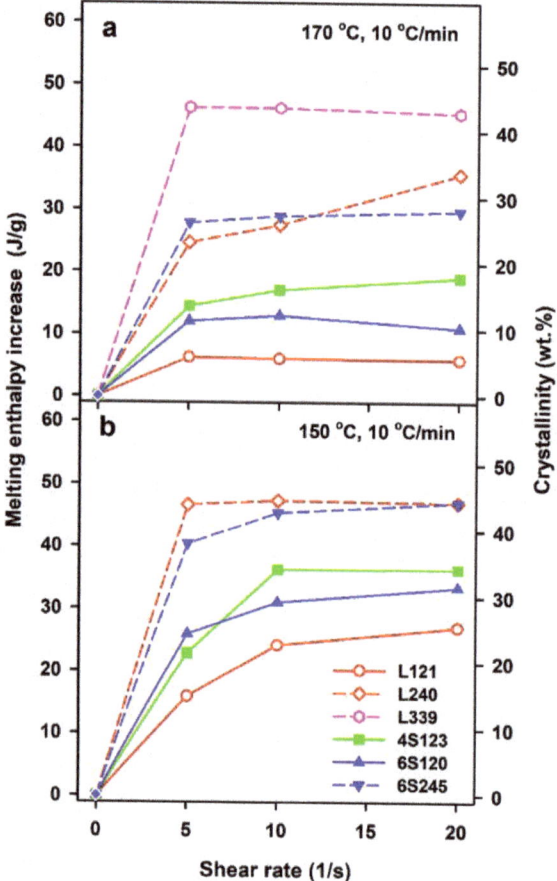

Figure 5. Increase of melting enthalpy, $\Delta H_{mc} - \Delta H_{mc}^q$, of crystalline phase formed in PLLAs during cooling at 10 °C/min caused by shearing at 170 °C (**a**) and 150 °C (**b**) versus shear rate, $\dot{\gamma}$. ΔH_{mc}^q denotes the melting enthalpy of crystals formed during cooling at 10 °C/min in control specimens.

The enhancement of the effect of shear in 6S120 and 4S123 as compared to L121 was undoubtedly caused by the star architecture of macromolecules, which hindered the relaxation of the stretched macromolecular chain network. In contrast to that, the shear-induced crystallization in 6S245 and L240, was similar, and even stronger in the latter, as shown in Figures 2 and 5. Although 6S245 and L240 had similar M_n and M_w, M_z of L240, 414 kg/mol, exceeded that of 6S245, 294 kg/mol, evidencing the higher content of larger macromolecules, which presence compensated the effect of 6S245 star architecture on the shear-induced crystallization. In the flow-induced crystallization of a polymer a high molar mass tail of its molar mass distribution plays a crucial role, due to long relaxation times, and at M_w of 240–245 kg/mol its effect compensated that of star architecture. It is also of importance that due to its higher molar mass, the number of branching points in 6S245 was smaller than in 6S120, hence their effect on the macromolecular mobility was reduced.

The crystallization, which was not completed during cooling continued during subsequent heating in DSC, resulting in cold-crystallization exotherms with peaks at T_{cc} of 97–109 °C, as shown in Figure 1. In many cases, pre-melting exotherms, with maxima at 159–164 °C, evidenced the alpha' to alpha form recrystallization. Usually, single melting peaks were observed, with T_m of 174–179 °C, although some of them with shoulders. As shown in Figure 6, ΔH_m values of the control specimens of 6S245 were equal to 37–38 J/g, whereas those of the other control PLLAs studied were higher, ranging from 43 to 51 J/g. The sheared PLLAs exhibited increased ΔH_m of 39–55 J/g. The lowest values of ΔH_m were those of 6S245 cooled at 10 °C/min after shearing at 170 °C, and cooled at 30 °C/min after shearing at 150 °C, as seen in Figure 6. In turn, among PLLAs sheared at 150 °C and cooled at 10 °C/min, L240 exhibited the highest ΔH_m, as evidenced in Figure 6c. These differences reflect the different ability of PLLAs studied to crystallize, as described in [24], and the different effect of shear influenced by the molar masses and macromolecular architecture of the studied PLLAs.

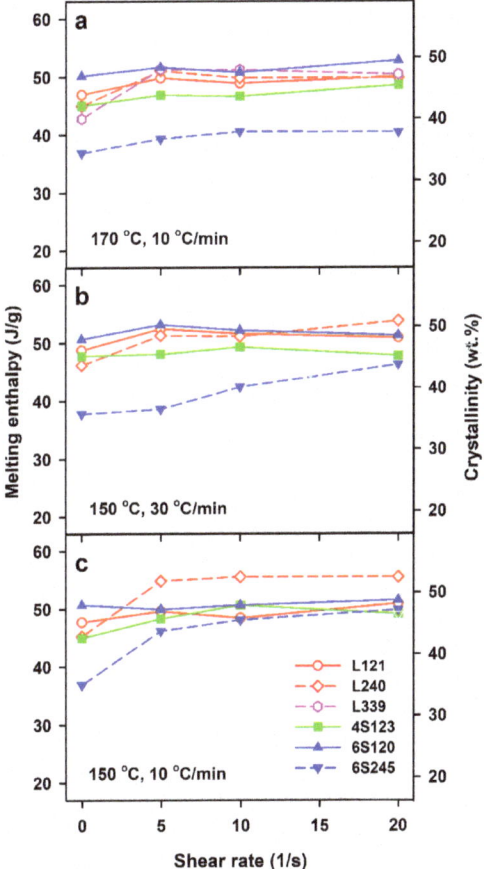

Figure 6. Melting enthalpy, ΔH_m, measured during DSC heating at 5 °C/min of PLLAs previously cooled at 10 and 30 °C/min, after shearing at 170 °C (**a**) and 150 °C (**b**,**c**) versus shear rate, $\dot{\gamma}$.

3.2. Structure

Examples of 2D-WAXS and 2D-SAXS patterns of sheared PLLA specimens are collected in Figures 7 and 8. Generally, the intensities of the reflections from the crystalline phase correlated with χ_c determined from DSC thermograms and plotted in Figure 2, and increased with increasing χ_c. (200)/(110) and (203) reflections (indicated by arrows in Figure S2 in SI) typical of both alpha and alpha' modifications were well visible on all patterns. Also, (210) reflection near 2θ of 22° characteristic of the alpha form was present in all the patterns, evidencing that the alpha phase was formed in all sheared specimens.

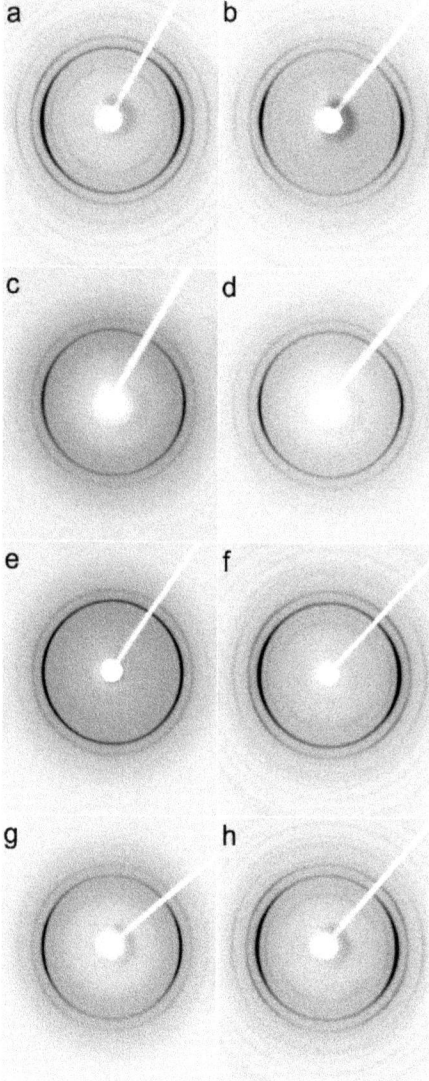

Figure 7. 2D-WAXS patterns of PLLAs: L339 sheared at 170 °C at 5/s for 20 s (**a**) and at 20/s for 5 s (**b**) and cooled at 10 °C/min, 4S123 sheared at 150 °C at 10/s for 10 s (**c**) and at 20/s for 5 s (**d**) cooled at 30 °C/min, L240 sheared at 150 °C at 10/s for 10 s (**e**) and at 20/s for 5 s (**f**) cooled at 30 °C/min, 6S245 sheared at 150 °C at 10/s for 10 s (**g**) and at 20/s for 5 s (**h**) cooled at 30 °C/min. Shearing direction -vertical.

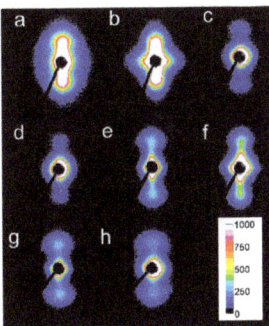

Figure 8. 2D-SAXS patterns of PLLAs: L339 sheared at 170 °C at 5/s for 20 s (**a**) and at 20/s for 5 s (**b**) and cooled at 10 °C/min, 4S123 sheared at 150 °C at 10/s for 10 s (**c**) and at 20/s for 5 s (**d**) cooled at 30 °C/min, L240 sheared at 150 °C at 10/s for 10 s (**e**) and at 20/s for 5 s (**f**) cooled at 30 °C/min, 6S245 sheared at 150 °C at 10/s for 10 s (**g**) and at 20/s for 5 s (**h**) cooled at 30 °C/min. Shearing direction-vertical.

This is in accordance with the T_c values shown in Figure 4, which indicate that the crystallization of most of the specimens occurred fully or partially above 110 °C; the higher the T_c the more intense was the (210) reflection. Only very weak (210) reflections were discernible in the patterns of PLLAs (not shown), which T_c was close to 100 °C, for instance, L121 and 6S120 sheared at 150 °C at 5/s and cooled at 30 °C/min, which indicated small alpha content.

In some of the 2D-WAXS patterns, intensities of (200)/(110) and (210) reflections were enhanced in equatorial regions, indicating the orientation of the respective crystallographic planes parallel to the shearing direction, and thus evidencing the orientation of polymer chain axes in the flow direction. This was corroborated by the strong polar reflections in the corresponding 2D-SAXS patterns reflecting the orientation of lamellae stacks perpendicular to the flow direction.

Among the specimens sheared at 170 °C and next cooled at 10 °C/min, L339 clearly exhibited such orientation, as shown in Figure 7a,b, and in Figure 8a,b. Weaker orientation was also detected in 6S245 and L240 sheared at 20/s (not shown). Shearing at 150 °C followed by cooling at 30 °C/min resulted in the crystal orientation in L240, 6S245, and 4S123 evidenced in Figures 7c–g and 8c–g. The orientation, although weaker, was also reflected in the patterns of L121 sheared at 20/s, 6S120 sheared at 10/s and 20/s (not shown). 2D-WAXS and 2D-SAXS patterns of PLLAs sheared at 150 °C and next cooled at 10 °C/min evidenced the same features of the morphology. It is well visible in 2D-SAXS patterns collected in Figure 9a–e, in which scattering from the crystalline phase was enhanced due to the higher χ_c developed during slower cooling.

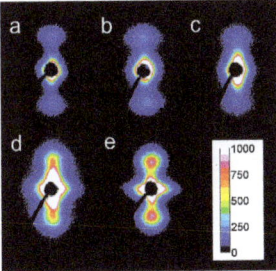

Figure 9. 2D-SAXS patterns of PLLAs: sheared at 150 °C at 20/s for 5 s and cooled at 10 °C/min: L121 (**a**), 4S123 (**b**), 6S120 (**c**), L240 (**d**), 6S245 (**e**). Shearing direction–vertical.

The reason for the orientation of crystals was the shear-induced formation of fibrillar nuclei aligned in the shearing direction on which grew lamella stacks perpendicular to the shearing direction. This is well seen in SEM micrographs of cross-section surfaces of sheared PLLAs, presented in Figure 10. The micrograph of L339 sheared at 170 °C at 20/s and next cooled at 10 °C/min, in Figure 9a, evidences the presence of lamellar stacks perpendicular or nearly perpendicular to the shearing direction, and lamella fans developed from the stacks. The stacks form cylindrical structures suggestive of nucleation on fibrillar nuclei. Similar morphology was found in L240 sheared at 150 °C at 20/s, and next cooled at 30 °C/min, as shown in Figure 10b, although spherulites between cylindrical structures were also discernible. Figure 10c presents 6S245 sheared at the same conditions and cooled at the same rate. In the micrograph, cylindrical structures are seen accompanied by spherulites, with radii of several micrometers. Amorphous areas are visible in few places between the spherulites, where crystallization was not accomplished. In L121, 6S120 and 4S123 sheared at 150 °C at 20/s crystallization during cooling at 30 °C/min was even less advanced, as shown in Figure 10d–f. The cylindrical structures and spherulites between them are well distinguishable, the latter more numerous than in PLLAs with higher M_w. The effect of slower cooling on PLLAs sheared at 150 °C at 20/s is illustrated in Figure 11. The morphology of L240 and 6S245 cooled at 10 °C/min was similar to that observed after faster cooing, as exemplified in Figure 11a showing L240. In PLLAs with M_w close to 120 kg/mol amorphous areas were not visible any longer because of high χ_c reached in these polymers during slower cooling. The specimens were completely filled with cylindrical structures and spherulites between them, as seen in Figure 11b showing L121.

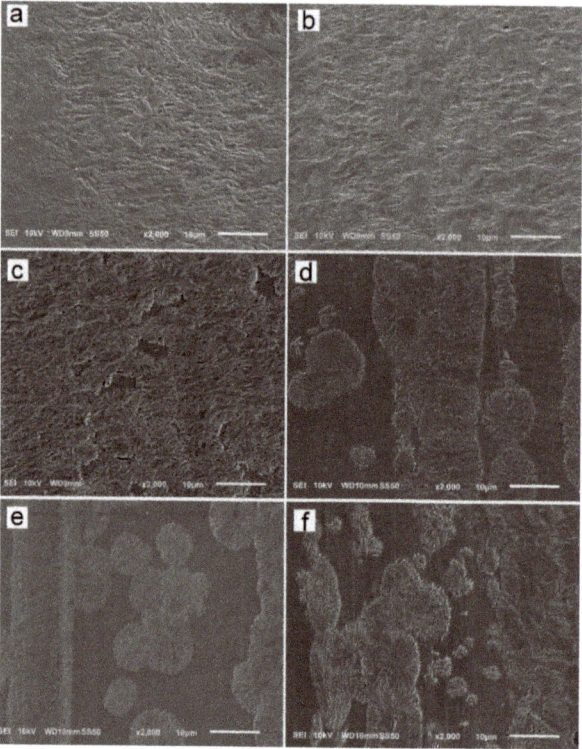

Figure 10. SEM micrographs of etched cross-section surfaces of PLLAs: L339 sheared at 170 °C at 5/s for 20 s, and next cooled at 10 °C/min (**a**), and L240 (**b**) 6S245 (**c**), L121 (**d**), 4S123 (**e**), 6S120 (**f**) sheared at 150 °C at 20/s for 5 s, and next cooled at 30 °C/min. Shearing direction–vertical.

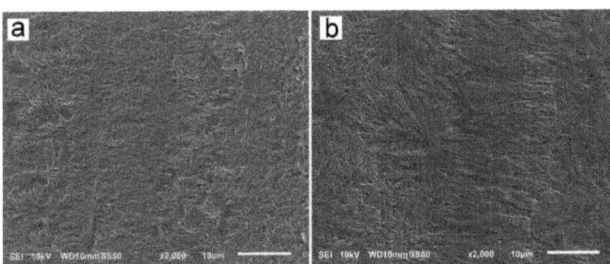

Figure 11. SEM micrographs of etched cross-section surfaces of L240 (**a**) and L121(**b**) sheared at 150 °C at 20/s for 5 s, and next cooled at 10 °C/min. Shearing direction—vertical.

2D-WAXS results evidenced orientation of the orthorhombic crystals of PLLAs with (200)/(110) and (210) crystallographic planes parallel to the shearing direction; hence, the orientation of polymer chain axes in the flow direction. It was accompanied by strong polar reflections in the corresponding 2D-SAXS patterns reflecting the orientation of lamellae stacks perpendicular to the flow direction. The orientation of crystals was stronger in PLLAs with M_w above 200 kg/mol than in those with M_w close to 120 kg/mol. However, among the latter, 4S123 exhibited the strongest orientation, whereas that of L121 was the weakest.

SEM analysis demonstrated the presence of lamellae stacks perpendicular to the shearing direction forming cylindrical structures, nucleated on fibrillar nuclei. These structures were accompanied by spherulites, especially in PLLAs with M_w close to 120 kg/mol. Obviously, the shear-induced fibrillar nucleation, although occurred, was less intense in these polymers than in PLLAs with higher M_w, which corroborated the conclusions drawn based on the results of X-ray scattering experiments.

In general, the strongest orientation of crystals was observed for the specimens, which crystallized at the highest temperatures and reached the highest χ_c. The crystal orientation resulted from the crystal growth on the shear-induced fibrillar nuclei; the more intense the nucleation, the higher T_c and χ_c. On the contrary, the weak or absent crystal orientation indicated the predominant point-like nucleation.

4. Conclusions

Three star PLLAs, 6-arm with M_w of 120 and 245 kg/mol, 4-arm with M_w of 123 kg/mol, and three linear PLLAs with M_w of 121, 240 and 339 kg/mol, were synthesized and their shear-induced crystallization was examined. The polymers were sheared at 170 °C and 150 °C at 5/s, 10/s and 20/s for 20 s, 10 s, and 5 s, respectively, and next cooled at 10 or 30 °C/min. The shear flow induced crystallization of the PLLAs during cooling at 30 °C/min and enhanced the crystallization at 10 °C/min, which was reflected in an increase of crystallization peak temperature, T_c, and crystallinity, χ_c. The flow-induced orientation of crystals was evidenced by 2D-WAXS and 2D-SAXS, although dependent on shearing conditions and molecular characteristics of the polymers. The lamellae stack perpendicular to the flow direction, suggestive of fibrillar nucleation, were observed by SEM, although spherulites were found between them, especially in PLLAs with M_w close to 120 kg/mol.

The results showed the crucial role of shearing temperature, T_s, and cooling rate, v, as the lower T_s increases the relaxation times of macromolecules and loweres the energy barrier for nucleation, whereas the slower cooling enabled a longer time for crystallization before too low temperature was reached. The shear-induced crystallization was also strongly influenced by molar mass of PLLA, as can be expected, but it was also affected by macromolecular architecture. It was well reflected in χ_c of PLLAs with M_w close to 120 kg/mol sheared at 150 °C and cooled at 30 °C/min. The effect of shear was the weakest on L121, stronger on 6S120, and the strongest on 4S123, despite the higher M_z of L121.

During cooling at 10 °C/min, T_c and ΔH_{mc} values of all PLLAs with M_w near 120 kg/mol were similar. However, this evidenced the stronger shear-induced nucleation in both star PLLAs than in L121, because the crystal growth in them was slower than in L121 [24]. The effect of molecular architecture on shear-induced crystallization during cooling at 10 °C/min was clearly seen in an increase of T_c and ΔH_{mc} in respect to the corresponding values for the control specimens, with similar thermal history but not subjected to shearing. The stronger effect of shear on 6S120 and 4S123 as compared to that on L121 undoubtedly resulted from the star architecture of macromolecules, which hindered the relaxation of the stretched macromolecular chain network. On the contrary, the shear-induced crystallization in 6S245 and L240 was similar, and even somewhat stronger in the latter. This can be understood taking into account that in the flow-induced crystallization of polymers, a high molar mass tail of molar mass distribution plays a crucial role, due to long relaxation times. M_z of L240, 414 kg/mol, exceeded that of 6S245, 294 kg/mol, evidencing the higher content of larger macromolecules, which at M_w of 240–245 kg/mol compensated the effect of star architecture on the shear-induced crystallization. Moreover, due to its higher molar mass, the number of branching points in 6S245 was smaller than in 6S120, which reduced their effect on macromolecular mobility.

Supplementary Materials: The following are available online. Figure S1: Increase of crystallization peak temperature, $T_c - T_c^q$, of PLLAs during cooling at 10 °C/min, caused by shearing at 170 and 150 °C, versus shear rate, $\dot{\gamma}$. T_c^q denotes the crystallization peak temperature of control specimens during cooling at 10 °C/min. Figure S2, 2D-WAXS pattern of PLLA L339 sheared at 170 °C at 5/s for 20 s and next cooled at 10 °C/min, with arrows indicating characteristic reflections.

Author Contributions: Conceptualization, J.B. and E.P.; methodology, J.B., A.M.; validation, J.B., E.P. and G.L.; formal analysis, J.B.; investigation, J.B.; writing—original draft preparation, J.B., E.P.; writing—review and editing, J.B., E.P.; visualization, J.B.; supervision, E.P.; funding acquisition, G.L. All authors have read and agreed to the published version of the manuscript.

Funding: This research was supported by the National Science Centre (Narodowe Centrum Nauki), Poland, grant No. 2013/09/B/ST5/03619, and statutory funds of CMMS PAS.

Institutional Review Board Statement: Not applicable.

Informed Consent Statement: Not applicable.

Data Availability Statement: The data presented in this study are available on request from the corresponding author.

Acknowledgments: The work was supported by the National Science Centre (Narodowe Centrum Nauki), Poland, grant No. 2013/09/B/ST5/03619 and statutory funds of CMMS PAS.

Conflicts of Interest: The authors declare no conflict of interest.

Sample Availability: Samples of the compounds are not available from the authors.

References

1. Piorkowska, E. Overview of biobased polymers. *Adv. Polym. Sci.* **2019**, *283*, 1–35. [CrossRef]
2. Masutani, K.; Kimura, Y. Present situation and future perspectives of poly(lactic acid). *Adv. Polym. Sci.* **2017**, *279*, 1–25. [CrossRef]
3. Babu, R.P.; O'Connor, K.; Seeram, R. Current progress on bio-based polymers and their future trends. *Prog. Biomater.* **2013**, *2*, 8. [CrossRef]
4. Castro-Aguirre, E.; Inigues-Franco, F.; Samsudin, H.; Fang, X.; Auras, R. Poly(lactic acid)-mass production, processing, industrial applications, and end of life. *Adv. Drug. Deliv. Rev.* **2016**, *107*, 333–366. [CrossRef]
5. Kost, B.; Svyntkivska, M.; Brzeziński, M.; Makowski, T.; Piorkowska, E.; Rajkowska, K.; Kunicka-Styczyńska, A.; Biela, T. PLA/β-CD-based fibres loaded with quercetin as potential antibacterial dressing materials. *Colloids Surf. B Biointerfaces* **2020**, *190*, 110949. [CrossRef]
6. Saeidlou, S.; Huneault, M.A.; Li, H.; Park, C.B. Poly(lactic acid) crystallization. *Prog. Polym. Sci.* **2012**, *37*, 1657–1677. [CrossRef]
7. Muller, A.J.; Avila, M.; Saenz, G.; Salazar, J. Crystallization of PLA-based materials. In *Poly(Lactic Acid) Science and Technology: Processing, Properties, Additives and Applications*; Jimenez, A., Peltzer, M., Ruseckaite, R., Eds.; Royal Society of Chemistry: Cambridge, UK, 2015; pp. 66–98. [CrossRef]

8. Zhang, J.; Tashiro, K.; Tsuji, H.; Domb, A.J. Disorder-to-order phase transition and multiple melting behavior of poly(L-lactide) Investigated by simultaneous measurements of WAXD and DSC. *Macromolecules* **2008**, *41*, 1352–1357. [CrossRef]
9. Kalish, J.P.; Aou, K.; Yang, X.; Hsu, S.H. Spectroscopic and thermal analyses of α′ and α crystalline forms of poly(L-lactic acid). *Polymer* **2011**, *52*, 814–821. [CrossRef]
10. Pan, P.; Yang, J.; Shan, G.; Bao, Y.; Weng, Z.; Cao, A.; Yazawa, K.; Inoue, Y. Temperature-variable FTIR and solid-state ^{13}C NMR investigations on crystalline structure and molecular dynamics of polymorphic poly(L-lactide) and poly(L-lactide)/poly(D-lactide) stereocomplex. *Macromolecules* **2012**, *45*, 189–197. [CrossRef]
11. Michalski, A.; Brzezinski, M.; Lapienis, G.; Biela, T. Star-shaped and branched polylactides: Synthesis, characterization, and properties. *Prog. Polym. Sci.* **2019**, *89*, 159–212. [CrossRef]
12. Cameron, D.J.A.; Shaver, M.P. Aliphatic polyester polymer stars: Synthesis, properties and applications in biomedicine and nanotechnology. *Chem. Soc. Rev.* **2011**, *40*, 1761–1776. [CrossRef] [PubMed]
13. Srisa-ard, M.; Baimark, Y. Effect of arm number and arm length on thermal properties of linear and star-shaped poly(D,L-lactides)s. *J. Appl. Sci.* **2010**, *10*, 1937–1943. [CrossRef]
14. Kim, S.H.; Han, Y.K.; Kim, Y.H.; Hong, S.I. Multifunctional initiation of lactide polymerization by stannous octoate/pentaerythritol. *Makromol. Chem.* **1992**, *193*, 1323–1631. [CrossRef]
15. Zhang, W.; Zheng, S. Synthesis and characterization of dendritic star poly(L-lactide)s. *Polym. Bull.* **2007**, *58*, 767–775. [CrossRef]
16. Zhang, C.X.; Wang, B.; Chen, Y.; Cheng, F.; Jiang, S.C. Amphiphilic multiarm star polylactide with hyperbranched polyethylenimine as core: A systematic reinvestigation. *Polymer* **2012**, *53*, 3900–3909. [CrossRef]
17. Biela, T.; Duda, A.; Penczek, S.; Rode, K.; Pasch, H. Well-defined star polylactides and their behavior in two-dimensional chromatography. *J. Polym. Sci. Part A Polym. Chem.* **2002**, *40*, 2884–2887. [CrossRef]
18. Kowalski, A.; Duda, A.; Penczek, S. Kinetics and mechanism of cyclic esters polymerization initiated with tin(II) octoate. 3. Polymerization of L,L-dilactide. *Macromolecules* **2000**, *33*, 7359–7370. [CrossRef]
19. Biela, T.; Kowalski, A.; Libiszowski, J.; Duda, A.; Penczek, S. Progress in polymerization of cyclic esters: Mechanisms and synthetic applications. *Macromol. Symp.* **2006**, *240*, 47–55. [CrossRef]
20. Tsuji, H. Quiescent crystallization of poly(lactic acid) and its copolymers-based materials. *Adv. Polym. Sci.* **2019**, *283*, 37–86. [CrossRef]
21. Tsuji, H.; Miyase, T.; Tezuka, Y.; Saha, S.K. Physical properties, crystallization, and spherulite growth of linear and 3-arm poly(L-lactide)s. *Biomacromolecules* **2005**, *6*, 244–254. [CrossRef]
22. Hao, Q.; Li, F.; Li, Q.; Li, Y.; Jia, L.; Yang, J.; Fang, Q.; Cao, A. Preparation and crystallization kinetics of new structurally well-defined star-shaped biodegradable poly(L-lactide)s initiated with diverse natural sugar alcohols. *Biomacromolecules* **2005**, *6*, 2236–2247. [CrossRef]
23. Wang, L.; Dong, C.M. Synthesis, crystallization kinetics, and spherulitic growth of linear and star-shaped poly(L-lactide)s with different numbers of arms. *J. Polym. Sci. Part A Polym. Chem.* **2006**, *44*, 2226–2236. [CrossRef]
24. Bojda, J.; Piorkowska, E.; Lapienis, G.; Michalski, A. Crystallization of star-shaped and linear poly(L-lactide)s. *Eur. Polym. J.* **2018**, *105*, 126–134. [CrossRef]
25. Lamberti, G. Flow induced crystallisation of polymers. *Chem. Soc. Rev.* **2014**, *43*, 2240–2252. [CrossRef]
26. Peters, G.W.M.; Balzano, L.; Steenbakkers, R.J.A. Flow-induced crystallization. In *Handbook of Polymer Crystallization*; Piorkowska, E., Rutledge, G.C., Eds.; John Wiley & Sons, Inc.: Hoboken, NJ, USA, 2013; pp. 399–432. [CrossRef]
27. Wang, Z.; Ma, Z.; Li, L. Flow-induced crystallization of polymers: Molecular and thermodynamic considerations. *Macromolecules* **2016**, *49*, 1505–1517. [CrossRef]
28. van Meerveld, J.; Peters, G.W.M.; Hutter, M. Towards a rheological classification of flow induced crystallization experiments of polymer melts. *Rheol. Acta* **2004**, *44*, 119–134. [CrossRef]
29. Janeschitz-Kriegl, H.; Ratajski, E.; Stadlbauer, M. Flow as an effective promotor of nucleation in polymer melts: A quantitative evaluation. *Rheol. Acta* **2003**, *42*, 355–364. [CrossRef]
30. Mykhaylyk, O.O.; Chambon, P.; Graham, R.S.; Fairclough, J.P.A.; Olmsted, P.D.; Ryan, A.J. The specific work of flow as a criterion for orientation in polymer crystallization. *Macromolecules* **2008**, *41*, 1901–1904. [CrossRef]
31. Hsiao, B.S. Role of chain entanglement network on formation of flow-induced crystallization precursor structure. In *Progress in Understanding of Polymer Crystallization*; Reiter, G., Strobl, G.R., Eds.; Springer: Berlin/Heidelberg, Germany, 2007; pp. 133–149. [CrossRef]
32. Somani, R.H.; Yang, L.; Zhu, L.; Hsiao, B.S. Shish-kebab precursor structures in entangled polymer melts. *Polymer* **2005**, *46*, 8587–8623. [CrossRef]
33. Rhoades, A.; Pantani, R. Poly(Lactic Acid): Flow-Induced Crystallization. *Adv. Polym. Sci.* **2019**, *283*, 87–117. [CrossRef]
34. Zhong, Y.; Fang, H.; Zhang, Y.; Wang, Z.; Yang, J.; Wang, Z. Rheologically determined critical shear rates for shear-induced nucleation rate enhancements of poly(lactic acid). *ACS Sustain. Chem. Eng.* **2013**, *1*, 663–672. [CrossRef]
35. Xu, H.; Xie, L.; Hakkarainen, M. Beyond a model of polymer processing-triggered shear: Reconciling shish-kebab formation and control of chain degradation in sheared poly(L-lactic acid). *ACS Sustain. Chem. Eng.* **2015**, *3*, 1443–1452. [CrossRef]
36. Bojda, J.; Piorkowska, E. Shear-induced nonisothermal crystallization of two grades of PLA. *Polym. Test.* **2016**, *50*, 172–181. [CrossRef]

37. Li, X.J.; Li, Z.M.; Zhong, G.J.; Li, L.B. Steady- shear- induced isothermal crystallization of poly(L-lactide) (PLLA). *J. Macrom. Sci. Part B Polym. Phys.* **2008**, *47*, 511–522. [CrossRef]
38. Kim, E.S.; Kim, B.C.; Kim, S.H. Structural effect of linear and star-shaped poly(L-lactic acid) on physical properties. *J. Polym. Sci. Part B Polym. Phys.* **2004**, *42*, 939–946. [CrossRef]
39. Fang, H.; Zhang, Y.; Bai, J.; Wang, Z. Shear-induced nucleation and morphological evolution for bimodal long chain branched polylactide. *Macromolecules* **2013**, *46*, 6555–6465. [CrossRef]
40. Wang, J.; Bai, J.; Zhang, Y.; Fang, H.; Wang, Z. Shear-induced enhancements of crystallization kinetics and morphological transformation for long chain branched polylactides with different branching degrees. *Sci. Rep.* **2016**, *6*, 26560. [CrossRef] [PubMed]
41. Michalski, A.; Łapienis, G. Synthesis and characterization of high-molar-mass star-shaped poly(L-lactide)s. *Polimery* **2018**, *63*, 488–494. [CrossRef]
42. Biela, T.; Duda, A.; Penczek, S. Enhanced melt stability of star-shaped stereocomplexes as compared with linear stereocomplexes. *Macromolecules* **2006**, *39*, 3710–3713. [CrossRef]
43. He, Y.; Wu, T.; Wie, J.; Fan, Z.; Li, S. Morphological investigation on melt crystallized polylactide homo- and stereocomplex by enzymatic degradation with proteinase K. *J. Polym. Sci. Part B Polym. Phys.* **2008**, *46*, 959–970. [CrossRef]
44. Huang, S.; Li, H.; Jiang, S.; Chen, X.; An, L. Crystal structure and morphology influenced by shear effect of poly(L-lactide) and its melting behaviour revealed by WAXS, DSC and in-situ POM. *Polymer* **2011**, *52*, 3478–3487. [CrossRef]
45. Righetti, M.C.; Gazzano, M.; Di Lorenzo, M.L.; Androsch, R. Enthalpy of melting of α'- and α- crystals of poly(L-lactic acid). *Eur. Polym. J.* **2015**, *70*, 215–220. [CrossRef]
46. Sarasua, J.R.; Prud'Homme, R.E.; Wisniewski, M.; Le Borgne, A.; Spassky, N. Crystallization and melting behaviour of polylactides. *Macromolecules* **1998**, *31*, 3895–3905. [CrossRef]
47. Garlotta, D. A literature review of poly(lactic acid). *J. Polym. Environ.* **2001**, *9*, 63–84. [CrossRef]
48. Piorkowska, E.; Galeski, A. Overall crystallization kinetics. In *Handbook of Polymer Crystallization*; Piorkowska, E., Rutledge, G.C., Eds.; John Wiley & Sons, Inc.: Hoboken, NJ, USA, 2013; pp. 215–236. [CrossRef]

Article

Modification of Polylactide Nonwovens with Carbon Nanotubes and Ladder Poly(silsesquioxane)

Mariia Svyntkivska, Tomasz Makowski *, Ewa Piorkowska, Marek Brzezinski, Agata Herc and Anna Kowalewska

Centre of Molecular and Macromolecular Studies Polish Academy of Sciences, Sienkiewicza 112, 90-363 Lodz, Poland; mariiasv@cbmm.lodz.pl (M.S.); epiorkow@cbmm.lodz.pl (E.P.); mbrzezin@cbmm.lodz.pl (M.B.); asherc@cbmm.lodz.pl (A.H.); anko@cbmm.lodz.pl (A.K.)
* Correspondence: tomekmak@cbmm.lodz.pl

Abstract: Electrospun nonwovens of poly(L-lactide) (PLLA) modified with multiwall carbon nanotubes (MWCNT) and linear ladder-like poly(silsesquioxane) with methoxycarbonyl side groups (LPSQ-COOMe) were obtained. MWCNT and LPSQ-COOMe were added to the polymer solution before the electrospinning. In addition, nonwovens of PLLA grafted to modified MWCNT were electrospun. All modified nonwovens exhibited higher tensile strength than the neat PLA nonwoven. The addition of 10 wt.% of LPSQ-COOMe and 0.1 wt.% of MWCNT to PLLA increased the tensile strength of the nonwovens 2.4 times, improving also the elongation at the maximum stress.

Keywords: polylactide nonwovens; electrospinning; reinforced fibers; MWCNT; linear ladder-like poly(silsesquioxane)

Citation: Svyntkivska, M.; Makowski, T.; Piorkowska, E.; Brzezinski, M.; Herc, A.; Kowalewska, A. Modification of Polylactide Nonwovens with Carbon Nanotubes and Ladder Poly(silsesquioxane). *Molecules* 2021, 26, 1353. https://doi.org/10.3390/molecules26051353

Academic Editors: Dimitrios Bikiaris, Marek Brzeziński and Małgorzata Baśko

Received: 1 February 2021
Accepted: 21 February 2021
Published: 3 March 2021

Publisher's Note: MDPI stays neutral with regard to jurisdictional claims in published maps and institutional affiliations.

Copyright: © 2021 by the authors. Licensee MDPI, Basel, Switzerland. This article is an open access article distributed under the terms and conditions of the Creative Commons Attribution (CC BY) license (https://creativecommons.org/licenses/by/4.0/).

1. Introduction

Electrospinning is a widely used technique to produce polymer fibers with different diameters and morphology. The nonwovens can be obtained via melt electrospinning or solution electrospinning [1]. The solution electrospinning does not require high temperature. Diameters and morphology of the electrospun fibers are determined by the type and molar mass of polymers, dispersity, and processing parameters, including applied voltage, tip-collector distance, type of solvent, feed rate, solution concentration, viscosity and conductivity. A wide range of electrospun polymer nonwovens has been obtained and studied for application in diverse fields: air [2] and water filtration [3], food packaging [4], tissue engineering [5] and reconstructive medicine [6], drug delivery systems [7], and antibacterial materials [8,9]. The drawback hindering the practical use of electrospun mats in some applications, for instance, filtration, is their poor tensile strength, as compared to woven fabrics. Fibers in nonwovens are loosely packed and weakly connected. The improvement of the mechanical performance of nonwovens can be achieved by modification of the fibers. Owing to their outstanding properties, including extremely high tensile strength and elastic modulus [10–12], carbon nanotubes (CNT) are attractive nanofillers for polymers [13]. The enhancement of properties of electrospun nonwovens by the introduction of CNT to electrospun polymer solutions was reported for different polymers, including poly(ethylene terephthalate), poly(vinylpyrrolidone), polyacrylonitrile/poly(vinyl chloride) blend, polyamide 6, and polylactide (PLA), among others [3,14–21]. Modification of electrospun nonwovens can be also achieved by incorporating other nanoparticles, for example, graphene oxide [14], cellulose nanocrystals [22], or hydroxyapatite nanoparticles [23].

PLA, a biodegradable polyester having good mechanical properties, is the most promising biobased polymer for replacement of conventional thermoplastics [24]. PLA's ability to crystallize depends on its enantiomeric composition; with the increasing content of units of different chirality, the ability to crystallize worsens. Slowly crystallizing PLAs can be cooled to the glassy state and crystallized during subsequent heating via "cold crystallization". The effective way to accelerate the crystallization of PLA is the addition of nucleating agents [25]. Additionally, shearing of PLA melt can enhance its crystallization [26,27]. To broaden its applications, PLA's properties are modified by various routes, including copolymerization, chain extension, plasticization, blending with other polymers, fillers, nanofillers, and fibers [28–37]. PLA modification with oligomeric linear ladder poly(silsesquioxane)s (LPSQ-R) with different sides substituents, including methoxycarbonyl (R = COOMe) groups was described recently by Herc et al. [38]. LPSQ-R are well-defined macromolecules with a double chain siloxane backbone, which makes them more rigid than typical polysiloxanes and limits their coiling in solutions. Interestingly, a significant increase of elongation at break was achieved for PLA blends with 5 wt.% of LPSQ-COOMe, with only a minor decrease of the yield strength, as compared to neat PLA. LPSQ-COOMe described in this report is a viscous amorphous substance of good thermal stability, well soluble in organic solvents. Herc et al. [39] recently demonstrated that LPSQ-COOMe enhanced the thermal stability of PLA.

Electrospinning allowed to obtain PLA fibers with diameters in the nanometer [40,41] and micrometer range [42–44]. To obtain modified PLA nonwovens, electrospinning of the polymer solutions with added plasticizers, nanofillers, or antibacterial substances, among others, was carried out, although post-electrospinning modification of fibers was also reported. For example, antibacterial activity of PLA nonwovens was achieved by introducing antibacterial agents to the solutions [8,9,42]. Plasticization of PLA/poly(hydroxybutyrate) blend nonwovens with oligomeric lactic acid (OLA) and acetyl(tributyl citrate) was described by Arriata et al. [45,46]. Leones et al. [47] used OLA to plasticize PLA nonwovens. In turn, the improvement of tensile strength of electrospun PLA mats was achieved by addition of nanofillers; for example, by cellulose nanocrystals [48], hydroxyapatite nanorods and graphene oxide [49]. PLA composite nonwovens with 5 wt.% of siliceous sponge spicules exhibited a tensile strength four times higher than that of neat PLA material [50].

The addition of nanocarbon materials to electrospun PLA solutions can result in an improvement of the tensile strength and elastic modulus [14,47,48,51]. A fourfold increase of the tensile strength and elastic modulus was reported for PLA nonwovens with 3 wt.% of multiwall CNT (MWCNT) relative to those of neat PLA nonwovens, although accompanied by a significant decrease of elongation at break [48]. Similar enhancement of the tensile strength of PLA nonwovens was achieved through the addition of 1 wt.% of functionalized MWCNT [51].

In this study, LPSQ-COOMe and MWCNT were used for modification of poly(L-lactide) (PLLA) electrospun nonwovens. PLLA grafted on MWCNT was prepared and used for electrospinning; MWCNT content was 0.1 wt.%. For comparison, nonwovens of commercial PLLA with the same content of MWCNT were also obtained. In addition, fibers of PLLA with 5 and 10 wt.% of LPSQ-COOMe were prepared, and also fibers with the same contents of LPSQ-COOMe and 0.1 wt.% of MWCNT with respect to PLLA. LPSQ-COOMe is not biodegradable, but like other silsesquioxanes [52] it is non-toxic and biocompatible. The influence of the additives on the morphology and diameters of the nonwovens as well as on their thermal and mechanical properties was studied. It was demonstrated that the modification with a small amount of MWCNT and 5–10 wt.% of LPSQ-COOMe significantly improved the tensile strength of the nonwovens. It is also worth noting that LPSQ-COOMe also improved thermal stability of the studied nonwovens.

2. Materials and Methods

2.1. Materials

Commercial PLLA Luminy® L175 was obtained from Total Corbion (The Netherlands) with D-lactide content below 1%, according to the supplier. Weight average molar mass M_w of 78 kg/mol and dispersity M_w/M_n of 1.8 were measured by size exclusion chromatography (SEC) with a multi-angle laser light scattering detector in dichloromethane using an Agilent Pump 1100 Series (preceded by an Agilent G1379A Degasser), equipped with a set of two PLGel 5 μ MIXED-C columns. A Wyatt Optilab Rex differential refractometer and a Dawn Eos (Wyatt Technology Corporation) laser photometer were used as detectors. Dichloromethane was used as eluent at a flow rate of 0.8 mL min^{-1} at room temperature (RT).

MWCNT Nanocyl NC-7000 (Nanocyl, Belgium) with an average diameter of 9.5 nm, length of 1.5 µm, and 90% purity, were used. PLLA covalently attached to MWCNT (PL-g-CNT) was synthesized by ring-opening polymerization of L,L-lactide according to a well-established procedure [53,54]. LPSQ-COOMe was synthesized as previously described [38,39]. Procedures for the synthesis of PL-g-CNT and LPSQ-COOMe are described below.

Dichloromethane purchased from StanLab (Poland) (purity 99.5%) and acetone purchased from POCH (Poland) (purity 99%) were used as received.

2.2. Synthesis of PL-g-CNT

Ring-opening polymerization of L,L-lactide was carried out as follows: $Sn(Oct)_2$ (1 mL of 0.25 mol L^{-1} solution in dry tetrahydrofuran (THF)) and L,L-Lactide (L,L-LA, 7.42 g, 51.5 mmol) were transferred under vacuum into break-seals and then sealed after being frozen in liquid N_2. MWCNT–OH (7.4 mg) was placed directly into the reaction ampule, dried under vacuum for 4 h, and sealed. Break-seals that contained the $Sn(Oct)_2$/THF solution and L,L-LA monomer and a tube that contained dry MWCNTs in a vial were sealed to the glass reaction vessel (~10 mL). Under vacuum, the break-seals and vial were broken, and all components were mixed at RT. THF was removed by distillation, and then the reaction vessel was sealed. The reaction vessel was placed into an ultrasonic bath for 60 min (130 °C) to disperse MWCNT–OH, then placed into a thermostated oil bath, and the polymerization was carried out at 130 °C for approx. 24 h. The resulting polymer was dissolved in $CHCl_3$ and precipitated into methanol, separated by filtration, and washed several times with methanol. The mass of the vacuum-dried product was 6.31 g (85% yield). The content of MWCNTs in the product was 0.1 wt.%. The molecular characteristics of the product were determined according to the procedure described previously [53]. It was found that PL-g-CNT was a mixture composed of PLLA grafted on MWCNT and free PLLA. Only approximately 15–20 wt.% of the PLLA chains were found to be polymerized from the MWCNT. M_w and M_w/M_n of the free PLLA fraction were 82 kg/mol and 1.4, respectively, as determined by SEC using the equipment described above. The molar masses of both PLLA fractions should be similar [55]. ^{1}H NMR of PL-g-CNT ($CDCl_3$): δ = 5.16 (q, 1H, CH-CH$_3$ polymer), 1.67 (d, 3H, CH$_3$ polymer) ppm.

2.3. Synthesis of LPSQ-COOMe

LPSQ-COOMe was synthesized as previously described [38,39]. Methyl thioglycolate (Acros Organics, 95%) was added into a solution of ladder-like poly(vinylsilsesquioxanes), M_n = 1000 g/mol, M_w/M_n = 1.4, obtained as described in [56], dissolved in dry THF (c = 8 wt.%) and placed in a quartz crucible. The ratio of reagents was $[HS]_0/[Vi]_0$ = 1.4). A photoinitiator (2,2-dimethoxy-2-phenylacetophenone, Acros Organics, 99%) was then added with stirring to the solution of reagents at $[Vi]_0/[DMPA]_0$ = 50. The mixture was irradiated for 15 min with UV light (λ = 365 nm). Volatiles were removed under reduced pressure and the residue was purified by precipitation into a large amount of hexane/ethyl acetate (1: 1 v/v). The product was dried under high vacuum at RT to the constant weight (reaction yield 91%) and characterized with ^{1}H NMR spectroscopy using a Bruker DRX-500 MHz spectrometer: δ [ppm] (THF-d8): 0.16 (s. OSiMe$_3$), 1.06 (m. SiCH$_2$), 2.77 (m. CH$_2$S), 3.25 (m. SCH$_2$), 3.65 (s. OCH$_3$). The ^{1}H NMR spectrum is shown in Figure S1 of Supporting Information (SI). Full NMR characteristics have been previously published in [38,39].

The obtained LPSQ-COOMe was essentially an amorphous substance with T_g at −41.2 °C as determined by DSC during heating at 10 °C/min from −100 °C.

2.4. Preparation of Nonwovens

Before further use, commercial PLLA pellets were dried under reduced pressure for 4 h at 100 °C. After drying, PLLA was dissolved in dichloromethane to obtain 12 wt.% solution. LPSQ-COOMe was added to PLLA solution to obtain 5:100 and 10:100 weight ratios of the modifier to PLLA (PL-LPSQ5, PL-LPSQ10). Furthermore, MWCNT were added to the solution containing neat PLLA and PLLA with LPSQ-COOMe to obtain 1:1000 weight ratio of MWCNT to PLLA (PL-CNT, PL-LPSQ5-CNT, PL-LPSQ10-CNT). All the solutions and dispersions were mixed with acetone in a volume ratio of 7:3. The dispersions were homogenized using an ultrasonic homogenizer, the Hielscher UP 200S (Hielscher, 130 Germany, power 200 W, amplitude 30%, frequency 24 kHz), at RT for 20 min.

PL-g-CNT was dissolved in dichloromethane to obtain a 6 wt.% solution. Then the solution was mixed with acetone in a volume ratio of 7:3 and homogenized, as described above.

PLLA-based nonwovens were electrospun at RT and relative humidity of 35–40% using a system consisting of a high voltage power supply CM5 (Simco-Ion, The Netherlands) and an aluminum plate 8 × 8 cm as a collector. Glass syringes and a step motor, model T-LLS105 from Zaber (Canada), were used for dosing of solutions. The tip-collector distance was 25 cm in all cases. The parameters of the electrospinning are listed in Table 1. The polymer concentrations in the solvent and other parameters of the electrospinning were selected based on preliminary experiments. The preparation scheme of the nonwovens is shown in Figure S2 in SI.

Table 1. Parameters of electrospinning.

Sample Code	Flow Rate (mL/min)	Voltage (kV)
PLLA	0.065	20
PL-LPSQ5	0.065	20
PL-LPSQ10	0.065	20
PL-g-CNT	0.065	18
PL-CNT	0.052	18
PL-LPSQ5-CNT	0.052	18
PL-LPSQ10-CNT	0.052	18

The obtained nonwovens were dried for 4 h at 35 °C under reduced pressure and then stored in a desiccator at RT. All measurements were carried out at least two days after the electrospinning.

2.5. Characterization

The thicknesses of all nonwovens were measured and their surface densities were determined by weighing. The surfaces of the nonwovens were examined using scanning electron microscopy (SEM) with energy dispersive spectroscopy (EDS) JEOL 6010LA (JEOL, Japan) at an accelerating voltage of 10 kV. Before the examination, the surfaces were vacuum sputtered with a 10 nm gold layer using a coater (Quorum EMS150R ES, UK). Diameter distributions of the fibers were determined from SEM micrographs.

Water contact angles (WCA) of 5 μl distilled water droplets with the nonwovens were determined at RT, using a goniometer 100-00-230 NRL Rame Hart (USA) with the Image Drop Analysis program. In each case, the WCA measurements were carried out five times and average values were calculated.

Thermal properties of the nonwovens were studied using differential scanning calorimetry (DSC 3 Mettler Toledo, Switzerland) during heating at 10 °C/min. Thermogravimetric analysis was performed using the TGA 550 from TA instruments (USA) in a nitrogen atmosphere at a heating rate of 20 °C/min.

Tensile drawing of nonwoven specimens was performed with the Linkam TST 350 Minitester (UK) at 25 °C. Three 10 mm wide strips of each material were drawn, with a distance between grips of 20 mm, at a rate of 2 mm/min (10%/min). The tests were carried out at least 48 h after preparation of the nonwovens.

Selected PLLA-based nonwovens were analyzed with Fourier-transform infrared spectroscopy (FTIR) using a Thermo Scientific Nicolet 6700 FT-IR instrument with an ATR (Attenuated Total Reflectance) (USA). FTIR spectra in the range of 500–4000 cm^{-1} were recorded with a 1 cm^{-1} resolution.

3. Results and Discussion

The structure of the electrospun PLLA-based nonwovens, fiber morphology, and the fiber size distributions are shown in Figures 1 and 2, whereas parameters of the nonwovens are collected in Table 2. All nonwovens were composed of uniform and defect-free fibers, randomly distributed. It is worth noting that TGA thermograms, shown in Figure S3 in SI, evidence that the solvents evaporated during the electrospinning and post-electrospinning drying; weight loss at 100 °C was below 0.1% in each case. Moreover, the TGA thermograms in Figure S3 in SI and data collected in Table S1 in SI show that all the materials with LPSQ-COOMe exhibited better thermal stability in comparison to neat PLLA, as already reported [39].

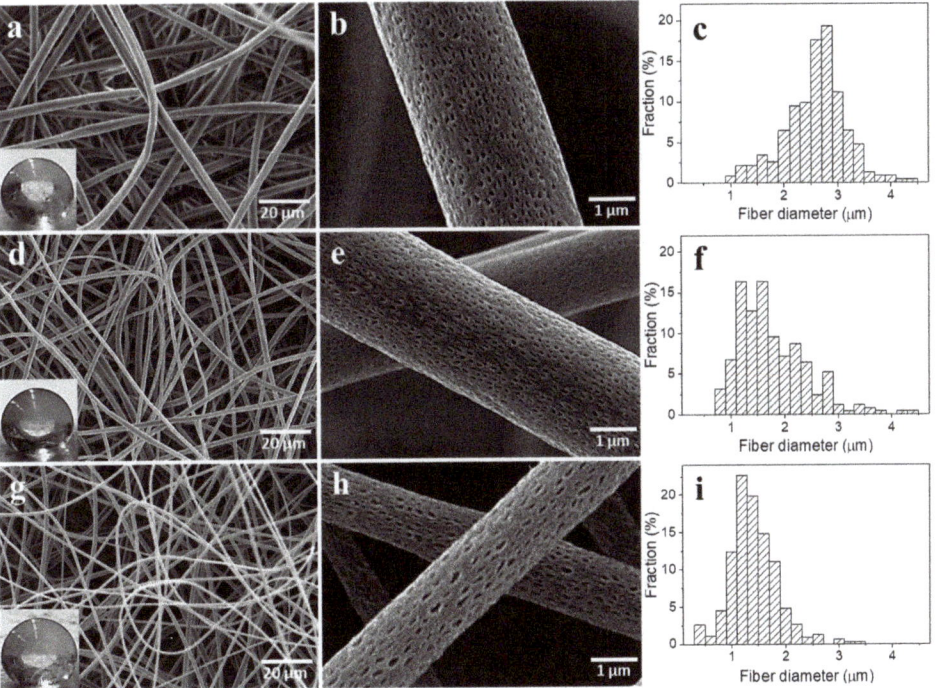

Figure 1. SEM micrographs of nonwovens: PLLA (**a,b**), PL-CNT (**d,e**), PL-g-CNT (**g,h**) and fiber diameter distributions; PLLA (**c**), PL-CNT (**f**), PL-g-CNT (**i**). Water drops on nonwoven surfaces are shown in the insets.

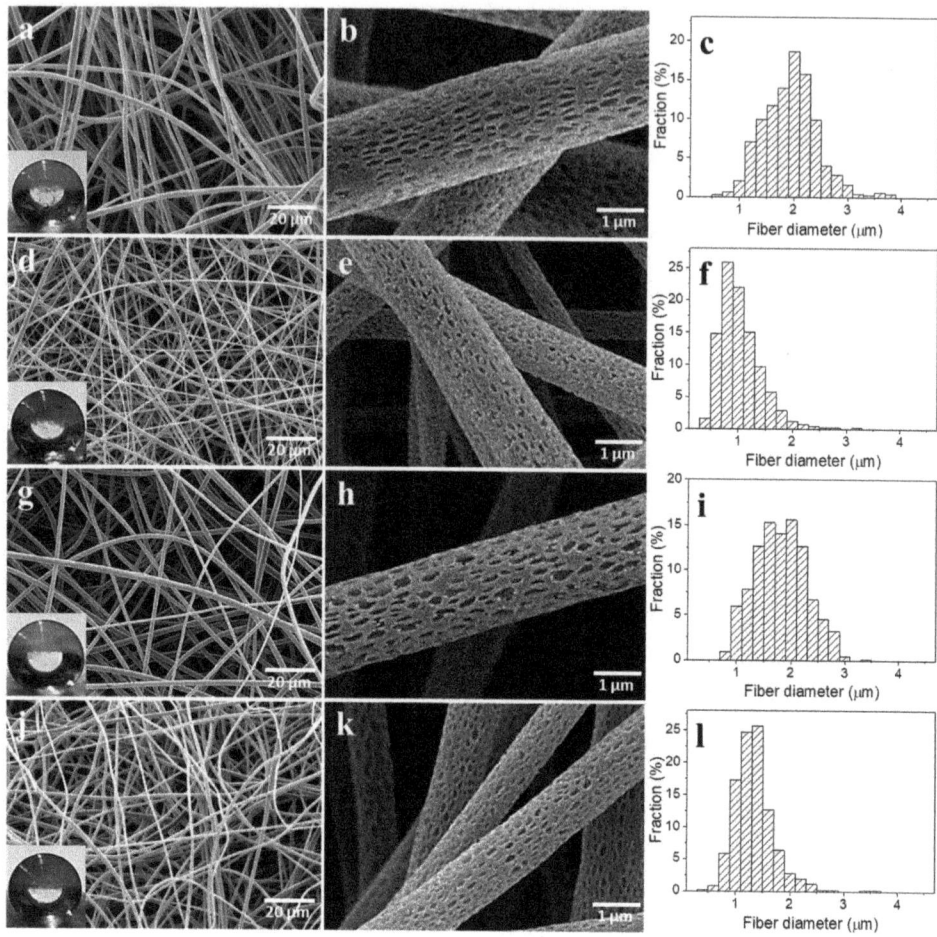

Figure 2. SEM micrographs of nonwovens: PL-LPSQ5 (**a**,**b**), PL-LPSQ5-CNT (**d**,**e**), PL-LPSQ10 (**g**,**h**), PL-LPSQ10-CNT (**j**,**k**) and fiber diameter distributions: PL-LPSQ5 (**c**), PL-LPSQ5-CNT (**f**), PL-LPSQ10 (**i**), PL-LPSQ10-CNT (**l**). Water drops on nonwoven surfaces are shown in the insets.

Table 2. Characteristics of PLLA-based nonwovens: thickness, surface density, average fiber diameter, the most probable diameter (at the histogram maximum) and water contact angle (WCA).

Sample Code	Thickness (mm)	Surface Density (mg/cm^2)	Average Fiber Diameter (μm)	Most Probable Fiber Diameter (μm)	WCA (°)
PLLA	0.84	6.8	2.5	2.7–2.9	155° ± 3
PL-CNT	0.58	5.9	1.7	1.5–1.7	154° ± 4
PL-g-CNT	0.68	6.0	1.8	1.1–1.3	156° ± 2
PL-LPSQ5	0.74	7.1	1.8	1.9–2.1	156° ± 2
PL-LPSQ5-CNT	0.68	6.3	0.9	0.7–0.9	155° ± 4
PL-LPSQ10	0.75	8.1	1.7	1.9–2.1	156° ± 3
PL-LPSQ10-CNT	0.68	7.0	1.2	1.2–1.4	156° ± 3

Diameters of a majority of neat PLLA fibers were in the 1.5–3.5 µm range, with an average of 2.5 µm, whereas diameters of PL-CNT and PL-g-CNT were smaller, mostly below 2 µm, with an average of 1.7 and 1.8 µm, respectively. The decrease of the PL-CNT fiber diameters was caused by an increase in the electrical conductivity of the solution. As a result, the solution underwent a greater extension under the influence of the electrostatic field, which caused the reduction of fiber diameters [6,57]. The same applies to PL-g-CNT electrospinning process, although it is worth noting that PL-g-CNT concentration in the solution had to be decreased in comparison to that of PL-CNT because of increased viscosity due to grafting of polymer chains on MWCNT. The pores on the fiber surfaces resemble those observed in [44,58] due to the breath figure mechanism. As a result of the evaporative cooling of the solvent, water vapor condensed and droplets formed on the surface of the fiber jets, causing the formation of pores. Moreover, the pores were elongated in the directions parallel to the fiber axes, as a result of deformation during jet stretching.

The addition of 5 and 10 wt.% of LPSQ-COOMe to PLLA resulted in thinner fibers, mostly with diameters ranging from 1 to 3 µm, and average diameters of 1.8 and 1.7 µm, respectively, as shown in Figure 2 and Table 2. The decreased average fibers diameters of PL-LPSQ5 and PL-LPSQ10 as compared to those of PLLA could result from a decrease of solution viscosity resulting from the presence of LPSQ-COOMe. Fibers of PL-LPSQ5-CNT and PL-LPSQ10-CNT were even thinner, mostly with diameters ranging from 0.5 to 1.5 µm and from 1 to 2 µm, with average values of 0.9 and 1.2 µm, respectively. The nonwoven thickness ranged from 0.68 to 0.84 mm. The thickest was the nonwoven of neat PLLA, whereas the thinnest were the nonwovens with MWCNT, composed of the markedly thinner fibers.

Figures 1 and 2 show also water droplets placed on the nonwoven surfaces, whereas the WCA values are listed in Table 2. All the obtained electrospun PLLA-based nonwovens were hydrophobic, with WCA values close to 155°. It is known that hydrophobicity of nonwovens results from their surface roughness and air entrapped between fibers [59].

DSC heating thermograms of PLLA-based nonwovens are compared in Figure 3, whereas the calorimetric parameters are collected in Table 3. Each thermogram shows a glass transition, a cold crystallization exotherms, and a melting endotherm. In addition, on all thermograms, pre-melting exotherms, attributed to the transition from the disordered orthorhombic alpha' form to the ordered orthorhombic alpha form [60], are visible, with a maximum at 157–159 °C. The cold crystallization exotherms, especially those of the materials with LPSQ-COOMe, exhibit long high-temperature tails, overlapping the pre-melting recrystallization exotherms, hence the latter are less pronounced and their enthalpies listed in Table 3 are only approximate. However, in each case the sum of enthalpies of the exothermic effects was equal to the melting enthalpy, evidencing that the nonwovens were amorphous before heating in DSC. T_g of neat PLLA nonwoven was at 59 °C, whereas the cold crystallization peak (T_{cc}) was at 97 °C. T_gs of both PL-CNT and PL-g-CNT were nearly the same, at 58 °C. T_gs of the nonwovens of PLLA blends with 5 and 10 wt.% of LPSQ-COOMe were lower, at 55 and 52–53 °C, respectively, due to the plasticizing effect of the additive. However, taking into account the T_gs of PLLA, LPSQ-COOMe and PLLA blends with LPSQ-COOMe, it can be calculated, using the Fox equation, that in PLLA with 5 and 10 wt.% of LPSQ-COOMe about 3 and 5 wt.% of the additive, respectively, is dispersed on a molecular level. This is in accordance with ref. [38], where phase separation was observed in PLA blends with LPSQ-COOMe. Figure S4 in SI shows SEM EDS mapping of silicon in PL-LPSQ5 and PL-LPSQ10 that confirms the presence of LPSQ-COOMe and its dispersion in the PLLA matrix. Inclusions of the additive were not discernible, most probably being too small, as in ref. [38]. It is also observed that T_{cc} of composite materials decreased in comparison to that of neat PLLA nonwoven, to 92 °C for PL-CNT and even more, to 87 °C, for PL-g-CNT. T_{cc}s of PL-LPSQ5 and PL-LPSQ10 nonwovens were even lower, at 86 °C and 82 °C, respectively. The presence of CNT further enhanced the decrease of T_{cc}, to 81 °C for both PL-LPSQ5-CNT and PL-LPSQ10-CNT. In turn, the melting peak temperatures (T_m) of the materials were at 173–175 °C. The only exception was PL-g-CNT, whose T_m of

179 °C was higher than that of PL-CNT, indicating the melting of thicker crystals, despite its low T_{cc}. CNT are known to nucleate crystallization of PLA [61], whereas LPSQ-COOMe increases the mobility of PLA chains and also enhances its cold crystallization [38]. It is worth noting that PLLA in PL-g-CNT was optically pure. Moreover, the additives can enhance the orientation of PLLA in the fibers during electrospinning, which also can promote cold-crystallization of the polymer.

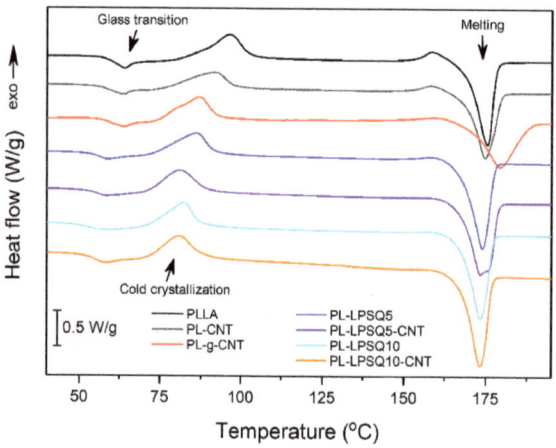

Figure 3. DSC heating thermograms of PLLA-based nonwovens.

Table 3. Calorimetric parameters of PLLA-based nonwovens: T_g–the glass transition temperature, T_{cc}, H_{cc}–the cold crystallization peak temperature and enthalpy, respectively, T_m and H_m–the melting peak and enthalpy, respectively, the enthalpy of the pre-melting recrystallization is given in brackets.

Sample Code	T_g (°C)	T_{cc} (°C)	H_{cc} (J/g)	T_m (°C)	H_m (J/g)
PLLA	59	97	34 (6)	175	41
PL-CNT	58	92	33 (4)	175	39
PL-g-CNT	58	87	36 (4)	179	41
PL-LPSQ5	55	86	44 (4)	174	48
PL-LPSQ5-CNT	55	81	49 (2)	173	51
PL-LPSQ10	52	82	48 (2)	173	50
PL-LPSQ10-CNT	53	81	50 (2)	173	52

Figure 4 shows the tensile behavior of the nonwovens, whereas the mechanical parameters are collected in Table 4. Fibers in the nonwovens were randomly distributed, without any preferred orientation. When the nonwovens were strained, the fibers tended to orient parallel to the stretching direction. Loosely connected structure without strong bonds between the fibers at their cross points facilitated the fiber alignment during drawing. However, some fibers broke at early stages of drawing, thus decreasing the tensile stress. In Figure 4 the engineering stress, calculated as a ratio of force to initial cross-section area of the nonwoven, is plotted vs. engineering strain. It is seen that with increasing strain the stress increases, passes through a maximum, and decreases, less or more sharply. The maximum stress value recorded during the drawing of neat PLLA nonwoven was 0.5 MPa. Similar values measured for electrospun neat PLA nonwovens were reported in [49,62]. The presence of 0.1 wt.% of CNT increased significantly the tensile strength to 0.8 and 0.95 MPa for PL-CNT and PL-g-CNT, respectively. The effect of 5 wt.% of LPSQ-COOMe on the PLLA nonwoven strength was weak, but PL-LPSQ10 exhibited a strength of 0.95 MPa.

However, the highest strength was achieved for PL-LPSQ5-CNT and PL-LPSQ10-CNT, 1.1 and 1.2 MPa, which exceeded more than two times the strength of neat PLLA nonwoven. It is worth noting that the maximum stress of PL-LPSQ10-CNT was achieved at an elongation of approx. 80%, significantly larger than in the case of the other materials.

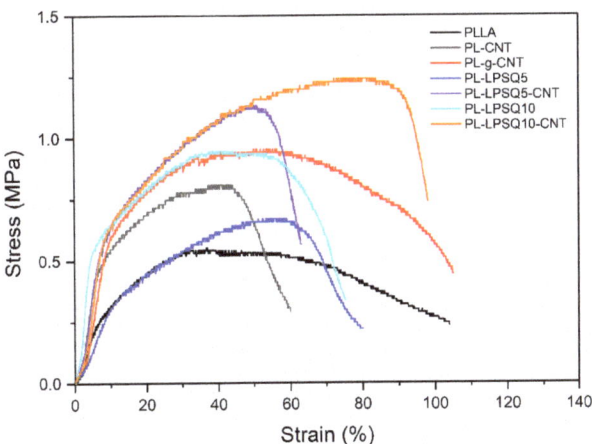

Figure 4. Stress-strain dependencies of PLLA-based nonwovens.

Table 4. Mechanical parameters of PLLA-based nonwovens.

Sample Code	Maximum Stress (MPa)	Strain at Maximum Stress (%)
PLLA	0.50	52
PL-CNT	0.81	40
PL-g-CNT	0.95	62
PL-LPSQ5	0.67	57
PL-LPSQ5-CNT	1.09	52
PL-LPSQ10	0.95	53
PL-LPSQ10-CNT	1.22	83

CNT-containing materials are known to exhibit outstanding mechanical properties, due to stress transfer from the weaker polymer to the stronger nanofiller, which allows a higher loading to be achieved, thus the improvement in strength [14]. However, in the nonwovens studied the weight ratio of MWCNT to PLLA was only 1:1000, therefore we attribute the increase in strength also to the enhancement of PLLA chain orientation during electrospinning. An increase in electrical conductivity of the solution due to the presence of CNT results in a greater extension under the influence of the electrostatic field, which can promote not only the fiber diameter reduction but also the orientation of polymer chains. It must be noted that CNT are highly anisotropic particles, which tend to be oriented during fiber jet stretching. It is also worth noting that the effect of LPSQ-COOMe on the tensile properties of the PLLA nonwovens was different than on PLA films, whereas at 5 wt.% content of the additive a large elongation at break, 230%, was achieved, accompanied by a minor, approx. 10%, decrease of the tensile strength [38]. LPSQ-COOMe double-strand structure hampers coiling, and orientation of its macromolecules during electrospinning can also promote the orientation of PLLA chains, especially that they are capable of supramolecular interactions with PLLA. Side ester groups of LPSQ-COOMe may participate in weak C-H···O=C hydrogen bonds, analogously to those in stereocomplex structures formed between enantiomeric chains of PLLA and poly(D-lactide) [63], and postulated for hybrid stereocomplex-PLA/LPSQ-COOMe blends [39]. The enhanced PLLA orientation can contribute to the improved strength of the modified nonwovens.

Figure 5 shows FTIR spectra of the selected nonwovens positioned perpendicular to the beam in the 2000–800 cm^{-1} range with the inset showing the enlarged peak at 1267 cm^{-1}.

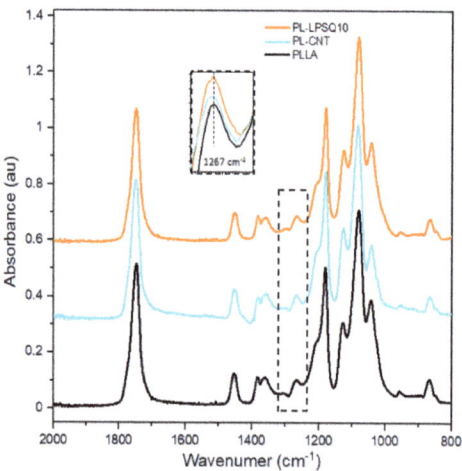

Figure 5. FTIR spectra of PLLAbased nonwovens in 2000–800 cm^{-1} range.

FTIR spectra of the selected nonwovens, shown in Figure 5, are typical for amorphous PLLA. 10 wt.% content of LPSQ-COOMe in PLLA corresponds to approx. 3 mol% content of functional groups, therefore the LPSQ-COOMe bands did not show up in the spectra, especially that some of them, e.g., those related to C=O or CH$_3$ groups, are overlapping with PLLA bands [39]. The band near 1750 cm^{-1} corresponds to C=O stretching, whereas the bands of the CH$_3$ asymmetric and symmetric bending and the first overtone of CH bending are seen near 1450, 1380, and 1360 cm^{-1} [64]. The bands near 1180, 1130, 1080, 1040 and 870 cm^{-1} are attributed to the asymmetric COC stretching and the asymmetric CH$_3$ rocking, the CH$_3$ symmetric rocking, the COC symmetric stretching, the C-CH$_3$ stretching, and the C-COO stretching, respectively [64]. The band near 1270 cm^{-1} is attributed to the CH bending and COC stretching. The spectra of the PLLA-based nonwovens shown in Figure 5 were nearly identical except for the 1266–1267 cm^{-1} band. The peaks were significantly higher for the modified PLLA nonwovens than for the neat PLLA nonwoven, as shown in the inset in Figure 5. This band is sensitive to the presence of gauche–gauche (gg) conformers in the PLLA chain, which are less energy-favorable than the gauche–trans (gt) conformers [65]. It was recently demonstrated that gg conformers can originate from the stretching of the PLLA macromolecule [66]. Thus, the increased intensity of 1266–1267 cm^{-1} band may evidence enhanced stretching and orientation of PLLA chains in the modified PLLA fibers, which contributed to the increased tensile strength of the modified nonwovens.

4. Conclusions

PLLA-based nonwovens were obtained by electrospinning. The PLLA nonwovens were modified with 0.1 wt.% of multiwall carbon nanotubes (MWCNT) and 5 and 10 wt.% of linear ladder poly(silsesquioxane) with methoxycarbonyl side groups (LPSQ-COOMe), which were added to the polymer solution before the electrospinning. In addition, nonwovens of PLLA grafted to modified MWCNT were electrospun; the content of MWCNT was 0.1 wt.%. The additives significantly reduced the fiber diameters. An average diameter decreased from 2.5 µm for neat PLLA fibers to about 1 µm for PL-LPSQ5-CNT and PL-LPSQ10-CNT. LPSQ-COOMe decreased T$_g$ of the fibers by up to 7 °C and increased their thermal stability. A decrease of cold crystallization temperature of modified fibers in comparison to those of neat PLLA was observed; the cold crystallization peak shifted from 97 °C to 81 °C. All modified nonwovens exhibited higher tensile strength than the neat PLLA nonwoven. The addition of 10 wt.% LPSQ-COOMe and 0.1 wt.% of MWCNT to PLLA increased the tensile strength of the nonwovens 2.4 times, improving also the elongation at maximum stress. FTIR analysis suggested the stronger orientation of PLLA chains in the modified fibers, which contributed to the improvement of their tensile strength.

Supplementary Materials: Figure S1: ^{1}H NMR spectrum of LPSQ-COOMe, recorded in THF-d$_8$ on a Bruker 200 spectrometer. Figure S2: Preparation scheme of PLLA-based nonwovens. Figure S3: TGA thermograms of PLLA-based nonwovens, in a nitrogen atmosphere at a heating rate of 20 °C/min. Table S1: Thermogravimetric data of PLLA-based nonwovens. Figure S4: EDS carbon and silicon mapping of (**a,b**) PL-LPSQ5 and (**c,d**) PL-LPSQ10 nonwovens.

Author Contributions: Conceptualization, M.S., T.M.; methodology, M.S., T.M., M.B., A.K.; validation, M.S., T.M., E.P.; formal analysis, M.S., T.M.; investigation, M.S., A.H., M.B.; writing—original draft preparation, M.S., T.M., M.B., A.K.; writing—review and editing, M.S., T.M., E.P.; visualization, M.S.; supervision, T.M.; funding acquisition, T.M. All authors have read and agreed to the published version of the manuscript.

Funding: This research was funded by National Science Centre (Narodowe Centrum Nauki), Poland, grant no. 2016/21/D/ST8/02908.

Data Availability Statement: The data presented in this study are available on request from the corresponding author.

Acknowledgments: Grant no. 2016/21/D/ST8/02908 from National Science Centre (Narodowe Centrum Nauki), Poland, and statutory funds of CMMS PAS. The authors are grateful to Total Corbion, The Netherlands, for providing PLLA *Luminy*® L175.

Conflicts of Interest: The authors declare no conflict of interest.

References

1. Xue, J.; Wu, T.; Dai, Y.; Xia, Y. Electrospinning and Electrospun Nanofibers: Methods, Materials, and Applications. *Chem. Rev.* **2019**, *119*, 5298–5415. [CrossRef]
2. Zhu, M.; Hua, D.; Pan, H.; Wang, F.; Manshian, B.; Soenen, S.J.; Xiong, R.; Huang, C. Green electrospun and crosslinked poly(vinyl alcohol)/poly(acrylic acid) composite membranes for antibacterial effective air filtration. *J. Colloid Interface Sci.* **2018**, *511*, 411–423. [CrossRef]
3. Namsaeng, J.; Punyodom, W.; Worajittiphon, P. Synergistic effect of welding electrospun fibers and MWCNT reinforcement on strength enhancement of PAN–PVC non-woven mats for water filtration. *Chem. Eng. Sci.* **2019**, *193*, 230–242. [CrossRef]
4. Zhao, L.; Duan, G.; Zhang, G.; Yang, H.; He, S.; Jiang, S. Electrospun Functional Materials toward Food Packaging Applications: A Review. *Nanomaterials* **2020**, *10*, 150. [CrossRef]
5. Scaffaro, R.; Maio, A.; Lopresti, F.; Botta, L. Nanocarbons in Electrospun Polymeric Nanomats for Tissue Engineering: A Review. *Polymer* **2017**, *9*, 76. [CrossRef] [PubMed]
6. Hudecki, A.; Łyko-Morawska, D.; Likus, W.; Skonieczna, M.; Markowski, J.; Wilk, R.; Kolano-Burian, A.; Maziarz, W.; Adamska, J.; Łos, M.J. Composite Nanofibers Containing Multiwall Carbon Nanotubes as Biodegradable Membranes in Reconstructive Medicine. *Nanomaterials* **2019**, *9*, 63. [CrossRef]
7. Nitanan, T.; Akkaramongkolporn, P.; Ngawhirunpat, T.; Rojanarata, T.; Panomsuk, S.; Opanasopit, P. Fabrication and evaluation of cationic exchange nanofibers for controlled drug delivery systems. *Int. J. Pharm.* **2013**, *450*, 345–353. [CrossRef] [PubMed]
8. Wang, P.; Mele, E. Effect of Antibacterial Plant Extracts on the Morphology of Electrospun Poly(Lactic Acid) Fibres. *Materials* **2018**, *11*, 923. [CrossRef] [PubMed]
9. Kost, B.; Svyntkivska, M.; Brzeziński, M.; Makowski, T.; Piorkowska, E.; Rajkowska, K.; Kunicka-Styczyńska, A.; Biela, T. PLA/β-CD-based fibres loaded with quercetin as potential antibacterial dressing materials. *Colloids Surf. B Biointerfaces* **2020**, *190*, 110949. [CrossRef]
10. Gupta, N.; Gupta, S.M.; Sharma, S.K. Carbon nanotubes: Synthesis, properties and engineering applications. *Carbon Lett.* **2019**, *29*, 419–447. [CrossRef]
11. Strozzi, M.; Pellicano, F. Linear vibrations of triple-walled carbon nanotubes. *Math. Mech. Solids* **2018**, *23*, 1456–1481. [CrossRef]
12. Venkataraman, A.; Amadi, E.V.; Chen, Y.; Papadopoulos, C. Carbon Nanotube Assembly and Integration for Applications. *Nanoscale Res. Lett.* **2019**, *14*, 1–47. [CrossRef]
13. Banerjee, R.; Ray, S.S. An overview of the recent advances in polylactide-based sustainable nanocomposites. *Polym. Eng. Sci.* **2021**, 1–33. [CrossRef]

14. Lee, J.K.Y.; Chen, N.; Peng, S.J.; Li, L.L.; Tian, L.L.; Thakor, N.; Ramakrishna, S. Polymer-based composites by electro-spinning: Preparation & functionalization with nanocarbons. *Prog. Polym. Sci.* **2018**, *86*, 40–84.
15. Mazinani, S.; Ajji, A.; Dubois, C. Fundamental Study of Crystallization, Orientation, and Electrical Conductivity of Elec-trospun PET/Carbon Nanotube Nanofibers. *J. Polym. Sci. B Polym. Phys.* **2010**, *48*, 2052–2064. [CrossRef]
16. Narvaez-Munoz, C.P.; Carrion-Matamoros, L.M.; Vizuete, K.; Debut, A.; Arroyo, C.R.; Guerrero, V.; Almeida-Naranjo, C.E.; Morales-Florez, V.; Mowbray, D.J.; Zamora-Ledezma, C. Tailoring Organic-Organic Poly(vinylpyrrolidone) Micro-particles and Fibers with Multiwalled Carbon Nanotubes for Reinforced Composites. *ACS Appl. Nano Mater.* **2019**, *2*, 4302–4312. [CrossRef]
17. Bazbouz, M.B.; Stylios, G.K. The tensile properties of electrospun nylon 6 single nanofibers. *J. Polym. Sci. Part B Polym. Phys.* **2010**, *48*, 1719–1731. [CrossRef]
18. Vicentini, N.; Gatti, T.; Salice, P.; Scapin, G.; Marega, C.; Filippini, F.; Menna, E. Covalent functionalization enables good dispersion and anisotropic orientation of multi-walled carbon nanotubes in a poly(l-lactic acid) electrospun nanofibrous matrix boosting neuronal differentiation. *Carbon* **2015**, *95*, 725–730. [CrossRef]
19. McCullen, S.D.; Stevens, D.R.; A Roberts, W.; I Clarke, L.; Bernacki, S.H.; E Gorga, R.; Loboa, E.G. Characterization of electrospun nanocomposite scaffolds and biocompatibility with adipose-derived human mesenchymal stem cells. *Int. J. Nanomed.* **2007**, *2*, 253–263.
20. Lu, P.; Hsieh, Y.-L. Multiwalled Carbon Nanotube (MWCNT) Reinforced Cellulose Fibers by Electrospinning. *ACS Appl. Mater. Interfaces* **2010**, *2*, 2413–2420. [CrossRef]
21. Baji, A.; Mai, Y.-W.; Abtahi, M.; Wong, S.-C.; Liu, Y.; Li, Q. Microstructure development in electrospun carbon nanotube reinforced polyvinylidene fluoride fibers and its influence on tensile strength and dielectric permittivity. *Compos. Sci. Technol.* **2013**, *88*, 1–8. [CrossRef]
22. Redondo, A.; Jang, D.; Korley, L.T.J.; Gunkel, I.; Steiner, U. Electrospinning of Cellulose Nanocrystal-Reinforced Polyure-thane Fibrous Mats. *Polymer* **2020**, *12*, 1021. [CrossRef]
23. Lopresti, F.; Pavia, F.C.; Vitrano, I.; Kersaudy-Kerhoas, M.; Brucato, V.; La Carrubba, V. Effect of hydroxyapatite con-centration and size on morpho-mechanical properties of PLA-based randomly oriented and aligned electrospun nano-fibrous mats. *J. Mech. Behav. Biomed. Mater.* **2020**, *101*, 103449. [CrossRef] [PubMed]
24. Piorkowska, E. Overview of Biobased Polymers. *Adv. Polym. Sci.* **2019**, *283*, 1–35. [CrossRef]
25. Saeidlou, S.; Huneault, M.A.; Li, H.; Park, C.B. Poly(lactic acid) crystallization. *Prog. Polym. Sci.* **2012**, *37*, 1657–1677. [CrossRef]
26. Rhoades, A.; Pantani, R. Poly(Lactic Acid): Flow-Induced Crystallization. *Adv. Polym. Sci.* **2019**, *283*, 87–117. [CrossRef]
27. Bojda, J.; Piorkowska, E. Shear-induced nonisothermal crystallization of two grades of PLA. *Polym. Test.* **2016**, *50*, 172–181. [CrossRef]
28. Liu, H.; Zhang, J. Research progress in toughening modification of poly(lactic acid). *J. Polym. Sci. Part B Polym. Phys.* **2011**, *49*, 1051–1083. [CrossRef]
29. Domenek, S.; Fernandes-Nassar, S.; Ducruet, V. Rheology, Mechanical Properties, and Barrier Properties of Poly(lactic acid). *Adv. Polym. Sci.* **2017**, *279*, 303–341. [CrossRef]
30. Lezak, E.; Kulinski, Z.; Masirek, R.; Piorkowska, E.; Pracella, M.; Gadzinowska, K. Mechanical and Thermal Properties of Green Polylactide Composites with Natural Fillers. *Macromol. Biosci.* **2008**, *8*, 1190–1200. [CrossRef]
31. Zubrowska, A.; Bojda, J.; Piorkowska, E. Novel Tough Crystalline Blends of Polylactide with Ethylene Glycol Derivative of POSS. *J. Polym. Environ.* **2017**, *26*, 145–151. [CrossRef]
32. Kowalczyk, M.; Piorkowska, E.; Dutkiewicz, S.; Sowinski, P. Toughening of polylactide by blending with a novel random aliphatic–aromatic copolyester. *Eur. Polym. J.* **2014**, *59*, 59–68. [CrossRef]
33. Siakeng, R.; Jawaid, M.; Ariffin, H.; Sapuan, S.M.; Asim, M.; Saba, N. Natural fiber reinforced polylactic acid composites: A review. *Polym. Compos.* **2019**, *40*, 446–463. [CrossRef]
34. Piekarska, K.; Sowinski, P.; Piorkowska, E.; Ul Haque, M.M.; Pracella, M. Structure and properties of hybrid PLA nano-composites with inorganic nanofillers and cellulose fibers. *Compos. Part A Appl. Sci. Manuf.* **2016**, *82*, 34–41. [CrossRef]
35. Goncalves, C.; Goncalves, I.C.; Magalhaes, F.D.; Pinto, A.M. Poly(lactic acid) Composites Containing Carbon-Based Nanomaterials: A Review. *Polymer* **2017**, *9*, 269. [CrossRef] [PubMed]
36. Murariu, M.; Dubois, P. PLA composites: From production to properties. *Adv. Drug Deliv. Rev.* **2016**, *107*, 17–46. [CrossRef]
37. Zhao, X.; Hu, H.; Wang, X.; Yu, X.; Zhou, W.; Peng, S. Super tough poly(lactic acid) blends: A comprehensive review. *RSC Adv.* **2020**, *10*, 13316–13368. [CrossRef]
38. Herc, A.S.; Bojda, J.; Nowacka, M.; Lewiński, P.; Maniukiewicz, W.; Piorkowska, E.; Kowalewska, A. Crystallization, structure and properties of polylactide/ladder poly(silsesquioxane) blends. *Polymer* **2020**, *201*, 122563. [CrossRef]
39. Herc, A.S.; Lewiński, P.; Kaźmierski, S.; Bojda, J.; Kowalewska, A. Hybrid SC-polylactide/poly(silsesquioxane) blends of improved thermal stability. *Thermochim. Acta* **2020**, *687*, 178592. [CrossRef]
40. Yang, T.; Wu, D.; Lu, L.; Zhou, W.; Zhang, M. Electrospinning of polylactide and its composites with carbon nanotubes. *Polym. Compos.* **2011**, *32*, 1280–1288. [CrossRef]
41. Zhou, Z.X.; Liu, L.J.; Yuan, W.Z. A superhydrophobic poly(lactic acid) electrospun nanofibrous membrane sur-face-functionalized with TiO_2 nanoparticles and methyltrichlorosilane for oil/water separation and dye adsorption. *New J. Chem.* **2019**, *43*, 15823–15831. [CrossRef]

42. Honarbakhsh, S.; Pourdeyhimi, B. Scaffolds for drug delivery, part I: Electrospun porous poly(lactic acid) and poly(lactic acid)/poly(ethylene oxide) hybrid scaffolds. *J. Mater. Sci.* **2010**, *46*, 2874–2881. [CrossRef]
43. Dasari, A.; Quirós, J.; Herrero, B.; Boltes, K.; García-Calvo, E.; Rosal, R. Antifouling membranes prepared by electrospinning polylactic acid containing biocidal nanoparticles. *J. Membr. Sci.* **2012**, *405*, 134–140. [CrossRef]
44. Makowski, T.; Svyntkivska, M.; Piorkowska, E.; Kregiel, D. Multifunctional polylactide nonwovens with 3D network of multiwall carbon nanotubes. *Appl. Surf. Sci.* **2020**, *527*, 146898. [CrossRef]
45. Arrieta, M.P.; Perdiguero, M.; Fiori, S.; Kenny, J.M.; Peponi, L. Biodegradable electrospun PLA-PHB fibers plasticized with oligomeric lactic acid. *Polym. Degrad. Stab.* **2020**, *179*, 109226. [CrossRef]
46. Arrieta, M.; López, J.; López, D.; Kenny, J.; Peponi, L. Biodegradable electrospun bionanocomposite fibers based on plasticized PLA–PHB blends reinforced with cellulose nanocrystals. *Ind. Crop. Prod.* **2016**, *93*, 290–301. [CrossRef]
47. Leonés, A.; Peponi, L.; Lieblich, M.; Benavente, R.; Fiori, S. In Vitro Degradation of Plasticized PLA Electrospun Fiber Mats: Morphological, Thermal and Crystalline Evolution. *Polymer* **2020**, *12*, 2975. [CrossRef] [PubMed]
48. Huan, S.Q.; Liu, G.X.; Cheng, W.L.; Han, G.P.; Bai, L. Electrospun Poly(lactic acid)-Based Fibrous Nanocomposite Rein-forced by Cellulose Nanocrystals: Impact of Fiber Uniaxial Alignment on Microstructure and Mechanical Properties. *Biomacromolecules* **2018**, *19*, 1037–1046. [CrossRef]
49. Liu, C.; Wong, H.M.; Yeung, K.W.K.; Tjong, S.C. Novel Electrospun Polylactic Acid Nanocomposite Fiber Mats with Hybrid Graphene Oxide and Nanohydroxyapatite Reinforcements Having Enhanced Biocompatibility. *Polymer* **2016**, *8*, 287. [CrossRef]
50. Wu, C.-S.; Wu, D.-Y.; Wang, S.-S. Bio-based polymer nanofiber with siliceous sponge spicules prepared by electrospinning: Preparation, characterisation, and functionalisation. *Mater. Sci. Eng. C* **2020**, *108*, 110506. [CrossRef]
51. Markowski, J.; Magiera, A.; Lesiak, M.; Sieron, A.L.; Pilch, J.; Blazewicz, S. Preparation and Characterization of Nano-fibrous Polymer Scaffolds for Cartilage Tissue Engineering. *J. Nanomater.* **2015**, *2015*, 564087. [CrossRef]
52. Janaszewska, A.; Gradzinska, K.; Marcinkowska, M.; Klajnert-Maculewicz, B.; Stanczyk, W.A. In Vitro Studies of Poly-hedral Oligo Silsesquioxanes: Evidence for Their Low Cytotoxicity. *Materials* **2015**, *8*, 6062–6070. [CrossRef]
53. Brzeziński, M.; Bogusławska, M.; Ilčíková, M.; Mosnáček, J.; Biela, T. Unusual Thermal Properties of Polylactides and Polylactide Stereocomplexes Containing Polylactide-Functionalized Multi-Walled Carbon Nanotubes. *Macromolecules* **2012**, *45*, 8714–8721. [CrossRef]
54. Boncel, S.; Brzezinski, M.; Mrowiec-Bialon, J.; Janas, D.; Koziol, K.K.K.; Walczak, K.Z. Oxidised multi-wall carbon nano-tubes-(R)-polylactide composite with a covalent β-D-uridine filler-matrix linker. *Mater. Lett.* **2013**, *91*, 50–54. [CrossRef]
55. Yoon, J.T.; Lee, S.C.; Jeong, Y.G. Effects of grafted chain length on mechanical and electrical properties of nanocomposites containing polylactide-grafted carbon nanotubes. *Compos. Sci. Technol.* **2010**, *70*, 776–782. [CrossRef]
56. Kowalewska, A.; Nowacka, M. Synthesis of Ladder Silsesquioxanes by in situ Polycondensation of Cyclic Tetravinylsiloxanetetraols. *Silicon* **2015**, *7*, 133–146. [CrossRef]
57. Saligheh, O.; Arasteh, R.; Forouharshad, M.; Farsani, R.E. Poly(Butylene Terephthalate)/Single Wall Carbon Nanotubes Composite Nanofibers by Electrospinning. *J. Macromol. Sci. Part B* **2011**, *50*, 1031–1041. [CrossRef]
58. Huang, C.; Thomas, N. Fabricating porous poly(lactic acid) fibres via electrospinning. *Eur. Polym. J.* **2018**, *99*, 464–476. [CrossRef]
59. Szewczyk, P.K.; Ura, D.P.; Metwally, S.; Knapczyk-Korczak, J.; Gajek, M.; Marzec, M.M.; Bernasik, A.; Stachewicz, U. Roughness and Fiber Fraction Dominated Wetting of Electrospun Fiber-Based Porous Meshes. *Polymer* **2018**, *11*, 34. [CrossRef]
60. Zhang, J.; Tashiro, K.; Tsuji, A.H.; Domb, A.J. Disorder-to-Order Phase Transition and Multiple Melting Behavior of Poly(l-lactide) Investigated by Simultaneous Measurements of WAXD and DSC. *Macromolecules* **2008**, *41*, 1352–1357. [CrossRef]
61. Barrau, S.; Vanmansart, C.; Moreau, M.; Addad, A.; Stoclet, G.; Lefebvre, J.-M.; Seguela, R. Crystallization Behavior of Carbon Nanotube−Polylactide Nanocomposites. *Macromolecules* **2011**, *44*, 6496–6502. [CrossRef]
62. Eang, C.; Opaprakasit, P. Electrospun Nanofibers with Superhydrophobicity Derived from Degradable Polylactide for Oil/Water Separation Applications. *J. Polym. Environ.* **2020**, *28*, 1484–1491. [CrossRef]
63. Tsuji, H. Poly(lactide) Stereocomplexes: Formation, Structure, Properties, Degradation, and Applications. *Macromol. Biosci.* **2005**, *5*, 569–597. [CrossRef]
64. Kister, G.; Cassanas, G.; Vert, M. Effects of morphology, conformation and configuration on the IR and Raman spectra of various poly(lactic acid)s. *Polymer* **1998**, *39*, 267–273. [CrossRef]
65. Pan, P.; Zhu, B.; Dong, T.; Yazawa, K.; Shimizu, T.; Tansho, M.; Inoue, Y. Conformational and microstructural charac-teristics of poly(L-lactide) during glass transition and physical aging. *J. Chem. Phys.* **2008**, *129*, 184902. [CrossRef]
66. Gao, X.-R.; Li, Y.; Huang, H.-D.; Xu, J.-Z.; Xu, L.; Ji, X.; Zhong, G.-J.; Li, Z.-M. Extensional Stress-Induced Orientation and Crystallization can Regulate the Balance of Toughness and Stiffness of Polylactide Films: Interplay of Oriented Amorphous Chains and Crystallites. *Macromolecules* **2019**, *52*, 5278–5288. [CrossRef]

Communication

Synergistic Flame Retardancy of Phosphatized Sesbania Gum/Ammonium Polyphosphate on Polylactic Acid

Qing Zhang *, Huiyuan Liu, Junxia Guan, Xiaochun Yang and Baojing Luo

College of Chemistry, Tangshan Normal University, Tangshan 063000, China; huiyuanliu123@sina.com (H.L.); gjxia1103@126.com (J.G.); yangxiaochun313@163.com (X.Y.); xiaoluobo_19881116@163.com (B.L.)
* Correspondence: qingzhang@tstc.edu.cn; Tel.: +86-3153863393

Abstract: Phosphating sesbania gum (DESG) was obtained by modifying sesbania gum (SG) with 9,10-dihydro-9-oxa-10-phosphaphenanthrene-10-oxide (DOPO) and endic anhydride (EA). The structure of DESG was determined using Fourier transform infrared (FTIR) spectroscopy and nuclear magnetic resonance spectroscopy (^{1}H-NMR). Flame-retardant polylactic acid (PLA) composites were prepared by melt-blending PLA with DESG, which acted as a carbon source, and ammonium polyphosphate (APP), which acted as an acid source and a gas source. The flame retardancy of the PLA composite was investigated using vertical combustion (UL-94), the limiting oxygen index (LOI) and the cone calorimeter (CONE) test. Thermal properties and morphology were characterized via thermogravimetric analysis (TGA) and field emission scanning electron microscopy (FESEM), respectively. Experimental results indicated that when the mass ratio of DESG/APP was equal to 12/8 the LOI value was 32.2%; a vertical burning test (UL-94) V-0 rating was achieved. Meanwhile, the sample showed a lowest total heat release (THR) value of 52.7 MJ/m^2, which is a 32.5% reduction compared to that of neat PLA. Using FESEM, the uniform distribution of DESG and APP in the PLA matrix was observed. The synergistic effect of DESG and APP effectively enhanced the flame retardancy of PLA. Additionally, the synergistic mechanism of DESG and APP in PLA was proposed.

Keywords: polylactic acid; sesbania gum; ammonium polyphosphate; flame retardancy

1. Introduction

The impact of worldwide production and consumption of petroleum-based plastics on economic and ecological sustainability has attracted widespread attention; these crises can be alleviated by developing bio-based plastics [1–3]. As an important environmentally friendly bio-based polymer, polylactic acid (PLA) can be obtained from corn, wheat, sugar beet, etc.; it completely degrades to carbon dioxide and water in a composting environment [4–6].

Recently, PLA materials have been commercialized in the fields of packaging materials, textile materials and medical materials because of their multiple advantages over traditional petro-polymers, including excellent mechanical properties, transparency, biocompatibility, biodegradability and processability [7–10]. However, the limiting oxygen index (LOI) of PLA is only 19%, which makes it easy to burn; a large number of molten droplets are formed during the burning process, which significantly restricts its commercial applications on a large scale [11,12].

Various flame-retardant additives have been incorporated into PLA matrices to solve this problem [13–18]. Polyphosphates, such as ammonium polyphosphate (APP), melamine polyphosphate, tris (hydroxymethyl)-aminomethane polyphosphate and isosorbide-based polyphosphonate, were introduced to PLA as halogen-free flame-retardants to increase its flame retardancy [19–22]; APP was often used in intumescent flame-retardants (IFR). As an eco-friendly flame-retardant, IFRs exhibit excellent flame retardancy on PLA due to their high effectiveness, low smoke production, low toxicity and low corrosiveness [23,24].

However, the char formation of pentaerythritol that usually acts as a carbon source in traditional IFRs is poor; a large amount of IFR needs to be added to achieve a vertical burning test (UL-94) V-0 rating, which leads to a significant decrease in the mechanical properties of PLA materials. Therefore, finding a carbon source with high flame-retardant efficiency is the key to improving the flame-retardant efficiency of IFRs in PLA.

With the development of bio-based materials and the increasing demand for ecologically sustainable development, research on bio-based flame-retardants has attracted the attention of many researchers. Bio-based polysaccharides such as starch, cellulose, lignin, cyclodextrin and chitosan have been used as "green" carbon sources in the flame-retardant modification of PLA [25–29]. Sesbania gum (SG) is a natural polysaccharide found in the seeds of sesbania, a native plant of China. It is a galactomannan; its main structure consists of one galactose on every other mannose unit. SG is an attractive biopolymer because of its abundance, low cost, biodegradability and potential application in the production of biodegradable films [30]. The large number of hydroxyl groups on the surface of these polysaccharides resulting in obvious phase separation from a hydrophobic matrix (such as PLA), resulting in poor interfacial compatibility between them and PLA. DOPO is commonly used as a reactive flame-retardant with high thermal stability, good antioxidant property, excellent flame-retardant ability and environmental compatibility. The reactive P–H bond in its structure may interact with unsaturated double bonds. Endic anhydride (EA), bicyclo[2.2.1]hept-5-ene-2,3-dicarboxylic anhydride, also has high reactivity, containing both unsaturated double bonds and anhydride bonds. The P–H bond of DOPO can undergo an addition reaction with the double bond of EA; the resulting product is further covalently bonded to the side chain of SG to reduce the polarity of SG and increase its compatibility with PLA. In the present work, SG was modified with DOPO and EA to prepare phosphorylated sesbania gum (DESG). Ammonium polyphosphate (APP), an acid source and gas source in IFRs, was combined with DESG and then added to PLA, resulting in a PLA-based composite material with flame retardancy. The synergistic flame retardancy of DESG and APP on PLA was investigated, thermal properties and morphologies of the resulting composites were characterized and the flame-retardant mechanism of DESG and APP in PLA was proposed.

2. Results

2.1. Structural Characterization of DESG

The preparation process of DESG consisted of two steps. In the first step, the P–H bond of DOPO and the double bond of EA underwent an addition reaction to prepare phosphatized EA (DEA). In the second step, the hydroxyl group of SG reacted with the anhydride group of DEA to obtain DESG. This process is shown in Figure 1.

The structures of SG, DEA and DESG were characterized using FTIR spectroscopy and ^{1}H-NMR. Figure 2a,b show the ^{1}H-NMR and FTIR spectra of SG, DEA and DESG, respectively. Compared with those of DOPO [31] and EA [32], in the ^{1}H-NMR spectrum of DEA the P–H signal (at δ 8.82 ppm in DOPO) and the peak (at approximately δ 6.5 ppm) attributed to the double bond of EA disappeared. However, new absorption peaks located at δ 3.6 and 2.1~2.8 ppm appeared, which demonstrated the occurrence of an addition reaction between DOPO and EA. For DESG, the signal that appeared in the range of δ 3.9~5.2 ppm was from SG; peaks at approximately δ 7~8 ppm could be assigned to the biphenyl group in DOPO, indicating that DEA was successfully grafted onto SG. All of this can also be confirmed in the FTIR spectra of DEA and DESG. For DEA, the stretching vibration band of double bond at 1628 cm^{-1} and the characteristic peak of P–H at 2436 cm^{-1} vanished. In the FTIR spectrum of DESG, the absorption peak of the anhydride group at 1864 cm^{-1} disappeared, but the absorption peaks of P–O–Ph (1159, 912, 758 cm^{-1}), P=O (1232, 1208 cm^{-1}) and P–Ph (1596, 1478 cm^{-1}) [33,34] were still present.

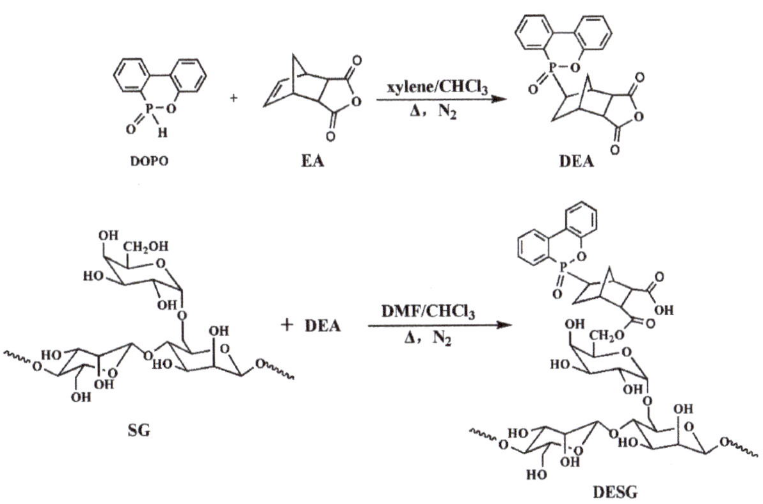

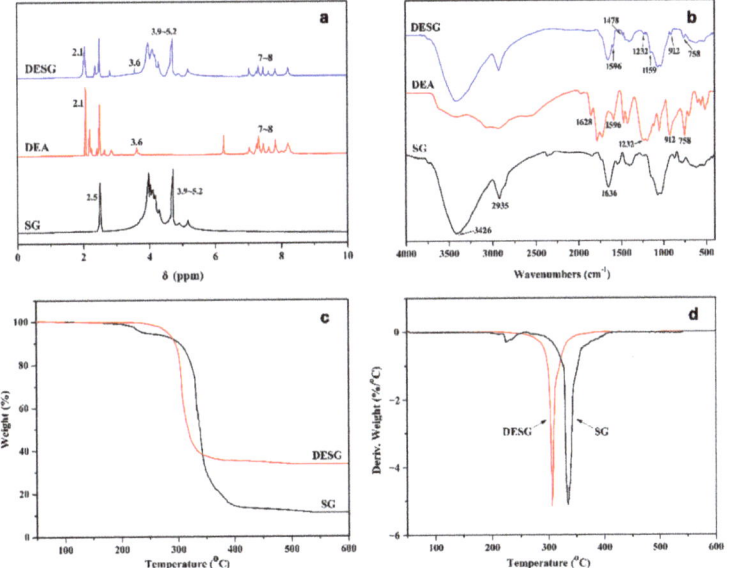

Figure 1. A DESG preparation route.

Figure 2. SG and DESG curves: (a) ¹HNMR, (b) FTIR, (c) TGA and (d) DTG.

The TGA and DTG curves of SG and DESG are shown in Figure 2c. Thermal stabilities were determined from the temperature at 5% weight loss, which was defined as the onset degradation temperature. From Figure 2c, we can see that the weight loss of SG occurred in two steps, including dehydration and thermolysis processes. The onset degradation temperature of SG was approximately 237.6 °C. DESG underwent only one stage of thermal degradation due to the thermolysis of the polysaccharide component. The onset degradation temperature shifted to a higher value (282.5 °C) compared with that of SG, which indicated that the introduction of the phosphorus element improved the early thermal stability of DESG. However, it can be found from the DTG curve that the maximum degradation rate temperature of DESG dropped approximately 28 °C. This was because the high temperature broke down the phospholipid bonds and the resulting acidic substances promoted the decomposition of SG, leading to a decrease in the maximum decomposition temperature. The final char residue at 800 °C was 33.5%, which was much higher than that of SG (11.0%); therefore, the high char residue and speed of carbon formation make DESG a potentially excellent carbon source for an IFR.

2.2. Flame Retardant Properties of DESG/APP on Polylactic Acid

DESG was combined with APP and then added to PLA, resulting in PLA-based composites with flame retardancy. Here, DESG was used as a carbon source and APP acted as both an acid source and a gas source. The flame retardancy of PLA composites was measured using UL-94 and LOI. The results are given in Table 1. The neat PLA sample showed a LOI value of 19.5%. For binary composites with 20 wt% SG or DESG added to PLA alone, the LOI did not change significantly. Similarly, SG or DESG alone were not helpful in upgrading the UL rating for PLA substrate. The incorporation of APP alone improved the LOI value of the A20 binary composite to 26.2%. However, the molten droplets produced ignited the degreasing cotton; the composite achieved only a UL V-2 rating. For a ternary composite, A12S8, containing 12 wt% APP and 8 wt% SG, the LOI value increased to 29.6%, but the UL rating was still V-2. This indicated that the synergistic flame retardancy of nonphosphatized SG and APP on PLA was weak. While 12 wt% APP and 8 wt% DESG were introduced into PLA, the A12D8 composite achieved a LOI value of 32.2% and a V-0 rating. The improvement in flame-retardant properties was due to the synergy effect between APP and DESG. As a result of this synergy, adding only 10 wt% flame-retardant, like A7D3 or A5D5, made the composite reach the V-0 rating. The effect of the addition ratio of APP and DESG on the flame-retardant properties of PLA composites was also investigated. The results from Table 1 show that the LOI value of the composites increased gradually and then decreased with an increase in APP content.

Table 1. Formulations and flammability of PLA and flame-retardant PLA composites.

Samples	PLA (wt%)	APP (wt%)	SG (wt%)	EDSG (wt%)	LOI (%)	UL-94	Cotton Ignition
PLA	100	0	0	0	19.5	NR [a]	Yes
S20	80	0	20	0	19.4	NR	Yes
D20	80	0	0	20	19.9	NR	Yes
A20	80	20	0	0	26.2	V-2	Yes
A12S8	80	12	8	0	29.6	V-2	Yes
A18D2	80	18	0	2	31.3	V-1	No
A15D5	80	15	0	5	31.9	V-0	No
A12D8	80	12	0	8	32.2	V-0	No
A9D11	80	9	0	11	29.2	V-0	No
A5D15	90	5	0	15	28.6	V-1	No
A9D1	90	9	0	1	25.8	V-1	No
A7D3	90	7	0	3	27.6	V-0	No
A5D5	90	5	0	5	26.1	V-0	No
A3D7	90	3	0	7	24.3	V-2	No

[a] NR = no rating.

In order to further investigate the synergistic effect of DESG and APP on PLA, a cone calorimeter (CONE) test, which is used to simulate the burning behavior of materials in a real fire, was conducted. The heat release rate (HRR) and total heat release (THR) curves of PLA and PLA composites are shown in Figure 3. In Figure 3a, one can see that neat PLA presented the largest peak heat release rate (pHRR) of 523.5 kW/m^2. The pHRR value of D20 and A20 were 501.6 and 482.5 kW/m^2, respectively. This indicated that the addition of DESG or APP alone cannot effectively reduce the HRR value. However, the pHRR value of the A12D8 sample decreased to 341.6 kW/m^2, exhibiting a 34.7% decrease compared to that of neat PLA. The HRR curve of A12D8 showed a double peak, while only a single peak was observed for other samples. This bimodal phenomenon was characteristic of the intumescent flame-retardants [29,35]. As shown in the THR curves of PLA and PLA composites (Figure 3b), the THR values of neat PLA, D20 and A20 were 78.1, 72.4 and 65.8 MJ/m^2, respectively. Similarly, although adding DESG or APP alone reduced the THR value compared with PLA, it was not as effective as adding both. The A12D8 sample showed the lowest THR value, 52.7 MJ/m^2, which was a 32.5% reduction compared to that of neat PLA.

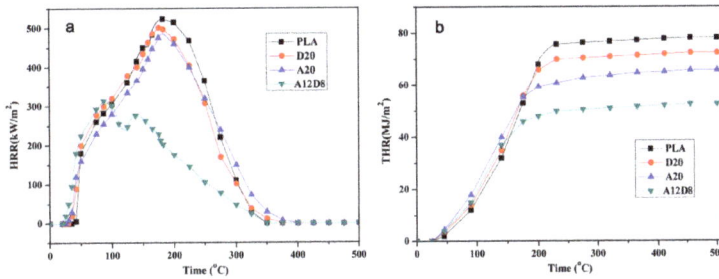

Figure 3. PLA, D20, A20 and A12D8 curves: (a) HRR and (b) THR.

Compared with the results reported for polysaccharide-added PLA flame-retardant composites [26,36], A12D8 composites achieved a higher LOI value as well as lower pHRR and THR values. For A12D8 composites, a V-0 rating can be achieved by adding only 10 wt% flame-retardants; however, 15–25 wt% was added in some studies [28,35,37].

2.3. Thermal Stabilities of Flame-Retardant PLA Composites

TGA was employed to evaluate the thermal stabilities of flame-retardant PLA composites. TGA and DTA thermograms of PLA and flame-retardant PLA composites in a nitrogen atmosphere (from 100 to 650 °C) are illustrated in Figure 4a,b. The initial decomposition temperature (T_i), defined as the 5% weight loss temperature, the temperature of maximal weight loss rate (T_{max}) and char yields at 400 and 600 °C are summarized in Table 2. For PLA, T_i and T_{max} were 309.4 °C and 338.7 °C, respectively. At approximately 400 °C, its decomposition was essentially complete; the char yield after 400 °C under N$_2$ was only 0.3%. As a result of SG's low decomposition temperature, the T_i of D20 decreased to 282.5 °C, with a char yield of 4.1% at 600 °C. The T_i of A20 was 296.5 °C and its char yield significantly increased at 400 °C. This occurred because the APP was decomposed in advance to generate phosphoric acid or polyphosphoric acid, which catalyzed the char formation of PLA itself. However, the char layer was unstable at high temperatures and underwent secondary decomposition at approximately 500 °C. For the A12D8 sample, thermal stability was higher after 300 °C than it was when adding APP or DESG alone; the presence of APP improved the thermal stability of DESG at a low temperature. Moreover, the weight loss of DESG at a low temperature was reduced compared with other samples. The char yield of A12D8 was as high as 32.1% and 16.8% at 400 °C and 600 °C, respectively. These results demonstrated the existence of the synergistic effect of APP and DESG combined in promoting char formation.

Table 2. TGA and DTG data for PLA, D20, A20 and A12D8.

Samples	T_i (°C)	T_{max} (°C)	Char Yield (%)	
			at 400 °C	at 600 °C
PLA	309.4	338.7	0.3	0.3
D20	282.5	319.2	4.2	4.1
A20	296.5	337.1	16.5	8.6
A12D8	273.0	343.6	32.1	16.8

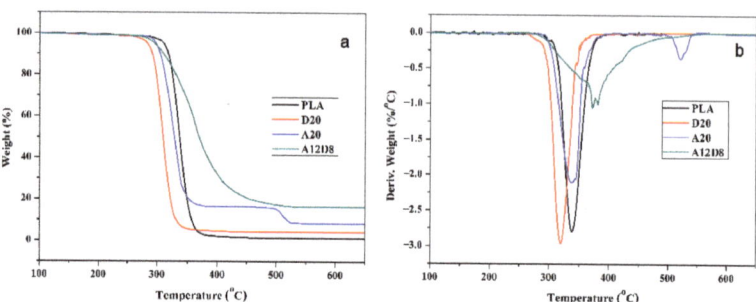

Figure 4. PLA, D20, A20 and A12D8 thermograms: (**a**) TGA and (**b**) DTG.

2.4. Dispersion of DESG/APP Flame-Retardant in PLA Matrix

To investigate the dispersion of DESG/APP flame-retardants in the PLA matrix, FESEM images of the fracture surfaces of D20, A20 and A12D8 samples are shown in Figure 5. It can be seen in Figure 5a,b that the fracture surfaces of the S20 and A20 samples were rough; there was an obvious phase interface between the SG or APP particles and the PLA matrix, indicating poor compatibility between the dispersed phase and the matrix. On one hand, because APP and SG are hydrophilic, there was a polarity difference between them and the hydrophobic PLA matrix. On the other hand, both APP and SG have strong intermolecular forces and are prone to agglomeration, which resulted in their poor dispersibility in PLA. After SG was phosphorylated by DEA, the hydrophobicity of SG improved significantly. Additionally, the carboxyl groups still existing on DESG could undergo esterification with the terminal hydroxyl groups of PLA during processing, which further improved the compatibility of DESG and PLA [38,39]. Therefore, homogeneous dispersion was found in the FESEM micrograph of the A12D8 sample (Figure 5c). In addition, the introduction of DESG also enhanced the compatibility of APP with the PLA matrix. This may be due to the reaction between the hydroxyl groups on DESG and the carboxyl groups on APP, which promoted better dispersion of the flame-retardant in the PLA matrix and further improved the synergistic flame-retardant efficiency of the flame-retardants.

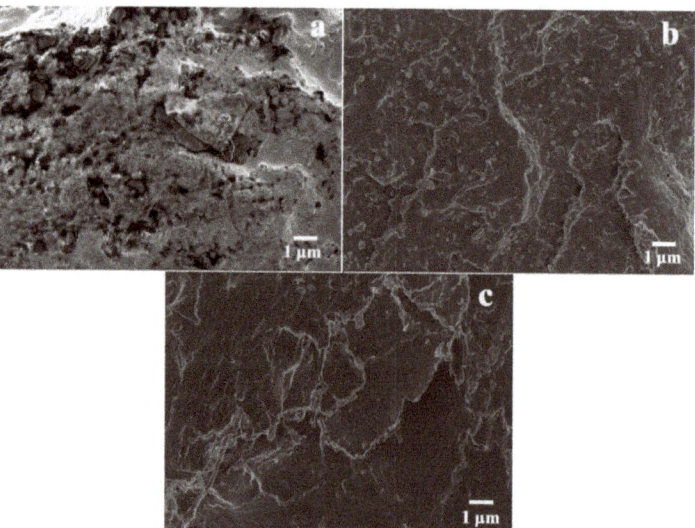

Figure 5. FESEM micrographs of PLA composites: (**a**) S20, (**b**) A20 and (**c**) A12D8.

It can be seen from the tensile properties of PLA and its composites (as shown in Table 3), that adding 20% DESG (sample D20) or APP (sample A20) alone significantly reduced the tensile strength and Young's modulus of the composites, which ranged from 67.6 to 48.1 MPa and from 1.75 to 1.08 GPa, respectively. This indicated that the interfacial compatibility between DESG or APP and PLA was very poor, resulting in a serious decline in the mechanical properties of composites. For A20, the elongation at break increased slightly, which was attributed to the dilution effect of APP on the PLA molecular chain [19,40]. After adding both DESG and APP (sample A12D8), the decrease in the strength and modulus of the composites was effectively alleviated; it was speculated that the incorporation of DESG could improve compatibility between the dispersed phase and the PLA matrix and reduce the influence on the mechanical properties of the composites.

Table 3. Tensile properties of PLA and its composites.

Samples	Tensile Strength (MPa)	Young's Modulus (GPa)	Elongation at Break (%)
PLA	67.6	1.75	7.6
D20	52.3	1.35	6.2
A20	48.1	1.08	8.2
A12D8	63.4	1.63	7.3

2.5. Flame-Retardant Mechanism

To understand the flame-retardant mechanism in the condensed phase, the Raman spectra of char residues of D20, A20 and A12D8 samples after cone calorimeter tests were investigated; the results are shown in Figure 6. The vibrations of disordered carbon and graphitic carbon were detected near 1340 cm^{-1} and 1585 cm^{-1}, which correspond to the D and G bands, respectively. The degree of graphitization was assessed using the value of the area ratio of D to G bands (AD/AG); a lower value suggested a higher graphitization degree of char layer, better shielding effect and higher thermal stability [41,42]. The AD/AG values of D20, A20 and A12D8 samples were 3.1, 2.9 and 2.3, respectively. The A12D8 sample had a lower AD/AG value compared to the others, indicating that the addition of both DESG and APP facilitated forming more graphitized char residue with a shielding effect for PLA.

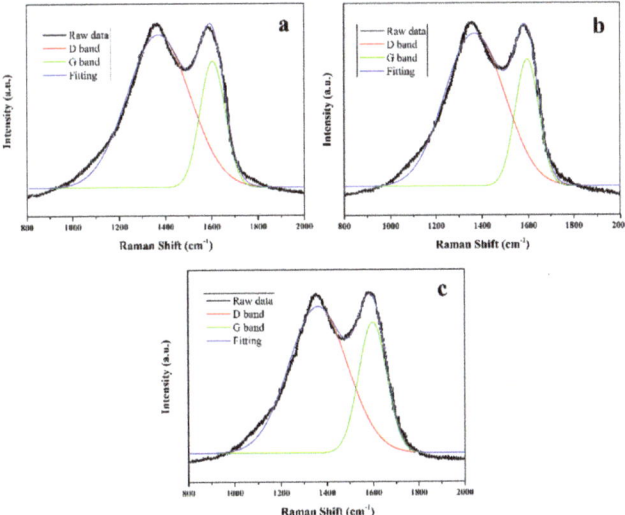

Figure 6. Raman spectra of (a) D20, (b) A20 and (c) A12D8 after cone calorimeter test.

Figure 7 presents a possible synergistic mechanism of the DESG/APP flame-retardant. When the composite material was heated and burned, APP, which was an acid source and a gas source, first decomposed to generate ammonia and polyphosphoric acid (or polymetaphosphoric acid). Ammonia, a non-flammable gas, could have a diluting effect in the gas phase. An esterification reaction between hydroxyl groups in the DESG molecules and polyphosphoric acid might occur. Meanwhile, the polyphosphoric acid, with strong dehydration, can accelerate the carbonization of the ester to form a phosphorus–carbon char layer containing cross-linked structures of P–O–C and P–O–P. With a continuous increase in the external temperature, the cross-linked structures then underwent various chemical reactions such as dehydrogenation, carbonization, breakage of the chemical bond, etc., to form char residues rich in phosphorus and carbon. These char residues appeared macroscopically as black carbonaceous skeletons. The water vapor generated by the reaction and some incombustible gases (such as NH_3) swelled the char layer, forming an intumescent protective layer. This intumescent layer, with high strength, good compactness and excellent thermal stability, covered the surface of the PLA matrix to produce a fine shielding effect, which further delayed the degradation rate of the material through effective thermal insulation and oxygen insulation and improved the flame-retardant performance. It is noteworthy that the esterification of DESG and polyphosphoric acid caused the initial weight loss of the composite, but at the same time a more stable carbon layer formed. This also explains why A12D8 had a 5% weight loss at a lower temperature of 273.0 °C, but less weight loss and a higher char residue at a higher temperature.

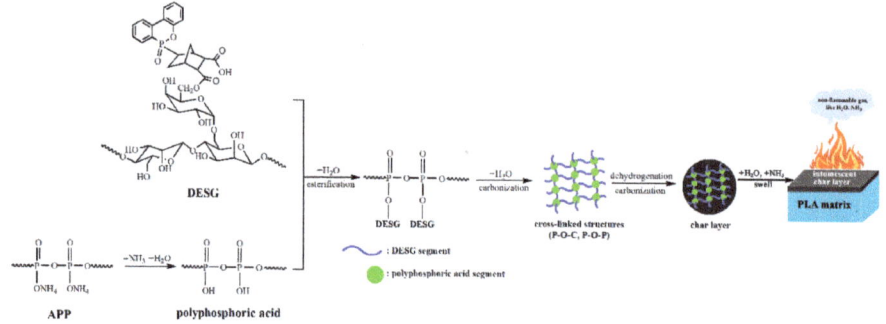

Figure 7. A possible synergistic mechanism of DESG/APP flame-retardant.

3. Materials and Methods

3.1. Materials

PLA (4032D) and SG were commercially obtained from Nature Work Company (Blair, NE, USA) and Ningbo Global Biomaterials Co., Ltd. (Ningbo, China), respectively. APP (analytical reagent) (AR) and DOPO (AR) were purchased from Sigma-Aldrich (Saint Louis, MO, USA). EA (AR), 4-dimethylaminopyridine (4DMAP) (AR), xylene (AR), chloroform (AR), anhydrous ethanol (AR) and N, N-Dimethylformamide (DMF) (AR), were supplied by Aladdin Industrial Co. (Beijing, China). All reagents were used without any further purification.

3.2. Preparation Procedures

3.2.1. Preparation of DESG

A total of 50 g of DOPO was dissolved in 100 mL of xylene at 75 °C in a three-neck flask equipped with a condenser; 46 g of EA dissolved in 80 mL of chloroform was added to this solution under a nitrogen atmosphere, followed by a reflux with magnetic stirring for 24 h. After the reaction was completed and cooled to room temperature, the solvent was removed via rotary evaporation at 75 °C to obtain DEA with a yield of 91.8%.

A total of 30 g of SG and 100 mL of DMF were added to a three-necked flask; the solution was heated to 90 °C, followed by a reflux for 4 h under a nitrogen atmosphere. This mixture was cooled to 75 °C. Next, 30 g of DEA dissolved in chloroform and an appropriate amount of 4DMAP was added. The reflux reaction was continued for 24 h under nitrogen protection. After the reaction was completed, the mixture was cooled to room temperature, washed with anhydrous ethanol 5 times and vacuum dried at 75 °C for 10 h to obtain DESG with a yield of 83.5%.

3.2.2. Blend Preparation of Flame-Retardant PLA Composites

PLA, APP, SG and DESG were dried in a vacuum drying oven at 80 °C for 24 h before use. The dried raw materials were weighed and premixed in a high-speed mixer for 8 min according to the formulations listed in Table 1. Next, the premixed blends were added to the mixing chamber of a torque rheometer to be melt-blended for 10 min at a temperature of 185 °C and a rotational speed of 45 rpm. The blends were compressed into plate specimens by a plate vulcanizer under 10 MPa for 5 min at 180 °C. Finally, these plates were cut and polished into standard splines for testing.

3.3. Analysis

The 1H-NMR spectra were measured on a Bruker AV II-400 MHz spectrometer (USA) using DMSO-d_6 as a solvent at 25 °C. The FTIR spectra were recorded on a Nicolet is 50 FTIR spectrometer within the 4000–400 cm^{-1} region with KBr matrix.

The morphology of the fractured surfaces of specimens was observed with a ZEISS Sigma 300 FESEM (Schnelldorf, Germany) with an accelerating voltage of 10 kV.

TGA and DTG analyses were carried out with a PerkinElmer TGA 4000 thermogravimetric analyzer (Waltham, MA, USA) at a heating rate of 10 °C/min from 30 to 650 °C under a nitrogen atmosphere.

LOI values were recorded on a Jiangning JF-3 oxygen index analyser (Nanjing, China) with sample dimensions of 130 mm × 6.5 mm × 3 mm according to ASTMD 2863-97. UL-94 vertical burning tests were performed using a Jiangning CZF-3 horizontal and vertical combustion tester (Najing, China) with sample dimensions of 130 mm × 13 mm × 3.0 mm in accordance with ASTMD 3801-2010.

CONE tests were carried out with a FTT cone calorimeter (East Grinstead, England) under an external heat flux of 50 kW/m^2. Sample dimensions were 100 mm × 100 mm × 3 mm according to ISO 5660.

Tensile properties were determined using a CMT4202 universal testing machine (Shenzhen, China) according to ASTM D638, at a test speed of 10 mm/min. Specimens were dumbbell-shaped with dimensions of 75 mm × 4 mm × 1 mm.

4. Conclusions

DESG was prepared by grafting SG with DOPO and EA. Flame-retardant PLA composites were obtained by melt-blending PLA with DESG and APP. Here, DESG acted as a carbon source due to its polysaccharide structure, while APP acted as an acid source and a gas source to catalyze the formation of char residue and release nonflammable gases, respectively. The flame retardancy, thermal properties, morphology and tensile properties of flame-retardant PLA composites were investigated. The results demonstrated that the synergistic effect of DESG and APP promoted the formation of an intumescent protective layer in PLA composites at a lower temperature, improved the high temperature stability of the composites and effectively enhanced the flame retardancy of PLA. When the mass ratio of DESG/APP is equal to 12/8, an LOI value of 32.2% and a UL-94 V-0 rating were achieved. A V-0 rating can also be reached at a lower addition ratio. Moreover, the addition of DESG could improve compatibility between the dispersed phase and the PLA matrix, as well as reduce the influence of the flame-retardant on the mechanical properties.

Author Contributions: Q.Z. and H.L. carried out the preparation work, analyzed the results and wrote the manuscript. J.G. performed the structural characterization and performance testing. X.Y. and B.L. discussed and analyzed the results. All authors have read and agreed to the published version of the manuscript.

Funding: This research was financially supported by the Program of Tangshan Normal University (2022C44).

Informed Consent Statement: Not applicable.

Data Availability Statement: Not applicable.

Conflicts of Interest: The authors declare no conflict of interest.

Sample Availability: Samples of the compounds are not available from the authors.

References

1. Gil-Castell, O.; Andres-Puche, R.; Dominguez, E.; Verdejo, E.; Monreal, L.; Ribes-Greus, A. Influence of substrate and temperature on the biodegradation of polyester-based materials: Polylactide and poly(3-hydroxybutyrateco-3-hydroxyhexanoate) as model cases. *Polym. Degrad. Stab.* **2020**, *180*, 109288. [CrossRef]
2. Zhou, Y.X.; Huang, Z.G.; Diao, X.Q.; Weng, Y.X.; Wang, Y.Z. Characterization of the effect of REC on the compatibility of PHBH and PLA. *Polym. Test.* **2015**, *42*, 17–25. [CrossRef]
3. Corre, Y.M.; Bruzaud, S.; Audic, J.L.; Grohens, Y. Morphology and functional properties of commercial polyhydroxyalkanoates: A comprehensive and comparative study. *Polym. Test.* **2012**, *31*, 226–235. [CrossRef]
4. Leluk, K.; Frackowiak, S.; Ludwiczak, J.; Rydzkowski, T.; Thakur, V.K. The impact of filler geometry on polylactic acid-based sustainable polymer composites. *Molecules* **2021**, *26*, 149. [CrossRef]
5. Zhang, J.; Lou, J.Z.; Ilias, S.; Krishnamachari, P.; Yan, J. Thermal properties of poly(lactic acid) fumed silica nanocomposites: Experiments and molecular dynamics simulations. *Polymer* **2008**, *49*, 2381–2386. [CrossRef]
6. Zhang, Q.; Li, D.; Zhang, H.; Su, G.; Li, G. Preparation and properties of poly(lactic acid)/sesbania gum/nano-TiO_2 composites. *Polym. Bull.* **2018**, *75*, 623–635. [CrossRef]
7. Wróblewska-Krepsztul, J.; Rydzkowski, T.; Michalska-Pożoga, I.; Thakur, V.K. Biopolymers for biomedical and pharmaceutical applications: Recent advances and overview of alginate electrospinning. *Nanomaterials* **2019**, *9*, 404. [CrossRef]
8. Castro-Aguirre, E.; Iniguez-Franco, F.; Samsudin, H.; Fang, X. Poly(lactic acid)—Mass production, processing, industrial applications, and end of life. *Adv. Drug Deliv. Rev.* **2016**, *107*, 333–336. [CrossRef]
9. Liu, S.; Qin, S.; He, M.; Zhou, D.; Qin, Q.; Wang, H. Current applications of poly(lactic acid) composites in tissue engineering and drug delivery. *Compos. Part B Eng.* **2020**, *199*, 108238. [CrossRef]
10. Benvenuta-Tapia, J.J.; Vivaldo-Lima, E. Reduction of molar mass loss and enhancement of thermal and rheological properties of recycled poly(lactic acid) by using chain extenders obtained from RAFT chemistry. *React. Funct. Polym.* **2020**, *153*, 104628. [CrossRef]
11. Chow, W.S.; Teoh, E.L.; Karger-Kocsis, J. Flame retarded poly(lactic acid): A review. *Express Polym. Lett.* **2018**, *12*, 396–417. [CrossRef]
12. Hamad, K.; Kaseem, M.; Yang, H.W.; Deri, F.; Ko, Y.G. Properties and medical applications of polylactic acid: A review. *Express Polym. Lett.* **2015**, *9*, 435–455. [CrossRef]
13. Tawiah, B.; Zhou, Y.Y.; Yuen, R.K.K.; Sun, J.; Fei, B. Microporous boron based intumescent macrocycle flame retardant for poly (lactic acid) with excellent UV protection. *Chem. Eng. J.* **2020**, *402*, 126209. [CrossRef]
14. Yang, W.; Tawiah, B.; Yu, C.; Qian, Y.F.; Wang, L.L.; Yuen, A.C.Y.; Zhu, S.E.; Hu, E.Z.; Chen, T.B.Y.; Yu, B. Manufacturing, mechanical and flame retardant properties of poly(lactic acid) biocomposites based on calcium magnesium phytate and carbon nanotubes. *Compos. Part A Appl. Sci.* **2018**, *110*, 227–236. [CrossRef]
15. Jin, X.B.; Xiang, E.L.; Zhang, R.; Qin, D.C.; Jiang, M.L.; Jiang, Z.H. Halloysite nanotubes immobilized by chitosan/tannic acid complex as a green flame retardant for bamboo fiber/poly(lactic acid) composites. *J. Appl. Polym. Sci.* **2021**, *138*, 49621. [CrossRef]
16. Mu, X.W.; Yuan, B.H.; Hu, W.Z.; Qiu, S.L.; Song, L.; Hu, Y. Flame retardant and anti-dripping properties of polylactic acid/poly(bis(phenoxy) phosphazene)/expandable graphite composite and its flame retardant mechanism. *RSC Adv.* **2015**, *5*, 76068–76078. [CrossRef]
17. Yang, W.; Yang, W.J.; Tawiah, B.; Zhang, Y.; Wang, L.L.; Zhu, S.E.; Chen, T.B.Y.; Chun, A.; Yuen, Y.; Yu, B.; et al. Synthesis of anhydrous manganese hypophosphite microtubes for simultaneous flame retardant and mechanical enhancement on poly(lactic acid). *Compos. Sci. Technol.* **2018**, *164*, 44–50. [CrossRef]
18. Tawiah, B.; Yu, B.; Yuen, R.K.K.; Hu, Y.; Wei, R.; Xin, J.H.; Fei, B. Highly efficient flame retardant and smoke suppression mechanism of boron modified graphene Oxide/Poly(Lactic acid) nanocomposites. *Carbon* **2019**, *150*, 8–20. [CrossRef]
19. Zhan, Y.; Wu, X.; Wang, S.; Yuan, B.; Fang, Q.; Shang, S.; Cao, C.; Chen, G. Synthesis of a bio-based flame retardant via a facile strategy and its synergistic effect with ammonium polyphosphate on the flame retardancy of polylactic acid. *Polym. Degrad. Stab.* **2021**, *191*, 109684. [CrossRef]

20. Guo, Y.; Chang, C.C.; Cuiffo, M.A.; Xue, Y.; Zuo, X.; Pack, S.; Zhang, L.; He, S.; Weil, E.; Rafailovich, M.H. Engineering flame retardant biodegradable polymer nanocomposites and their application in 3D printing. *Polym. Degrad. Stab.* **2017**, *137*, 205–215. [CrossRef]
21. Wang, X.; Wang, W.; Wang, S.; Yang, Y.; Li, H.; Sun, J.; Gu, X.; Zhang, S. Self-intumescent polyelectrolyte for flame retardant poly(lactic acid) nonwovens. *J. Clean. Prod.* **2021**, *282*, 124497. [CrossRef]
22. Mauldin, T.C.; Zammarano, M.; Gilman, J.W.; Shields, J.R.; Boday, D.J. Synthesis and characterization of isosorbide-based polyphosphonates as biobased flame-retardants. *Polym. Chem.* **2014**, *5*, 5139–5146. [CrossRef]
23. Wang, X.; Hu, Y.; Song, L.; Xuan, S.; Xing, W.; Bai, Z.; Lu, H. Flame retardancy and thermal degradation of intumescent flame retardant poly(lactic acid)/starch biocomposites. *Ind. Eng. Chem. Res.* **2011**, *50*, 713–720. [CrossRef]
24. Feng, C.; Liang, M.; Jiang, J.; Huang, J.; Liu, H. Flame retardant properties and mechanism of an efficient intumescent flame retardant PLA composites. *Polym. Adv. Technol.* **2016**, *27*, 693–700. [CrossRef]
25. Feng, J.X.; Su, S.P.; Zhu, J. An intumescent flame retardant system using b-cyclodextrin as a carbon source in polylactic acid (PLA). *Polym. Adv. Technol.* **2011**, *22*, 1115–1122. [CrossRef]
26. Zhu, T.; Guo, J.; Fei, B.; Feng, Z.Y.; Gu, X.Y.; Li, H.F.; Sun, J.; Zhang, S. Preparation of methacrylic acid modified microcrystalline cellulose and their applications in polylactic acid: Flame retardancy, mechanical properties, thermal stability and crystallization behavior. *Cellulose* **2020**, *27*, 2309–2323. [CrossRef]
27. Cayla, A.; Rault, F.; Giraud, S.; Salaun, F.; Fierro, V.; Celzard, A. PLA with intumescent system containing lignin and ammonium polyphosphate for flame retardant textile. *Polymers* **2016**, *8*, 331. [CrossRef]
28. Wang, J.J.; Ren, Q.; Zheng, W.G.; Zhai, W.T. Improved flame-retardant properties of poly(lactic acid) foams using starch as a natural charring agent. *Ind. Eng. Chem. Res.* **2014**, *53*, 1422–1430. [CrossRef]
29. Chen, C.; Gu, X.; Jin, X.; Sun, J.; Zhang, S. The effect of chitosan on the flammability and thermal stability of polylactic acid/ammonium polyphosphate biocomposites. *Carbohydr. Polym.* **2017**, *157*, 1586–1593. [CrossRef]
30. Zhang, Q.; Gao, Y.; Zhai, Y.A.; Liu, F.Q.; Gao, G. Synthesis of sesbania gum supported dithiocarbamate chelating resin and studies on its adsorption performance for metal ions. *Carbohydr. Polym.* **2008**, *73*, 359–363. [CrossRef]
31. Liu, B.Z.; Gao, X.Y.; Zhao, Y.F.; Dai, L.N.; Xie, Z.M.; Zhang, Z.J. 9,10-Dihydro-9-oxa-10-phosphaphenanthrene 10-oxide-based oligosiloxane as a promising damping additive for methyl vinyl silicone rubber (VMQ). *J. Mater. Sci.* **2017**, *52*, 8603–8617. [CrossRef]
32. Kas'yan, L.I.; Tarabara, I.N.; Pal'chikov, V.A.; Krishchik, O.V.; Isaev, A.K.; Kas'yan, A.O. Acylation of aminopyridines and related compounds with endic anhydride. *Russ. J. Org. Chem.* **2005**, *41*, 1530–1538. [CrossRef]
33. Xiong, Y.Q.; Jiang, Z.J.; Xie, Y.Y.; Zhang, X.Y.; Xu, W.J. Development of a DOPO-containing melamine epoxy hardeners and its thermal and flame-retardant properties of cured products. *J. Appl. Polym. Sci.* **2013**, *127*, 4352–4358. [CrossRef]
34. Wang, X.D.; Zhang, Q. Synthesis, characterization, and cure properties of phosphorus-containing epoxy resins for flame retardance. *Eur. Polym. J.* **2004**, *40*, 385–395. [CrossRef]
35. Jin, X.D.; Cui, S.P.; Sun, S.B.; Sun, J.; Zhang, S. The preparation and characterization of polylactic acid composites with chitin-based intumescent flame retardants. *Polymers* **2021**, *13*, 3513. [CrossRef]
36. Wang, B.; Qian, X.; Shi, Y.; Yu, B.; Hong, N.; Song, L.; Hu, Y. Cyclodextrin microencapsulated ammonium polyphosphate: Preparation and its performance on the thermal, flame retardancy and mechanical properties of ethylene vinyl acetate copolymer. *Compos. Part B Eng.* **2015**, *69*, 22–30. [CrossRef]
37. Wang, S.; Zhang, L.; Semple, K.; Zhang, M.; Zhang, W.; Dai, C. Development of biodegradable flame-retardant bamboo charcoal composites, part I: Thermal and elemental analyses. *Polymers* **2020**, *12*, 2217. [CrossRef]
38. Kucharczyk, P.; Zednik, J.; Sedlarik, V. Synthesis and characterization of star-shaped carboxyl group functionalized poly(lactic acid) through polycondensation reaction. *Macromol. Res.* **2017**, *25*, 180–189. [CrossRef]
39. Bishai, M.; De, S.; Banerjee, R. Copolymerization of lactic acid for cost-effective PLA synthesis and studies on its improved characteristics. *Food Sci. Biotechnol.* **2013**, *22*, 73–77. [CrossRef]
40. Jia, Y.W.; Zhao, X.; Fu, T.; Li, D.F.; Guo, Y.; Wang, X.L.; Wang, Y.Z. Synergy effect between quaternary phosphonium ionic liquid and ammonium polyphosphate toward flame retardant PLA with improved toughness. *Compos. Part B Eng.* **2020**, *197*, 108192. [CrossRef]
41. Ferrari, A.C.; Basko, D.M. Raman spectroscopy as a versatile tool for studying the properties of grapheme. *Nat. Nanotechnol.* **2013**, *8*, 235–246. [CrossRef]
42. Wu, J.N.; Chen, L.; Fu, T.; Zhao, H.B.; Guo, D.M.; Wang, X.L. New application for aromatic Schiff base: High efficient flame-retardant and anti-dripping action for polyesters. *Chem. Eng. J.* **2018**, *336*, 622–632. [CrossRef]

Review

Strategies for Enhancing Polyester-Based Materials for Bone Fixation Applications

Raasti Naseem [1,*], Charalampos Tzivelekis [2], Matthew J. German [2], Piergiorgio Gentile [1], Ana M. Ferreira [1] and Kenny Dalgarno [1]

[1] School of Engineering, Newcastle University, Newcastle upon Tyne NE1 7RU, UK; piergiorgio.gentile@newcastle.ac.uk (P.G.); ana.ferreira-duarte@newcastle.ac.uk (A.M.F.); kenny.dalgarno@newcastle.ac.uk (K.D.)

[2] School of Dental Sciences, Translational and Clinical Research Institute, Faculty of Medical Sciences, Newcastle University, Newcastle upon Tyne NE1 7RU, UK; babis.tzivelekis@newcastle.ac.uk (C.T.); matthew.german@newcastle.ac.uk (M.J.G.)

* Correspondence: raasti.naseem@newcastle.ac.uk

Abstract: Polyester-based materials are established options, regarding the manufacturing of bone fixation devices and devices in routine clinical use. This paper reviews the approaches researchers have taken to develop these materials to improve their mechanical and biological performances. Polymer blending, copolymerisation, and the use of particulates and fibre bioceramic materials to make composite materials and surface modifications have all been studied. Polymer blending, copolymerisation, and particulate composite approaches have been adopted commercially, with the primary focus on influencing the in vivo degradation rate. There are emerging opportunities in novel polymer blends and nanoscale particulate systems, to tune bulk properties, and, in terms of surface functionalisation, to optimise the initial interaction of devices with the implanted environment, offering the potential to improve the clinical performances of fracture fixation devices.

Keywords: biomaterials; polyesters; polymer blends; copolymers; biodegradable materials; bone regeneration; mechanical properties; composites; glass fibres

1. Introduction

Fractured bones require stabilisation to allow healing to occur. For complex fractures, and in anatomical areas not conducive to external fixation, internal fracture fixation plates are required. Currently, titanium alloys are the most commonly used materials for these plates due to their strength and stiffness being high enough to protect the damaged bone during healing. However, once healing is complete, these high mechanical properties can become problematic, in particular, disrupting the normal remodelling processes of the bone, leading to a reduction in bone quality under the plates, termed stress shielding [1,2]. In fact, stress shielding can become so sufficiently severe that a significant number of revision operations are required annually to remove the fracture plates—along with the associated costs to health care funders and health risks for patients [3].

To reduce the incidence of post-operative stress-shielding research has focused on developing alternative materials for use in fracture plates. In particular, attractive materials are polymers that dissolve, or resorb, in aqueous environments, such as those based on polylactic acid (PLA) and polyglycolic acid (PGA) [4]. This resorption behaviour offers the potential advantage of the fracture plate disappearing once healing is complete. By combining PLA and PGA as copolymers or blends, and by controlling the concentrations of each component, materials with a range of mechanical properties have been developed, with differences in dissolution times, ranging from months to years [5]. Consequently, fracture plates have been developed that are suitable for use in a number of anatomical areas (e.g., the craniofacial region). However, the inferior mechanical properties of these polymers, in respect to titanium alloys and the bone means that applications for these polymeric fracture plates are limited [4]. The plates often need to be much thicker than Ti-alloy plates designed for similar applications due to the lower stiffness of the polymer plates being insufficient to protect the healing bone without the increased polymer bulk. Additionally, these plates have several (potential) post-operative complications (e.g., acid degradation products can cause inflammation) [6–9]. Considerable research has been conducted to improve these polymers, to extend the range of clinical applications of these materials and reduce the incidence of post-operative complications.

This review describes the properties of PLA and PGA homopolymers, their copolymers, and blends, together with a review of other polymers that have been added to improve properties. It also describes the methods used to improve the fracture plate's mechanical and physical properties, by producing composite materials, which are often made using the addition of glasses and ceramics. Further, we discuss attempts to improve the biocompatibility of these materials by adding bioactive components or functionalisation of already existing components of the polymer glass composites.

2. Polymer Enhancements

2.1. Copolymers and Polymer Blends

Copolymers are favourable as they combine the beneficial properties of the monomeric species whilst simultaneously negating their disadvantageous properties. The same applies for polymer blends and the polymers that constitute these blends. This results in the possibility of a range of materials with desirable properties, including tailored material degradation rate, tunable biophysical, and biochemical properties [10]. Over the years, there has been a wide range of copolymers investigated for bone fixation applications, with copolymerisation being the key feature. Some of the most common polymers investigated include poly-L-lactide (PLLA), poly(D-lactic acid) (PDLA), poly(D,L-lactic acid) (PDLLA), and polyglycolic acid (PGA). These are summarised in Table 1.

Table 1. Summary of Polymer Blends.

Polymer	Properties	Study Type	Degradation	Clinical Application	Reference
Poly-L/DL-Lactide [(P[L/DL]LA)] (70/30) Copolymer	Sufficient to support fractures, bendable	In vivo, human	Consistent with bone healing	Orbital fractures	[11]
P(D(2%),L(98%))lactide Copolymer	Sufficient to support fractures, bendable	In vivo, human	Consistent with bone healing	Interference fixation screws for anterior cruciate ligament surgery	[12]
PLGA (L-lactide 82: glycolide 18) Copolymer	7 GPa Young's modulus 50% by 12 weeks	In vitro	50% decline in mechanical properties by 12 weeks, peak retention at 8 weeks	Choice of material in foot surgery	[13]
PLGA/PLA (100:0, 75:25, 50:50, 25:75, 0:100) Copolymer	N/A	In vivo, rodent	2 weeks–6 months	Oral resorbable implants	[14]
Poly(lactic acid)-b-poly(lactide-co-caprolactone) (PLA-b-PLCL) 30 wt% PLCL Copolymer	173 MPa tensile strength 5.4 GPa Young's modulus)	In vitro	N/A	Smart bone fixation material with shape memory effect	[15]
PLLA/PHBV (40:60) Blend	Improved elasticity compared to PLLA	In vitro	PLLA: 12 weeks, PHBV: 53 weeks	Orthopaedics	[16]
P(L/D,L)lactide/TMC (56:24:20 and 49:21:30) Copolymer	Decrease in Young's modulus and tensile strength compared to P(L/D,L)LA (0.9 GPa from 3.1, 27 MPa from 50 MPa)	In vitro	N/A	Soft tissue engineering	[17]

Table 1. Cont.

Polymer	Properties	Study Type	Degradation	Clinical Application	Reference
Poly-e-caprolactone-co-L-lactide (100:0, 90:10, 80:20, 60:40) compatibilised with 2.0 phr Joncryl® Blend	Young's modulus/stress at break: 100:0—1.5 GPa/57.6 MPa 90:10—1.2 GPa/44.8 MPa 80:20—1.1 GPa/41.8 MPa 60:40—0.32 GPa/14.6 MPa	In vitro	N/A	Long term implantable devices, tissue engineering, drug delivery	[18]
Poly(D,L-lactide-co-glycolide)/(L-lactide-co-E-caprolactone) (PDLGA/PLCL) PDLGA(DL-lactide/glycolide, 53/47 M ratio), 70/30 L-lactide/ε-caprolactone M ratio PDLGA/PLCL (80:20, 60:40, 40:60, 20:80) Blend	Young's modulus/Yield strength PDLGA—1.2 GPa/36 MPa PDLGA:PLCL(80:20)—1.1 GPa/28 MPa PDLGA:PLCL (60:40)—0.6 GPa/19 MPa PDLGA:PLCL (40:60)—0.02 GPa/5.6 MPa PDLGA:PLCL (20:80)—7.1 MPa/-	In vitro	Degradation accelerated by larger amounts of PLDGA. PDLGA has a lower molecular weight compared with PLCL; therefore, favours an increased hydrolytic degradation rate	Minimally invasive surgery, shape memory polymer	[19]
PDLLA/P(TMC-CL) (Poly(L/D-lactide) (85:15)/20% wt (50/50 trimethylene carbonate-co-e-caprolactone) Blend	Decrease in tensile strength (50 MPa (PLDLA) in comparison to 30 MPa (PLDLA20%P(TMC)CL)), bending modulus increase (2.7 GPa to 4.9 GPa), elongation increase (7.5% to 130%), increase in impact strength	In vitro and in vivo, canine	No significant mass loss up to 45 weeks in vitro, in vivo healing within 12 weeks, screws and plates loosened after 18 weeks	Single fractures of the mandible	[20]

Polylactic acid (PLA) is an aliphatic, biodegradable polyester used in a range of biomedical applications [21–23]. PLA polymers are used in bone fixation applications as it is possible to tailor the rate and degree of crystallinity, and, thereby, mechanical properties, degradation behaviour, and processing temperatures [5]. These are achieved through controlling the stereochemical architecture and the molecular weight of the polymer. Two stereoisomers of PLA are PLLA (poly(L-lactide) and PDLA poly(D-lactide) (Figure 1). PDLA is amorphous because of the racemic mixture of monomer, which disturbs crystallinity; consequently, erosion occurs at a significantly faster rate compared with PLLA [24–27]. To compare the two, a highly crystalline PLLA will take 2–5 years [28] to degrade in a normal physiological environment in vivo and in vitro (pH 7.4, phosphate-buffered solution (PBS), and incubated at 37 °C) [29]. An amorphous poly(D,L-lactic acid) (PDLLA) shows loss of integrity in 2 months and complete degradation in 12 months under similar conditions [30,31].

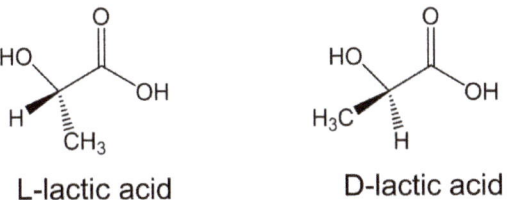

Figure 1. Enantiomer structures of PLA monomer [32].

PLLA and PDLA have been used extensively as copolymers to create materials for bone fixation applications [20,22,23,33,34]. They can also be combined to form poly(D,L-lactic acid) PDLLA, with ratios of each polymer being varied. By controlling L/D ratios, a variety of different mechanical properties and degradation profiles can be obtained and implemented for bone fixation at various sites [35]. Poly-L/DL-lactide (70/30) plates have been successfully used in maxillofacial applications where an enhanced degradation rate is necessary to eliminate oedema formation; which can occur as a result of the slow degrading PLLA [11]. The strength of these plates is comparable to those of PLLA. Poly-D (2%),L (98%)-lactide interference fixation screws have been used in clinical applications, displaying no adverse side effects and a good rate of degradation due to the D isomer in the copolymer [12]. Superior thermal and mechanical properties, compared to 100% pure L- or D- lactide polymers, have been reported from the production of stereo-complexes of enantiomeric PLAs. A polymer stereocomplex can be defined as "a stereoselective interaction between two complementing stereoregular polymers, that interlock and form a new composite" [36]. This results in a material with enhanced physical properties due in part to the formation of crystallites with intermolecular crosslinks [5,26,37–39].

Self-reinforcement (SR) is a form of copolymerisation where a composite structure is formed from a polymer matrix being reinforced with oriented reinforcing units, such as fibres, fibrils, or extended chain crystals. These units are composed of the same material as the matrix [40]. This technique can produce a range of polymeric devices of very high strength, including poly-L/DL-Lactide (SR-P(L/DL)LA) 70/30, which was developed for use as miniplates and miniscrews for anterior mandibular fracture fixation [40]. The matrix material, PLDLLA, was formed into a composite structure by reinforcing with the same material (PLDLLA) and was reported to have bending strengths of up to 400 MPa [40]. Self-reinforced materials were developed and enhanced through controlling copolymer ratios (L/D). Through blending the L/D isomers, it is possible to obtain fracture fixation plates that are bendable at room temperature without the application of heat, which can be required for other types of non-ductile resorbable materials used in fracture fixation [22,41–43]. SR-PDLLA/PLLA (40 PDLLA:60 PLLA) showed faster absorption in vivo compared with SR-PLLA, possesses sufficient mechanical properties for fixation of osteotomies, and displays no adverse foreign body reactions at 48 weeks post implantation [44]. Highly oriented polyester fibres were used as reinforcement in self-reinforced polymers or single polymer composites [45]. PLLA matrices with PLLA fibre reinforcement produced materials with increased strength values compared to unreinforced PLLA [46,47] (bending strength of 200 MPa and shear strength of 94–98 MPa), appropriate for treatment of the cortical bone in certain fixation sites [48]. Self-reinforced P(L/DL)LA plates and screws of 70:30 P(L/DL)LA (BioSorb Fx, BionX Ltd., Bedford, MA, USA) were clinically assessed for bone fixation with minimal post-operative complications [49] in craniomaxillofacial [40] and limb surgery [50]. Self-reinforced PLA was successfully used as fixation plates and screws in the treatment of children's forearm fractures, demonstrating stability in a long arm cast, and supporting osteosynthesis whilst eliminating the need for a second surgery [51].

Polymer blends involve the physical entanglement of at least two polymers in the mix. PLLA has also commonly been combined with PGA for the purpose of bone fixation applications. PGA is a rigid material with high crystallinity, which translates to a high strength and modulus, but also has a faster degradation rate than PLA, on the scale of 6–12 months [52]. When combined with PLLA, a desirable blend with good mechanical properties is formed. PLLA/PGA (82:18) has been used in widespread craniofacial applications due to its ability to maintain sufficient mechanical strength for 6–8 weeks, with in vitro studies showing mass loss being complete in 9–15 months [13]. When PGA is combined with PLA, the copolymer has a balance of the hydrophilic properties of PGA and PLLA, leading to an intermediate rate of hydrolysis. Higher incorporation of PGA may elicit a faster degradation rate, but this is not always the case. One study found that a content of PLA between 75 and 100% and PGA 0–25% was the most desirable in terms of degradation characteristics [14]. Clinically, LactoSorb copolymer pins, composed of PLLA/PGA (82/18) are used in an array of orthopaedic applications [13]. This material has a strength loss profile that better mimics the healing process in comparison the homopolymer PGA or PLLA. A degradation rate slower than that of PGA reduces the occurrence of premature fixation strength loss before bony union occurs, whilst a faster degradation rate than PLLA allows for a faster transfer of load to the surrounding healing tissue, promoting remodelling. The PLLA/PGA pins were used successfully in distal chevron osteotomies, providing fixation on a level comparable with metallic implants [53].

Blending lactide polymers with non-lactide polymers were investigated in order to alleviate the brittle nature of the former materials. Poly-L-lactide/ε-caprolactone (PLCL), trimethylene carbonate (TMC), and poly(3-hydroxybutyrate-co-3-hydroxyvalerate) (PHBV) have all shown promise in reducing the brittle character of PLLA specifically. Poly DL-lactide/glycolide copolymer (PDLGA) and poly-L-lactide/ε-caprolactone (PLCL) were combined to form a set of PDLGA/PLCL blends. Greater incorporation of PLCL into the blend provided a decline in elastic modulus and yield stress and an increase in yield strain. Characteristics were provided from the elastic polymer, PLCL. When PLCL makes up 60% of the polymer blend, the brittle nature of the polymer is eradicated [19]. A novel terpolymer based on L-lactide, D,L-lactide, and trimethylene carbonate (TMC) displays brilliant mechanical and biocompatibility properties of PLDLA. TMC, an elastomeric component, was incorporated in with the lactide polymers to alleviate the brittle nature. The three components were first copolymerised and then moulded into pins for evaluation. Incorporation of TMC into the blend reduced mechanical properties of the material however make it suitable for use in soft tissue applications [17]. PDLLA was combined with copolymer of trimethylene carbonate (TMC) and e-caprolactone (CL) to form a PDLLA/P(TMC-CL) blend, assessing its suitability for fracture fixation by studying degradation. The material, manipulated into bone plates and screws, was tested in vitro and in vivo (mandibular fractures). In vitro, the initial tensile strength was maintained for a time-period after bone healing occurred at 6–12 weeks. At 45 weeks of the study, it was seen that minimal water was absorbed by the degrading blend, with no observation of significant mass loss. In vivo, the implants devised enabled uneventful bone healing and no premature failure of the devices were seen; however, although undisturbed bone healing occurred around 12 weeks, the implanted screws had broken and bone plates loosened [20]. Inion (Finland) at present has clinically available products on the market that are composed of L-Lactide, D,L-Lactide, polyglycolide, and TMC [54]. Clinical studies using the material in the form of miniplate conclude that the resorbable blend can be used in the same circumstances as titanium miniplate, with exception in maxillary elongation and mandibular setback [55], due to higher relapse in these circumstances and one degradable plate not being stable enough to prevent such an occurrence. Pins composed of PLLA/PHBV were assessed for their mechanical and degradation properties, where the addition of PHBV improved thermal properties and dampened the brittle character of PLLA. Intermediate ratios of the PLLA/PHBV (60/40, 50/50, 40/60) copolymer degraded faster than pure PHBV whilst maintaining mechanical properties for longer than pure PLLA [16].

2.2. Orientation

Higher polymer crystallinity and chain orientation can be used to enhance mechanical properties in the direction of orientation [56]. Polymer chain orientation is a way to produce polymers with high strength and modulus. This can be achieved typically through cold- and/or hot-drawing of the polymer materials by melt processing above a polymer glass transition temperature (Tg), but below the melting temperature (Tm) [57].

Processing methods, such as extrusion, compression, and injection moulding can be used to achieve chain orientation. It has been found that processing poly(D-lactide) by solid-state extrusion improved the bending strength and bending modulus of the material. The draw rate and temperature of extrusion are the key parameters that determine the mechanical properties of the polymer. Annealing of the polymers prior to solid state extrusion also allowed for relief of stress concentration and chain orientation [57], and the crystallinity of compression-moulded or extruded PLLA samples were increased through annealing, and sterilisation by ethylene oxide gas processing. Increased crystallinity is associated with an increase in the Young modulus; however, it is also accompanied with a decrease in tensile strength and elongation at break [58]. Optimisation of the injection moulding process and use of nucleating agents can result in an increase of crystalline content of commercial grade PLA from 5 to 42%. An added benefit is the concomitant decrease in processing time [59]. Despite the advantages of processing methods on some of the mechanical properties of material, the molecular weight does diminish with heat treatment of polymers and should be taken into consideration, as this would, in turn, impact the mechanical behaviour of the polymer overall.

Orientation of polymer chains can also be achieved through enhanced entanglement between the long-branched chains. A poly(lactic acid)-branched-poly(lactide-co-caprolactone) (PLA-b-PLCL) was developed through two phase separated structures using long chained branches and demonstrated high tensile strength and modulus (172 MPa and 5.4 GPa respectively) showing potential for use in bone fixation [15].

3. Composite Materials

The limited mechanical properties, the lack of bioactive behaviour that allow bone apposition and bonding on the polyester implant, and complications associated with the acidic degradation products are major drawbacks of polyester implants. Ceramic materials are high strength, biocompatible, corrosion resistant, and possess a high level of bioactivity. For this purpose, composite polyester implants with ceramic fillers (hydroxyapatite (HA), tricalcium phosphate (TCP)), silicate, and phosphate glasses have been used to improve the overall mechanical properties and endow bioactivity in bioresorbable osteofixation devices. In addition, alkaline dissolution from bioactive ceramic fillers counteracts the decrease of pH in the ambient implantation site by the acidic degradation products of the polyester matrix.

3.1. Particulate Bioceramics

Hydroxyapatite (HA) and β-tricalcium phosphate (TCP), have undergone intensive research for their use in bone applications [60]. HA ($Ca_{10}(PO_4)_6(OH)_2$) has a molar ratio of 1.67 Ca/P and is a bioactive ceramic that is typically used in coatings and can be incorporated into materials in particulate form. HA has been shown to improve the biocompatibility of biomaterials due to its similarity in structure and composition to bone and enamel [61]. It is osteoconductive and osteoinductive [60]. Tricalcium phosphate has three polymorphs; α-TCP, β-TCP, and α'-TCP. In particular, β-TCP displays excellent biocompatibility, bioactivity, and bio-resorbability. Although HA has greater mechanical strength than β-TCP, the latter is more resorbable. This encourages faster growth of new bone around the implant.

Polyesters, such as PLLA, PCL, and PHBV can take up to 2 years to degrade, and ceramic incorporation has been found to decrease this. HA enhances the degradation rate of PCL, with mass loss in vitro increasing with HA content. In addition, the compressive yield strength and modulus of the material also increased almost linearly with HA content [62]. Up to 3 wt% nano-hydroxyapatite (n-HA) incorporated into a poly-L-lactic-co-glycolic acid (PLGA) matrix has been found to significantly increase the mechanical properties of the polymer [63,64]. It was concluded that a higher content of n-HA in the composite promoting enhanced crystallisation, but also caused greater agglomeration in the PLGA matrix, which resulted in a decrease in mechanical properties. The addition of 3 wt% n-HA enhanced the degradation performance of the material, making the material promising for use in clinical applications in comparison to pure PLGA in bone fracture internal fixation materials. Porous polymer-hydroxyapatite scaffolds for femur fracture treatment produced by 3D printing have presented promising results. Hydroxyapatite nanoparticles were used to enhance PLA polymer and subsequently 3D-printed into porous cylindrical structures. Comparing 5% to 25% HA content in PLA, the compressive modulus and elastic modulus increased by ~38% and ~92%, respectively, a 11% decrease in porosity of the scaffolds was also seen. These results illustrate the reinforcing behaviour of HA and the mechanical properties being a function of porosity [65]. The mechanical and thermal properties of injection-moulded poly(3-hydroxybutyrate-co-3hydroxyhexanoate) (P(3HB-co-3HHx))/hydroxyapatite nanoparticles (n-HA) parts for use in bone tissue engineering and bone fixation were assessed [66]. The addition of 20% nHA improved the mechanical properties; specifically, the tensile moduli and flexural moduli by approximately 60%. A larger content of n-HA leads to agglomeration, with the ductility, toughness, and thermal stability of the material declining. The addition of nHA up to 10% most closely resembled natural bone in terms of strength and ductility and demonstrated controlled agglomeration. Considering its biocompatibility, the copolymer composite can be appropriate as a bioresorbable implant for low-stress bone fixation sites. An alternative way to simple melt mixing of ceramics with polyesters is covalently linking the two. A novel study surface first grafted nano-HA to PLLA. This nano-HA/PLLA composite was then blended with PLLA. Surface-grafting HA to PLLA led to an improvement in mechanical properties, compared to similar non-grafted PLLA/HA composites due to the grafting providing strong linkages between the HA particles and the PLLA matrix [67]. Further works by the researchers saw surface modification of carbonate hydroxyapatite particles with PLGA, with processing as previously described [68]. The PLGA/g-CHAP nanocomposites displayed improved mechanical properties in comparison to unmodified PLGA. This was attributed to the strong interfacial bonding between PLGA and g-CHAP particles. At 2% of g-CHAP content, the fracture strain was increased from ~5% (for neat PLGA) to 20%. When g-CHAP content was between 2 and 15%, the composites showed enhanced tensile strength and fracture strain. The tensile strength decreased linearly with filler content beyond 20%.

In vitro, TCP incorporation into PLLA showed good biocompatibility and a rate of degradation consistent with bone healing [69]. In particular, material strength was maintained for 16 weeks, after which a decline was observed with no measurable strength of the material at 40 weeks. There was a decrease in tensile strength of the material over the investigated period. In vivo, these results were seen on a shorter time scale. The addition of 10% volume TCP filler to PLLA gave a biocomposite material with an extended strength and molecular weight retention period, both in vitro and in vivo [69]. The same experiments conducted on PHBV showed slower degradation rates, which was attributed to the hydrophobic nature of PHBV. Ternary blends of PCL/PGA/tricalcium phosphate (TCP) (80/10/10 and 70/10/20) displayed success for use in low load bearing applications, such as maxillofacial surgery [70]. Adhesion strength of the materials was tested using a previously established protocol, where investigated biomaterials were melted and applied to two bone sections before curing. The mechanical properties of the set fixation was then tested through compressive and tensile force [71]. The blends retain the adhesive strength of PCL whist having improved hydrophilicity. A higher incorporation of TCP also results in enhanced degradation and support for osteoblast growth. PLLA was strengthened with a poly(ε-caprolactone-co-L-lactide) copolyester [18]. A higher quantity of copolyester (PCL/LLA) in the blend provided a greater elongation at break with a concurrent decrease in the Young's modulus and strength. Use of a chain extender (Joncryl® ADR 4368, BASF, Thailand) enhanced phase compatibility of PLLA and the interspersed copolyester phase.

3.2. Glass Fibres

Bioactive glasses were first developed in 1969 and represent a group of materials that have the capacity to bond with bone in physiological environments. They have been greatly studied due to their desirable characteristics, which include biocompatibility, degradation, and mechanical strength [72]. By changing the composition of glass, it is possible to obtain a range of mechanical properties and high controllability over the resorption and ion dissolution that resemble and complement the mineral content of bone. For instance, it is possible to decrease degradation rates in a phosphate glass composition by addition of hydration-resistant metal oxides, such as Al_2O_3, Fe_2O_3, and TiO_2 [73,74]. The use of high aspect ratio fibrous bioactive glass structures, as reinforcing phases, has produced fully bioresorbable polyester composites with higher mechanical properties compared to composites with particulate reinforcement [74]; however, studies have shown variable results, with some showing a rapid loss of mechanical properties with degradation [75]. The use of fibres in bioresorbable composites has also been associated with an immediate osteoinductive effect owed to the presence of the filler in the outer surface of the polyester implant [76]. Glass fibre reinforced polyester composites have also been found to act protectively to the polyester matrix against gamma radiation from deterioration of its mechanical properties [74].

The properties of the composite for a given polyester matrix and glass fibre reinforcement depend on the geometry (i.e., aspect ratio) of the filler, distribution within the matrix, volume fraction, and surface area of the composite and strength of the fibre–matrix interface. Therefore, fibre and composite manufacturing methods significantly influence the properties of the composites. Glass fibres are typically fabricated into continuous fibres via melt spinning or preform drawing manufacturing processes, with single filament production being more common [77]. The mechanical, thermal, and degradation properties of glass fibres vary from those of bulk glasses depending on processing temperatures, drawing speed and ratio, and viscosity of molten glass, which ultimately determine the fibre diameter. A smaller fibre diameter leads to increased dissolution times, a fact that is associated with the active surface area of the filler [74].

The fibres can be appropriately aligned within the polyester matrix to create long-fibre composites, or chopped and dispersed within the matrix to create short fibre composites. The use of long fibre reinforcement is associated with improved mechanical properties in the composites when measured along the axis of the fibre. During manufacturing, the fibrous preform has to be positioned within a mould cavity and infiltrated with polyester matrix, which might require complex tooling. Short fibre composites are easier to process with extrusion/injection or casting processes, with the filler is melt-mixed or dissolved with the polyester matrix prior to processing. Solvent casting methods have also been used to create short glass fibre polyester composites. The method of incorporation of glass fibres into the polyester matrix plays an important role in the obtained mechanical and degradation properties.

Long-fibre unidirectional (UD) woven mats and randomly orientated short fibre non-woven mats (RM) of iron doped glass phosphate fibres as reinforcements in a PLA matrix have produced bioresorbable composites with enhanced mechanical properties [78]. UD matt reinforced composites with filler volume fraction of 20% revealed a faster depletion of mechanical properties during degradation compared to the randomly oriented short fibre matt composites of 30–40% volume fraction. A maximum modulus and strength for the RM and UD were 10 GPa/120 MPa and 11.5 GPa/130 MPa, respectively, falling short of the ideal properties for cortical bone. Increased concentration of fibres, despite allowing for enhancement of mechanical properties of the composite material in comparison to polymer alone, also lead to an increased degradation time. Further work demonstrated that ~30% fibre volume fraction of unidirectionally and randomly aligned fibres into PLA rods imparted the composite with a higher initial modulus, which succumbed to degradation faster than PLA alone [79]. The RM and UD-filled PLLA was manufactured into intramedullary nails and the mechanical properties were assessed [80]. The composite reinforced with unidirectionally aligned fibres provided enhanced mechanical properties compared to pure PLA rod. The method of processing via forging at 100 °C also improved the properties of the PLA matrices by influencing chain orientation. This method has a similar effect to the drawing process at low draw ratios. Fibre incorporation and material processing were jointly responsible for the property enhancement. The materials were also processed into bioresorbable screws with promising results [81,82].

Different treatment processes performed on phosphate glass fibres further impact their performance in composites. When short glass fibres were incorporated into PLA in the form of randomly oriented non-woven mats, there was a mass loss of 14% and 10%, respectively, for non-treated and heat-treated fibre composites over 6 weeks, in comparison to no mass loss seen for pure PLA. Incorporation of glass fibres enhanced the material modulus (2.5 GPa → ~ 5 GPa for both composites) and flexural strength significantly, with the latter matching that for cortical bone. Concerning the retention of mechanical properties over 6 weeks of in vitro degradation, the strength of PLA declined slightly, while the modulus was maintained. For the non-treated and treated samples, there was a significant decline in both modulus and flexural strength; 0.5–1 GPa and ~40 MPa, respectively. There was no mass loss for PLA over the 6 weeks compared with ~12% and ~14% for heat-treated and non-treated, respectively. Heat treatment of the fibres led to a decreased dissolution rate. The PLA alone and heat-treated composites displayed higher cell viability due to their slower degradation [83]. Incorporation of short non-treated and heat-treated glass fibres in the form of randomly oriented non-woven mats was also investigated in polycaprolactone. The composite materials presented with a flexural strength and modulus of up to 30 MPa and 2.5 GPa, respectively, values that are comparable to those of the human trabecular bone. A higher mass loss was seen in the composites with a higher volume fraction (V_f 17/18%), 20% compared with 8% for (V_f 6.4%) over 5 weeks [84]. The rapid decrease in mechanical properties in glass fibre reinforced polyesters is attributed mainly to early hydration of the reinforcement due to weak interfacial interactions and polymer swelling during degradation that increases the internal stress of the system inducing early cracking and failure [85,86]. Coupling agents, including silanes, acids, and other agents that can create covalent bonds between filler and matrix material have been used to enhance the interfacial properties between phosphate glass and PLA matrices. The improved interfacial shear strength between the phosphate glasses and PLA matrix prevents early hydration of the filler and assists with the fibre/matrix load transfer, thereby improving the overall mechanical properties of the composite material [87].

Two main points can be seen from assessing composite material. Firstly, the method of incorporation of bioceramic fillers into the polymer matrices must be considered, as studies have shown certain processes can be disadvantageous to mechanical properties due to improper blending and homogeneity across the composite material [88]. Secondly, the volume fraction of the filler must be optimised, as excessive content can lead to agglomeration and depletion of mechanical properties.

4. Surface Enhancement

4.1. Overview of Surface Enhancements

Surface modifications for bone fixation has been an active area of research for decades and is commonly carried out in order to enhance physiological bone fixation, assist the healing process, and improve biocompatibility, functionality, and biological efficacy. The success or failure of the implant is dependent on the device and surrounding tissue at the implant interface [89]. Fabrication processes can change the surface composition in comparison to the material bulk. This may be due to oxidation or hydrolysis of surface groups and/or preferred molecular orientation of surface groups in order to minimise surface free energy. Such effects typically occur unevenly over a surface, which can impact the performance of the material [90]. It is thereby crucial to alter polymer surfaces to be able to regulate concurrent surface interactions and responses.

Controlling biocompatibility is an ongoing challenge with biomaterials, as synthetic and naturally occurring polymers quite often do not have the surface properties, which are required for specific applications [91]. Generally, surface enhancements are made with certain objectives. When considering polyester surface enhancements, they are challenging to modify due to their ease of degradation with chain scission, solvent sensitivity, and low heat stability [92,93]. Surface enhancements are typically conducted to increase or reduce [90]:

- Hydrophilicity;
- ionic charge/pH;
- adhesion of microorganisms;
- adsorption of molecules;
- permeation of molecules;
- roughness;
- impurities;
- chemical/biological reaction kinetics.

Surface engineering generally includes alteration of topographical (i.e., roughness) and chemical (i.e., coating) characteristics of a medical device. An increased roughness is desirable as it leads to an increased surface area, which in turn gives a larger area for cell adhesion. It additionally enhances biomechanical interlocking between bone tissue and implant [89]. Calcium phosphates are typically the group of materials used for coating orthopaedic implants due to the excellent bioactivity of these ceramics. Together with the mechanical advantages of the substrate implant, there is an improvement of the implants overall performance [94].

4.2. Surface Enhancements of Polyesters for Bone Fixation

From reviewing literature and the commercial products, which are available on the market for bone fixation (Table 2), it is apparent that surface enhancements of polyesters used in bone fixation are not yet employed. A vast array of in vivo and in vitro work has been conducted on bone tissue engineering applications, but this has not yet been thoroughly investigated in bone fixation materials and devices. Figure 2 illustrates the array of techniques that can be used to alter surface properties. These techniques have been divided into three categories: (i) roughening, (ii) coatings, adhesions and depositions, and (iii) grafting. Each group of techniques allows for different topographical surface enhancements. It can be difficult to assess which technique is the best fit, for the purpose of enhancement, for a specific application. Work conducted on plasma, chemical, or laser methodologies investigated what treatment was most appropriate for the modification of PLA surfaces. The effects of each treatment was looked into mechanically (with surface roughness analysis), surface wettability, and chemically (via XPS). Chemical treatment caused the most drastic increase in surface roughness. Plasma treatment lead to an increase in roughness and was found to be dependent upon exposure time. Laser treatment appeared to decrease surface roughness when compared to the untreated PLA. Given these results, chemical modification may be an appropriate method to be used on PLA joint implant surfaces, as an increased surface roughness leads to an increased strength of the joints. Materials processed with plasma exhibit an oligomeric layer on the surface, which could be detrimental to the adhesive joint formation process. As laser-treated surfaces showed a decrease in surface roughness, this method was not deemed suitable to use with regards to material mechanical properties. Chemical modifications cause the least change in water contact angle, but plasma and laser treatments show a significant increase. Plasma treatment overall induces the higher surface energy of the three treatments. Oxygen content of the surfaces increases with all three modes, with plasma processing giving the highest. Plasma modifications are deemed to be the most beneficial of the three processes for polymer implants [95]. Plasma modifications are known to be an effective method to treat the surface of polymers for biomedical applications as these treatments can be selective, yet not affect the bulk polymer characteristics [96,97]. As shown in Figure 2, plasma can be used to introduce surface roughness, graft surfaces, and deposit material onto a surface. Plasma surface modifications often offer a shorter processing time in comparison to other surface modification methods. Although both plasma treatment and plasma coating technologies are commercially available for modifying polymeric surface properties, their use in fracture fixation applications has not been significant to date.

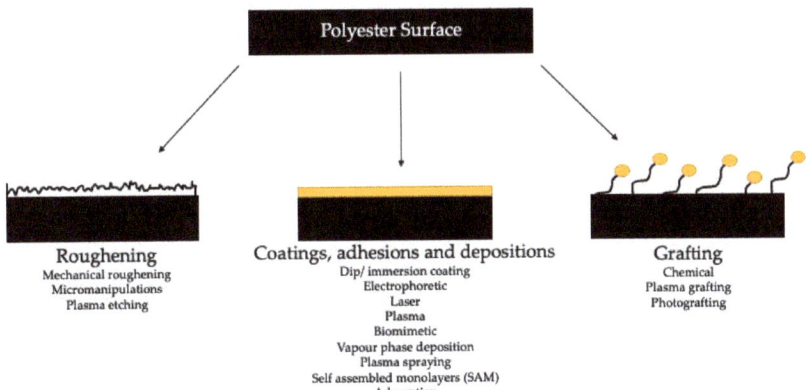

Figure 2. Polymer surface modification methods (for biodegradable polymers) [89,91,98–101].

There is a large scope of data regarding the biocompatibility of polyester materials, specifically with regards to surface enhancement [101–105]. As the polymer biomaterials have interaction directly with an extracellular environment, it is vital that an adverse immune response is not provoked [91,100]. A few important surface characteristics can be altered to obtain the right biocompatibility control for the intended application. These include surface morphology, chemical structure and functional groups, interfacial free energy, wettability, cytotoxicity, and adhesion [91,105]. Plasma treated PDLLA was combined with collagen anchorage. The plasma surface property alterations included improved surface hydrophilicity and increased surface energy. As a result of plasma pre-treatment, there was more collagen fixated to the surface. Mouse fibroblast cells were used to assess cell affinity and showed good affinity with the plasma treated, collagen anchored samples in comparison to untreated surfaces [106]. Similar enhanced cell affinity behaviour on PDLLA has also been reported after treatment with anhydrous ammonia plasma treatment [107]. In contrast, PLA modified with medium pressure dielectric barrier discharge (DBD) plasma treatment illustrated an increase in hydrophilicity along with an increase in oxygen content, measured by contact angle measurements and XPS, respectively. Biologically, plasma modification of PLA led to increased initial cell attachment; however, after 7 days, there was no significant difference in activity on the untreated and treated samples. It was concluded that cell proliferation was not influenced by the application of plasma treatment to the surface [98]. The influence of plasma treatments on these polymer surfaces is dependent on the experiment parameters and materials, which will give a variation in cell attachment success.

A challenge with altering material surfaces is the side effect that may be imposed on the bulk material [91], however; there are surface enhancement techniques that can limit or even avoid this. A method investigated on polyester materials, which may limit the influence on the bulk, is extreme ultraviolet radiation (EUV). Degradation of the bulk material can be limited through using short wavelength radiation, which is in the extreme ultraviolet range; which, in turn, is only absorbed by a very thin layer of the polymer surface (<100 nm) [91]. Vapour phase (VP) grafting of N-vinylpyrrolidone was conducted on four types of biodegradable polymers; PLA, PCL, PLGA, and PTMC. Wettability of all materials was enhanced by the process and it was possible for the surface topographies not to be altered due to the thin graft layer applied. Additionally, film surfaces for the grafted materials with VP after 30 min was rougher than the original polymers and enhanced good cellular adhesion was noted with PLLA, PLGA, and PTMC [99]. Radio frequency (RF) plasma is an additional process for modification without bulk properties being affected [108]. Characteristics, including advantages and disadvantages of some commonly used surface enhancement treatments on polyester surfaces, are summarised in Table 2.

Table 2. Summary of surface enhancement treatments on polymers [89,102,105,109,110].

Process	Advantages	Disadvantages
Chemical grafting	Exposure of functional groups on material surface. Long and stable effects produced.	Limited by the functional groups on the surface. Chemical modifications (i.e., aminolysis/hydrolysis) may be required prior to fixation of biomolecules, causing destruction of the topological structure of the surface.
Pulsed Laser Deposition (PLD)	Simple, versatile, rapid, cost effective. Precise control of the thickness and morphology of the films deposited.	Small area of structural and thickness uniformity.
Photografting	Solvent free approach. Non-destructive Surface topography maintains a thin graft layer.	May affect the material bulk properties and induce material degradation.
Extreme UV radiation (EUV)	Penetration depth limited (<100 nm in upper layer of polymers), affect surface layers only. Strong interaction of EUV photons with material.	Lack of commercially available lab sources of EUV radiation. High equipment costs.
Plasma Modification	Simple and widely used. Bulk properties are not affected.	Size of treated material restricted by the size of the treatment chamber.
Physical coating	Simple and effective methodology.	Bonding relatively weak, specifically in aqueous environments.

The methods discussed highlight some of the breakthroughs, which have enabled triggering of specific responses, recruitment of the correct cells and stimulus for these cells to perform. These are additional functionalities to otherwise inert polyester biomaterials [89].

5. Current Market Products

Table 3 summarises the current polyester-based biomedical devices used in bone fixation applications. It can be seen that the majority of devices are composed of PLA and its composites, which is to be expected, as PLA in clinical applications is the most commonly used biodegradable polymer [111]. This is predominantly due to its excellent biocompatibility, mechanical properties, and history of use in medical applications since the 1970s [112]. The majority of products that are detailed are used in small load bearing applications, which suggests that material enhancements to biodegradable polymers are still a requirement in order to be used in larger load applications. This may in part be due to the biodegradability and mechanical properties, prior to and during degradation, of these materials, when scaled into larger devices.

Table 3. Orthopaedic fracture fixation devices on the market.

Company	Device	Application	Material
CONMED	SmartPin/SmartPin PDX	Foot and ankle.	PLA
	SmartNail	Foot and ankle.	PLA
	BioScrew	Knee (tibial/femoral applications). Bioabsorbable interference screw.	PLA
	BioMini-Revo	Shoulder.	PLA
J&J	RAPIDSORB	Resorbable plates, meshes and screws intended for use in fracture repair, and reconstructive procedure of the craniofacial skeleton. Implants resorbed in 12 m.	85:15 poly(L-lactide-co-glycolide)
	ORTHOMESH	Resorbable graft containment system.	85:15 poly(L-lactide-co-glycolide)
Stryker	SonicPin	Austin/chevron osteotomy. Maintain alignment and fixation of bone fractures, osteotomy, or bone grafts in hallux valgus applications in the presence of appropriate immobilisation (e.g. rigid fixation implants, cast and brace.)	PLDLLA
	Delta System	8–13 months, craniofacial and mid-facial skeleton fixation.	P-L/D-LA/GA

Table 3. Cont.

Company	Device	Application	Material
Smith & Nephew	Regenesorb	Material used in orthopaedic applications.	PLGA with calcium sulphate and β-TCP
	SureTac III system	Shoulder.	
Teijin Medical Technologies	OSTEOTRANS-OT	Orthopaedic and thoracic surgery. Products include screw, pin, washer, interference screw, rib/sternum pin.	µ-HA and PLLA
	OSTEOTRANS-MX	Products include meshes and screws. Used in cranio, oral, and maxillofacial, plastic and reconstructive surgeries.	µ-HA and PLLA
	FIXSORB	Used in cranial, oral, maxillofacial, plastic, and reconstructive surgeries. Products include screws, washers, pins, rods.	PLLA
	FIXSORB MX	Used in cranial, oral, maxillofacial, plastic, and reconstructive surgeries. Products include plates and screws.	PLLA
Gunze (Japan)	GRAND FIX	Oral, craniomaxillofacial, and plastic surgery. Products include plates and screws (mini for plate locking and cortical full thread screws), rib pins (rib and sternum fixation) and pins, screws, and ACL screws.	PLLA
Acumed (USA)	Biotrak Screws	Fixation for small bones and bone fragments in the upper and lower extremities, including fractures, fusions, and osteotomies. Composed of Biotrak helical nail, pin, standard, and mini screw.	PLLA
Arthrex (USA)	Trim it spin Pin	Pins. Foot and ankle.	PLLA
Takiron	Fixsorb	Fracture fixation,	PLLA/HA
Biomet Arthrotek	Bio-Phase Reunite Screws, pins plates	Fracture fixation,	PLLA/PLG
	LactoSorb		PLLA/PGA

6. Final Considerations and Perspective

Table 4 summarises the enhancement strategies, material/s, and the resultant modified properties, as discussed in this review.

The main reasons for seeking to enhance the properties of materials for fixation are:

- Modifying the modulus, to allow either the stiffness of the device to be changed, or to allow stiffness to be maintained whilst using less material.
- Modifying the strength, again to change the strength of the device or to allow the strength to be maintained whilst using less material.
- Making the material less brittle, to avoid brittle failure in use or when fitting the device.
- Changing the degradation rate of the material to alter the time for a device to resorb.
- Changing the degradation chemistry to avoid an excess of acidic degradation products.
- Improving the interaction between the device and native tissue.

Table 4. Polyester enhancement strategies for bone fixation bioresorbable polyester materials.

Enhancement Strategy	Enhanced Polyester Material	Reference	Modified Properties
Self-reinforcement (oriented units)	SR-P(L/DL)LA SR-PLLA, SR-PDLLA/PLLA SR-PLA	[40,49–51] [22,44] [46–48]	
Copolymers/blends	PLA stereocomplexation P(L/DL)LA (70:30) PDLA (2:98) PLLA/PGA PLA-b-PLCL PLLA/PHBV P(L,DL)LA/TMC PLLA/(PCL/LLA) PLDGA/PLCL PDLLA/P(TMC-CL)	[37–39] [11] [12] [13] [15] [16] [17] [18] [19] [20]	Strength and elasticity, thermal properties, degradation/absorption profile, retention of mechanical integrity
Tuning of thermoforming parameters, nucleating agents, thermal post-processing	PDLA	[57]	Chain orientation and crystallinity
	PLLA	[7,58]	
Bioceramic reinforced co/polyester composites	PLLA/TCP, PHBV/TCP PCL/HA PLGA/nHA PLLA/PLLA-grafted HA PLGA/PLGA grafted CHA P(3HB-co-3HHx)/nHA PLGA/nHA, TCP, Mg-CP, Sr-CP PLLA/phosphate glass fibres PCL/CP glass fibres PLA/CP (coupling agents)	[69] [62] [63] [67] [68] [66] [88] [78–83] [84] [87]	Strength and elasticity, thermal properties, degradation/absorption profile, retention of mechanical integrity, endowment of bioactivity (osteoinduction)
Surface functionalisation	Covalent grafting techniques on PLLA, PCL, PLGA, PTMC substrates	[96]	Surface topography and roughness, surface free energy, and chemistry to improve cell adhesion and proliferation and/or induce specific responses
	Chemical/plasma/laser on PLA substrates	[95,99]	

For the most part, enhancements have been applied using PLA or PLLA as a comparator material, aiming to retain strength, reduce brittleness, increase degradation to promote resorption, mitigate the acidic degradation products, and promote better integration between the polymer and the native tissue.

The complexity arises from the inter-related nature of the properties, which are being manipulated: most of the enhancement methods previously described affect more than one of the properties of interest, and the device design can be used to mitigate some effects. For example, most devices are relatively thin, which enhances degradation and limits the volume of degradation products released at the implantation site.

From Table 2 it is clear that two key material enhancement strategies have been adopted broadly adopted clinically, and so can be seen to offer benefits with clinical value:

- Copolymerisation and blending of polymers for greater flexibility and increased degradation rates, with use of glycolic acid and variations in the use of L- and D-lactic monomers or polymers, the most common approaches in clinically applied devices.
- Composite materials with particulate bioceramics offer benefits in terms of mitigating acidic degradation, and increasing degradation rate, whilst having a relatively small effect on the other properties, although care must be taken to ensure that the amount of particulate loading is not so high as to make the material brittle.

The use of materials with anisotropic mechanical properties, through orientation of the polymer or through long fibre reinforcement, has not been adopted clinically, perhaps suggesting that the loading that the devices are put under in practice is too complex to be addressed by enhancement in one orientation.

In terms of future developments, there is ongoing interest in blending new polymers with polyhydroxyalkanoate-based materials, popular for their green production route, biocompatibility, and benign degradation products [113]. There is also ongoing interest in refinements to particulate bioceramic additives, in terms of the additive itself, with Mg and Sr substituted HA systems explored, and in terms of the form of the reinforcement, with both nanoscale additives and short fibres being explored for enhanced biocompatibility and degradation profiles.

Surprisingly, surface modification techniques have not been significantly explored for fixation devices, despite the more general growth of interest in using surface functionalisation methods to enhance medical devices [114]. Reported work to date has predominantly focussed on promoting hydrophilicity on surfaces, but it is possible to apply specific molecules onto surfaces to elicit specific responses in vivo. The reasons for the application may be to generate a more biomimetic surface for attachment to the native tissue, or to release stimulatory or therapeutic molecules to the fracture site, and exploration of the potential for this in fixation applications could offer a way to further improve the performance of this important class of medical devices.

Author Contributions: All authors have read and agreed to the published version of the manuscript.

Funding: R.N. is funded by the European Union's Horizon 2020 research and innovation program under grant agreement no. 814410. C.T. is funded by the UK Medical Research Council Confidence in Concept fund (MC_PC_18057).

Conflicts of Interest: The authors declare no conflict of interest.

Abbreviations

Polylactic acid	PLA
Poly (L-lactic acid)m	PLLA
Poly (L,D lactic acid)	PLDLA
Poly (D-lactic acid)	PDLA
Poly (D,L-lactic acid)	PDLLA
Poly(L, DL-lactic acid)	PLDLLA
Poly(glycolic acid)	PGA
Poly-L-lactic-co-glycolic acid	PLGA
Poly(lactic acid)- b-poly(lactide- co-caprolactone)	PLA-b-PLCL
Poly(3-Hydroxybutyrate-Co-Hydroxyvalerate)	PHBV
Tri-calcium phosphate	TCP
Calcium phosphate	CP
nano/carbonated/hydroxyapatite	n/C/HA
Poly(trimethylene carbonate)	PTMC

References

1. Alexander, R.; Theodos, L. Fracture of the bone-grafted mandible secondary to stress shielding: Report of a case and review of the literature. *J. Oral Maxillofac. Surg.* **1993**, *51*, 695–697. [CrossRef]
2. Kennady, M.C.; Tucker, M.R.; Lester, G.E.; Buckley, M.J. Stress shielding effect of rigid internal fixation plates on mandibular bone grafts. A photon absorption densitometry and quantitative computerized tomographic evaluation. *Int. J. Oral Maxillofac. Surg.* **1989**, *18*, 307–310. [CrossRef]
3. Hanson, J.; Lovald, S.; Cowgill, I.; Erdman, M.; Diamond, B. National hardware removal rate associated with internal fixation of facial fractures. *J. Oral Maxillofac. Surg.* **2011**, *69*, 1152–1158. [CrossRef]
4. On, S.-W.; Cho, S.-W.; Byun, S.-H.; Yang, B.-E. Bioabsorbable Osteofixation Materials for Maxillofacial Bone Surgery: A Review on Polymers and Magnesium-Based Materials. *Biomedicines* **2020**, *8*, 300. [CrossRef] [PubMed]
5. Farah, S.; Anderson, D.G.; Langer, R. Physical and mechanical properties of PLA, and their functions in widespread applications—A comprehensive review. *Adv. Drug Deliv. Rev.* **2016**, *107*, 367–392. [CrossRef]
6. Geddes, L.; Carson, L.; Themistou, E.; Buchanan, F. In Vitro Inflammatory Response Evaluation Of Pre-Ddegraded Bioresorbable Polymers Used In Trauma Fixation And Tissue Regeneration Applications. *Orthop. Proc.* **2018**, *100-B*, 62.
7. Böstman, O.; Pihlajamäki, H. Clinical biocompatibility of biodegradable orthopaedic implants for internal fixation: A review. *Biomaterials* **2000**, *21*, 2615–2621. [CrossRef]
8. Athanasiou, K.A.; Niederauer, G.G.; Agrawal, C.M. Sterilization, toxicity, biocompatibility and clinical applications of polylactic acid/polyglycolic acid copolymers. *Biomaterials* **1996**, *17*, 93–102. [CrossRef]

9. Wu, D.; Chen, X.; Chen, T.; Ding, C.; Wu, W.; Li, J. Substrate-anchored and degradation-sensitive anti-inflammatory coatings for implant materials. *Sci. Rep.* **2015**, *5*, 1–12. [CrossRef]
10. Francois, E.; Dorcemus, D.; Nukavarapu, S. Biomaterials and scaffolds for musculoskeletal tissue engineering. In *Regenerative Engineering of Musculoskeletal Tissues and Interfaces*; Elsevier Inc.: Amsterdam, The Netherlands, 2015; pp. 3–23. [CrossRef]
11. Al-Sukhun, J.; Törnwall, J.; Lindqvist, C.; Kontio, R. Bioresorbable poly-L/DL-lactide (P[L/DL]LA 70/30) plates are reliable for repairing large inferior orbital wall bony defects: A pilot study. *J. Oral Maxillofac. Surg.* **2006**, *64*, 47–55. [CrossRef]
12. Barber, F.A. Poly-D,L-lactide interference screws for anterior cruciate ligament reconstruction. *Arthrosc. J. Arthrosc. Relat. Surg.* **2005**, *21*, 804–808. [CrossRef]
13. Pietrzak, W.S.; Caminear, D.S.; Perns, S.V. Mechanical characteristics of an absorbable copolymer internal fixation pin. *J. Foot Ankle Surg.* **2002**, *41*, 379–388. [CrossRef]
14. Miller, R.A.; Brady, J.M.; Cutright, D.E. Degradation rates of oral resorbable implants (polylactates and polyglycolates): Rate modification with changes in PLA/PGA copolymer ratios. *J. Biomed. Mater. Res.* **1977**, *11*, 711–719. [CrossRef] [PubMed]
15. Liu, Y.; Cao, H.; Ye, L.; Coates, P.; Caton-Rose, F.; Zhao, X. Long-chain branched poly(lactic acid)-b-poly(lactide-co-caprolactone): Structure, viscoelastic behavior, and triple-shape memory effect as smart bone fixation material. *Ind. Eng. Chem. Res.* **2020**, *59*, 4524–4532. [CrossRef]
16. Ferreira, B.M.P.; Duek, E.A.R. Pins Composed of Poly(L-Lactic Acid)/Poly(3-Hydroxybutyrate-Co-Hydroxyvalerate) PLLA/PHBV Blends: Degradation in Vitro. *J. Appl. Biomater. Biomech.* **2005**, *3*, 50–60. [PubMed]
17. Motta, A.C.; De Rezende Duek, E.A. Synthesis and characterization of a novel terpolymer based on L-lactide, D,L-lactide and trimethylene carbonate. *Mater. Res.* **2014**, *17*, 619–626. [CrossRef]
18. Srisuwan, Y.; Baimark, Y.; Suttiruengwong, S. Toughening of Poly(L-lactide) with Blends of Poly(ε-caprolactone-*co*-L-lactide) in the Presence of Chain Extender. *Int. J. Biomater.* **2018**. [CrossRef]
19. Petisco-Ferrero, S.; Etxeberria, A.; Sarasua, J.R. Mechanical properties and state of miscibility in poly(racD,L-lactide-co-glycolide)/(L-lactide-co-ε-caprolactone) blends. *J. Mech. Behav. Biomed. Mater.* **2017**, *71*, 372–382. [CrossRef]
20. Tams, J.; Joziasse, C.A.P.; Bos, R.R.M.; Rozema, F.R.; Grijpma, D.W.; Pennings, A.J. High-impact poly(L/D-lactide) for fracture fixation: In vitro degradation and animal pilot study. *Biomaterials* **1995**, *16*, 1409–1415. [CrossRef]
21. Naseem, R.; Zhao, L.; Liu, Y.; Silberschmidt, V.V. Experimental and computational studies of poly-L-lactic acid for cardiovascular applications: Recent progress. *Mech. Adv. Mater. Mod. Process.* **2017**, *3*, 1–18. [CrossRef]
22. Haers, P.E.; Suuronen, R.; Lindqvist, C.; Sailer, H. Biodegradable polylactide plates and screws in orthognathic surgery: Technical note. *J. Cranio-Maxillofac. Surg.* **1998**, *26*, 87–91. [CrossRef]
23. Bos, R.R.M.; Rozema, F.R.; Boering, G.; Nijenhuis, A.J.; Pennings, A.J.; Jansen, H.W.B. Bone-plates and screws of bioabsorbable poly (L-lactide) An animal pilot study. *Br. J. Oral Maxillofac. Surg.* **1989**, *27*, 467–476. [CrossRef]
24. Pruitt, L.A.; Chakravartula, A.M. *Mechanics of Biomaterials: Fundamental Principles for Implant Design*; Cambridge University Press: Cambridge, UK, 2011.
25. Röhm, E.; Aldrich, S. *RESOMER® Biodegradable Polymers for Medical Device Applications*; Merck KGaA: Darmstadt, Germany, 2015.
26. Quynh, T.M.; Mitomo, H.; Nagasawa, N.; Wada, Y.; Yoshii, F.; Tamada, M. Properties of crosslinked polylactides (PLLA & PDLA) by radiation and its biodegradability. *Eur. Polym. J.* **2007**, *43*, 1779–1785.
27. Wang, Y.; Zhang, X. Vascular restoration therapy and bioresorbable vascular scaffold. *Regen. Biomater.* **2014**, *1*, 49–55. [CrossRef] [PubMed]
28. Narayanan, G.; Vernekar, V.N.; Kuyinu, E.L.; Laurencin, C.T. Poly (lactic acid)-based biomaterials for orthopaedic regenerative engineering. *Adv. Drug Deliv. Rev.* **2016**, *107*, 247–276. [CrossRef] [PubMed]
29. Weir, N.A.; Buchanan, F.J.; Orr, J.F.; Farrar, D.F.; Dickson, G.R. Degradation of poly-L-lactide. Part 2: Increased temperature accelerated degradation. *Proc. Inst. Mech. Eng. Part H J. Eng. Med.* **2004**, *218*, 321–330. [CrossRef]
30. Nair, L.S.; Laurencin, C.T. Biodegradable polymers as biomaterials. *Prog. Polym. Sci.* **2007**, *32*, 762–798. [CrossRef]
31. Pina, S.; Ferreira, J.M.F. Bioresorbable plates and screws for clinical applications: A review. *J. Healthc. Eng.* **2012**, *3*, 243–260. [CrossRef]
32. NPTEL. Stereoisomers of Lactic Acid. Available online: http://nptel.ac.in/courses/116102006/Flash/6.1%20figure%202.jpg (accessed on 13 September 2020).
33. Barnes, G. *REGENESORB Absorbable Biocomposite Material A Unique Formulation of Materials with Long Histories of Clinical Use*; Smith & Nephew, Inc.: Andover, MA, USA, 2013.
34. Smit, T.H.; Engels, T.A.P.; Wuisman, P.I.; Govaert, L.E. Time-Dependent Mechanical Strength of 70/30 Poly(L,DL-lactide) Shedding Light on the Premature Failure of Degradable Spinal Cages. *Spine* **2008**, *33*, 14–18. [CrossRef]
35. Kanno, T.; Sukegawa, S.; Furuki, Y.; Nariai, Y.; Sekine, J. Overview of innovative advances in bioresorbable plate systems for oral and maxillofacial surgery. *Jpn. Dent. Sci. Rev.* **2018**, *54*, 127–138. [CrossRef]
36. Slager, J.; Domb, A.J. Biopolymer stereocomplexes. *Adv. Drug Deliv. Rev.* **2003**, *55*, 549–583. [CrossRef]
37. Tsuji, H.; Horii, F.; Hyon, S.H.; Ikada, Y. Stereocomplex Formation between Enantiomeric Poly(lactic acid)s. 2. Stereocomplex Formation in Concentrated Solutions. *Macromolecules* **1991**, *24*, 2719–2724. [CrossRef]
38. Yamane, H.; Sasai, K. Effect of the addition of poly(D-lactic acid) on the thermal property of poly(L-lactic acid). *Polymer (Guildf)* **2003**, *44*, 2569–2575. [CrossRef]

39. Furuhashi, Y.; Kimura, Y.; Yamane, H. Higher order structural analysis of stereocomplex-type poly(lactic acid) melt-spun fibers. *J. Polym. Sci. Part B Polym. Phys.* **2007**, *45*, 218–228. [CrossRef]
40. Ylikontiola, L.; Sundqvuist, K.; Sàndor, G.K.B.; Törmälä, P.; Ashammakhi, N. Self-reinforced bioresorbable poly-L/DL-Lactide [SR-P(L/DL)LA] 70/30 miniplates and miniscrews are reliable for fixation of anterior mandibular fractures: A pilot study. *Oral Surg. Oral Med. Oral Pathol. Oral Radiol. Endod.* **2004**, *97*, 312–317. [CrossRef]
41. Weisberger, E.C.; Eppley, B.L. Resorbable Fixation Plates in Head and Neck Surgery. *Laryngoscope* **1997**, *107*, 716–719. [CrossRef]
42. Lim, H.-Y.; Jung, C.-H.; Kim, S.-Y.; Cho, J.-Y.; Ryu, J.-Y.; Kim, H.-M. Comparison of resorbable plates and titanium plates for fixation stability of combined mandibular symphysis and angle fractures. *J. Korean Assoc. Oral Maxillofac. Surg.* **2014**, *40*, 285. [CrossRef]
43. Ongodia, D.; Li, Z.; Xing, W.Z.; Li, Z. Resorbable Plates for Fixation of Complicated Mandibular Fractures in Children. *J. Maxillofac. Oral Surg.* **2014**, *13*, 99–103. [CrossRef]
44. Majola, A.; Vainionpää, S.; Vihtonen, K.; Vasenius, J.; Törmälä, P.; Rokkanen, P. Intramedullary fixation of cortical bone osteotomies with self-reinforced polylactic rods in rabbits. *Int. Orthop.* **1992**, *16*, 101–108. [CrossRef] [PubMed]
45. Gao, C.; Yu, L.; Liu, H.; Chen, L. Development of self-reinforced polymer composites. *Progress Polymer Sci.* **2012**, *37*, 767–780. [CrossRef]
46. Tormala, P.; Rokkanen, P.; Laiho, J.; Tamminmaki, M.; Vainionpaa, S. Material for Osteosynthesis Devices. U.S. Patent 4,743,257, 8 May 1985.
47. Wright-Charlesworth, D.D.; Miller, D.M.; Miskioglu, I.; King, J.A. Nanoindentation of injection molded PLA and self-reinforced composite PLA after in vitro conditioning for three months. *J. Biomed. Mater. Res. Part A* **2005**, *74*, 388–396. [CrossRef] [PubMed]
48. Rokkanen, P. Absorbable self-reinforced polylactide (SR-PLA) composite rods for fracture fixation: Strength and strength retention in the bone and subcutaneous tissue of rabbits. *J. Mater. Sci. Mater. Med.* **1992**, *3*, 43–47.
49. Moure, C.; Qassemyar, Q.; Dunaud, O.; Neiva, C.; Testelin, S.; Devauchelle, B. Skeletal stability and morbidity with self-reinforced P (l/dL) la resorbable osteosynthesis in bimaxillary orthognathic surgery. *J. Cranio-Maxillofac. Surg.* **2012**, *40*, 55–60. [CrossRef] [PubMed]
50. Waris, E.; Ninkovic, M.; Harpf, C.; Ninkovic, M.; Ashammakhi, N. Self-reinforced bioabsorbable miniplates for skeletal fixation in complex hand injury: Three case reports. *J. Hand Surg. Am.* **2004**, *29*, 452–457. [CrossRef]
51. Beslikas, T.; Papavasiliou, K.; Sideridis, A.; Kapetanos, G.; Papavasiliou, V. The use of bio-absorbable self-reinforced polylactic acid co-polymer fixation plates and screws in the treatment of children's forearm fractures. *Orthop. Proc.* **2018**, *86-B*, 162.
52. Shetye, S.S.; Miller, K.S.; Hsu, J.E.; Soslowsky, L.J. 7.18 Materials in tendon and ligament repair. In *Comprehensive Biomaterials II*; Elsevier: Amsterdam, The Netherlands, 2017; pp. 314–340. [CrossRef]
53. Caminear, D.S.; Pavlovich, R.; Pietrzak, W.S. Fixation of the chevron osteotomy with an absorbable copolymer pin for treatment of hallux valgus deformity. *J. Foot Ankle Surg.* **2005**, *44*, 203–210. [CrossRef]
54. Inion CPSTM Fixation System—Inion. Available online: https://www.inion.com/product/inion-cps-fixation-system/ (accessed on 29 June 2020).
55. Ballon, A.; Laudemann, K.; Sader, R.; Landes, C.A. Segmental stability of resorbable P(L/DL)LA-TMC osteosynthesis versus titanium miniplates in orthognatic surgery. *J. Cranio-Maxillofac. Surg.* **2012**, *40*, e408–e414. [CrossRef]
56. Gogolewvki, S. Resorbable polymers for internal fixation. *Clin. Mater.* **1992**, *10*, 13–20. [CrossRef]
57. Weiler, W.; Gogolewski, S. Enhancement of the mechanical properties of polylactides by solid-state extrusion: I. Poly(D-lactide). *Biomaterials* **1996**, *17*, 529–535. [CrossRef]
58. Weir, N.A.; Buchanan, F.J.; Orr, J.F.; Farrar, D.F.; Boyd, A. Processing, annealing and sterilisation of poly-L-lactide. *Biomaterials* **2004**, *25*, 3939–3949. [CrossRef]
59. Harris, A.M.; Lee, E.C. Improving mechanical performance of injection molded PLA by controlling crystallinity. *J. Appl. Polym. Sci.* **2008**, *107*, 2246–2255. [CrossRef]
60. Winkler, T.; Sass, F.A.; Duda, G.N.; Schmidt-Bleek, K. A review of biomaterials in bone defect healing, remaining shortcomings and future opportunities for bone tissue engineering: The unsolved challenge. *Bone Jt. Res.* **2018**, *7*, 232–243. [CrossRef]
61. Surmenev, R.A.; Surmeneva, M.A.; Ivanova, A.A. Significance of calcium phosphate coatings for the enhancement of new bone osteogenesis—A review. *Acta Biomater.* **2014**, *10*, 557–579. [CrossRef]
62. Ang, K.C.; Leong, K.F.; Chua, C.K.; Chandrasekaran, M. Compressive properties and degradability of poly(ε-caprolactone)/hydroxyapatite composites under accelerated hydrolytic degradation. *J. Biomed. Mater. Res. Part A* **2007**, *80A*, 655–660. [CrossRef]
63. Liuyun, J.; Chengdong, X.; Dongliang, C.; Lixin, J.; Xiubing, P. Effect of n-HA with different surface-modified on the properties of n-HA/PLGA composite. *Appl. Surf. Sci.* **2012**, *259*, 72–78. [CrossRef]
64. Liuyun, J.; Chengdong, X.; Lixin, J.; Dongliang, C.; Qing, L. Effect of n-HA content on the isothermal crystallization, morphology and mechanical property of n-HA/PLGA composites. *Mater. Res. Bull.* **2013**, *48*, 1233–1238. [CrossRef]
65. Esmaeili, S.; Hossein, A.A.; Motififard, M.; Saber-Samandari, S.; Montazeran, A.H.; Bigonah, M.; Sheikhbahaei, E.; Khandan, A. A porous polymeric-hydroxyapatite scaffold used for femur fractures treatment: Fabrication, analysis, and simulation. *Eur. J. Orthop. Surg. Traumatol.* **2020**, *30*, 123–131. [CrossRef]
66. Ivorra-Martinez, J.; Quiles-Carrillo, L.; Boronat, T.; Torres-Giner, S.; Covas, J.A. Assessment of the Mechanical and Thermal Properties of Injection-Molded Poly(3-hydroxybutyrate-co-3-hydroxyhexanoate)/Hydroxyapatite Nanoparticles Parts for Use in Bone Tissue Engineering. *Polymers* **2020**, *12*, 1389. [CrossRef] [PubMed]

67. Hong, Z.; Zhang, P.; He, C.; Qiu, X.; Liu, A.; Chen, L.; Chen, X.; Jing, X. Nano-composite of poly(L-lactide) and surface grafted hydroxyapatite: Mechanical properties and biocompatibility. *Biomaterials* **2005**, *26*, 6296–6304. [CrossRef] [PubMed]
68. Hong, Z.; Zhang, P.; Liu, A.; Chen, L.; Chen, X.; Jing, X. Composites of poly(lactide-co-glycolide) and the surface modified carbonated hydroxyapatite nanoparticles. *J. Biomed. Mater. Res. Part A* **2007**, *81A*, 515–522. [CrossRef]
69. Jones, N.L.; Cooper, J.J.; Waters, R.D.; Williams, D.F.; Agrawal, C.M.; Parr, J.E.; Lin, S.T. Resorption Profile and Biological Response of Calcium Phosphate filled PLLA and PHB7V. In *Synthetic Bioabsorbable Polymers for Implants*; ASTM STP 1396; ASTM International: West Conshohocken, PA, USA, 2000.
70. Bou-Francis, A.; Piercey, M.; Al-Qatami, O.; Mazzanti, G.; Khattab, R.; Ghanem, A. Polycaprolactone blends for fracture fixation in low load-bearing applications. *J. Appl. Polym. Sci.* **2020**, *137*, 48940. [CrossRef]
71. Bou-Francis, A.; Ghanem, A. Standardized methodology for in vitro assessment of bone-to-bone adhesion strength. *Int. J. Adhes. Adhes.* **2017**, *77*, 96–101. [CrossRef]
72. Carvalho, S.M.; Oliveira, A.A.R.; Lemos, E.M.F.; Pereira, M.M. Bioactive Glass Nanoparticles for Periodontal Regeneration and Applications in Dentistry. In *Nanobiomaterials in Clinical Dentistry*; Elsevier Inc.: Amsterdam, The Netherlands, 2012; pp. 299–322. [CrossRef]
73. Navarro, M.; Ginebra, M.P.; Planell, J.A. Cellular response to calcium phosphate glasses with controlled solubility. *J. Biomed. Mater. Res. Part A* **2003**, *67*, 1009–1015. [CrossRef]
74. Zakir Hossain, K.M.; Felfel, R.M.; Grant, D.M.; Ahmed, I. Chapter 11: Phosphate Glass Fibres and Their Composites. In *RSC Smart Materials*; 2017-January; Royal Society of Chemistry: Cambridge, UK, 2017; pp. 257–285.
75. Parsons, A.J.; Ahmed, I.; Haque, P.; Fitzpatrick, B.; Niazi, M.I.K.; Walker, G.S.; Rudd, C.D. Phosphate Glass Fibre Composites for Bone Repair. *J. Bionic Eng.* **2009**, *6*, 318–323. [CrossRef]
76. Melo, P.; Tarrant, E.; Swift, T.; Townshend, A.; German, M.; Ferreira, A.M.; Gentile, P.; Dalgarno, K. Short phosphate glass fiber—PLLA composite to promote bone mineralization. *Mater. Sci. Eng. C* **2019**, *104*, 109929. [CrossRef]
77. Zhu, C.; Ahmed, I.; Parsons, A.; Wang, Y.; Tan, C.; Liu, J.; Rudd, C.; Liu, X. Novel bioresorbable phosphate glass fiber textile composites for medical applications. *Polym. Compos.* **2018**, *39*, E140–E151. [CrossRef]
78. Ahmed, I.; Jones, I.A.; Parsons, A.J.; Bernard, J.; Farmer, J.; Scotchford, C.A.; Walker, G.S.; Rudd, C.D. Composites for bone repair: Phosphate glass fibre reinforced PLA with varying fibre architecture. *J. Mater. Sci. Mater. Med.* **2011**, *22*, 1825–1834. [CrossRef]
79. Felfel, R.M.; Ahmed, I.; Parsons, A.J.; Walker, G.S.; Rudd, C.D. In vitro degradation, flexural, compressive and shear properties of fully bioresorbable composite rods. *J. Mech. Behav. Biomed. Mater.* **2011**, *4*, 1462–1472. [CrossRef] [PubMed]
80. Felfel, R.M.; Ahmed, I.; Parsons, A.J.; Harper, L.T.; Rudd, C.D.; Ahmed, I.; Parsons, A.J.; Harper, L.T.; Rudd, C.D. Initial mechanical properties of phosphate-glass fibre-reinforced rods for use as resorbable intramedullary nails. *J. Mater. Sci.* **2012**, *47*, 4884–4894. [CrossRef]
81. Felfel, R.M.; Ahmed, I.; Parsons, A.J.; Rudd, C.D. Bioresorbable screws reinforced with phosphate glass fibre: Manufacturing and mechanical property characterisation. *J. Mech. Behav. Biomed. Mater.* **2013**, *17*, 76–88. [CrossRef]
82. Felfel, R.M.; Ahmed, I.; Parsons, A.J.; Rudd, C.D. Bioresorbable composite screws manufactured via forging process: Pull-out, shear, flexural and degradation characteristics. *J. Mech. Behav. Biomed. Mater.* **2013**, *18*, 108–122. [CrossRef] [PubMed]
83. Ahmed, I.; Cronin, P.S.; Neel, E.A.; Parsons, A.J.; Knowles, J.C.; Rudd, C.D. Retention of mechanical properties and cytocompatibility of a phosphate-based glass fiber/polylactic acid composite. *J. Biomed. Mater. Res. Part B Appl. Biomater.* **2009**, *89*, 18–27. [CrossRef]
84. Ahmed, I.; Parsons, A.J.; Palmer, G.; Knowles, J.C.; Walker, G.S.; Rudd, C.D. Weight loss, ion release and initial mechanical properties of a binary calcium phosphate glass fibre/PCL composite. *Acta Biomater.* **2008**, *4*, 1307–1314. [CrossRef] [PubMed]
85. Felfel, R.M.; Ahmed, I.; Parsons, A.J.; Haque, P.; Walker, G.S.; Rudd, C.D. Investigation of crystallinity, molecular weight change, and mechanical properties of PLA/PBG bioresorbable composites as bone fracture fixation plates. *J. Biomater. Appl.* **2012**, *26*, 765–789. [CrossRef] [PubMed]
86. Ramsay, S.D.; Pilliar, R.M.; Santerre, J.P. Fabrication of a biodegradable calcium polyphosphate/polyvinyl-urethane carbonate composite for high load bearing osteosynthesis applications. *J. Biomed. Mater. Res. Part B Appl. Biomater.* **2010**, *94*, 178–186. [CrossRef]
87. Hasan, M.S.; Ahmed, I.; Parsons, A.J.; Rudd, C.D.; Walker, G.S.; Scotchford, C.A. Investigating the use of coupling agents to improve the interfacial properties between a resorbable phosphate glass and polylactic acid matrix. *J. Biomater. Appl.* **2012**. [CrossRef]
88. dos Santos, T.M.B.K.; Merlini, C.; Aragones, Á.; Fredel, M.C. Manufacturing and characterization of plates for fracture fixation of bone with biocomposites of poly (lactic acid-co-glycolic acid) (PLGA) with calcium phosphates bioceramics. *Mater. Sci. Eng. C* **2019**, *103*, 109728. [CrossRef] [PubMed]
89. Bosco, R.; Van Den Beucken, J.V.; Leeuwenburgh, S.; Jansen, J. Surface engineering for bone implants: A trend from passive to active surfaces. *Coatings* **2012**, *2*, 95–119. [CrossRef]
90. Hoffman, A.S. Surface modification of polymers: Physical, chemical, mechanical and biological methods. *Macromol. Symp.* **1996**, *101*, 443–454. [CrossRef]
91. Ul Ahad, I.; Bartnik, A.; Fiedorowicz, H.; Kostecki, J.; Korczyc, B.; Ciach, T.; Brabazon, D. Surface modification of polymers for biocompatibility via exposure to extreme ultraviolet radiation. *J. Biomed. Mater. Res. Part A* **2014**, *102*, 3298–3310. [CrossRef]

92. Mayekar, P.C.; Castro-Aguirre, E.; Auras, R.; Selke, S.; Narayan, R. Effect of nano-clay and surfactant on the biodegradation of poly(lactic acid) films. *Polymers* **2020**, *12*, 311. [CrossRef]
93. Low, Y.J.; Andriyana, A.; Ang, B.C.; Zainal Abidin, N.I. Bioresorbable and degradable behaviors of PGA: Current state and future prospects. *Polym. Eng. Sci.* **2020**, *60*, 25508. [CrossRef]
94. Duta, L.; Popescu, A. Current Status on Pulsed Laser Deposition of Coatings from Animal-Origin Calcium Phosphate Sources. *Coatings* **2019**, *9*, 335. [CrossRef]
95. Moraczewski, K.; Rytlewski, P.; Malinowski, R.; Zenkiewicz, M. Comparison of some effects of modification of a polylactide surface layer by chemical, plasma, and laser methods. *Appl. Surf. Sci.* **2015**, *346*, 11–17. [CrossRef]
96. Yoshida, S.; Hagiwara, K.; Hasebe, T.; Hotta, A. Surface modification of polymers by plasma treatments for the enhancement of biocompatibility and controlled drug release. *Surf. Coat. Technol.* **2013**, *233*, 99–107. [CrossRef]
97. Favia, P.; D'Agostino, R. Plasma treatments and plasma deposition of polymers for biomedical applications. *Surf. Coat. Technol.* **1998**, *98*, 1102–1106. [CrossRef]
98. Jacobs, T.; Declercq, H.; De Geyter, N.; Cornelissen, R.; Dubruel, P.; Leys, C.; Beaurain, A.; Payen, E.; Morent, R. Plasma surface modification of polylactic acid to promote interaction with fibroblasts. *J. Mater. Sci. Mater. Med.* **2013**, *24*, 469–478. [CrossRef]
99. Källrot, M.; Edlund, U.; Albertsson, A.C. Surface functionalization of degradable polymers by covalent grafting. *Biomaterials* **2006**, *27*, 1788–1796. [CrossRef]
100. Mutch, A.L.; Grøndahl, L. Challenges for the development of surface modified biodegradable polyester biomaterials: A chemistry perspective. *Biointerphases* **2018**, *13*, 06D501. [CrossRef]
101. Chu, P.K.; Chen, J.Y.; Wang, L.P.; Huang, N. Plasma-surface modification of biomaterials. *Mater. Sci. Eng. R Rep.* **2002**, *36*, 143–206. [CrossRef]
102. Lech, A.; Butruk-Raszeja, B.A.; Ciach, T.; Lawniczak-Jablonska, K.; Kuzmiuk, P.; Bartnik, A.; Wachulak, P.; Fiedorowicz, H. Surface modification of plla, ptfe and pvdf with extreme ultraviolet (Euv) to enhance cell adhesion. *Int. J. Mol. Sci.* **2020**, *21*, 9679. [CrossRef] [PubMed]
103. Nedĕla, O.; Slepička, P.; Švorčík, V. Surface Modification of Polymer Substrates for Biomedical Applications. *Materials* **2017**, *10*, 1115. [CrossRef]
104. Rudolph, A.; Teske, M.; Illner, S.; Kiefel, V.; Sternberg, K.; Grabow, N. Surface Modification of Biodegradable Polymers towards Better Biocompatibility and Lower Thrombogenicity. *PLoS ONE* **2015**, *10*, 142075. [CrossRef]
105. Bu, Y.; Ma, J.; Bei, J.; Wang, S. Surface Modification of Aliphatic Polyester to Enhance Biocompatibility. *Front. Bioeng. Biotechnol.* **2019**, *7*, 98. [CrossRef]
106. Yang, J.; Bei, J.; Wang, S. Enhanced cell affinity of poly (D,L-lactide) by combining plasma treatment with collagen anchorage. *Biomaterials* **2002**, *23*, 2607–2614. [CrossRef]
107. Yang, J.; Bei, J.; Wang, S. Improving cell affinity of poly(D,L-lactide) film modified by anhydrous ammonia plasma treatment. *Polym. Adv. Technol.* **2002**, *13*, 220–226. [CrossRef]
108. Gomathi, N.; Sureshkumar, A.; Neogi, S. RF plasma-treated polymers for biomedical applications. *Curr. Sci.* **2008**, *94*, 1478–1486.
109. Morintale, E.; Constantinescu, C.-D.; Dinescu, M. Thin films development by pulsed laser-assisted deposition. *Ann. Univ. Craiova Phys.* **2010**, *20*, 43–56.
110. Janorkar, A.V.; Metters, A.T.; Hirt, D.E. Degradation of poly(L-lactide) films under ultraviolet-induced photografting and sterilization conditions. *J. Appl. Polym. Sci.* **2007**, *106*, 1042–1047. [CrossRef]
111. Da Silva, D.; Kaduri, M.; Poley, M.; Adir, O.; Krinsky, N.; Shainsky-Roitman, J.; Schroeder, A. Biocompatibility, biodegradation and excretion of polylactic acid (PLA) in medical implants and theranostic systems. *Chem. Eng. J.* **2018**, *340*, 9–14. [CrossRef]
112. Singhvi, M.S.; Zinjarde, S.S.; Gokhale, D.V. Polylactic acid: Synthesis and biomedical applications. *J. Appl. Microbiol.* **2019**, *127*, 1612–1626. [CrossRef]
113. Bonartsev, A.P.; Bonartseva, G.A.; Reshetov, I.V.; Kirpichnikov, M.P.; Shaitan, K.V. Application of polyhydroxyalkanoates in medicine and the biological activity of natural poly(3-hydroxybutyrate). *Acta Nat.* **2019**, *11*, 4–16. [CrossRef] [PubMed]
114. Stewart, C.; Akhavan, B.; Wise, S.G.; Bilek, M.M.M. A review of biomimetic surface functionalization for bone-integrating orthopedic implants: Mechanisms, current approaches, and future directions. *Prog. Mater. Sci.* **2019**, *106*, 100588. [CrossRef]

Review

Poly(Lactic Acid)-Based Graft Copolymers: Syntheses Strategies and Improvement of Properties for Biomedical and Environmentally Friendly Applications: A Review

Jean Coudane [1,*], Hélène Van Den Berghe [1], Julia Mouton [2,3], Xavier Garric [1] and Benjamin Nottelet [1]

1. Department of Polymers for Health and Biomaterials, Institut des Biomolecules Max Mousseron, UMR 5247, University of Montpellier, CNRS, ENSCM, 34000 Montpellier, France; helene.van-den-berghe@umontpellier.fr (H.V.D.B.); xavier.garric@umontpellier.fr (X.G.); benjamin.nottelet@umontpellier.fr (B.N.)
2. Polymers Composites and Hybrids, IMT Mines d'Alès, 30100 Alès, France; julia.mouton@epf.fr
3. EPF Graduate School of Engineering, 34000 Montpellier, France
* Correspondence: jean.coudane@umontpellier.fr

Abstract: As a potential replacement for petroleum-based plastics, biodegradable bio-based polymers such as poly(lactic acid) (PLA) have received much attention in recent years. PLA is a biodegradable polymer with major applications in packaging and medicine. Unfortunately, PLA is less flexible and has less impact resistance than petroleum-based plastics. To improve the mechanical properties of PLA, PLA-based blends are very often used, but the outcome does not meet expectations because of the non-compatibility of the polymer blends. From a chemical point of view, the use of graft copolymers as a compatibilizer with a PLA backbone bearing side chains is an interesting option for improving the compatibility of these blends, which remains challenging. This review article reports on the various graft copolymers based on a PLA backbone and their syntheses following two chemical strategies: the synthesis and polymerization of modified lactide or direct chemical post-polymerization modification of PLA. The main applications of these PLA graft copolymers in the environmental and biomedical fields are presented.

Keywords: poly(lactic acid); chemical modification; graft copolymers; compatibilization; biomedical and environmental applications

1. Introduction

Today, bioplastics, compounds derived from sustainable sources, are one of the best alternatives to petroleum-based plastics. They are natural or synthetic biopolymers, and include poly(lactic acid) (PLA), which is of great commercial interest due to various factors. First, PLA is produced by polymerizing lactide, a derivative of lactic acid industrially produced from plants, making it a biosourced thermoplastic. As such, extrusion, molding, injection molding, thermoforming, and fiber spinning are largely used to process PLA for many industrial applications. Lastly, it forms an intrinsically biocompatible system in a living environment and is biodegradable, with a tunable degradability as a function of its molecular weight and tacticity, which makes it suitable for many applications in biomedical and environmental fields such as tissue engineering, drug delivery, "green" packaging, textiles, etc. [1,2].

Despite these clear advantages, PLA suffers from some limitations. From an economical point of view, it remains more expensive than many non-biodegradable commodity polymers. Moreover, regarding its thermomechanical properties, it has a low toughness and poor impact strength. To overcome these limitations, polymer blends can provide the desired properties at a low cost through simple physical processes, rather than chemical approaches such as copolymerization reactions. Polymer blends and composites [3], as well as plasticizers [4], are used to improve the mechanical properties of polymers. The melt blending of dissimilar polymers is a classic method for obtaining new enhanced properties. Unfortunately, PLA-based blends exhibit an insufficient performance because the blended polymers are often thermodynamically immiscible, resulting in a poor compatibility between the blended components [5]. This phenomenon is particularly important in high-molecular-weight polymers commonly found in the field of orthopedics.

In order to circumvent the compatibility problem, compatibilizers have been proposed [6]. Compatibilizers are used to improve the properties of immiscible or partially miscible polymer blends. They improve the adhesion between the blended polymers. Compatibilizers can be "reactive" (they chemically react with at least one of the two blended polymers) or "non-reactive" (they have secondary interactions with both polymers). More generally, these PLA compatibilizers may consist of a copolymer comprising the PLA and the polymer to be compatibilized. PLA-based copolymers can be of a "block" or "graft" architecture. It is necessary to have at least one reactive function on the PLA to obtain these blocks or graft copolymers. However, PLA only has reactive functions at its chain ends, typically alcohol and carboxylic acid functions. Therefore, it is quite easy to prepare block copolymers (di-, tri- or multi-blocks) from PLA. The most common PLA-based block copolymers are probably the amphiphilic PLA-b-Poly(ethylene glycol) (PEG) di-block copolymers and PLA-b-PEG-b-PLA triblock copolymers, in which the hydrophobicity of PLA is decreased and its toughness is improved [7]. There are many review articles on the formation of PLA-based block copolymers and their applications, especially in the biomedical field [7–9].

The reactive functions at the chain ends of PLA can also react with the reactive functions in the chain of certain polymers, such as polysaccharides, to give polymer-g-PLA graft copolymers in a so-called "classic" structure (Figure 1), where the polymer backbone is grafted with PLA side chains [10]. The synthesis of "reverse" structures, i.e., with a PLA main chain grafted with other polymer side chains is more challenging because, unlike polysaccharides, the PLA backbone is not functionalized. Therefore, it is necessary to first functionalize the PLA chain before subsequent grafting of polymer segments onto the PLA backbone. From a theoretical point of view, two methods can be used to obtain a functionalized PLA backbone: (i) copolymerization of a lactide with a pre-functionalized lactide and (ii) direct chemical modification of a preformed PLA chain. It is therefore the aim of the present review to focus on these relatively uncommon reverse PLA-g-polymer structures.

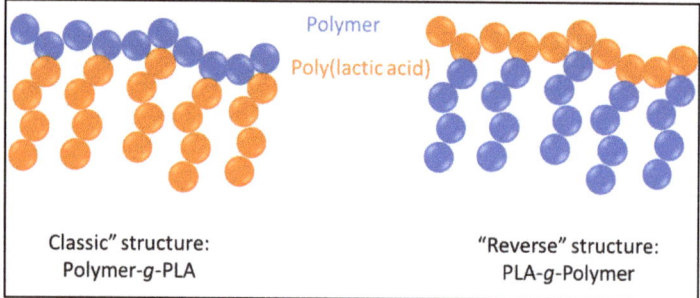

Figure 1. Illustration of "classic" and "reverse" structures of PLA-based graft copolymers.

2. Results

2.1. Copolymerization of Lactide with a Functionalized Lactide for the Preparation of PLA Graft Copolymers

To the best of our knowledge, according to the first method mentioned above, few functionalized lactides are described in the literature, and only a small proportion of these substituted lactides has been copolymerized with lactide to yield functionalized

PLA backbones [11–14]. For example, some lactides functionalized with benzyl, allyl and propargyl groups have also been prepared and copolymerized with lactide, but no polymer chains were subsequently grafted onto the PLA backbone [15].

Yu et al. prepared an alkyne-functionalized lactide that was copolymerized with lactide to give an alkyne-functionalized PLA. An azide-paclitaxel-PEG was then reacted with the alkyne-functionalized PLA via a Cu(I)-catalyzed azide–alkyne cycloaddition (CuAAC) click reaction to give a novel graft polymer–drug conjugate (GPDC): PLA-g-Paclitaxel-PEG (Figure 2) targeting sustained release of Paclitaxel [16]. Zhang et al. prepared and (co)polymerized a dipropargyloxylactide by ring opening polymerization (ROP) in the presence of Sn(Oct)$_2$ and TBBA (tertiobutylbenzyl alcohol). An azido-PEG (PEG-N$_3$) was then grafted via CuAAC click chemistry (Figure 3) [17]. Turbidity, dynamic light scattering and NMR results suggested that these grafted copolymers exhibit a reversible thermo-responsive property, with LCST ranging from 22 to 69 °C depending on the molecular weight of the PEG.

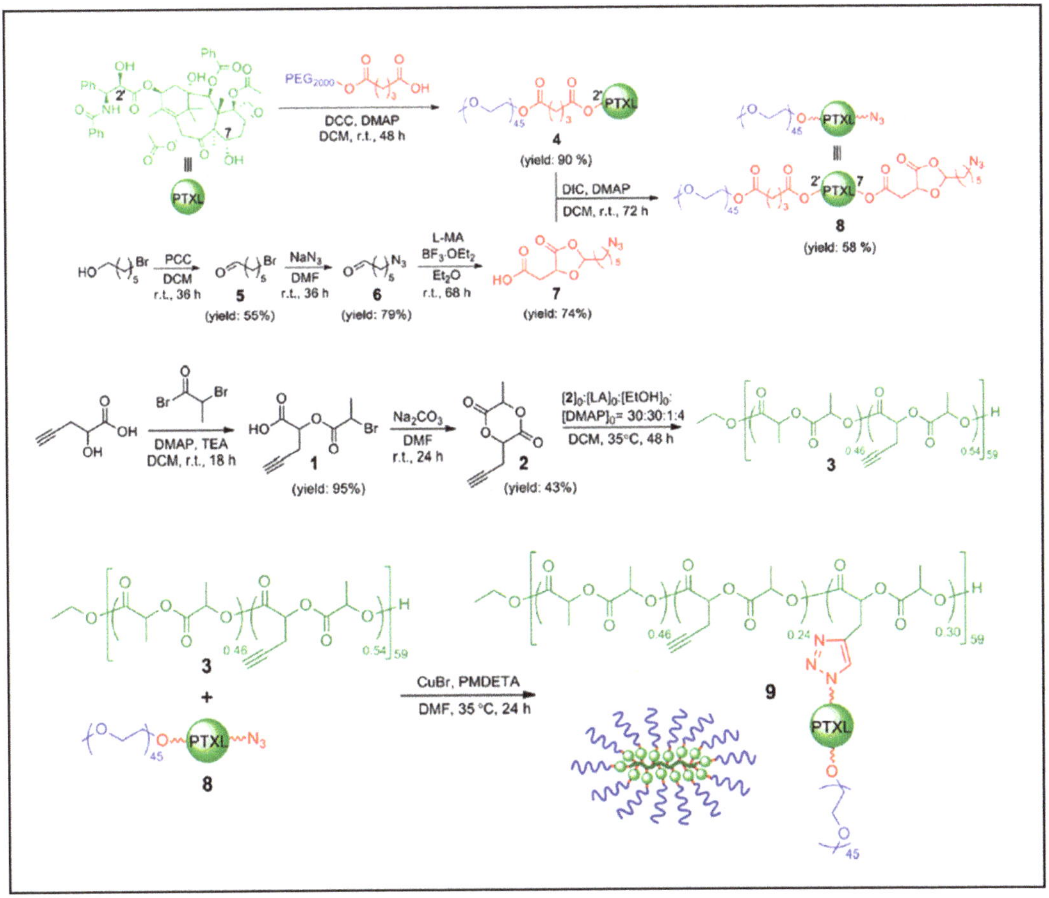

Figure 2. Reaction scheme for the synthesis of PLA-g-Paclitaxel-PEG (adapted from Yu et al. [16], Copyright American Chemical Society, reproduced with permission).

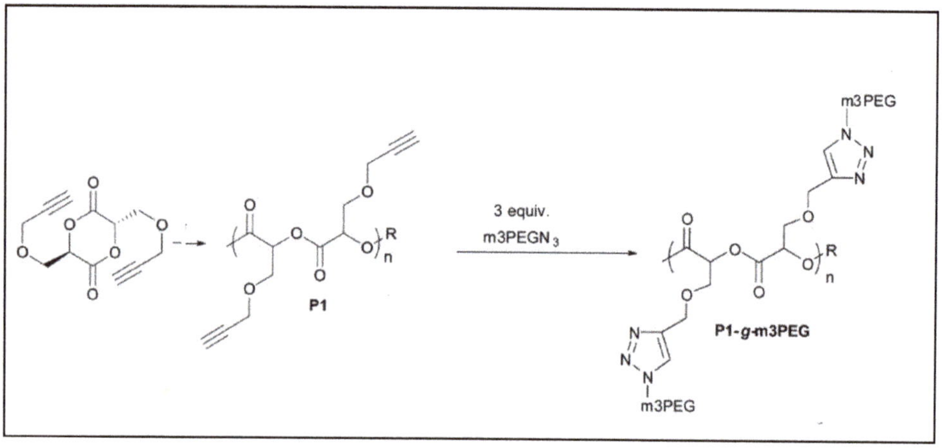

Figure 3. Reaction scheme for the synthesis of dipropargyloxymethyl PLA and the grafting of PEG via click a reaction (adapted from Zhang et al. [17], Copyright Royal Society of Chemistry, reproduced with permission).

In another synthetic method, Castillo et al. prepared a spirolactide–heptene monomer that yielded a PEG-grafted lactide after reaction with an azido-PEG (Figure 4) [18]. This PEG-grafted lactide was then polymerized with molecular weights above 10kDa. Typically, polymerization was carried out in anhydrous CH_2Cl_2 with 1,5,7-triazabicyclo [4.4.0]dec-5-ene (TBD)/benzyl alcohol as the catalyst/initiator system. Preliminary biological studies showed that PLA-g-PEG reduced non-specific protein adsorption and cell adhesion compared to the original PLA.

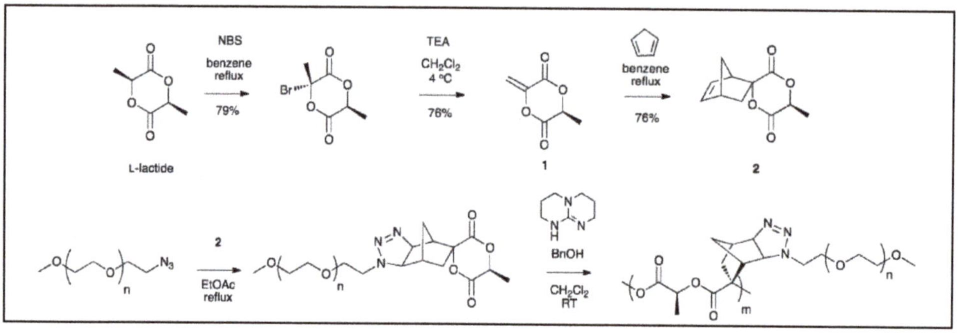

Figure 4. Reaction scheme for the preparation of PLA-g-PEG (from Castillo et al. [18], copyright American Chemical Society, reproduced with permission).

This limited number of examples illustrates that this approach based on functional-lactides is time-consuming, requires many steps and often gives low yields. For example, the global yields of the syntheses are 20% for dipropargyloxylactide [17] and 10% for PLA-g-Paclitaxel-PEG [16]. These copolymers are therefore expensive, and while this is not a fundamental problem for biomedical applications, it is a real drawback for environmental applications that use much larger amounts of product. Therefore, the second method, chemical modification of a preformed PLA chain, is the most widely used to obtain a functionalized PLA backbone allowing for the grafting of a second polymer.

2.2. Direct Functionalization of PLA Backbone: Towards PLA-Based Graft Copolymers

PLA is an aliphatic polyester known for being highly sensitive to chain breakings, especially in acid or alkaline environments. As a result, unlike poly ε-caprolactone, for which a general method of chain modification in anionic media is described [19], almost all substitution reactions on the PLA backbone are carried out under free radical conditions. In most cases, a functional small molecule is grafted on the PLA backbone to give it functionality and, in a second step, a polymer chain is grafted through this small molecule. This grafting can be carried out by polymerization of a monomer from this functional group ("grafting from") or by the direct grafting of a polymer chain through its reactive chain end ("grafting onto"). From an experimental point of view, the functionalization is carried out in blenders or extruders in the presence of a radical precursor. Very frequently, the modification of the PLA chain aims to obtain a compatibilizing agent for blends of PLA and another polymer in order to improve the mechanical properties of the blends. Zeng et al. described the main basic strategies for the compatibilization of the PLA blends, among them functionalization of the PLA backbone with a reactive compound [20].

2.2.1. PLA-g-Maleic Anhydride (PLA-g-MA)

The most widely used reagent for functionalizing the PLA chain is maleic anhydride (MA) due to its good chemical reactivity, low toxicity and good stability to the experimental synthesis conditions [21–23]. The grafting of MA onto the PLA backbone results in a reactive compatibilizer due to the presence of the anhydride function. PLA graft copolymers are therefore obtained by a reaction of the anhydride group with some reactive functions of the blended polymer.

Functionalization with MA is typically carried out between 120 °C and 200 °C with stirring at 50–200 rpm in an extruder in the presence of a radical initiators, such as dicumyl peroxide (DCP) or benzyl peroxide (BPO). A typical reaction scheme is shown in Figure 5 [21]. The role of different reaction parameters (MA concentration, nature of the initiator, temperature, molar mass of PLA) on the percentage of grafting is also described [23]. It was found that the molecular weights of PLA decreased during the reaction due to chain scissions. The grafting percentage of low-molecular-weight PLA is greater than that of the high-molecular-weight PLA because of steric hindrance (Table 1). It should be noted that the grafting percentage of MA increased with the initial concentration of MA, but remained low (<1.25%) regardless of the different parameters of the reaction, which shows the low efficiency of these radical reactions [23]. If one wishes to graft at least two molecules of anhydride per PLA chain, the molar mass of the latter must be higher than 14,500 g.mole^{-1} if substitution degree = 1%, which is not always the case. Nevertheless, the results obtained for the compatibility of various PLA-based blends are significant. Examples of variations in mechanical properties are shown in Table 2 in blends of PLA/cellulose nanofibers (CNF) [24].

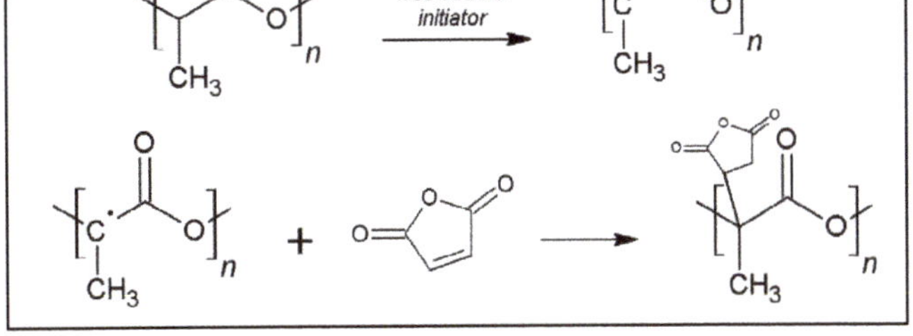

Figure 5. Reaction scheme for the grafting of MA on PLA (from González-López et al. [21], copyright Taylor and Francis, reproduced with permission).

Table 1. Variations in molecular weight and grafting percentage during functionalization of PLA by MA (extract of data from Muenprasat et al. [23]).

PLA Molecular Weight	PLA-g-MA Molecular Weight	Grafting %
294,000	196,000	0.65
92,000	76,000	1.25

Table 2. Mechanical properties of PLA, PLA/CNF and PLA/CNF/PLA-g-MA (extract of data from Ghasemi et al. [24]).

	Tensile Strength (MPa)	Strain at Break (%)	Tensile Modulus (GPa)
Original PLA	22.4	2.3	1.1
PLA/CNF(5%)	34.8	3.0	1.3
PLA/CNF(5%)/PLA-g-MA (5%)	60.3	5.5	1.5

The following paragraphs show the main PLA graft copolymers obtained from PLA-g-MA with their principal properties and applications.

PLA-g-Cellulosic Derivatives

Cellulose fibers derived from renewable biomass have attracted interest as microscale reinforcements in composite materials. Natural fibers have many advantages—low density, low cost, renewability, biodegradability—that make them excellent candidates for the design of biodegradable materials. They can advantageously replace mineral reinforcements in PLA matrices [3,4]. However, a poor compatibility between the fiber and the polymer matrix leads to materials with poor performances. In particular, nanocelluloses, due to their polar surfaces, are difficult to uniformly disperse in a non-polar medium. The consequences of this poor interfacial compatibility between polymer and filler are poor properties of the final blend. The compatibilization of PLA/cellulosic derivatives blends to improve many of the blends' properties, especially mechanical properties, while maintaining a natural source, are the most widely described in the literature [24,25]. Because PLA-g-MA acts as a reactive compatibilizer, a PLA-g-cellulose copolymer is formed as a result of the reaction between the anhydride of PLA-g-MA, and alcohol functions of the cellulosic derivative (Figure 6) [21]. The presence of a very low percentage of PLA-g-MA (<1%) in the original blend significantly improved the properties of the blend [24]. Tensile strength, tensile modulus and strain at break were increased by 55.3%, 15.45% and 30.4%, respectively, over neat PLA by adding 5 wt.% of cellulose nanofibers (CNFs), and by 169.2%, 36.3% and 139.1%, respectively, by adding 5% of PLA-g-MA to the blend PLA/CNF.

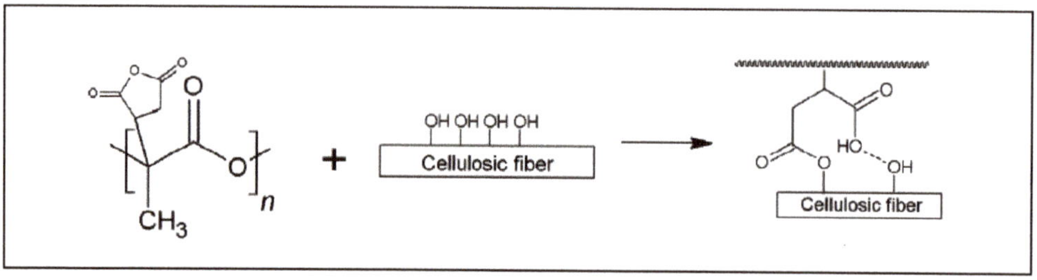

Figure 6. Reaction scheme for the grafting of a cellulosic fiber on PLA-g-MA (from González-López et al. [21], copyright Taylor and Francis, reproduced with permission).

Many cellulosic derivatives were introduced into PLA matrices in the presence of PLA-g-MA to improve the mechanical properties of blends such as Luffa [26], flax [27], coffee grounds [28], wood flour or rice husk [29,30], sisal fibers [31], straw [32], bamboo fiber [33], cassava starch [34,35], starch [36], and lemongrass fiber [37]. For example, the use of PLA-g-MA in a PLA/cassava starch blend has a significant impact on elongation at break but not on Young's modulus or tensile strength. However, it was noted that PLA-g-MA with a higher proportion of grafted MA (0.52 wt.%) had a lower molecular weight and higher dispersity value, showing some degradation of the polymer backbone [34,38].

Many other properties are improved in PLA-g-MA compatible blends, such as morphological, rheological, thermal, tensile and moisture sorption properties as well as thermal degradation [39–41]. For example, Figure 7 shows SEM micrographs of PLA/TPS (thermoplastic starch) blends (70/30 w/w) without (Figure 7a) and with two parts per hundred rubber (phr) PLA-g-MA (Figure 7b), highlighting the compatibilization of the blend of PLA-g-starch formed in situ.

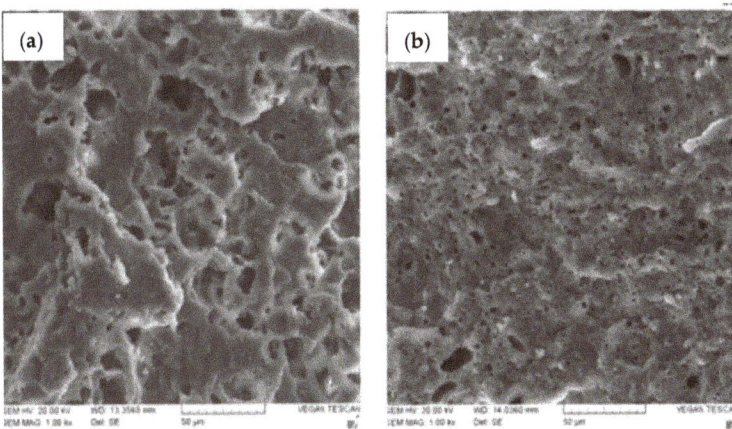

Figure 7. SEM micrographs of PLA/TPS blends (70/30 w/w) (**a**) without and (**b**) with 2 phr PLA-g-MA (adapted from Moghaddam et al. [39], copyright Springer Science, reproduced with permission).

To obtain a PLA-g-starch copolymer, another method is to react maleic anhydride with starch to obtain a maleated thermoplastic starch (MTPS), which is then mixed with PLA in the presence of Luperox 101 (2,5-bis(tert-butylperoxy)-2,5-dimethylhexane) in a Brabender at 180 °C for 5 min. The reaction scheme is shown in Figure 8 [42].

Direct grafting of cellulose nanocrystals (CNC) on PLA, without the addition of PLA-g-MA, is also described, following the reaction scheme of Figure 9 [43]. could his DCP was sprayed onto PLA beads, and the DCP-coated PLA pellets were mixed with CNC and extruded in a twin-screw extruder at 180 °C at 50 rpm for 5 min. The effective grafting of CNC onto PLA was identified by SEC, FTIR and NMR, but NMR showed a very low proportion of CNC in the copolymer. Some mechanical and structural properties were significantly impacted (increased Young modulus, decreased elongation, increased crystallinity).

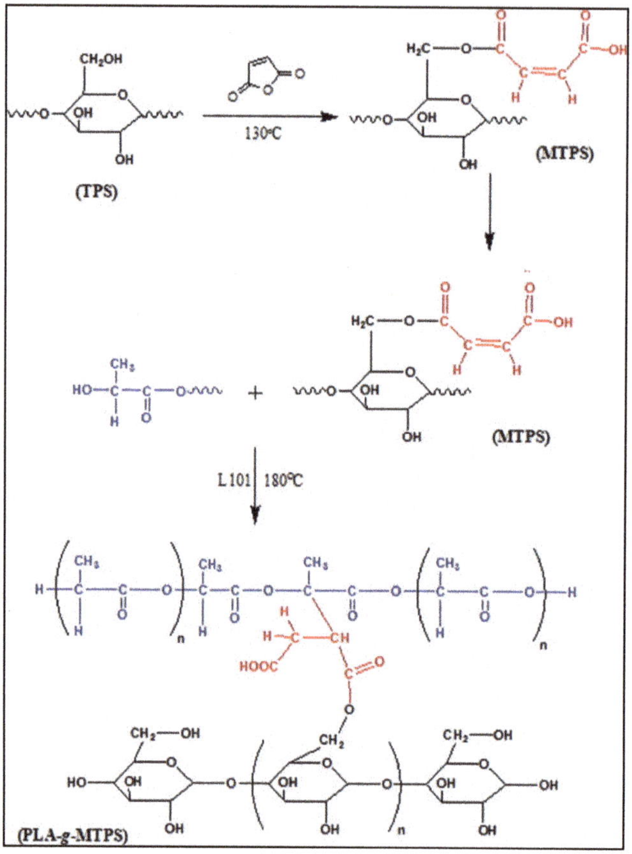

Figure 8. Reaction scheme of the MTPS formation and coupling to PLA (from Wootthikanokkhan et al. [42], copyright Wiley-VCH GmbH. Reproduced with permission).

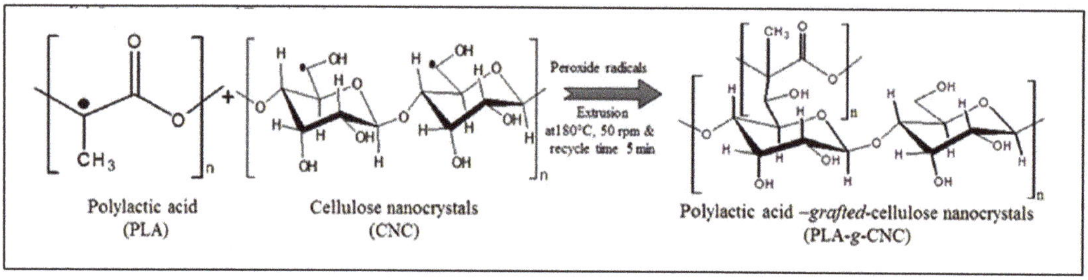

Figure 9. Reaction scheme for the synthesis of PLA-g-CNC (from Dhar et al. [43], copyright Elsevier, reproduced with permission).

PLA-g-Natural Rubber (PLA-g-NR)

Among the drawbacks of PLA materials, we can also highlight their fragility. Natural rubbers (NR), on the other hand, are highly flexible, environmentally friendly and derived from a renewable resource. They are good toughness agents due to their high molecular weight and very low glass transition temperature. However, due to the non-polarity of NRs, PLA/NR blends are immiscible and not compatible. To improve the interfacial interaction between PLA and NR, the reactive compatibilizer PLA-g-MA is used to form a graft copolymer PLA-g-NR [44,45]. Typically, the compatibilized blend is made in a twin screw extruder at a temperature between 160 and 180 °C and a screw speed of around 30 rpm. With the addition of PLA-g-MA, the mechanical properties of the material were significantly improved. It was found that a 3% PLA-g-MA was the best compatibilizer composition to achieve the best performance of the material [44]. The reverse reaction of a maleic anhydride on NR (NR-MA), followed by reaction on PLA in a radical medium, was also performed with similar results regarding mechanical properties [46,47]. However, in this case, the proposed mechanism does not involve a reaction on the PLA backbone but only the alcohol chain end, which reacts on the backbone of NR-MA (Figure 10).

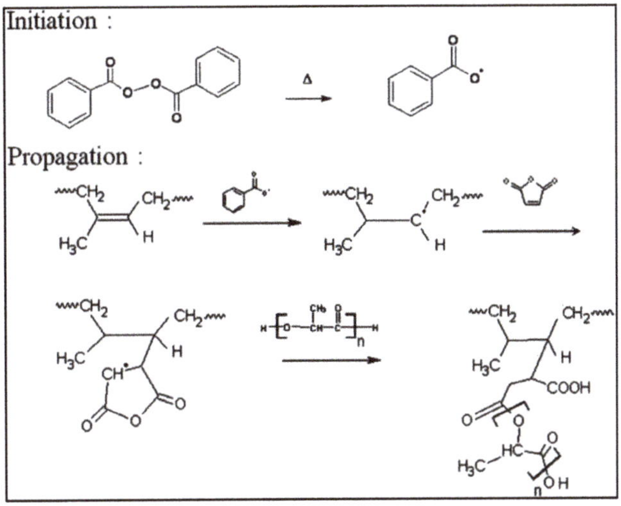

Figure 10. Proposed mechanism for the grafting of NR-MA on PLA (from Thepthawat et al. [47], copyright Wiley-VCH GmbH. Reproduced with permission).

PLA-g-Polyester

Other degradable and flexible polyesters are now being developed on an industrial scale, such as polybutylene adipate-co-terephthalate (PBAT), polybutylene succinate (PBS) or poly ε−caprolactone (PCL); the first two examples are derived from renewable resources. Blends of PLA with these polyesters are good candidates for improving the brittleness and toughness of PLA while maintaining the biodegradability of the blends. Unfortunately, these blends are largely incompatible, especially with PBAT, due to its structure, which contains aromatic rings [48]. As with PLA/NR blends, PLA-g-MA was used as a reactive compatibilizer whose possible reaction of the alcohol chain end of PBAT with the grafted anhydride is shown in Figure 11.

Typically, the reaction is carried out in a co-rotating twin screw extruder at a temperature between 160 and 180 °C and a screw speed of 25–100 rpm [49,50]. Effective grafting of PBAT was demonstrated by a drastic increase in the molecular mass of the copolymer, from 10 kDa to ca. 20 kDa. The mechanical properties of the PLA/PBAT blend were only slightly improved despite the incorporation of PLA-g-MA. However, the improvement of the interfacial adhesion between PLA and PBAT was evidenced by SEM micrographs. A very limited compatibility effect of the addition of PLA-g-MA in PLA/PBAT/TiO$_2$ blends was also observed, with TiO$_2$ acting as a nucleating agent [51]. The addition of CaCO$_3$ to PLA/PBAT/PLA-g-MA blends resulted in an increase in Young's modulus [52].

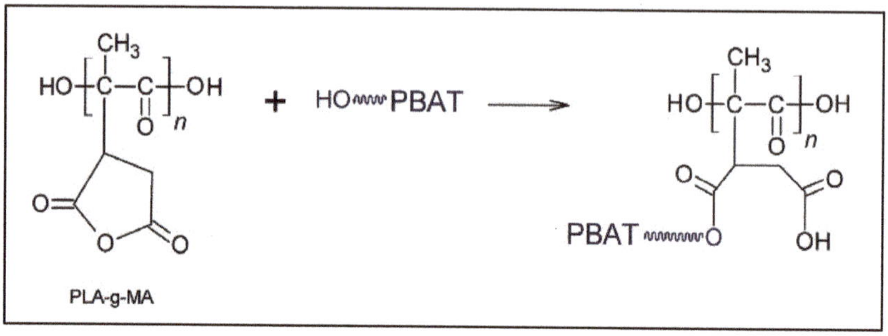

Figure 11. Possible reaction between PLA-g-MA and PBAT (from Rigolin et al. [50], copyright Elsevier, reproduced with permission).

Phetwarotai et al. studied PLA and PBS blends in the presence of PLA-g-MA or toluene diisocyante (TDI) as compatibilizers [53]. They showed that TDI is a more effective compatibilizer than PLA-g-MA for PLA/PBS blend films, their failure mode changed from brittle to ductile due to the improved compatibility. The effect of PLA-g-MA was also studied on PLA/PBAT/thermoplastic starch (TPS) ternary blend films. The thermal stability, tensile properties and compatibility of the PLA, PBAT, and TPS blends were slightly improved with the addition of the compatibilizer [54].

In the case of PLA/PCL with non-compatible blends, a reaction of MA with PCL is more likely to occur between the anhydride and OH groups on the PCL chain ends. As in the case of PBAT, it is the terminal alcohol of the PCL that reacts with the grafted anhydride of the PLA chain (Figure 12A) [55]. According to the authors, the compatibility effect of PLA-g-MA was shown by SEM micrographs (Figure 12B) and by a significant increase in the elongation at break of the material, from 7% in PLA or in a mixture of PLA and PCL to 53% in a compatibilized blend.

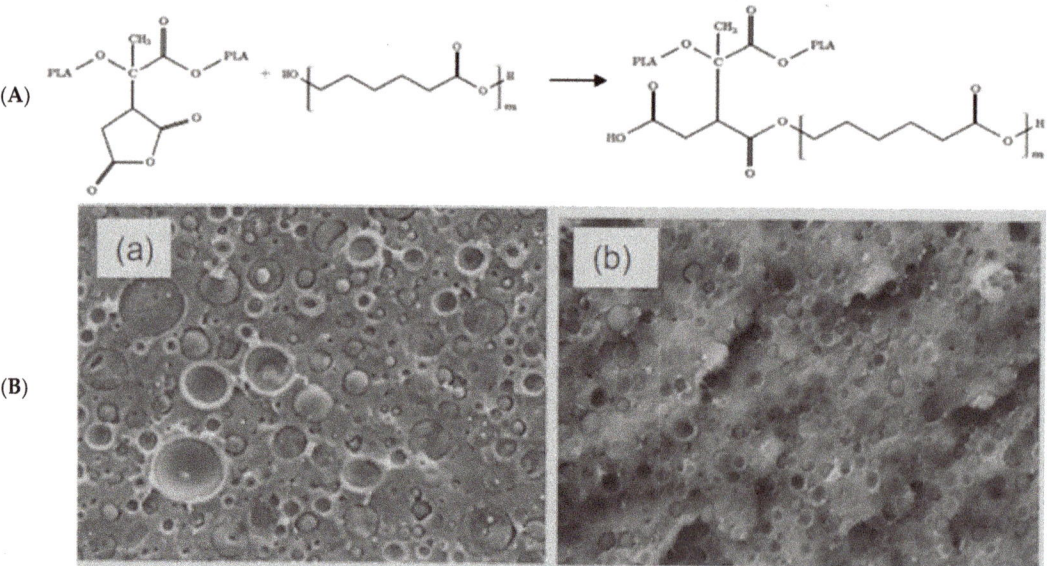

Figure 12. (**A**) Proposed reaction scheme between PLA-g-MA and PCL (**B**) SEM micrographs of PLA/PCL blends (**a**) without and (**b**) with 10% PLA-g-MA (adapted from Gardella et al. [55], copyright Springer, reproduced with permission).

By reactive extrusion of the bio-based poly(glycerol succinate-co-maleate) (PGSMA) with PLA at 150–180 °C in the presence of a free radical initiator, a PLA-g-PGSMA graft copolymer was obtained that acts as an interfacial compatibilizer due to the double bond of the maleate unit in PGSMA [56]. The tensile strength of PLA/PGSMA blends was improved by almost 400% compared to that of pure PLA. This increase was caused by (i) the in situ formation of PLA-g-PGSMA graft copolymers and (ii) the crosslinking of PGSMA within the PLA matrix, which act as interfacial compatibilizers. Two-dimensional NMR and FTIR confirmed the formation of PLA-g-PGSMA, but the substitution degree on the PLA backbone was not evaluated. It is important to note that this work was performed with very-low-molecular-weight PGSMA (Mn < 1200 Da).

PLA-g-MA is also used to compatibilize PLA with thermoplastic polyurethane elastomers (TPU) and thermoplastic polyester elastomers (TPE). Charpy impact, toughness and fracture toughness of brittle PLA were improved when the blends were compatibilized by addition of PLA-g-MA without adversely damaging effects on other mechanical and thermal properties of the PLA blends [57].

Other PLA-Based Blends

PLA-g-MA serves as reactive compatibilizer for other polymer blends by reacting the anhydride functions on the second constituent of the blend. Examples include polyamides 11 and 12, where anhydride function reacts in the amine groups [58,59]. The ductility and impact strength of the compatibilized blends were increased by a factor two compared to non-compatibilized blends. A crosslinking agent (trimethylolpropane trimethacrylate) was also used to enhance the impact strength of the PLA-g-MA-containing blend. PLA/Polyamide6 (PA6) blends were compatibilized with PLA-g-IA (IA = itaconic anhydride), a compound similar to PLA-g-MA [60]. It was reported that IA can react with amine and amide functions of PA6. Unfortunately, a weak compatibility effect was obtained, likely due to the low concentration of IA moieties grafted onto PLA backbone (0.4 wt.%). PLA-g-IA is also a compatibilizer of PLA/Novatein (a protein-based thermoplastic) as shown in the SEM micrographs of Figure 13 [61].

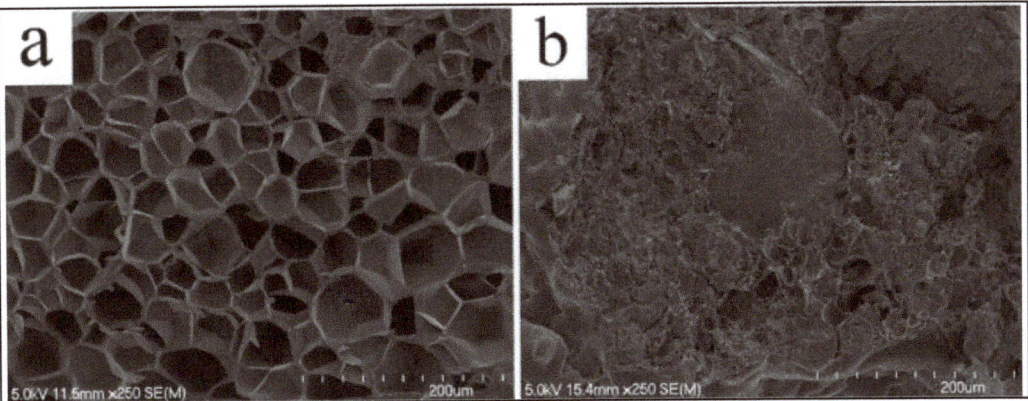

Figure 13. SEM micrographs of PLA/novatein 90/10 blends (**a**) uncompatibilized, (**b**) compatibilized (adapted from Walallavita et al. [61], copyright Wiley-VCH GmbH. Reproduced with permission).

Compatibilizing effects of PLA-g-MA on the properties of PLA/Soy Protein concentrate (SPC) blends were demonstrated by a 19% increase in tensile strength compared to the non-compatibilized blend [62]. Finally, PLA-g-MA also has the ability to provide some compatibility to PLA/mineral mixtures such as PLA/carbon nanotubes [22], PLA/Titanium oxide [63], PLA/halloysite [64], PLA/hydroxyapatite [65], PLA/talc [66], PLA polyhedral oligomeric silsesquioxane (POSS) [67].

2.2.2. PLA-g-Glycidyl Methacrylate (PLA-g-GMA)

A second type of compound that allows for the reactive functionalization of the PLA chain is glycidyl methacrylate (GMA). Grafting is a free-radical reaction performed in a mixer (80 rpm) at 160 °C for 12 min in the presence of BPO [68]. A proposed scheme of the reaction is shown in Figure 14 [69].

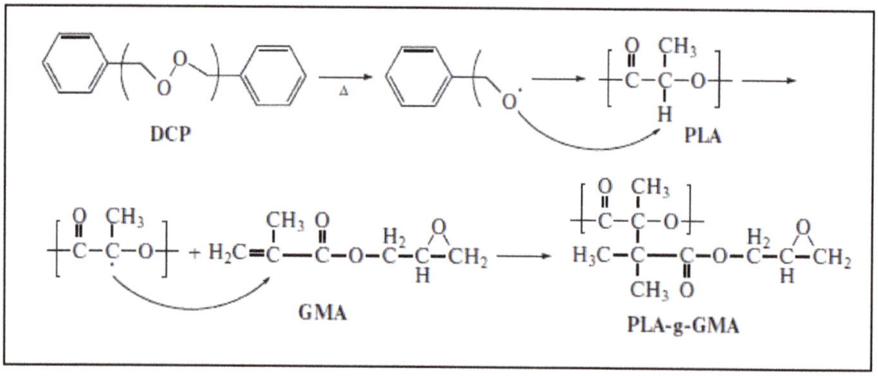

Figure 14. Reaction scheme for the grafting of GMA on PLA backbone (from Gu et al. [69], copyright Springer Nature, reproduced with permission).

The possibility of further functionalization is ensured by the presence of the new oxirane group on the grafted chain. As for PLA-g-MA, the main application lies in the compatibility of polymer blends. The graft content increased from 1.8 to 11.0 wt.% as the GMA concentration in the feed is varied from 5 to 20 wt.%. [68]. The degree of substitution, which is significantly higher than that of PLA-g-MA, is therefore the main advantage of GMA over MA. The characterization and properties of PLA-g-GMA (crystallization, characteristics, tensile stress, stress–strain curve, brittleness, thermal properties) compared to the one of PLA, are described by Kangwanwatthanasiri et al. [70].

The main graft copolymers obtained from PLA-g-GMA and their principal properties and applications are described in the following paragraphs. We found almost the same types of structures as those obtained with PLA-g-MA; therefore, they will not be described in detail.

PLA-g-Cellulosic Derivatives

PLA-g-GA is blended with cellulosic derivatives, fully renewable and degradable resources, to obtain a fully biodegradable product. For example, when used with starch, a small percentage of PLA-g-GMA plays the role of a compatibilizer, which is incorporated into both PLA and starch phases [68]. Essentially, when PLA-g-GMA is incorporated into PLA/cellulosic blends, the mechanical properties are improved: the starch tensile strength at break increased from 18.6 ± 3.8 to 29.3 ± 5.8 MPa, the tensile modulus from 510 ± 62 to 901 ± 62 MPa, and elongation at break from 1.8 ± 0.4 to $3.4 \pm 0.6\%$ [68].

Based on the reaction with PLA-g-GMA, the various PLA/cellulosic derivatives blends that were compatibilized are PLA–starch copolymers [68], PLA-treated arrowroot fiber [71], PLA–lignin [72,73] PLA–cassava pulp [74], PLA–cellulose [75], PLA–bamboo flour [76], PLA–rice straw fiber [77].

PLA-g-Polyesters

For the same reasons as described with MA, PLA/PBAT and PLA/PBS blends were compatibilized by the addition of PLA-g-GMA, with the main results of improved mechanical and thermal properties [69]. Specifically, the presence of PLA-g-GMA in a blend PLA/PBAT led to a decrease in crystallization rate, an increase in melt strength and viscosity, an improvement of tensile strength, and elongation at break, which are dependent on the proportion of PLA-g-GMA [78]. In a typical operating procedure, PLA, PBS or PBAT and PLA-g-GMA were melt blended together in a twin-screw extruder at a rotational speed of 30 rpm for several minutes. The temperature was between 150 and 170 °C [79]. The PLA/PBAT/PLA-g-GMA blends were successfully printed by 3D printing. A reaction mechanism between PBAT and PLA-g-GMA is described in Figure 15 [78]. With 10 wt.% of compatibilizer, the viscosity of the PBAT/PLA blend increased, and there was no longer a crystalline region of PBAT, showing an improved compatibility of PLA and PBAT.

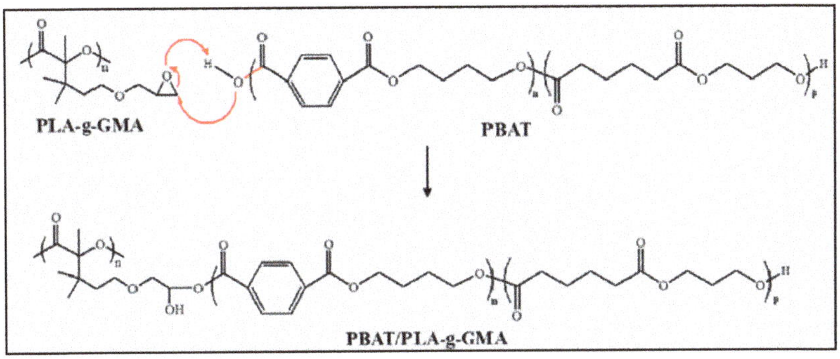

Figure 15. Mechanism of the reaction between PLA-g-GPA and PBAT (from Lyu et al. [78], copyright Elsevier, reproduced with permission).

PLA/cassava pulp/PBS ternary biocomposites were also compatibilized by PLA-g-GPA, with the mechanical properties of the PLA/cassava pulp/PBS composites being improved with the addition of PLA-g-GMA [80]. Similar to polyesters, PLA/thermoplastic polyurethane (TPU) blends were also compatibilized in the presence of PLA-g-GMA [81]. In this example, PLA-g-TPU acted as a compatibilizer for the blend PLA/TPU.

2.2.3. PLA-g-Acrylic Acid (PLA-g-AA)

Another functionalization of the PLA chain is used in the field of the compatibilization of PLA-based blends, namely the grafting of acrylic acid (AA) to give a PLA-g-AA graft copolymer. Typically, AA grafting is performed under free radical conditions, by adding a mixture of AA and BPO to molten PLA in a mixer at 95 °C for a period of 6 h [82]. A PLA-g-PAA copolymer is formed, which can then react with alcohol functions of the cellulosic derivatives by esterifying the alcohol functions of the cellulosic compound (Figure 16).

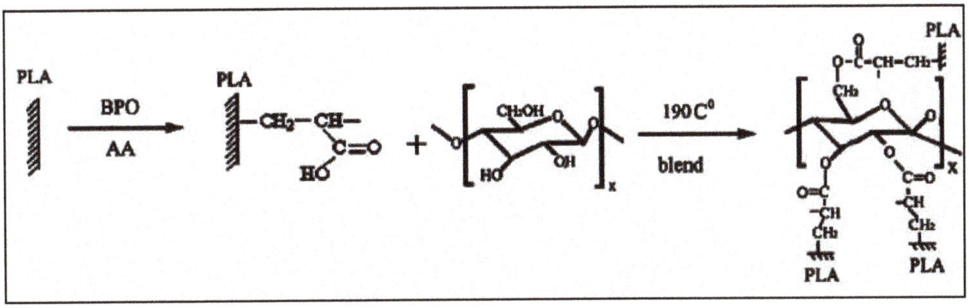

Figure 16. Reaction scheme for the grafting of acrylic acid on PLA and reaction with starch (from Wu, [82], copyright Wiley-VCH GmbH. Reproduced with permission).

As with PLA-g-MA, PLA-g-AA can compatibilize blends with natural cellulosic compounds, such as sisal fiber [31], wood flour [83], corn starch [84], rice husk [85], and hyaluronic acid [86]. In all cases, improvements in mechanical properties and/or biodegradation are obtained in the compatibilized mixtures. The compatibilization effect is shown by the size of the corn starch (CS) phase in PLA/CS and PLA/CS/PLA-g-corn starch. In a PLA/CS blend (50/50 w/w), the CS phase size decreased from 17.5 µm to 7.3 µm when PLA-g-corn starch was added in the blend [84].

Acrylic acid can also be graft-polymerized onto PLA chains in a solution using a photoinitiator, typically benzophenone, under UV irradiation at 254 nm [87]. Finally, the grafting of PAA onto the PLA backbone was also obtained by a free-radical reaction of BPO onto a solution of PLA in chloroform, followed by a reaction and polymerization of AA at 100 °C for 10 min under pressure. A drastic decrease in toughness and an increase in tensile modulus were observed in PLA-g-PAA as compared to PLA [88].

Inorganic–organic hybrid composites, based on mixtures of PLA and SiO_2 [89] and TiO_2 [90] generated via a sol–gel process, also showed improved mechanical and thermal properties when PLA was replaced by PLA-g-AA. This was attributed to stronger interfacial forces between carboxylic acid groups of PLA-g-AA and the residual Si-OH and Ti-OH groups [89].

2.2.4. PLA-g-Halogen

The halogenation, in particular bromination, of the PLA chain is another method for reactive functionalization of the PLA chain. Usually, bromination is achieved by a free-radical mechanism. Typically, PLA is treated with N-bromo succinimide (NBS) in the presence of BPO over a period of 5 days [91]. There is no detectable chain degradation or crosslinking based on SEC results. Similarly, authors prepared chlorinated and iodinated PLAs. They used short-chain PLA (Mn = 2 kDa) to facilitate polymer characterization. The substitution degree depended on the halogen: from 3.2% for bromination and chlorination to 0.5% for iodination. However, the degree of bromination can be increased to 8% with microwave activation.

PLA-g-Br was used as a multisite macroinitiator for the ATRP of methyl methacrylate (MMA) and oligo ethylene glycol methacrylate (OEGMA) (Figure 17). Depending on the PLA-g-Br/MMA ratio, the side chains had different lengths. Complete bromine consumption was achieved during polymerization [92].

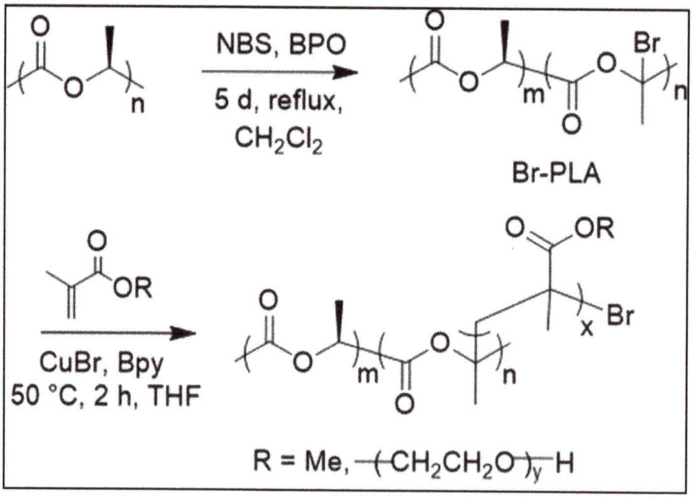

Figure 17. Formation of PLA-g-PMMA and PLA-g-POEGMA by ATRP (from Kalelkar et al. [92], copyright American Chemical Society, reproduced with permission).

The bromination reaction was also performed on the surface of PLA films by treating the surface with NBS in H_2O under UV irradiation. The incorporation of up to 3.7% bromine on the surface was achieved. Surface-initiated ATRP of quaternary ammonium methacrylate (QMA) chloride in the presence of CuBr and 2,2′-bipyridyl (bpy) was then performed, as shown in Figure 18. The cationic grafted surfaces are significantly more toxic to E.coli cells than genuine PLA, but no toxicity to HeLa cells upon contact was found [93].

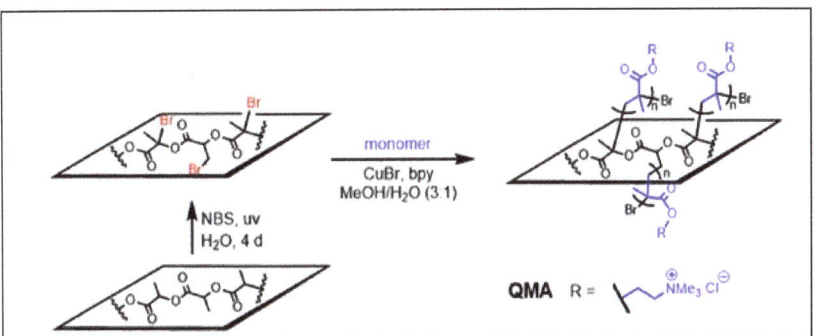

Figure 18. Surface-initiated ATRP of methacrylate monomers (from Kalelkar et al. [93], copyright Elsevier, reproduced with permission).

A direct surface grafting (without any pre-halogenation) of poly(methacrylic acid) (PMAA) onto PLA was obtained after activation of the surface of a PLA film via photo-oxidation followed by the UV-induced polymerization of methacrylic acid. The grafting was confirmed in particular by FTIR analysis [94]. This method was employed to prepare a nano hydroxyapatite/g-PLA composite.

PLA nanofibers obtained by electrospinning were coated with PMMA by plasma polymerization [95]. The coated PLA fibers showed an increase in diameter from 250 nm to 700 nm. MTT assays and cells count showed that the PLA-*g*-PMMA copolymers form intrinsically biocompatible systems.

2.2.5. Other PLA-Based Graft Copolymers

PLA-*g*-Nitrilotriacetic Acid (PLA-*g*-NTA)

In the field of "green packaging", it is desirable to have non-migratory, metal-chelating and biodegradable materials. To this end, metal-chelated nitrilotriacetic acid (NTA) was grafted onto PLA using a classical radical mechanism, as shown in Figure 19 [96]. The grafting was evidenced by ATR-FTIR and XPS. Significant radical scavenging and metal-chelating efficacies as well as the ability to delay the degradation of ascorbic acid showed the antioxidant capacity of PLA-*g*-NTA. However, this NTA grafting was not followed by any polymer grafting, even if the grafted carboxylic acids are functional groups.

PLA-*g*-Vinyltrimethoxysilane (PLA-*g*-VTMS)

PLA-*g*-VTMS was prepared by a free radical reaction of vinyltrimethoxysilane (VTMS) on PLA in the presence of DCP in a Brabender at 190 °C for 5 min. The trimethoxysilane was then hydrolyzed to allow the crosslinking of the PLA chains to improve the mechanical properties of the PLA nanofibers (Figure 20) [97]. Electrospun nanofibrous mats based on PLA/NCC and PLA-*g*-silane/NCC nanocomposites were fabricated and compared. PLA-*g*-VTMS was used to improve the mechanical properties of PLA/nanocrystalline cellulose (NCC). In particular, the impact of NCC on improving tensile strength was notable, even though no chemical reaction of PLA-*g*-VTMS with NCC is reported. In addition, a cytotoxicity assessment showed the biocompatibility of the modified nanofibers, making them good candidates for tissue engineering applications.

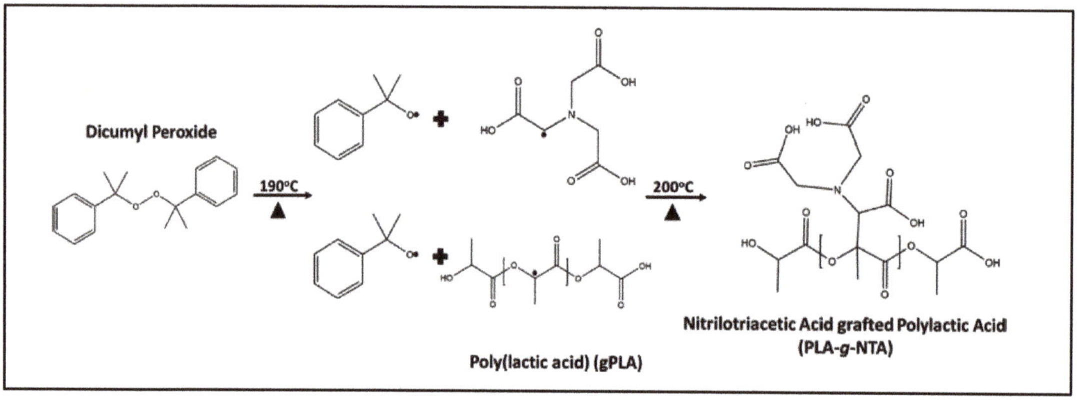

Figure 19. Reaction scheme for the grafting of NTA on PLA (from Herskovitz et al. [96], copyright American Chemical Society, reproduced with permission).

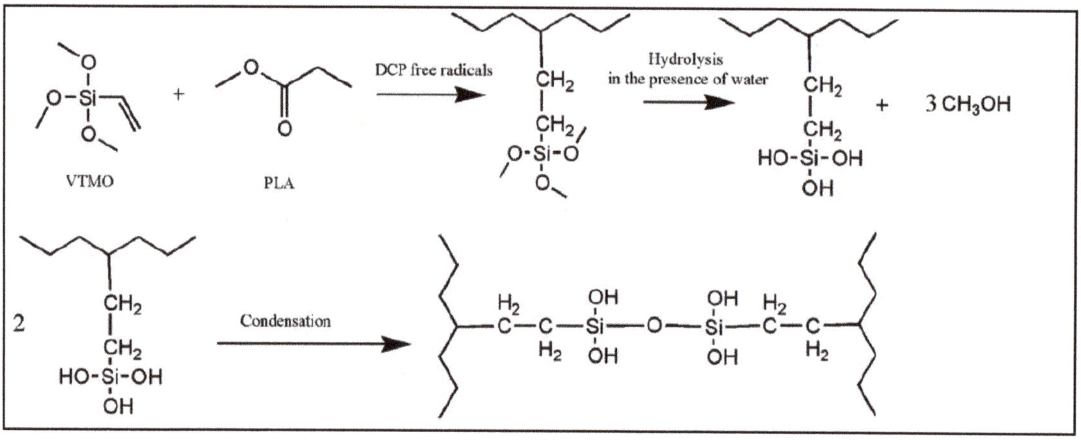

Figure 20. Synthesis scheme of PLA-g-VTMS, hydrolysis of methoxysilane, and crosslinking of PLA chains (adapted from Rahmat et al. [97], copyright Elsevier, reproduced with permission).

PLA-g-Poly(Vinyl Pyrrolidone) (PLA-g-PVP)

A PLA film was treated with a solution of N-vinyl pyrrolidone (NVP) in methanol and AgNO$_3$ using ^{60}Co γ−radiation polymerization at a dose of 1–30 kGy at room temperature. After washings, a PLA-g-PVP film was formed on the surface. Silver nanoparticles were also immobilized on the film surface. A surface grafting ratio, in the range of 25–49%, is assessed by the FTIR ratio of the bands at 1660 cm^{-1} of PVP and the sum of the bands at 1660 cm^{-1} and 1750 cm^{-1} of PLA [98]. There is no indication of the grafting degree of PVP onto the PLA backbone. It is noted, however, that PVP grafting significantly accelerated PLA degradation and does not impede cell proliferation [99]. Controlled variation in the grafting ratio could broaden the applications of this material in tissue engineering scaffolds, drug delivery, and the prevention of post-surgical adhesion.

2.2.6. Anionic Derivatization

The anionic derivatization of PCL was described by Ponsart et al. [19]. It is a remarkably powerful one-pot two-step method for grafting many types of substituents on the PCL backbone. This method is theoretically applicable to many polyesters, but it leads to varying degrees of chain cleavage depending on the nature of the polyester, due to the anionic medium caused by the presence of lithium diisopropylamide. Even though PCL is relatively resistant to this basic medium, which is not the case with PLA, There are still many chain breakings. Nevertheless, El Habnouni et al. applied the method to the surface of the PLA film and nanofibers in a non-solvent medium that causes only moderate-chain scissions and allows for the preparation of functional PLA surfaces [100]. In particular, this approach was exploited with propargylated PLA allowing the grafting of bioactive polymers through CuAAC or thiol-yne click reactions. Therefore, anti-biofilm and bactericidal PLA surfaces were obtained by the reaction of α-azido QPDMAEMA (quaternized poly(2-(dimethylamino)ethyl methacrylate)) or thiol-functional polyaspartamide derivatives (Figure 21) [101,102].

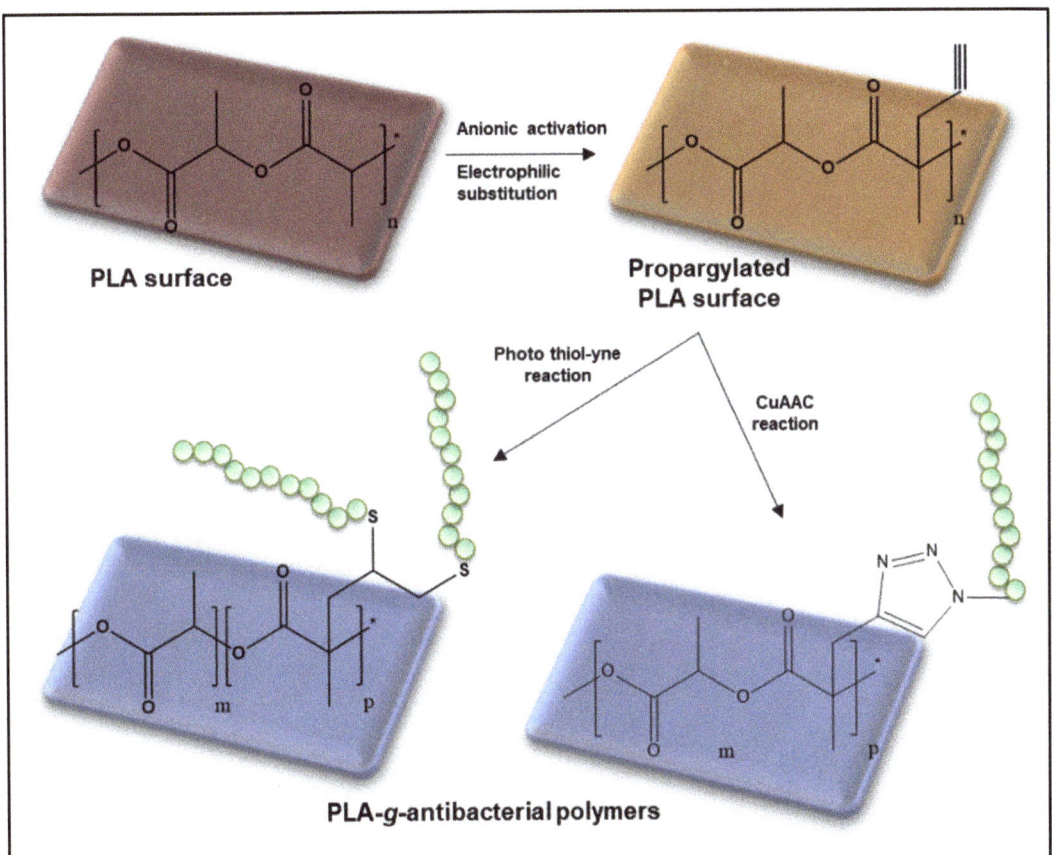

Figure 21. General synthesis scheme of PLA-g-antibacterial polymers from propargylated PLA surface (adapted from Sardo et al. [102] and El Habnouni et al. [101]).

The main copolymers grafted onto the PLA chain, precursors, copolymers and literature references are summarized in Table 3.

Table 3. Main PLA-based grafted copolymers according to literature.

Precursor: Functionalized PLA	PLA-g-Copolymer	Refs.
PLA-g-MA	cellulosic derivatives	[24,40,41]
	luffa	[26]
	flax	[27]
	coffee grounds	[103]
	wood flour	[29,30]
	rice husk	[29,30]
	sisal	[31]
	straw	[32]
	bamboo	[33]
	cassava	[34,35]
	starch	[36,39,42]
	lemongrass	[37]
	natural rubber	[44,45]
	polyesters	
	PBAT	[48–50]
	PBAT/TiO$_2$	[51]
	PBAT/CaCO$_3$	[52]
	PBAT/starch	[54]
	PBS	[53]
	PCL	[55]
	PGSMA	[56]
	TPU	[57]
	PA	[58,59]
	soy protein	[62]
	mineral compounds	[22,63–67]
PLA-g-IA	polyamide	[60,61]
PLA-g-GMA	cellulosic derivatives	
	starch	[68]
	arrowroot	[71]
	lignin	[72,73]
	cassava	[74]
	cellulose	[75]
	bamboo	[76]
	rice-straw	[77]
	polyesters	
	PBAT	[69,78,79]
	PBAT/cassava	[80]
	TPU	[81]
PLA-g-AA	cellulosic derivatives	
	starch	[82,84,90]
	sisal	[31]
	wood flour	[83]
	rice husk	[85]
	mineral compounds	[89]
	hyaluronic acid	[86]
PLA-g-halogen	PMMA and POEGMA	[92–94]
	PVP	[98,99]
PLA-g-alkyne	PEG	[17]
Direct grafting	cellulose nanocrystals	[43]
	natural rubber	[46,47]
	PMMA	[95]
Surface-anionic derivatization	QPDMAEMA	[101]
	α,β-poly(N-2-hydroxyethyl)-D,L-aspartamide	[102]

3. Conclusions

Research on the chemical modifications of the PLA backbone to yield PLA-*g*-polymer graft copolymers is scarce. These modifications mostly occur via a radical mechanism in the presence of a peroxide, leading to the covalent substitution of a reagent on the methine proton of the PLA chain. These reactions are essentially carried out in mass at high temperature in a mixer or an extruder. The main substituents are anhydride or epoxy groups that allow a reactive compatibilization of PLA-based polymer blends. The degree of substitution remains low (<2%) but allows for significant improvements in properties, mainly in mechanical properties. The reactions of the anhydride or epoxide functions grafted on the PLA chain with other polymers (cellulose derivatives, polyesters, polyamides, natural gums, PMMA) lead to the formation of numerous graft copolymers whose backbone is PLA. If these PLA-*g*-polymers are mostly described for the compatibilization of PLA-containing blends, it appears that the applications of these PLA-based grafted copolymers could cross over into biomedical and environmental fields, because they are intrinsically biocompatible systems. In any case, the low degree of grafting obtained in these grafting reactions highlights the importance of finding new grafting approaches to develop functionalization on the PLA chain in order to obtain new PLA-based graft copolymers.

Author Contributions: Conceptualization, J.C. and B.N.; Writing—original draft preparation: J.C.; Writing & review, J.M. and H.V.D.B.; Validation: X.G., H.V.D.B., B.N. and J.C. All authors have read and agreed to the published version of the manuscript.

Funding: This research received no external funding.

Institutional Review Board Statement: Not applicable.

Informed Consent Statement: Not applicable.

Data Availability Statement: Not applicable.

Conflicts of Interest: The authors declare no conflict of interest.

References

1. Farah, S.; Anderson, D.G.; Langer, R. Physical and Mechanical Properties of PLA, and Their Functions in Widespread Applications—A Comprehensive Review. *Adv. Drug Deliv. Rev.* **2016**, *107*, 367–392. [CrossRef] [PubMed]
2. Naser, A.Z.; Deiab, I.; Defersha, F.; Yang, S. Expanding Poly(Lactic Acid) (PLA) and Polyhydroxyalkanoates (PHAs) Applications: A Review on Modifications and Effects. *Polymers* **2021**, *13*, 4271. [CrossRef] [PubMed]
3. Zhao, X.; Hu, H.; Wang, X.; Yu, X.; Zhou, W.; Peng, S. Super Tough Poly(Lactic Acid) Blends: A Comprehensive Review. *RSC Adv.* **2020**, *10*, 13316–13368. [CrossRef]
4. Alhanish, A.; Abu Ghalia, M. Developments of Biobased Plasticizers for Compostable Polymers in the Green Packaging Applications: A Review. *Biotechnol. Prog.* **2021**, *37*, e3210. [CrossRef] [PubMed]
5. Muthuraj, R.; Misra, M.; Mohanty, A.K. Biodegradable Compatibilized Polymer Blends for Packaging Applications: A Literature Review. *J. Appl. Polym. Sci.* **2018**, *135*, 45726. [CrossRef]
6. Anderson, K.S.; Schreck, K.M.; Hillmyer, M.A. Toughening Polylactide. *Polym. Rev.* **2008**, *48*, 85–108. [CrossRef]
7. Oh, J.K. Polylactide (PLA)-Based Amphiphilic Block Copolymers: Synthesis, Self-Assembly, and Biomedical Applications. *Soft Matter* **2011**, *7*, 5096. [CrossRef]
8. Stefaniak, K.; Masek, A. Green Copolymers Based on Poly(Lactic Acid)—Short Review. *Materials* **2021**, *14*, 5254. [CrossRef]
9. Bawa, K.K.; Oh, J.K. Stimulus-Responsive Degradable Polylactide-Based Block Copolymer Nanoassemblies for Controlled/Enhanced Drug Delivery. *Mol. Pharm.* **2017**, *14*, 2460–2474. [CrossRef]
10. Nouvel, C.; Dubois, P.; Dellacherie, E.; Six, J.-L. Controlled Synthesis of Amphiphilic Biodegradable Polylactide-Grafted Dextran Copolymers. *J. Polym. Sci. Part Polym. Chem.* **2004**, *42*, 2577–2588. [CrossRef]
11. Yin, M.; Baker, G.L. Preparation and Characterization of Substituted Polylactides. *Macromolecules* **1999**, *32*, 7711–7718. [CrossRef]
12. He, J.; Wang, W.; Zhou, H.; She, P.; Zhang, P.; Cao, Y.; Zhang, X. A Novel PH-Sensitive Polymeric Prodrug Was Prepared by SPAAC Click Chemistry for Intracellular Delivery of Doxorubicin and Evaluation of Its Anti-Cancer Activity in Vitro. *J. Drug Deliv. Sci. Technol.* **2019**, *53*, 101130. [CrossRef]
13. Trimaille, T.; Gurny, R.; Möller, M. Synthesis and Properties of Novel Poly(Hexyl-Substituted Lactides) for Pharmaceutical Applications. *CHIMIA* **2005**, *59*, 348. [CrossRef]
14. du Boullay, O.T.; Saffon, N.; Diehl, J.-P.; Martin-Vaca, B.; Bourissou, D. Organo-Catalyzed Ring Opening Polymerization of a 1,4-Dioxane-2,5-Dione Deriving from Glutamic Acid. *Biomacromolecules* **2010**, *11*, 1921–1929. [CrossRef] [PubMed]

15. Pound-Lana, G.; Rabanel, J.-M.; Hildgen, P.; Mosqueira, V.C.F. Functional Polylactide via Ring-Opening Copolymerisation with Allyl, Benzyl and Propargyl Glycidyl Ethers. *Eur. Polym. J.* **2017**, *90*, 344–353. [CrossRef]
16. Yu, Y.; Zou, J.; Yu, L.; Ji, W.; Li, Y.; Law, W.-C.; Cheng, C. Functional Polylactide-*g*-Paclitaxel–Poly(Ethylene Glycol) by Azide–Alkyne Click Chemistry. *Macromolecules* **2011**, *44*, 4793–4800. [CrossRef]
17. Zhang, Q.; Ren, H.; Baker, G.L. Synthesis and Click Chemistry of a New Class of Biodegradable Polylactide towards Tunable Thermo-Responsive Biomaterials. *Polym. Chem.* **2015**, *6*, 1275–1285. [CrossRef] [PubMed]
18. Castillo, J.A.; Borchmann, D.E.; Cheng, A.Y.; Wang, Y.; Hu, C.; García, A.J.; Weck, M. Well-Defined Poly(Lactic Acid)s Containing Poly(Ethylene Glycol) Side Chains. *Macromolecules* **2012**, *45*, 62–69. [CrossRef]
19. Ponsart, S.; Coudane, J.; Vert, M. A Novel Route to Poly(ε-Caprolactone)-Based Copolymers via Anionic Derivatization. *Biomacromolecules* **2000**, *1*, 275–281. [CrossRef]
20. Zeng, J.-B.; Li, K.-A.; Du, A.-K. Compatibilization Strategies in Poly(Lactic Acid)-Based Blends. *RSC Adv.* **2015**, *5*, 32546–32565. [CrossRef]
21. González-López, M.E.; Robledo-Ortíz, J.R.; Manríquez-González, R.; Silva-Guzmán, J.A.; Pérez-Fonseca, A.A. Polylactic Acid Functionalization with Maleic Anhydride and Its Use as Coupling Agent in Natural Fiber Biocomposites: A Review. *Compos. Interfaces* **2018**, *25*, 515–538. [CrossRef]
22. Verginio, G.E.A.; Montanheiro, T.L.d.A.; Montagna, L.S.; Marini, J.; Passador, F.R. Effectiveness of the Preparation of Maleic Anhydride Grafted Poly(lactic Acid) by Reactive Processing for Poly(lactic Acid)/Carbon Nanotubes Nanocomposites. *J. Appl. Polym. Sci.* **2021**, *138*, 50087. [CrossRef]
23. Muenprasat, D.; Suttireungwong, S.; Tongpin, C. Functionalization of Poly(Lactic Acid) with Maleic Anhydride for Biomedical Application. *J. Met. Mater. Miner.* **2010**, *20*, 189–192.
24. Ghasemi, S.; Behrooz, R.; Ghasemi, I. Investigating the Properties of Maleated Poly(Lactic Acid) and Its Effect on Poly(Lactic Acid)/Cellulose Nanofiber Composites. *J. Polym. Eng.* **2018**, *38*, 391–398. [CrossRef]
25. Wang, H.-M.; Chou, Y.-T.; Wu, C.-S.; Yeh, J.-T. Polyester/Cellulose Acetate Composites: Preparation, Characterization and Biocompatible. *J. Appl. Polym. Sci.* **2012**, *126*, E242–E251. [CrossRef]
26. Rahem, Z.; Mayouf, I.; Guessoum, M.; Delaite, C.; Douibi, A.; Lallam, A. Compatibilization of Biocomposites Based on Sponge-Gourd Natural Fiber Reinforced Poly(Lactic Acid). *Polym. Compos.* **2019**, *40*, 4489–4499. [CrossRef]
27. Mihai, M.; Ton-That, M.-T. Novel Bio-Nanocomposite Hybrids Made from Polylactide/Nanoclay Nanocomposites and Short Flax Fibers. *J. Thermoplast. Compos. Mater.* **2019**, *32*, 3–28. [CrossRef]
28. Wu, C.-S. Renewable Resource-Based Green Composites of Surface-Treated Spent Coffee Grounds and Polylactide: Characterisation and Biodegradability. *Polym. Degrad. Stab.* **2015**, *121*, 51–59. [CrossRef]
29. Tsou, C.-H.; Hung, W.-S.; Wu, C.-S.; Chen, J.-C.; Huang, C.-Y.; Chiu, S.-H.; Tsou, C.-Y.; Yao, W.-H.; Lin, S.-M.; Chu, C.-K.; et al. New Composition of Maleic-Anhydride-Grafted Poly(Lactic Acid)/Rice Husk with Methylenediphenyl Diisocyanate. *Mater. Sci.* **2014**, *20*, 446–451. [CrossRef]
30. Tsou, C.-Y.; Wu, C.-L.; Tsou, C.-H.; Chiu, S.-H.; Suen, M.-C.; Hung, W.-S. Biodegradable Composition of Poly(Lactic Acid) from Renewable Wood Flour. *Polym. Sci. Ser. B* **2015**, *57*, 473–480. [CrossRef]
31. Wu, C.-S. Preparation, Characterization, and Biodegradability of Renewable Resource-Based Composites from Recycled Polylactide Bioplastic and Sisal Fibers. *J. Appl. Polym. Sci.* **2012**, *123*, 347–355. [CrossRef]
32. Mihai, M.; Ton-That, M.-T. Novel Polylactide/Triticale Straw Biocomposites: Processing, Formulation, and Properties. *Polym. Eng. Sci.* **2014**, *54*, 446–458. [CrossRef]
33. Rigolin, T.R.; Takahashi, M.C.; Kondo, D.L.; Bettini, S.H.P. Compatibilizer Acidity in Coir-Reinforced PLA Composites: Matrix Degradation and Composite Properties. *J. Polym. Environ.* **2019**, *27*, 1096–1104. [CrossRef]
34. Detyothin, S.; Selke, S.E.M.; Narayan, R.; Rubino, M.; Auras, R.A. Effects of Molecular Weight and Grafted Maleic Anhydride of Functionalized Polylactic Acid Used in Reactive Compatibilized Binary and Ternary Blends of Polylactic Acid and Thermoplastic Cassava Starch. *J. Appl. Polym. Sci.* **2015**, *132*. [CrossRef]
35. Bher, A.; Uysal Unalan, I.; Auras, R.; Rubino, M.; Schvezov, C. Toughening of Poly(Lactic Acid) and Thermoplastic Cassava Starch Reactive Blends Using Graphene Nanoplatelets. *Polymers* **2018**, *10*, 95. [CrossRef] [PubMed]
36. Orozco, V.H.; Brostow, W.; Chonkaew, W.; López, B.L. Preparation and Characterization of Poly(Lactic Acid)-g-Maleic Anhydride + Starch Blends. *Macromol. Symp.* **2009**, *277*, 69–80. [CrossRef]
37. Jing, H.; He, H.; Liu, H.; Huang, B.; Zhang, C. Study on Properties of Polylactic Acid/Lemongrass Fiber Biocomposites Prepared by Fused Deposition Modeling. *Polym. Compos.* **2021**, *42*, 973–986. [CrossRef]
38. Detyothin, S.; Selke, S.E.M.; Narayan, R.; Rubino, M.; Auras, R. Reactive Functionalization of Poly(Lactic Acid), PLA: Effects of the Reactive Modifier, Initiator and Processing Conditions on the Final Grafted Maleic Anhydride Content and Molecular Weight of PLA. *Polym. Degrad. Stab.* **2013**, *98*, 2697–2708. [CrossRef]
39. Moghaddam, M.R.A.; Razavi, S.M.A.; Jahani, Y. Effects of Compatibilizer and Thermoplastic Starch (TPS) Concentration on Morphological, Rheological, Tensile, Thermal and Moisture Sorption Properties of Plasticized Polylactic Acid/TPS Blends. *J. Polym. Environ.* **2018**, *26*, 3202–3215. [CrossRef]
40. Aouat, T.; Kaci, M.; Devaux, E.; Campagne, C.; Cayla, A.; Dumazert, L.; Lopez-Cuesta, J.-M. Morphological, Mechanical, and Thermal Characterization of Poly(Lactic Acid)/Cellulose Multifilament Fibers Prepared by Melt Spinning. *Adv. Polym. Technol.* **2018**, *37*, 1193–1205. [CrossRef]

41. Aouat, T.; Kaci, M.; Lopez-Cuesta, J.-M.; Devaux, E. Investigation on the Durability of PLA Bionanocomposite Fibers Under Hygrothermal Conditions. *Front. Mater.* **2019**, *6*, 323. [CrossRef]
42. Wootthikanokkhan, J.; Kasemwananimit, P.; Sombatsompop, N.; Kositchaiyong, A.; Isarankura na Ayutthaya, S.; Kaabbuathong, N. Preparation of Modified Starch-Grafted Poly(Lactic Acid) and a Study on Compatibilizing Efficacy of the Copolymers in Poly(Lactic Acid)/Thermoplastic Starch Blends. *J. Appl. Polym. Sci.* **2012**, *126*, E389–E396. [CrossRef]
43. Dhar, P.; Tarafder, D.; Kumar, A.; Katiyar, V. Thermally Recyclable Polylactic Acid/Cellulose Nanocrystal Films through Reactive Extrusion Process. *Polymer* **2016**, *87*, 268–282. [CrossRef]
44. Mohammad, N.N.B.; Arsad, A.; Rahmat, A.R.; Talib, M.S.; Mad Desa, M.S.Z. Influence of Compatibilizer on Mechanical Properties of Polylactic Acid/Natural Rubber Blends. *Appl. Mech. Mater.* **2014**, *554*, 81–85. [CrossRef]
45. Ruf, M.F.H.M.; Ahmad, S.; Chen, R.S.; Shahdan, D.; Zailan, F.D. *Tensile and Morphology Properties of PLA/LNR Blends Modified with Maleic Anhydride Grafted-Polylactic Acid and -Natural Rubber*; AIP Publishing LLC: Selangor, Malaysia, 2018; p. 020015. [CrossRef]
46. Abdullah Sani, N.S.; Arsad, A.; Rahmat, A.R. Synthesis of a Compatibilizer and the Effects of Monomer Concentrations. *Appl. Mech. Mater.* **2014**, *554*, 96–100. [CrossRef]
47. Thepthawat, A.; Srikulkit, K. Improving the Properties of Polylactic Acid by Blending with Low Molecular Weight Polylactic Acid-g-Natural Rubber. *Polym. Eng. Sci.* **2014**, *54*, 2770–2776. [CrossRef]
48. Jiang, L.; Wolcott, M.P.; Zhang, J. Study of Biodegradable Polylactide/Poly(Butylene Adipate-Co-Terephthalate) Blends. *Biomacromolecules* **2006**, *7*, 199–207. [CrossRef]
49. Teamsinsungvon, A.; Jarapanyacheep, R.; Ruksakulpiwat, Y.; Jarukumjorn, K. Melt Processing of Maleic Anhydride Grafted Poly(Lactic Acid) and Its Compatibilizing Effect on Poly(Lactic Acid)/Poly(Butylene Adipate-Co-Terephthalate) Blend and Their Composite. *Polym. Sci. Ser. A* **2017**, *59*, 384–396. [CrossRef]
50. Rigolin, T.R.; Costa, L.C.; Chinelatto, M.A.; Muñoz, P.A.R.; Bettini, S.H.P. Chemical Modification of Poly(Lactic Acid) and Its Use as Matrix in Poly(Lactic Acid) Poly(Butylene Adipate-Co-Terephthalate) Blends. *Polym. Test.* **2017**, *63*, 542–549. [CrossRef]
51. Phetwarotai, W.; Tanrattanakul, V.; Phusunti, N. Synergistic Effect of Nucleation and Compatibilization on the Polylactide and Poly(Butylene Adipate-Co-Terephthalate) Blend Films. *Chin. J. Polym. Sci.* **2016**, *34*, 1129–1140. [CrossRef]
52. Teamsinsungvon, A.; Ruksakulpiwat, Y.; Jarukumjorn, K. Properties of Biodegradable Poly(Lactic Acid)/Poly(Butylene Adipate-Co-Terephthalate)/Calcium Carbonate Composites. *Adv. Mater. Res.* **2010**, *123–125*, 193–196. [CrossRef]
53. Phetwarotai, W.; Maneechot, H.; Kalkornsurapranee, E.; Phusunti, N. Thermal Behaviors and Characteristics of Polylactide/Poly(Butylene Succinate) Blend Films via Reactive Compatibilization and Plasticization. *Polym. Adv. Technol.* **2018**, *29*, 2121–2133. [CrossRef]
54. Phetwarotai, W.; Aht-Ong, D. Characterization and Properties of Nucleated Polylactide, Poly(Butylene Adipate-Co-Terephthalate), and Thermoplastic Starch Ternary Blend Films: Effects of Compatibilizer and Starch. *Adv. Mater. Res.* **2013**, *747*, 673–677. [CrossRef]
55. Gardella, L.; Calabrese, M.; Monticelli, O. PLA Maleation: An Easy and Effective Method to Modify the Properties of PLA/PCL Immiscible Blends. *Colloid Polym. Sci.* **2014**, *292*, 2391–2398. [CrossRef]
56. Valerio, O.; Pin, J.M.; Misra, M.; Mohanty, A.K. Synthesis of Glycerol-Based Biopolyesters as Toughness Enhancers for Polylactic Acid Bioplastic through Reactive Extrusion. *ACS Omega* **2016**, *1*, 1284–1295. [CrossRef]
57. Kaynak, C.; Meyva, Y. Use of Maleic Anhydride Compatibilization to Improve Toughness and Other Properties of Polylactide Blended with Thermoplastic Elastomers. *Polym. Adv. Technol.* **2014**, *25*, 1622–1632. [CrossRef]
58. Raj, A.; Samuel, C.; Prashantha, K. Role of Compatibilizer in Improving the Properties of PLA/PA12 Blends. *Front. Mater.* **2020**, *7*, 193. [CrossRef]
59. Nam, B.-U.; Son, Y. Enhanced Impact Strength of Compatibilized Poly(Lactic Acid)/Polyamide 11 Blends by a Crosslinking Agent. *J. Appl. Polym. Sci.* **2020**, *137*, 49011. [CrossRef]
60. Petruš, J.; Kučera, F.; Jančář, J. Online Monitoring of Reactive Compatiblization of Poly(Lactic Acid)/Polyamide 6 Blend with Different Compatibilizers. *J. Appl. Polym. Sci.* **2019**, *136*, 48005. [CrossRef]
61. Walallavita, A.S.; Verbeek, C.J.R.; Lay, M.C. Biopolymer Foams from Novatein Thermoplastic Protein and Poly(Lactic Acid). *J. Appl. Polym. Sci.* **2017**, *134*, 45561. [CrossRef]
62. Zhu, R.; Liu, H.; Zhang, J. Compatibilizing Effects of Maleated Poly(Lactic Acid) (PLA) on Properties of PLA/Soy Protein Composites. *Ind. Eng. Chem. Res.* **2012**, *51*, 7786–7792. [CrossRef]
63. Can, U.; Kaynak, C. Effects of Micro-Nano Titania Contents and Maleic Anhydride Compatibilization on the Mechanical Performance of Polylactide. *Polym. Compos.* **2020**, *41*, 600–613. [CrossRef]
64. Chow, W.; Tham, W.; Seow, P. Effects of Maleated-PLA Compatibilizer on the Properties of Poly(Lactic Acid)/Halloysite Clay Composites. *J. Thermoplast. Compos. Mater.* **2013**, *26*, 1349–1363. [CrossRef]
65. Liu, Z.; Chen, Y.; Ding, W. Preparation, Dynamic Rheological Behavior, Crystallization, and Mechanical Properties of Inorganic Whiskers Reinforced Polylactic Acid/Hydroxyapatite Nanocomposites. *J. Appl. Polym. Sci.* **2016**, *133*, 43381. [CrossRef]
66. Park, S.B.; Lee, Y.J.; Ku, K.H.; Kim, B.J. Triallyl Isocyanurate-Assisted Grafting of Maleic Anhydride to Poly(Lactic Acid): Efficient Compatibilizers for Poly(Lactic Acid)/Talc Composites with Enhanced Mechanical Properties. *J. Appl. Polym. Sci.* **2022**, *139*, 51488. [CrossRef]
67. Gardella, L.; Colonna, S.; Fina, A.; Monticelli, O. On Novel Bio-Hybrid System Based on PLA and POSS. *Colloid Polym. Sci.* **2014**, *292*, 3271–3278. [CrossRef]

68. Liu, J.; Jiang, H.; Chen, L. Grafting of Glycidyl Methacrylate onto Poly(Lactide) and Properties of PLA/Starch Blends Compatibilized by the Grafted Copolymer. *J. Polym. Environ.* **2012**, *20*, 810–816. [CrossRef]
69. Ting, G.; Zhu, D.; Lu, Y.; Lu, S. Effect of PLA-g-GMA on the Thermal, Rheological and Physical Behavior of PLA/PBAT Blends. *Polym. Sci. Ser. A* **2019**, *61*, 317–324. [CrossRef]
70. Kangwanwatthanasiri, P.; Suppakarn, N.; Ruksakulpiwat, C.; Ruksakulpiwat, Y. Glycidyl Methacrylate Grafted Polylactic Acid: Morphological Properties and Crystallization Behavior. *Macromol. Symp.* **2015**, *354*, 237–243. [CrossRef]
71. Tarique, J.; Sapuan, S.M.; Khalina, A.; Sherwani, S.F.K.; Yusuf, J.; Ilyas, R.A. Recent Developments in Sustainable Arrowroot (Maranta Arundinacea Linn) Starch Biopolymers, Fibres, Biopolymer Composites and Their Potential Industrial Applications: A Review. *J. Mater. Res. Technol.* **2021**, *13*, 1191–1219. [CrossRef]
72. Wang, N.; Zhang, C.; Weng, Y. Enhancing Gas Barrier Performance of Polylactic Acid/Lignin Composite Films through Cooperative Effect of Compatibilization and Nucleation. *J. Appl. Polym. Sci.* **2021**, *138*, 50199. [CrossRef]
73. Yang, W.; Dominici, F.; Fortunati, E.; Kenny, J.M.; Puglia, D. Effect of Lignin Nanoparticles and Masterbatch Procedures on the Final Properties of Glycidyl Methacrylate-g-Poly(lactic Acid) Films before and after Accelerated UV Weathering. *Ind. Crops Prod.* **2015**, *77*, 833–844. [CrossRef]
74. Nguyen, T.C.; Ruksakulpiwat, C.; Ruksakulpiwat, Y. Effect of Cellulose Nanofibers from Cassava Pulp on Physical Properties of Poly(Lactic Acid) Biocomposites. *J. Thermoplast. Compos. Mater.* **2020**, *33*, 1094–1108. [CrossRef]
75. Nguyen, T.C.; Ruksakulpiwat, C.; Rugmai, S.; Soontaranon, S.; Ruksakulpiwat, Y. Crystallization Behavior Studied by Synchrotron Small-Angle X-Ray Scattering of Poly(lactic Acid)/Cellulose Nanofibers Composites. *Compos. Sci. Technol.* **2017**, *143*, 106–115. [CrossRef]
76. Wang, Y.; Weng, Y.; Wang, L. Characterization of Interfacial Compatibility of Polylactic Acid and Bamboo Flour (PLA/BF) in Biocomposites. *Polym. Test.* **2014**, *36*, 119–125. [CrossRef]
77. Wu, C.-S.; Liao, H.-T.; Jhang, J.-J.; Yeh, J.-T.; Huang, C.-Y.; Wang, S.-L. Thermal Properties and Characterization of Surface-Treated RSF-Reinforced Polylactide Composites. *Polym. Bull.* **2013**, *70*, 3221–3239. [CrossRef]
78. Lyu, Y.; Chen, Y.; Lin, Z.; Zhang, J.; Shi, X. Manipulating Phase Structure of Biodegradable PLA/PBAT System: Effects on Dynamic Rheological Responses and 3D Printing. *Compos. Sci. Technol.* **2020**, *200*, 108399. [CrossRef]
79. Chuai, C.Z.; Zhao, N.; Li, S.; Sun, B.X. Study on PLA/PBS Blends. *Adv. Mater. Res.* **2011**, *197–198*, 1149–1152. [CrossRef]
80. Kangwanwatthanasiri, P.; Suppakarn, N.; Ruksakulpiwat, C.; Yupaporn, R. Biocomposites from Cassava Pulp/Polylactic Acid/Poly(Butylene Succinate). *Adv. Mater. Res.* **2013**, *747*, 367–370. [CrossRef]
81. Mo, X.-Z.; Wei, F.-X.; Tan, D.-F.; Pang, J.-Y.; Lan, C.-B. The Compatibilization of PLA-g-TPU Graft Copolymer on Polylactide/Thermoplastic Polyurethane Blends. *J. Polym. Res.* **2020**, *27*, 33. [CrossRef]
82. Wu, C.-S. Improving Polylactide/Starch Biocomposites by Grafting Polylactide with Acrylic Acid—Characterization and Biodegradability Assessment. *Macromol. Biosci.* **2005**, *5*, 352–361. [CrossRef] [PubMed]
83. Wu, C.-S.; Liao, H.-T. Modification of Biodegradable Polylactide by Silica and Wood Flour through a Sol–Gel Process. *J. Appl. Polym. Sci.* **2008**, *109*, 2128–2138. [CrossRef]
84. Wu, C.-S. Polylactide-Based Renewable Composites from Natural Products Residues by Encapsulated Film Bag: Characterization and Biodegradability. *Carbohydr. Polym.* **2012**, *90*, 583–591. [CrossRef] [PubMed]
85. Wu, C.-S.; Tsou, C.-H. Fabrication, Characterization, and Application of Biocomposites from Poly(Lactic Acid) with Renewable Rice Husk as Reinforcement. *J. Polym. Res.* **2019**, *26*, 44. [CrossRef]
86. Wu, C.-S.; Liao, H.-T. A New Biodegradable Blends Prepared from Polylactide and Hyaluronic Acid. *Polymer* **2005**, *46*, 10017–10026. [CrossRef]
87. Wang, J.-X.; Huang, Y.-B.; Yang, W.-T. Photo-Grafting Poly(Acrylic Acid) onto Poly(Lactic Acid) Chains in Solution. *Chin. J. Polym. Sci.* **2020**, *38*, 137–142. [CrossRef]
88. Orellana, J.L.; Mauhar, M.; Kitchens, C.L. Cellulose Nanocrystals versus Polyethylene Glycol as Toughening Agents for Poly(Lactic Acid)-Poly(Acrylic Acid) Graft Copolymer. *J. Renew. Mater.* **2016**, *4*, 340–350. [CrossRef]
89. Yeh, J.-T.; Chai, W.-L.; Wu, C.-S. Study on the Preparation and Characterization of Biodegradable Polylactide/SiO_2–TiO_2 Hybrids. *Polym.-Plast. Technol. Eng.* **2008**, *47*, 887–894. [CrossRef]
90. Liao, H.-T.; Wu, C.-S. New Biodegradable Blends Prepared from Polylactide, Titanium Tetraisopropylate, and Starch. *J. Appl. Polym. Sci.* **2008**, *108*, 2280–2289. [CrossRef]
91. Beuille, E.; Darcos, V.; Coudane, J.; Lacroix-Desmazes, P.; Nottelet, B. Regioselective Halogenation of Poly(Lactide) by Free-Radical Process. *Macromol. React. Eng.* **2014**, *8*, 141–148. [CrossRef]
92. Kalelkar, P.P.; Collard, D.M. Tricomponent Amphiphilic Poly(Oligo(Ethylene Glycol) Methacrylate) Brush-Grafted Poly(Lactic Acid): Synthesis, Nanoparticle Formation, and *In Vitro* Uptake and Release of Hydrophobic Dyes. *Macromolecules* **2020**, *53*, 4274–4283. [CrossRef]
93. Kalelkar, P.P.; Geng, Z.; Cox, B.; Finn, M.G.; Collard, D.M. Surface-Initiated Atom-Transfer Radical Polymerization (SI-ATRP) of Bactericidal Polymer Brushes on Poly(Lactic Acid) Surfaces. *Colloids Surf. B Biointerfaces* **2022**, *211*, 112242. [CrossRef] [PubMed]
94. Xiao, Y.M.; Li, X.D.; Fan, H.S.; Zhu, X.D.; Teng, L.Z.; Gu, Z.W.; Zhang, X.D. A Novel Preparation Method for Nano-HA/PLA Composite with Grafted PLA. *Key Eng. Mater.* **2006**, *309–311*, 1105–1108. [CrossRef]
95. Li, C.; Hsieh, J.H.; Wang, H.Y. Plasma Polymerised Poly(Methyl Methacrylate) and Cyclopropylamine Films on Polylactic Acid Nanofibres by Electrospinning. *Int. J. Nanotechnol.* **2017**, *14*, 977. [CrossRef]

96. Herskovitz, J.E.; Goddard, J.M. Reactive Extrusion of Nonmigratory Antioxidant Poly(Lactic Acid) Packaging. *J. Agric. Food Chem.* **2020**, *68*, 2164–2173. [CrossRef] [PubMed]
97. Rahmat, M.; Karrabi, M.; Ghasemi, I.; Zandi, M.; Azizi, H. Silane Crosslinking of Electrospun Poly(lactic Acid)/Nanocrystalline Cellulose Bionanocomposite. *Mater. Sci. Eng. C* **2016**, *68*, 397–405. [CrossRef]
98. Wang, J.; Chen, H.; Chen, Z.; Chen, Y.; Guo, D.; Ni, M.; Liu, S.; Peng, C. In-Situ Formation of Silver Nanoparticles on Poly(lactic Acid) Film by γ-Radiation Induced Grafting of N-Vinyl Pyrrolidone. *Mater. Sci. Eng. C* **2016**, *63*, 142–149. [CrossRef]
99. Peng, C.; Chen, H.; Wang, J.; Chen, Z.; Ni, M.; Chen, Y.; Zhang, J.; Yuan, T. Controlled Degradation of Polylactic Acid Grafting N-Vinyl Pyrrolidone Induced by Gamma Ray Radiation. *J. Appl. Polym. Sci.* **2013**, *130*, 704–709. [CrossRef]
100. El Habnouni, S.; Darcos, V.; Garric, X.; Lavigne, J.-P.; Nottelet, B.; Coudane, J. Mild Methodology for the Versatile Chemical Modification of Polylactide Surfaces: Original Combination of Anionic and Click Chemistry for Biomedical Applications. *Adv. Funct. Mater.* **2011**, *21*, 3321–3330. [CrossRef]
101. El Habnouni, S.; Lavigne, J.-P.; Darcos, V.; Porsio, B.; Garric, X.; Coudane, J.; Nottelet, B. Toward Potent Antibiofilm Degradable Medical Devices: A Generic Method for the Antibacterial Surface Modification of Polylactide. *Acta Biomater.* **2013**, *9*, 7709–7718. [CrossRef]
102. Sardo, C.; Nottelet, B.; Triolo, D.; Giammona, G.; Garric, X.; Lavigne, J.-P.; Cavallaro, G.; Coudane, J. When Functionalization of PLA Surfaces Meets Thiol–Yne Photochemistry: Case Study with Antibacterial Polyaspartamide Derivatives. *Biomacromolecules* **2014**, *15*, 4351–4362. [CrossRef] [PubMed]
103. Wu, C.-S. Renewable Resource-Based Composites of Recycled Natural Fibers and Maleated Polylactide Bioplastic: Characterization and Biodegradability. *Polym. Degrad. Stab.* **2009**, *94*, 1076–1084. [CrossRef]

Review

Biocompatible Polymers Combined with Cyclodextrins: Fascinating Materials for Drug Delivery Applications

Bartłomiej Kost *, Marek Brzeziński *, Marta Socka, Małgorzata Baśko and Tadeusz Biela

Centre of Molecular and Macromolecular Studies, Polish Academy of Sciences, Sienkiewicza 112, 90-363 Lodz, Poland; msocka@cbmm.lodz.pl (M.S.); baskomeg@cbmm.lodz.pl (M.B.); tadek@cbmm.lodz.pl (T.B.)
* Correspondence: kost@cbmm.lodz.pl (B.K.); mbrzezin@cbmm.lodz.pl (M.B.)

Academic Editor: Ivan Gitsov
Received: 30 June 2020; Accepted: 24 July 2020; Published: 28 July 2020

Abstract: Cyclodextrins (CD) are a group of cyclic oligosaccharides with a cavity/specific structure that enables to form inclusion complexes (IC) with a variety of molecules through non-covalent host-guest interactions. By an elegant combination of CD with biocompatible, synthetic and natural polymers, different types of universal drug delivery systems with dynamic/reversible properties have been generated. This review presents the design of nano- and micro-carriers, hydrogels, and fibres based on the polymer/CD supramolecular systems highlighting their possible biomedical applications. Application of the most prominent hydrophobic aliphatic polyesters that exhibit biodegradability, represented by polylactide and polycaprolactone, is described first. Subsequently, particular attention is focused on materials obtained from hydrophilic polyethylene oxide. Moreover, examples are also presented for grafting of CD on polysaccharides. In summary, we show the application of host-guest interactions in multi-component functional biomaterials for controlled drug delivery.

Keywords: polylactide; cyclodextrin; drug delivery systems; controlled release

1. Introduction

Supramolecular chemistry aims to construct highly complex, functional systems through a variety of noncovalent interactions (e.g., multiple hydrogen bonding, metal coordination, host–guest interactions, and aromatic stacking) [1]. The discovery of host-guest recognition that mimics biological structure formation and their functions [2,3], led to the development of numerous supramolecular systems. For instance, the host-guest interactions provide a possibility to integrate two or more chemical moieties in a supramolecular system [4,5]. In this review, among the many investigated host molecules, we will focus on the cyclodextrins (CDs), which belongs to the family of cyclic oligomers composed of α-(1→4)-linked glucose units in $4C_{10}$ chair conformation [4]. Depending on the number of glucose units, the ones containing 6, 7, and 8 are named as α-, β-, and γ-CD as shown in Figure 1 [6], respectively. The structure of CDs is described as a truncated cone with a hydrophilic outer surface and hydrophobic core that allows for the encapsulation of a variety of bioactive guest molecules. The formation of inclusion complex is based on non-covalent bonds such as hydrogen or van der Waals bonds. The main driving force of IC formation is the release of the water molecules from the hydrophilic cavity. The water is replaced by more hydrophilic compounds present in the solution and the apolar-apolar interaction is formed. The formation of this type of interaction causes decreasing of CDs ring strain and resulting in more favourable and lower energy state. The list of potential guest which could be encapsulated inside the CDs is quite varied and included low molecular weight molecules such as aldehydes, organic acid, ketones, fatty acids, amines and linear or branched polymers. Due to the size of the inner cavity of CDs, different CDs can complex different molecules, for instance: α-CD can typically complex low molecular weight molecules or aliphatic polymer chains, β-CD will complex aromatics or heterocycles compounds, and γ-CD can entrap the largest molecules such as macrocycles or steroids. Moreover, the hydroxyl functionalities present at CDs surface can be used as a initiating groups in ring-opening polymerization (ROP) of cyclic esters, leading to functionalized polyesters with the core able to form IC [7]. In addition, the grafting to methods (e.g., click reaction [8], coupling [9]) can be used to combine polyesters or polyethers with CDs. Moreover, grafting from and grafting to methods are also used to combine the CDs with saccharides and polysaccharides [10]. All these methods lead to the formation of well-defined and relatively uniform polymers with the ability to form host-guest interactions. The growing interest in the application of macrocyclic molecules in combination with biodegradable polymers, especially with CDs, is related with their biocompatibility and reversibility of host-guest interactions which leads to the stimuli-responsive supramolecular systems [11,12]. This feature is closely related to the kinetic and thermodynamic properties of host–guest complexes and it is determined by the sizes of host and guest, as well as the environmental conditions such as pH and temperature [13]. Therefore, a large amount of efforts was focused on the utilization of this phenomenon in the control drug release of CD-based supramolecular systems. This review focuses on the highly organized drug delivery systems which combines biocompatible synthetic polymers/biopolymers with CDs and discusses the fabrication and biomedical applications of supramolecular systems based on host–guest interactions (with the exclusion of non-degradable polymers and cellulose since the excellent review about cellulose/CDs is already existing in the literature [14]. Therefore, we group the research work into four sections which describe the drug delivery systems containing CD combined with polylactide, poly(ε-caprolactone), poly(ethylene glycol), and polysaccharides. The specific biomedical applications of the host–guest systems based on these polymers are discussed with the focus on several leading directions, that is, drug delivery, gene delivery, antibacterial activity, bone regeneration, photodynamic therapy, tumour targeting, wound healing, and removal of micropollutants. The design and functions of nano- and micro-carriers, hydrogels, and fibres based on the CD-based supramolecular systems have been highlighted.

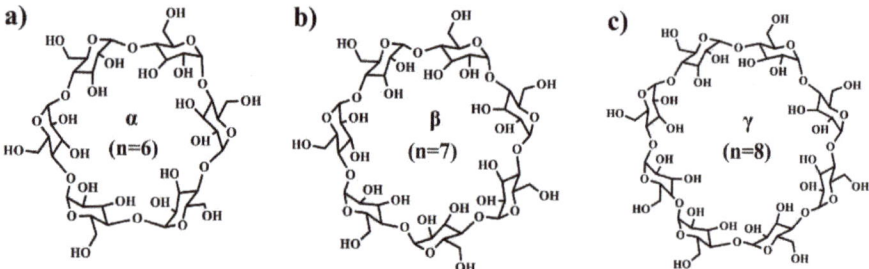

Figure 1. The structure of (a) α-CD, (b) β-CD, (c) γ-CD, n-number of glucopiranose units.

2. Polylactide Systems Based on Cyclodextrin for Controlled Drug Delivery

Polylactide (PLA) (Figure 2) is biodegradable aliphatic polyester that can be produced from naturally occurring renewable resources, such as corn or sugar beets. Due to its biocompatibility and ability to degradation to non-toxic products, PLA can be excellent platform for the preparation of various polymeric drug delivery systems [15]. However, some applications of PLA are limited because of its low solubility in water, long degradation time, and weak encapsulation of polar drug. To overcome these drawbacks, LA is copolymerized with chosen kind of monomers [16] or PLA is connected with polyethylene glycol (PEG) [17]. Despite these disadvantages, PLA and its copolymers are frequently used for biomedical applications for nanoparticles (NPs), microparticles (MPs), fibres or hydrogels preparation. In this part of review, the application of polymeric drug delivery systems such as NPs, MPs, hydrogels and fibres using a combination of cyclodextrin, polylactide, and its copolymers is described and summarized in Table 1.

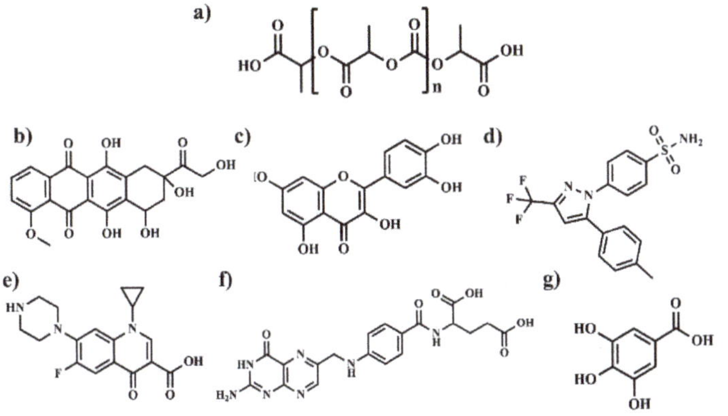

Figure 2. The chemical structure of (**a**) the carrier polymer-PLA, and chemical structure of (**b**) DOX, (**c**) quercetin, (**d**) celecoxib, (**e**) ciprofloxacine, (**f**) folic acid, (**g**) gallic acid.

Supramolecular hydrogels are extensively used as a promising tools for drug delivery systems due to their ability to control release of drug, absorb large amount of water and low toxicity. Polyglycolide (PGA) or polyethylene glycol (PEG) are hydrophilic polymers widely used for preparation PLA-based hydrogels [18]. Also triblock copolymer of poly(lactide-co-glycolide-co-ethylene glycol) (PLGA-PEG-PLGA) was used to prepare supramolecular hydrogel due to its good solubility in water. However, the high M_n of PEG and low LA/GA ratio is crucial to maintain hydrophilic nature of PLGA-PEG-PLGA copolymer. After mixing of a solution of triblock copolymer with solution of α-CD in water, the gelation process occurs as a result of IC formation between PLGA chain and cyclodextrin cavity. The gelation time can be reduced by increasing the α-CD concentration. In addition, the increasing of hydrophilic-lipophilic balance causes shorter gelation time and faster model drug e.g., vitamin B_{12} (B_{12}) releases. Moreover, the alteration of the ratio of PLGA-PEG-PLGA to α-CD leads to controlled release of B_{12} from supramolecular hydrogel [19]. Therefore, the hydrogels could be also obtained by treating the solution of diblock copolymer PLA-b-PEG with an α-CD solution (Figure 3b). The core-shell structure of hydrogel was achieved due to the amphiphilic nature of PLA-b-PEG copolymer and their micellar aggregation. The aggregation process is driven by the spontaneous self-assembly of the polymer in water. Therefore, as a result of this process hydrophobic part of copolymer (PLA) creates inner core while hydrophilic part (PEG) forms outer corona. However, the two types of physical interactions are required for gelation: association of PLA-PEG micelles and formation of IC microcrystals. The formation of micellar hydrogels depend on several factors such as α-CD concentration, polymer concentration, and temperature of the process. The selection of proper parameters is essential to prepare high-quality hydrogels. The hydrogels were also loaded with doxorubicin (DOX) and its release profiles were evaluated in PBS at 37 °C with varying type of hydrogels formulation. The release rate of DOX from hydrogels decreases as the concentration of α-CD increases because of enhancement in hydrogel strength. It is worth noting that blank micellar hydrogel is non-toxic against HeLa cells, and after DOX encapsulation the hydrogel efficiently delivers cargo to the desired target. It was confirmed by the uptake analysis, which showed that DOX can be located in nucleus and cytoplasm through the carrier–mediated endocytosis pathway [20].

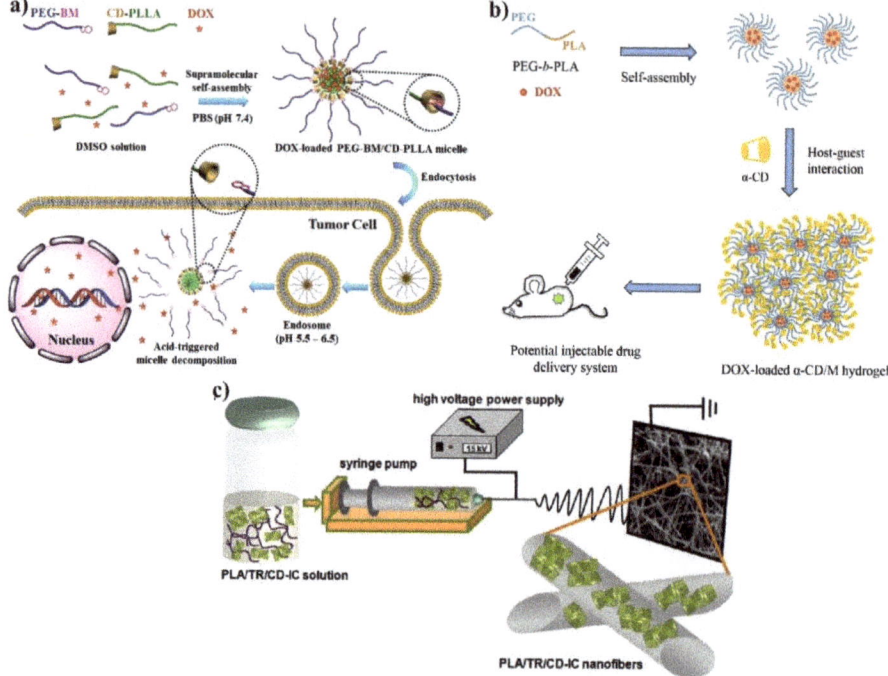

Figure 3. Schematic illustration of formation (**a**) nanoparticles, (**b**) hydrogels and, (**c**) fibres based on PLA and its copolymers for drug delivery applications. (Copyrights 2019 ACS (1A), Copyrights 2018 Elsevier (1B), Copyrights 2013 ACS (1C)).

Table 1. The summary of described hydrogels, microparticles and fibres for drug delivery based on combination of PLA and its copolymers with different CDs.

Type of Drug Delivery System	Platform	Type of CD	Drug	Release Medium	In Vitro Studies	Ref.
Hydrogel	PLGA-PEG	α-CD	Vitamin B12	PBS	nd	[19]
Hydrogel	PLA-PEG	α-CD	DOX	PBS	HeLa cells	[20]
Fibre	PLA	γ-CD	Gallic acid	10% or 95% EtOH	nd	[21]
Fiber	PLA	Mβ-CD	Quercetin	PBS	nd	[22]
Fibre	PLA	polyCD	Ciprofloxaxin	PBS	NIH3T3 cells	[23]
Microparticle	PLGA	γ-CD	Dexamethasone	PBS	nd	[24]
Microparticle	PLGA	DMβ-CD	Celecoxib	PBS	Human chondrocytes	[25]
Microparticle	PLGA	HPβ-CD	Prostaglandin	PBS	Calu-3 cells	[26]
Nanoparicle	PLGA	HPβ-CD	Triamcinolone acetonide	PBS	Rabbit eyes	[27]
Nanoparicle	PLA-PEG	β-CD	Folic acid	nd	HEK293T cells	[28]
Nanoparticle	PLA/PEG	β-CD	DOX	pH 6.0, pH 5.5, pH 7.4	HepG2 cells	[29]

The well-known electrospinning process can be useful method for production of microfibers (MF) or nanofibers (NF) for biomedical application as shown in Figure 3c. Uyar and co-workers used linear poly(lactide) (PLA) modified with CD to prepare nanofibers loaded with two different antioxidant agents: gallic acid (GA), tocopherol (TC) or an antibiotic: triclosan (TR). The electrospinning allows for the preparation PLA nanofibers with incorporated IC between GA or TC and γ-CD (Figure 4a). The molecular modelling was employed to confirm the ability of GA to IC formation. The penetration of cyclodextrin cavity by GA leads to the formation of two energetically stable structures with relatively low energy. The release studies showed that dosage control of GA and TC strongly depends on release media, nanofiber diameter and solubility of GA. Additionally, due to the antioxidant properties of GA or TC, the radical scavenging assay was performed. However, the GA incorporated in γ-CD showed slightly lower activity than free GA due to the specific orientations of GA hydroxyl group in γ-CD cavity. Therefore, despite faster release of tocopherol (α-TC) from PLA/α-TC/γ-CD-IC-NF, in comparison to α-TC-loaded PLA nanofibres, there was no significant difference in antioxidant activity and lipid oxidation inhibition between both systems. The TR- loaded fibres showed the broad spectrum of activity against gram-positive and gram-negative bacteria. Due to the well-known antibacterial properties of TR, the PLA-TR nanofibres cause the inhibition in the growth of both *E. coli* and *S. aureus* bacteria strains. The observed inhibition zone around the fibres is wider in the case of polylactide NF with IC (PLA-CD/TR) than with free TR what can be related to better solubility of CD/TR systems in agar media. Most importantly, the better inhibitions level against bacteria was observed for NF with the β-CD IC. The difference between β-CD and γ-CD systems may be caused by the partly uncomplexed TR in the case of β-CD, and uncomplexed TR can affect the bacteria at the initial stage before the complexed drug release from NF [21,30,31].

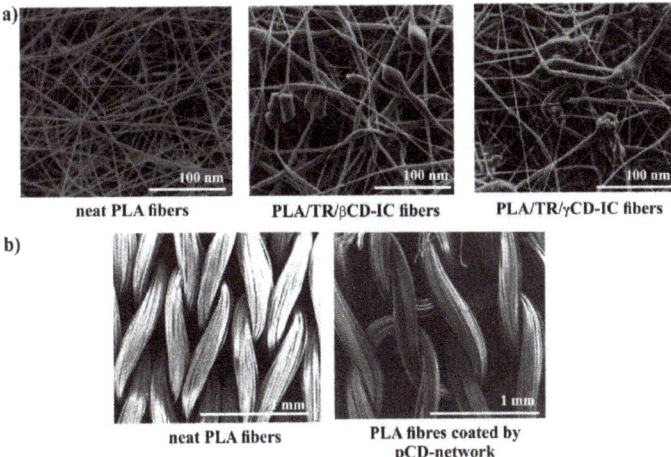

Figure 4. The SEM imaging of PLA fibres with (**a**) TR inclusion complex and (**b**) coated by pCD-citrate network for antibacterial drugs delivery. (Copyrights 2013 ACS (**a**), Copyrights 2017 Acta Materialia (**b**)).

Microfibers (MF) with β-CD were used as an excellent platform for quercetin (Q) delivery. Kost et al. have prepared PLAs with ability to formation of IC due to presence of β-CD in the polymer core. The presence of β-CD covalently built into polymer structure allow for creation of IC between β-CD-PLA and Q. Electrospinning of such modified polylactides leads to formation of various types of Q-loaded supramolecular MF with antibacterial activity. Importantly, MF forms Q protective scaffold against the destructive effects of light and oxygen. In contrast to TR-loaded microfibers, in this case the release of Q to the phosphate buffer saline (PBS) or agar plate was not observed. Moreover, the inhibition zone appears only where the materials come into direct contact with the agar medium. The delayed release of Q is associated with a strong entrapment of Q in the fibres or in the cyclodextrin cavity, poor solubility of Q in water and the hydrophobic nature of PLA. Despite the fact that all nonwovens with Q loading were yellow, after rubbing against the skin or white paper they did not leave yellow marks, which is important from a practical point of view when we intend to use them as dressing material [22]. To prepare PLA fabrics covered by a β-CD network with prolonged antibacterial activity, the pad/dry/cure technique was successfully employed (Figure 4b). The PLA was impregnated in a water solution containing CDs, citric acid and catalyst. After extraction of unreacted components (130 °C), the PLA textiles covered with β-CD network were prepared (PLA/β-CD). The ciprofloxacin-loaded PLA/β-CD textiles after exposition on two strains (*E. coli*, *S. aureus*) sufficiently reduced the number of bacteria. Moreover, the increasing of β-CD ratio in PLA/β-CD textiles from 8% to 33%, provided prolonged antibacterial activity up to 24 h and 120 h for *S. aureus* and *E. coli*, respectively. Additionally, PLA/β-CD textiles were estimated as a non-toxic against mouse fibroblast after 3 h. After 6 days, the viability of cells was reduced to 65% and decreased with increasing of cyclodextrin concertation. The possible degradation of cyclodextrin network by hydrolysis of esters bond of PLA produces free carboxylic acid and thus faster cell proliferation [23]. Due to their biocompatibility and biodegradability, poly(lactide-*co*-caprolactone) (PLA-*co*-PCL) was chosen to produce nanofibres scaffold for the inhibition of growth MCF-7 cells. The different combinations of MgO nanoparticles (MgO NPs), curcumin (Cur) and aloe vera (AV) allow for producing various types of drug delivery nanofibres scaffold via electrospinning. The MgO NPs can be mixed with β-CD and Cur to form multitask, anticancer system. Additionally, MgO NPs were chosen due to their ability to inhibit the uncontrolled growth of cancer cells, and AV was added to the NF to improve the hydrophilicity of scaffold. After blending the PLACL with the AV the contact angle was reduced from 127° to 54° due to the presence of polyphenols in AV structure.

The presence of MgO NPs causes increasing of contact angel because of the replacement of polyphenols group to MgO which is more hydrophobic. The morphology of NF scaffolds were uniform, with rough fibre surfaces. The diameter of the scaffolds was 786 ± 286, 507 ± 171, 334 ± 95, 360 ± 94 and 326 ± 80 nm for PLACL, PLACL/AV, PLACL/AV/MgO, PLACL/AV/MgO/CUR and PLACL/AV/MgO/β-CD, respectively. The obtained materials were tested on the MCF-7 cells to estimate their cytotoxicity properties by measurement the number and the shape of breast cancer cells (MCF-7). PLACL/AV/MgO/Cur and PLACL/AV/MgO/β-CD nanofibres showed the highest apoptotic effect, destroyed polygonal morphologies and reduced the number of MTC-7 cells. It is worth noticing that PLACL/AV/MgO also possessed apoptotic effect but its activity significantly increased in combination with Cur or β-CD. The obtained results showed that PLACL/AV/MgO NF scaffolds supported cell adhesion and proliferation but combination of CUR or β-CD showed only a slight cytotoxicity effect on the cell line. Moreover the reduction of the concentration of active ingredients from 5% to 1% show that the scaffold still possess anti-tumor properties. The in vitro studies showed that PLA-*co*-PCL nanofiber scaffold with natural ingredients can be effective against breast cancer cells [32].

The emulsion/solvent evaporation method is one of the common techniques to prepare polymeric microparticles (MPs) for the encapsulation of bioactive molecules [33]. The preparation of biodegradable PLA microcapsules has been extensively investigated in recent years [34–36]. The PLGA was used as a platform for encapsulating dexamethasone sodium phosphate (DE). DE is sensitive to degradation and requires stabilization to prolong its activity. Before encapsulation in PLGA MPs, DE was firstly physically complexed by three types of molecules: hydroxypropyl-β-CD (HPCD), γ-CD, or water-soluble polyethylenimine (PEI). The CDs were selected as entrapping agents due to their ability to form the IC, while the complexation of DE by PEI is based on electrostatic interaction between phosphoryl part of DE and nitrogen atoms of PEI. Therefore, PLGA exhibit lower encapsulation efficiency of DE, when MPs are obtained without complexion agents. The complexation of DM leads to increase encapsulation efficiency, however, PEI was found to be the most effective complexing agent since it possesses a large number of imine groups in one polymer chain. The presumptive protective effect of DE complexation by CDs and EPI against UV irradiation showed the negative result. After irritation, the free DE, as well as complexed one, underwent photodegradation to the same extent. Unexpectedly, the more stable complex between DE and γ-CD showed the highest photodegradation degree. The formation of intermolecular hydrogen bond between DE and γ-CD have destructing effect on its photochemical stability. However, the presence of any secondary interaction slows the DE release from PLGA MPs in PBS at 37 °C. Moreover, the release of DE from PLGA MPs with EPI is lower than those of the drugs from PLGA with HPCD and γ-CD. The delay of DE release from MPs with EPI is associated with stronger electrostatic interactions in comparison to weak van der Waals forces in IC [24]. PLGA microspheres were prepared also by Cannava et al. as a sustainable drug delivery system for anti-inflammatory agent Celecoxibe

(CE). They evaluated the influence of IC between β-CD and CE on the size of MPs, the release of CE and anti-inflammatory activity. The presence of the IC causes the size reduction of MPs because of its ability to act as a surfactant and prevent coalescence during the evaporation process. Unexpectedly, the creation of an IC reduces the encapsulation efficiency of CE from 80% to 54%. The formation of stable IC with high solubility in water limits its ability to diffusion to the oil phase and reduce affinity to the PLGA matrix during MPs preparation. However, the formation of IC also enhances the solubility of CE in PBS thus the release of CE from all formulation with IC was much faster than from MPs without β-CD. Only 15% of CE was released from MPs without β-CD within 15 days. In addition, the presence of β-CD/CE in the PLGA matrix introduces the porosity and affects the release rate. The prepared MPs with β-CD were more effective than free drug as an anti-inflammatory drug on human chondrocyte cultures. This higher degree of activity was correlated with the better solubility of CE in the culture media and to the fact that β-CD/CE can act as a penetration enhancer [25]. The same release behaviour was observed by Gupta and co-workers for prostaglandin. The presence of IC enhances the solubility of hydrophilic prostaglandin (1.6×10^{-2} M at 25°C) from 10 to 60 times after addition the appropriate CDs [37], and accelerate its release and improve bioavailability. The prostaglandin administrated via pulmonary route is rapidly metabolised within 2 h at 37°C, and thus possesses short therapeutic activity. However, the encapsulation of prostaglandin into the PLGA microparticles induces the sustainable and prolonged release and thus extending the prostaglandin activity. It is well-known that all components, which build MPs are non-toxic and approved for FDA to biomedical application thus, the obtained MPs with prostaglandin were found to be safe against Calu-3 human airway epithelial cells [26].

The IC is widely used to solve the problems with solubility, degradability or bioavailability of drugs in physiological condition. The simplest method to introduce IC to polymeric NPs is based on the physical mixing of components before NPs formation [27,38]. However, the most attractive approach is to attach β-CD to the polymer chain before the formation of IC and NPs. To introduce β-CD to polymeric chain researchers used different reactions such as: azide-alkyne cycloaddition reaction (CuAAC) [39], esterification [40], carbodiimidazole coupling reaction [41] or polymerization [42]. To produce polymeric NPs with cyclodextrin moiety, Gao et al. used the well-known reaction between the carboxyl group of PLA [43,44] or PLGA with amino terminated β-CD [45]. This multistep process leads to introducing to the polymeric chain one or two β-CD moieties and obtaining the polymers with the ability to form IC. In addition, the hydrophobicity of polymers decreases as a result of attachment of β-CD moieties to the PLA chain and decreases with increasing concentration of β-CD or decreasing the molecular weight of the polymer. The NPs were fabricated by using nanoprecipitation method. The slowly instillation of acetone polymeric solution to the deionised water allows for production of spherical, uniform nanoparticles with narrow dispersity. However, the obtained NPs size and stability strongly depend on the molecular weight of polymer, solution viscosity, hydrophilic-lipophilic balance, added surfactant or encapsulated drug. The employed strategy allowed to use these nanoparticles to deliver BSA protein. The encapsulation efficiency (EE) of polymeric NPs with β-CD moiety was much higher than for non-functionalized NPs. The high EE is associated with the three types of phenomena: the creation of IC between BSA and β-CD cavity, ionic interaction of BSA carboxyl groups with protonated –NH– group of modified β-CD or amphiphilic nature of β-CD-terminated PLA. Moreover, the method used allows to control the amount of encapsulated drug. In double emulsion method, the water solution of BSA is used as an inner water phase and thus, BSA could be effectively encapsulated inside the NPs [45]. However, the influence of β-CD in the polymer chain on release behavior was not observed for all prepared NPs. As suggested by Gao et al. the degree of degradation of the polymer matrix may be responsible for the observed differences in the course of drug release curves [43–45]. To introduce more than two cyclodextrin moieties to the polymer, the 4-arms PEG-(PLA)$_4$ β-CD end-capped copolymers were synthesized. It is worth noting that β-CD-functionalized PLGA can undergo self-assembly into reversible micelles in an aqueous solution. However, the presence of four β-CD could not induce higher encapsulation efficiency (EE) of BSA. The formation of the IC does not determine the encapsulation process and only occurs when the acrylic part of the BSA chain penetrates the cyclodextrin cavity. The encapsulation of BSA in reversible micelles is mainly based on hydrophobic interaction between PLA and BSA chain and increase with increasing of polylactide molecular weight. The release studies showed that the release of BSA from star-shaped or linear PLA-PEG NPs with β-CD moiety is faster than from PLA-PEG without β-CD. It is probably due to a faster diffusion of BSA from NPs with larger pores formed into particles in the presence of β-CD [40]. Y-shaped PDLA with β-CD moiety (β-CD-b-[PDLA]$_2$) and Y-shaped poly(L-lactide-co-dimethylaminoethyl methacrylate) (PDMAEMA-b-[PLLA]$_2$) were used to produce stereocomplexed micelles for doxorubicin (DOX) delivery. The combination of ROP, ATRP and click reaction allows for preparing this Y, unique architecture of copolymers. The presence of β-CD in copolymers allows for formation of the IC with adamantane terminated FA. Due to these untypical architectures (Y-like structure), the copolymers can create stable, uniform NPs with folic acid (FA) on the surface reinforced by stereocomplexation between enantiomeric chain of PLLA and PDLA. The cytotoxicity assay showed no obvious toxicity against human embryonic kidney cells (HEK 293 cells) no-adverse-effect and biocompatible nature of stereocomplexed micelles [46].

To prepare stereocomplex micelles another approach also was applied by Kost et al. The partly methylated-β-CD was introduced to the polymers (PLLA and PDLA) by using it as an initiator of lactides polymerization. The prepared in this way PLLA and PDLA homopolymers are still able to form host-guest inclusion complex but also stereocomplex in addition. The slow precipitation of DMF polymeric solution to distilled water allows for formation uniform, stable enantiomeric or stereocomplexed NPs. The stability of NPs in PBS at 37 °C was achieved due to the addition of poly(D,L-lactide-co-ethylene glycol methyl ether) to the solution during nanoprecipitation. Cholesteryl-terminated PLLA-PEG or PDLA copolymers and poly(cyclodextrin) (PCD) were used for preparation three polyester platforms containing different amount of PCD moieties for DOX delivery. The linear PCD was prepared from β-CD by using epichlorohydrin (EPI) as a linking agent [47]. The host-guest interaction between cholesterol groups and PCD based on weak van der Waals interaction and stereocomplexation are crucial interactions that led to the formation of PDLA/PLLA/PCD cross-linked micelles (CSMs). A careful selection of components allows for preparation uniform, stable dimension rage of NPs from 60 to 100 nm. Moreover, the creation of IC and increasing of PLA/PCD ratio (PLA: PCD, 1:2) improved the compactness and stability of NPs in water making it appropriate vehicle for DOX delivery A slight decrease in the DOX release profile from crosslinked stereocomplex micelles (CSMs) was observed as compared to stereocomplex micelles without PCD. The IC between PCD and cholesterol-terminated PLA forms an extra barrier, which could slow down the release of DOX from NPs under physiological condition. The cellular-uptake efficiency showed that the presence of the highest amount of PCD in CSMs causes the highest release ratio of DOX. The PCD promote the drug delivery carries to swell, dissolve and faster release of DOX from CSMs with the highest degree of PCD. The half-maximal inhibitory concentration for CSMs with the highest degree of PCD, showed the most effective proliferation activity. However, no increased toxicity was observed for all DOX loaded micelles compared to free DOX [48].

Hydroxyapatite (HA) could also be coated by PCD in a simple reaction with citric acid. This modification leads to introducing PLA to HA via ROP of LA and thus the materials composed of HA/PCD/PLA may be potentially used for bone tissue engineering. It is well-known that toxicity of materials based on HA might be correlated with Ca^{2+} release from HA. The release of calcium ions disrupts the body's calcium homeostasis and may be the main cause of its toxicity. On the other hand, PLA chains form a hydrophobic layer on the HA surface and inhibit the free release of Ca^{2+} into the human body during application. It was also reported that the β-CD could complex some active lipophilic compounds (cholesterin) and ensure its prolonged activity. In addition, the introduction of HA/PCD/PLA into mesenchymal stem cells induces a local anabolic reaction and significantly increases its cell adhesion, mineralization, biocompatibility, and osteoinductive activity. The combination of HA/PCD/PLA reduces the cytotoxicity of HA and makes it promising material for tissue engineering [49]. The pH-sensitive micelles based on associated/disassociated IC between β-CD-terminated PLA and benzimidazole (BM) terminated-PEG were successfully synthesized as shown in Figure 3a and classic micelles, independently synthesized, based on PLA-*b*-PEG copolymer were used in the control test. The change in pH value from 5.5 to 7.4 allows for controlling the size and stability of NPs due to associated/disassociated of IC. The low pH leads to the creation of an anionic form of BM and causes dissociation of IC thereby the pH-responsive micelles become larger and more unstable. The release studies showed that by pH adjustment, 3 times faster DOX release could be achieved in pH 5.5 in comparison to the physiological condition. Furthermore, the in vitro studies were investigated on two cells line: human breast cancer cell line (HeLa) and human liver cancer cell line (HepG2). It was not observed the significant difference in the antitumor activity of pH-responsive micelles, PLA-*b*-PEG micelles and free DOX. However, all DOX-loaded micelles showed lower drug accumulation in the livers and kidney in comparison to free DOX. This results suggested that DOX-loaded micelles slightly enhance the antitumor activity and reduce systemic toxicity of DOX [29]. Additionally, the improving of targeting properties of polymeric micelles with β-CD might be achieved by conjugating folic acid to the polymer chain. The FA-modified micelles exhibit better targeting properties and could specifically delivery encapsulated drug to the tumor side due to the ability of FA to bind to folate receptors over-expressed in the cells [28,41].

The biodegradable and biocompatible PLA materials are widely used for the preparation of different types of drug carriers in the form of NP, MPs or fibres. Unfortunately, the blank polymers and their copolymers usually do not have appropriate functions enabling their direct application as drug carriers. Only the desired modifications provide them with new functions, structures and morphology allowing for controlled drug release under external stimuli. The α-, β- or γ-CD are extensively applied to improve the application properties of the drug, especially solubility in water or their penetration through cellular membranes. Although, mixing polymers with CDs is still the simplest method to improve the functionality of the polymer this operation does not always affect the drug release behavior. A promising approach to creating useful drug delivery systems is based on covalent incorporation of CDs moiety to the polymer structure by means of well-described organic reactions. It should be mentioned, however, that some of the potential drug carriers presented in the literature are very complicated in synthesis and the sophisticated materials produced in this way are usually very expensive. This limits their future use.

3. Polymeric Drug Delivery Systems from Cyclodextrin and Poly(ε-Caprolactone)

Aliphatic semi-crystalline poly(ε-caprolactone) (PCL) (Figure 5) is approved by the U.S. Food and Drug Administration (FDA) for biomedical and pharmaceutical applications. It possesses good biocompatibility and slow biodegradability, high flexibility, as well as good solubility in many the organic solvents. Therefore, PCL is a polymer frequently used as a matrix for preparation of the biodegradable and bioresorbable materials. For instance, PCL based systems were applied in the production of wound dressings, surgical sutures, prosthetics materials, as well as carriers for a variety of drugs [50–53]. However, due to the hydrophobicity and lack of active sites (similar as in PLA) needed to attach bioactive molecules, its biological applications are limited. Among the methods used to modify PCL and expanding its applications, the following strategies have been proposed: chemical modification of PCL and surface modification its particles [54,55] the formation of self-assembling copolymers with hydrophilic polymers [56,57], and coating of PCL nanofibers with bioactive materials [58,59]. The alternative way of bioactive sites addition is the preparation of non-covalent ICs of PCL with CDs (α, β, or γ-CD) [60]. Recently, PCL/CD-based systems in the form of micro- [61,62] and nanoparticles [63], micelles [64–66], nanofibres [67,68] or hydrogels [64,69,70] are promising drug delivery carriers due to their unique properties in terms of drug release and/or stimuli-responsiveness [64] as shown in Table 2. Therefore, in this part of the review, the application of polymeric drug delivery systems based on the combination of PCL or its copolymers and CD, are described.

Table 2. The summary of described micro- and nanoparticles, fibers, and hydrogels for drug delivery based on a combination of PCL and its copolymers with different CDs.

Type of Drug Delivery System	Platform	Type of CD	Drug/Biomolecule	Release Medium	In Vitro/In Vivo Studies	Ref.
Microparticle	PCL	HP-β-CD	Diffractaic acid	pH 7.4	Vero cells	[61]
Nanoparticle	PCL	α-CD	Camptothecin/DOX	PBS	HepG2 cells	[63]
Hydrogel	PCL-b-PEG-b-PCL	β-CD	Indomethacin	PBS	Rat paws	[64]
Nanoparticle	PCL-b-PEG	β-CD	DOX	PBS	SKOV3 cells/ HEK293T cells	[65]
Nanofiber	PCL	HP-β-CD	Sulfisoxazole	PBS	nd	[68]
Hydrogel	PCL-b-PEG	α-CD	Diclofenac	PBS	L-929 cells/HCEC cells/ rabbit eyes	[69]
Hydrogel	(mPEG acetal-PCL-acetal-)₃	α-CD	DOX hydrochloride	pH 5.0, pH 7.4	nd	[70]
Nanofiber	PCL	β-CD	α-Tocopherol	PBS	nd	[71]
Nanofiber	PCL	β-CD	Naproxen	PBS	nd	[72]
Nanofiber	PCL	β-CD	Ciprofloxacin	pH 7.2	nd	[73]
Nanofiber	PCL	γ-CD	Oxidoreductase	pH 7.5	Catalase enzyme/	[74]
Nanofiber	PCL	γ-CD	Oxidoreductase	pH 7.5	Laccase enzyme	[75]
Nanofiber	PCL	β-CD	Silver sulfadiazine	PBS	E. coli/Klebsiella pneumoniae/ S. epidermidis/S. aureus	[76]
Nanofiber	PCL	β-CD	Tetracycline	PBS	Aggregatibacteractinomycetem comitans/ Porphyromonas gingivalis	[77]
Nanoparticle	PCL-b-PEG	HP-β-CD	Docetaxel	PBS	MCF-7 cells	[78]
Nanoparticle	PCL-b-PEG	HP-β-CD	Paclitaxel/ Cidofovir	pH 5.5	L929 cells,	[79]
Nanoparticle	PCL-b-PEG	β-CD	Paclitaxel	PBS	HepG2 cells	[80]
Nanoparticle	methoxy-PEG-b-PCL	Triacetyl-β-CD	Sepiapterin	PBS	nd	[81]
Nanoparticle	PCL-b-PEG	β-CD	DOX	pH 7.0–2.0	nd	[82]
Nanoparticle	Star-PCL-PEG	DMβ-CD	DOX	PBS/ 1,4-dithio-threitol solution	Skov cells/HEK293 T cells/ tumor-bearing mouse	[83]
Nanoparticle	Star-PCL-PEG	β-CD	DOX	PBS	A549 cells	[84]
Nanoparticle	Star-PCL-PEG	β-CD	DOX	PBS	HeLa cells	[85]
Hydrogel	PCL-b-PEG-b-PCL	γ-CD	Insulin	PBS	nd	[86]
Hydrogel	PCL-b-PEG-b-PCL	α-CD	B₁₂ vitamin	PBS	nd	[87]
Hydrogel	PCL-b-PEG and poly(ethylene glycol)-b-poly(acrylic acid)	α-CD	DOX/Cisplatin	PBS	Human bladder carcinoma EJ cells	[88]
Hydrogel	methoxy PEG)-b-PCL-co-1,4,8-trioxa-[4.6]spiro-9-un-decanone	α-CD	Paclitaxel	PBS	HeLa cells/7703 cells	[89]
Hydrogel	mPEG–PCL–mPEG	α-CD	Erythropoietin	PBS	Rats heart muscle	[90]
Hydrogel	mPEG–b-PCL-b-poly[2-(dimethylamino) ethyl methacrylate	α-CD	DNA Polyplexes	PBS	COS7 cells	[91]
Hydrogel	PCL-b-PEG-b-PCL	γ-CD	Dexamethasone	PBS	nd	[92]
Nanoparticle	PCL-b-PEG	HP-β-CD	Docetaxel	PBS	HeLa cells	[93]

Biodegradable drug-loaded polymeric microparticles (MPs) (e.g., microspheres or microcapsules) based on aliphatic polyesters are interesting and promising carriers for developing an oral-dosed controlled release [94]. The bioactive particles encapsulated in polymeric MP are usually obtained by emulsion-solvent evaporation [95]. For instance, Silva et al. reported the incorporation of CD/diffractaic acid ICs into PCL microspheres using the multiple W/O/W emulsion-solvent evaporation technique [61]. As an active substance, the diffractaic acid was used, which is known from its antiulcerogenic and gastroprotective properties. The 35-fold increase in solubility of used biologically active compound was achieved by the application of 2-hydroxy-propyl-β-CD and the formation of ICs between this drug and CD cavity. The formation of ICs during the PCL-microparticles precipitation reduced drug cytotoxicity against Vero cells, what increased safety and therapeutic efficacy. A novel class of β-CD and pH-sensitive polymer-based polymersome MPs, designed for potential therapeutic tools in the treatment of cholesterol-associated neurodegenerative diseases, was presented by the Yagci group [62]. Polymer structure was specifically designed for lysosomal-targeting through the incorporation of a benzoic imine bond between the β-CD and the PCL backbone. At physiological conditions (PBS, pH = 7.4) the obtained polymer self-assembled into stable polymersomes with a detachable β-CD core and negatively charged surface due to the presence of carboxylic-groups incorporated into PCL macromolecules. Under weakly acidic pH conditions (pH 5.5), the formed nanostructure undergoes hydrolysis of the imine linkage, with consequent disassembling and release of monomeric CDs. The obtained MPs were non-toxic according to the cellular viability test on the HUVECs line. This was a promising therapeutic approach for CD-PMs delivery to endosome and lysosome, due to rapid hydrolysis at pH 5.5 and fast cellular uptake. Besides, the presence of the carboxylic groups at the polymer termini provided the possibility of further surface modification of the microparticles for improved delivery.

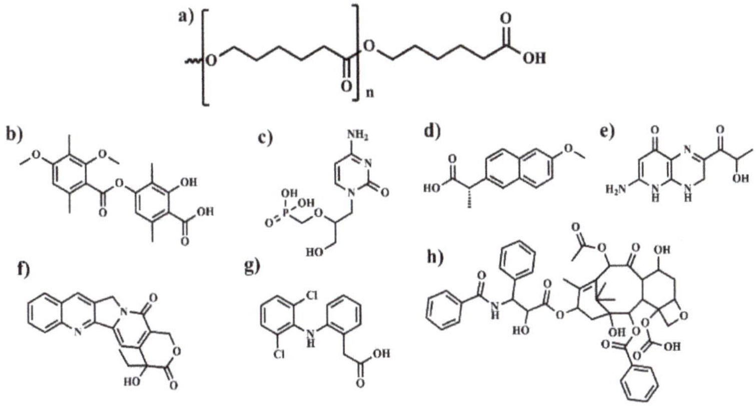

Figure 5. The chemical structure of (**a**) the carrier polymer -PCL, and chemical structure of (**b**) diffractaic acid, (**c**) cidofovir, (**d**) naproxen, (**e**) sepiapterin, (**f**) camptothecin, (**g**) diclofenac, (**h**) paclitaxel.

The nanosized-based carriers are extensively studied as promising tools for drug delivery systems which can be passively accumulated at tumor sites due to the enhanced permeability and retention effect (EPR effect) [96]. The incorporation of amphiphilic CDs into polymers can alter the amphiphilicity of the polymer, resulting in the formation of supramolecular nanostructures self-assembled as polyrotaxanes (PRXs). For instance, CDs and PCL are able to form of PRXs structures in a solid-state, termed as "molecular necklaces", what induces the crystallization of CDs in channel-like structures and keeps CD molecules threaded onto the polymer backbone during the post-functionalization step [97,98]. The polyester-based PRXs are described in the literature as drug delivery systems with drugs loaded by physical entrapment via hydrophobic interactions, or with drugs conjugated to the backbone of PRXs via stimulus-responsive conjugation. For example, Li et al. developed α-CD-based nanoscale micelles of PRXs structure with enzyme-regulated release behavior [63]. The novel supramolecular micelles were loaded with the anticancer camptothecin (CPT) chemically bonded with PCL through a hydrolyzable linkage. The steady-state release of drug was observed without lipase, it was shown that less than 30% CPT released over 120 h, with the negligible initial burst release. On the contrary, 50% CPT was released with 5 U (μmol/min) of lipase and 70% was released with 10 U of lipase within 10 h. To evaluate the drug encapsulation capacity of the drug in the formed micelles, DOX, another hydrophobic anticancer drug, was used as a model. DOX was loaded by dialysis method with a drug entrapment efficiency of 43.7% and drug loading efficiency of 17.3%. The release of the physically loaded DOX was compared with that of chemically conjugated camptothecin. The DOX was released much faster through diffusion, than CPT through hydrolysis of the ester bond. The enzyme-induced drug release behavior and cytotoxicity against HepG2 cells were evaluated, confirming the utility of obtained micelles for controlled drug delivery.

The unique advantages of materials obtained by combining electrospinning of aliphatic polyesters and enriching them in CDs have been discovered in the last decade [67]. For example, the nanofibers capable of capturing small molecules, such as environmental toxins in water and air, as well as hydrophilic drugs, can be obtained by this approach. The electrospinning of the aliphatic polyesters combined with CDs leads to the formation of nanofibers with improved hydrophilicity and/or crystallinity. Narayanan group first reported the successful fabrication of PCL nanofibers containing α- and γ-CDs [99], as well as β-CDs [100]. Those PCL-based nanofibers containing α- or γ-CD were prepared using electrospinning from 60:40 chloroform/N,N-dimethylformamide [99]. The average diameter of the obtained nanofibers increased only slightly with increasing loading of CD. However, a significant reduction in water contact angle was observed even with additions of a small percentage of CD (~5%). The phenolphthalein absorption tests showed that γ-CD-functionalized nanofibers absorbed faster than α-CD-functionalized at all CD loadings. This indicated that γ-CD was more available in unthreaded form than α-CD in their PCL nanowebs. Narayanan et al. obtained also the PCL/β-CD functional nanofibers in a similar way [100]. The efficiency of wound odor absorbance by prepared nanofibers was studied using a simulated wound odor solution, consisting of butyric and propionic acids in ethanol. Immersion tests indicated that the nanofibers containing β-CDs were very efficient in masking the odor.

In addition, the CD enriched PCL fibers were used for encapsulation of small molecules like α-tocopherol [71], naproxen [72], sulfisoxazole [68], ciprofloxacin [73], as well as even large molecule (e.g., polymer or enzyme) [74,75,98,101,102]. Those studies were initiated to enhance the stability of active compounds against various environmental factors during delivery by entrapment them in PCL nanofibers. For example, the Uyar group has chosen α-tocopherol (TC), a form of vitamin-E frequently applied as a healing factor in wound dressings, and used it as an active agent for the formation of β-CD-ICs encapsulated into electrospun nanofibers (Figure 6a–e) [71]. The marginal increase in fiber diameter (345 ± 140 nm vs. 205 ± 115 nm) was observed for PCL/α-TC nanofibers after β-CD application. An increase in the released amount of α-TC was found when it was delivered from PCL/α-TC/β-CD-IC-nanofibers, in comparison to PCL/α-TC nanofibers (Figure 6f) The SEM images of UV-treated PCL/α-TC nanofibers and PCL/α-TC/β-CD-IC-nanofibres presented in Figure 6g,h showed that nanofibers maintained their fibrous structure up to 65 min under the applied conditions. Authors proved that inclusion complexation between α-TC and β-CD during the formation of PCL electrospun nanofibers improves its photostability (~6% higher than un-encapsulated form) and antioxidant activity.

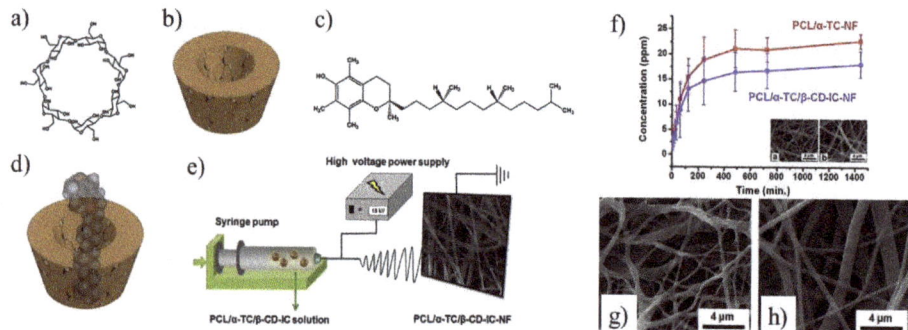

Figure 6. α-TC enriched PCL nanofibers preparation: chemical structure of β-CD (**a**), schematic representation of β-CD (**b**), the chemical structure of α-TC (**c**), the formation of α-TC/β-CD-IC (**d**), electrospinning of nanofibers from PCL/α-TC/β-CD-IC solution (**e**), the release of α-TC from PCL/α- TC-nanofibers and PCL/α-TC/β-CD-IC-nanofibers (**f**), SEM images of UV-treated (**g**) PCL/α -TC-NF and (**h**) PCL/α-TC/b-CD-IC-NF [71]. (Copyrights 2016 Elsevier).

Moreover, wound dressing material based on β-CD and PCL was also obtained by Souza et al. [76]. In this work, the bioactive electrospun fibers containing silver sulfadiazine complexed with β-CD in the PCL nanofibers matrix were synthesized to modulate the drug release as well as to reduce the direct contact between silver and skin. Although complexation promoted a decrease in hemolytic index and slowed drug release, no negative effect on antimicrobial activity was observed. Among the drugs incorporated into PCL nanofibers, naproxen, and sulfisoxazole can also form the ICs with CDs, and these hydrophobic drugs are widely used for relieving pain. The Uyar group encapsulated naproxen in β-CD-cavities and embedded in a PCL nanofibres matrix [72]. Observation by means the XRD technique proved the successful incorporation of naproxen (NAP)/β-CD-complex into electrospun PCL nanofibers. The SEM imaging of the electrospun PCL/NAP and PCL/NAP/β-CD-complex nanofibers showed that the average diameter of the nanofibers was around 300 nm. In addition, the aggregates of β-CD-IC nanofibers were also observed. HPLC analysis revealed that the β-CD-complex based nanofibres releases NAP two times higher than PCL nanofibers with neat NAP. This is a very promising result for the future of drug delivery systems. Uyar group also described sulfisoxazole/hydroxypropyl-β-CD inclusion complex, incorporated in hydroxypropyl cellulose nanofibers via electrospinning [68]. Sandwich configurations were prepared by placing IC-enriched cellulose nanofibers between electrospun PCL nanofibers. As a result, PCL-nanofibers enriched structures exhibited a slower release of sulfisoxazole as compared with neat PCL-nanofibers. Tetracycline, a biocidal drug with poor water solubility, was also embedded in PCL nanofibers and encapsulated in β-CD cavities to be used in the regeneration of periodontal ligaments [77]. An antimicrobial diffusion test was performed for a set of nanofibers with the microorganisms like *Aggregatibacter actinomycetemcomitans* (A.a.) and *Porphyromonas gingivalis* (P.g.). Tests revealed significantly higher halos of bacterial inhibition against both oral bacteria in PCL nanofibers containing tetracycline/β-CD (34 ± 3 and 30 ± 3 mm for A.a. and P.g. respectively), compared to PCL nanofibers with non-complexed tetracycline (28 ± 4 and 26 ± 3 mm). The collected data indicated that nanofibers containing tetracycline/β-CD promote the adhesion and slower dentine demineralization enhance the potential of this formulation for clinical application.

The hydrophobic biocidal drug, ciprofloxacin was combined with PCL nanofiber. The preparation of supramolecular CD/PCL containers allowed for efficient encapsulation of ciprofloxacin and use in this form The antibacterial application of free drug is impaired due to poor solubility and limited stability [73]. α-CD/ciprofloxacin and β-CD/ciprofloxacin ICs formation was carried out under two different conditions: at room temperature or with sonic energy. Larger amount of ciprofloxacin was trapped in the CDs cavity when ultra-sonication was applied. SEM analysis indicated that the incorporation of CDs/ciprofloxacin ICs inside PCL nanofibers did not affect the morphology of electrospun nanofibers. After incorporation of the ICs into PCL nanofibers, the release of ciprofloxacin was followed at pH 7.2. The release of ciprofloxacin from PCL nanofibers increased with increasing solubility of the drug via ICs formation. The drug release from nanofibers was mainly controlled by the diffusion, and this process did not affect nanofiber structures. The increase in the released amount of drug was found in stimulated physiologic environment, when it was delivered from β-CD/PCL nanofibers, compared to α-CD/PCL nanofibers. The key advantage of CDs is their facile capability to form ICs not only with small but also with large molecules. The enzymes are a type of large biomacromolecules incorporated into CD enriched PCL fibers. The introduction of CD-ICs into electrospun nanofibers having high surface area and highly porous nanostructure make them suitable substrate for biocatalyst immobilization [74,75]. For instance, catalase, an anti-free radical enzyme, was successfully immobilized onto poly(ethylene oxide) nanofibers containing γ-CDs, sandwiched between PCL nanofibers [74]. The positive influence of CDs on enzyme activity and the stability of the catalase was showed. Similar to the catalase enzyme, laccase was immobilized on γ-CD/PCL nanofibers and showed higher catalytic activity (96.48 U/mg), compared to enzymes immobilized on PCL nanofibers without CDs (23.2 U/mg) or γ-CD/laccase physical mixtures in PCL nanofibers (71.6 U/mg) [75]. During the formation of the CD-enzyme complex, the enzyme used did not lose activity and no denaturation was observed.

The core-shell nanocarriers based on amphiphilic copolymers have attracted great attention as a potential agents for cancer chemotherapy since the hydrophilic shell can ensure prolonged circulation of the carrier in blood, whereas the hydrophobic core can enhance drug loading efficiency [103,104]. Core-shell NPs from aliphatic polyesters are extensively used as promising materials for drug and gene delivery due to its useful properties, biocompatibility and excellent degradability [105,106]. Polymeric micelles made of amphiphilic copolymers, such as poly (ε-caprolactone)-block-poly (ethylene glycol) (PCL-b-PEG), are of great interest, especially in recent years [107]. The combination of CD-based ICs with PCL/PEG copolymers of various microstructure was widely used to solve the problem with solubility, degradability, or bioavailability of many important drugs. For instance, Varan et al. designed hydroxypropyl-β-CD coated and docetaxel-loaded nanoparticles composed of PCL and PCL-b-PEG to be applied as implants to site following after surgical operation of tumor [78]. The coating with CD significantly increased the drug encapsulation and anticancer efficacy against MCF-7 human breast adenocarcinoma cell lines, however, it did not change particle size and polydispersity. Those PEG-b-PCL and hydroxypropyl-β-CD-based systems were used farther in inkjet printing of antiviral/anticancer combination dosage forms. As a result of that research, a combination product consisting of anticancer paclitaxel and antiviral cidofovir drugs was manufactured as an adhesive film for local treatment of cervical cancers [79]. Characterization studies of obtained material showed that the printing process did not influence neither the structure of nanoparticles nor inclusion complex. The paclitaxel and cidofovir containing ink and film formulations have higher anticancer efficacy as compared with drugs solution. Incorporation of paclitaxel into PCL/PEG copolymer nanoparticles was also reported by Ahmed et al. [80]. In their report, the β-CD grafted poly(acrylic acid) was synthesized by radical polymerization, and then embedded on the surface of PCL-b-PEG-b-PCL nanoparticles through host-guest interaction and hydrogen bonding between the oxygen atom of PEG and hydrogen atom of carboxyl group of poly(acrylic acid). Paclitaxel was released smoothly without remarkable initial burst release during the in vitro drug release experiments (i.e., only 20% drug was released in the first 12h). After drug loading, the NPs displayed significant cytotoxicity against HepG2 cells. Kuplennik et al. applied 2,3,6-triacetyl-β-cyclodextrin within methoxy-PEG-b-PCL nanoparticles to improve the encapsulation efficiency of sepiapterin, the natural precursor of the essential cofactor tetrahydrobiopterin [81]. For this purpose, sepiapterin/cyclodextrin complexes were produced by spray-drying of binary solutions in ethanol and encapsulated within methoxy-PEG-b-PCL nanoparticles by nano-precipitation. The encapsulation efficiency and drug loading were 85% and 2.6%, respectively, as opposed to the much lower values (14% and 0.6%, respectively) achieved with pristine drug. Moreover, the sustained release of the sepiapterin from nanoparticles was observed, with a relatively low burst effect of 20%.

The CDs enriched core-shell nanoparticles reveal the ability for drug delivery through the skin by enhancing solubilization of lipophilic drugs as well as increasing the amount of solubilized species at the absorption site by promoting drug transport through passive diffusion [108]. The drug carrier system designed for the delivery of lipophilic drug through the skin was described by the Quaglia group. For this purpose, the core-shell nanoparticles based on PEG-b-PCL associated with 2-hydroxypropyl- β-CD were employed [109]. The NPs entrapping the second generation of photosensitizer Zn^{2+} phthalocyanine (ZnPc), highly lipophilic and fluorescent model molecule, were formed. The transport of ZnPc through porcine ear skin was evaluated on Franz-type diffusion cells. The confocal Raman spectroscopy demonstrated that 2-hydroxypropyl-β-CD caused an alteration of water profile in the skin and a high reduction in the degree of hydration at stratum corneum/viable epidermis interface which can promote NPs transport.

To enhance the controlled release of anticancer drugs and minimize the side effects of those drugs, the rational approach is to use stimuli-responsive PCL/PEG/CD micelles which show the response to pH [82], temperature change [110], light [65], ionic strength or enzymes. The pH-sensitive polymeric micelles, stable at the physiological pH, which can dissociate to release drugs in the acidic environment of solid tumor tissues, are playing an important role in controlled cancer treatment [111]. In this regard, the pH-sensitive PCL-b-PEG micelles containing a polymeric form of β-cyclodextrin were also used by

Gao et al. as a copolymer block [82]. The complex micelles were formed via host-guest interactions between poly(β-CD) in diblock PEG-b-PCD copolymer and BM groups in BM-PCL. The DOX encapsulation efficiency of complexed micelles was up to 74.77%. The release of DOX from PEG-b-PCD/BM-PCL polymeric micelles was suppressed at neutral pH solutions and accelerated at acidic solutions or high temperatures. The cumulative release of DOX increased from 70.3% to 98.6% with a decrease of pH from 7.0 to 2.0. At weakly acidic conditions (pH 5.2), BM groups were protonated resulting in the partial disruption of complexation with β-CD in PEG-b-PCD, so the drug molecules could be slowly released from micelles, and the cumulative release reached approximately 80%.

Li et al. presented glutathione (GSH)/light dual-responsive supramolecular drug carriers based on the CD modified by PEG and azobenzene-PCL fabricated for intracellular delivery of DOX (Figure 7a) [65]. The azobenzene groups/CD complexes are typical supramolecular assemblies and have been intensively studied for their unique photo-responsive properties induced by the photochemical trans-cis isomerization of the azobenzene units. The obtained spherical carriers exhibited glutathione sensitivity attributed to disulfide bonds between PEG and β-CD, and the light sensitivity response was achieved by a simple host-guest interaction between β-CD with azobenzene groups. The DOX was selected to evaluate the drug loading capacity and therapeutic effect of the carriers. The total drug loading was determined as 30.4% by UV. After 48 h of drug release experiment, the cumulative drug release rate was less than 20% without stimulation. When a stimulus (365 nm light wavelength or 10 mM GSH) or double stimulus (365 nm light wavelength and 10 mM GSH) were applied, the drug release rate was significantly accelerated, the cumulative drug release rate reached more than 55% after 48 h of drug release experiment (Figure 7b). The cytotoxicity of ICs and drug-loaded carriers were explored against normal cells (HEK293T cells) and tumor cells (SKOV3 cells). According to cytotoxicity studies of blank drug carriers, the cell viabilities were more than 85% at a wide range of concentration (0–600 mg·mL^{-1}) indicating that the drug carriers exhibit lower cytotoxicity and good biocompatibility (Figure 7c). It was shown that the drug-loaded carriers have a better pharmacodynamics performance to tumor cells, but less toxicity to normal cells, since the DOX can be released under the trigger of light and glutathione.

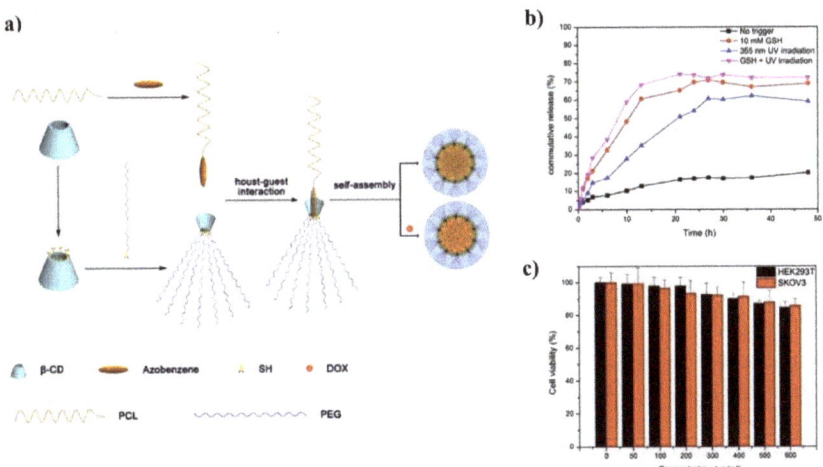

Figure 7. Synthesis route of glutathione/light dual-responsive supramolecular DOX loaded PCL-b-PEG-based carriers (**a**) drug release curves in different stimulations (**b**) cell viability of SKOV3, and HEK293T cells following incubation with as-prepared carriers for 24 h (**c**) [65]. (Copyrights 2019 Elsevier).

Star-shaped copolymers attracted much attention because their branched structures can form unimolecular micelles with better stability than the micelles self-assembled from conventional linear copolymers [112]. It is well known that those copolymers provide a stable environment for drug loading and its sustained release [113,114]. Those particular features made amphiphilic star-shaped copolymers especially useful for the formation of drug delivery systems, e.g., supramolecular NPs or hydrogel preparation. To simplify the preparation of star-shaped structures with a precisely controlled degree of branching, the supramolecular host-guest pair can be used as the block junction. For instance, Gou et al. reported the synthesis of novel drug-conjugated amphiphilic PCL star copolymers containing β-CD as core moiety. In this work, PCL/poly(acrylic acid) [115] and PCL/PEG multimiktoarm [116] copolymers were synthesized by the combination of controlled ring-opening polymerization with "click" chemistry and atom transfer radical polymerization, respectively. These new types of amphiphilic copolymers, which were composed of biocompatible poly(acrylic acid) or PEG corona surrounding both biodegradable CD core and PCL arms could self-assemble into multimorphological aggregates in aqueous solution. In addition, the hydrophobic ibuprofen-loaded nanoparticles fabricated from these drug-conjugated PCL/PEG/CD copolymers were investigated [116]. The hydrophobic ibuprofen was incorporated into the chain ends of the PCL by the reaction in the presence of dicyclohexylcarbodiimide and 4-dimethylaminopyridine. The drug-loading efficiency and drug-encapsulation efficiency of the ibuprofen-conjugated miktoarm copolymers were significantly higher than those of the corresponding non-drug conjugated counterpart. It was a result of the conjugated ibuprofen influence on the hydrophobicity of the miktoarm star-shaped copolymer, leading to an increase of drug loading amount, as well as interactions (such as π-π aromatic stacking force) between covalently bonded ibuprofen with the free ibuprofen, which forces the free ibuprofen to incorporate into the micellar core.

DOX loaded nanocarriers based on the star-shaped amphiphilic mPEG-b-PCL copolymers and β-CD were described by Li et al. [83]. Authors developed smart, reductive stimulus-responsive nanosystems using modified β-CD molecules. The secondary hydroxyl groups of CD were methylated to improve solubility, whereas the primary hydroxyl groups were conjugated with mPEG-b-PCL-SH through a disulfide linkage to amplify the hydrophobic cavity and enhance the stability of the nanocarriers. The DOX-loaded micelles were prepared with the highest drug loading capacity (LC) of 31.9 wt.% and encapsulation efficiency (EE) of 83.9%. DOX release from the micelles under a reductive stimulus was carried out in PBS (pH 7.4) and in 1,4-dithio-threitol solution which simulated the reductive tumor microenvironment. DOX was released significantly faster in the presence of DTT than in its absence, and the cumulative release rate of DOX loaded micelles containing methylated β-CD in the DTT solution was > 50% within 8 h. In contrast, the cumulative release rate in PBS was < 20% even after 100 h. It confirmed the usability of disulfide bonds, which can rapidly be broken under reducing conditions and the accelerated macromolecules dissociation is observed. It was shown that synthesized nanocarriers accumulated at the tumor site via EPR and released the drug in a controlled manner in the reductive tumor microenvironment, with negligible premature leakage, and side effects on the healthy tissues.

The release of drugs from the star-shaped copolymeric micelles can be also induced by the reactive oxygen species. The example of those carriers are systems with oxidation-sensitiveness due to non-covalent, combination of β-CD/ferrocene. Those systems are able to produce an excess amount of reactive oxygen species in the specific tumor cell lines [117]. In this regard, the electrochemical redox stimulus systems based on β-CD and ferrocene linker were described by Yuan [84] and Wei [85] groups. Yuan and co-workers applied 4-arms PCL terminated with β-CD and linear polymer polyethylene glycol terminated with ferrocene to improve the biocompatibility and efficiency of DOX delivery (Figure 8a) [84]. The electrochemically-responsive supramolecular micelles were obtained, which exhibited faster release and better biocompatibility compared with their linear analogues, namely linear ferrocene-terminated PEGs (Fc-PEGs). The cyclic voltammetry and 2D NOE NMR were used to confirm the host-guest interaction between these two polymers. Cytotoxicity experiment of the supramolecular micelles, conducted on A549 cells, proved their biocompatibility (Figure 8b). Through electrochemical control, a reversible assembly-disassembly transition of the micelles was realized, which was investigated by TEM. In order to confirm the high efficiency of star polymers as drug carriers, UV-vis spectra were used to calculate the drug loading content and drug loading efficiency respectively, which turned out to be 11.0% and 67.7% (8.0% and 49.4% in case of linear analogues). In vitro drug release experiments under electrochemical stimuli showed that DOX could be released from drug carriers in several hours (Figure 8c). Upon applying a potential of +0.8 V, DOX was released, however, along with the decomposition of micelles caused by the oxidation of ferrocene attached with PEG macromolecules. Those comparative studies revealed the advantages in drug loading of the star-shaped copolymers over linear analogues for use as drug carriers.

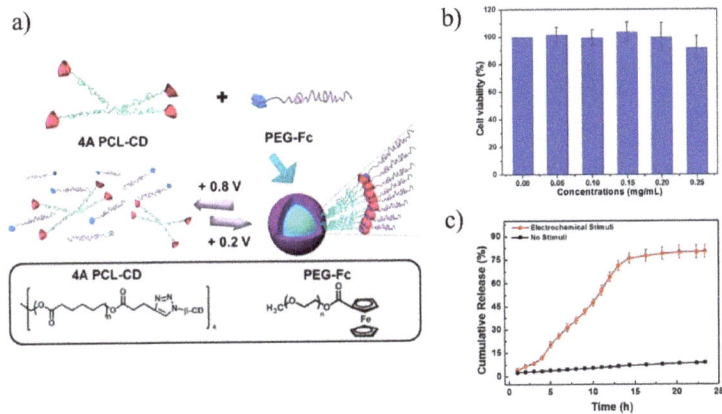

Figure 8. Structures of 4-arm PCL-CD and PEG-Fc and schematic representation of the potential-responsive controlled assembly and disassembly of the 4-arm PCL-CD/Fc-PEG micelles (a), cytotoxicity evaluation of the 4-arm PCL-CD/Fc-PEG micellar solutions, measured by comparing the cell viability of A549 cells (b), the dependence of the amount of DOX released from 4-arm PCL-CD/Fc-PEG micelles and time under no stimuli and +0.8 V stimuli (c) [84]. (Copyright 2019 Elsevier).

In addition, the Wei group described the supramolecular structures obtained from 3, 4, and 6 arm star-shaped PCLs with ferrocene end-capped arms and 3-arm poly(oligo ethylene glycol) methacrylates terminated by β-CD [85]. The micelles obtained from star-shaped exhibited the highest drug loading content and the encapsulation efficiency, most likely due to its highest stability reflected by its critical aggregation concentration value. The in vitro drug release profiles, at the physiological conditions (PBS, pH = 7.4) and in an oxidizing medium (PBS, pH = 7.4, 0.2 mM NaClO) at 37 °C, proved that NaClO significantly promoted the drug release with 20–30% increase for all formulations, confirming the oxidation-triggered dissociation of β-CD/Fc complexation and the structural deformation of supramolecular micelles. Finally, the cytotoxicity tests of all supramolecular star-shaped micelle constructs for HeLa cells, revealed that DOX-loaded micelle formulations exhibit lower cytotoxic activity than the free DOX, most likely due to the slower internalization mechanism (endocytosis vs direct membrane permeation). The cytotoxicity studies proved also that higher degree branching of the obtained star-shaped copolymer and growing hydrophilic arm lengths enhanced the therapeutic efficacy of the DOX-loaded nanocarriers.

The PCL/PEG copolymers and CD-based supramolecular hydrogels (SMGels) with their reversible sol-gel transition properties were widely explored as injectable biomaterials capable of establishing versatile drug delivery systems [70,86,118]. The SMGel based on α-CD [87,119], β-CD [64], or γ-CD [86] and PCL/PEG copolymers of various microstructure were developed and investigated due to their controlled drug release and site-specific drug delivery triggered by various stimuli. The advantage of SMGel is the dynamic nature of their structure that can be easily broken by shear forces because it is composed of weak noncovalent interaction. Tabassi et al. presented the example of SMGel with shear-thinning thixotropic behaviour, showing that hydrogel composed of copolymers with a PCL to PEG ratio of 1:4 are suitable for syringeable SMGel preparation [87]. The mixing of α-CD (12%) and PCL-b-PEG-b-PCL (10%) induces gel formation in less than one minute and enables sustained release of vitamin B12 for at least 20 days. Moreover, their thixotropic behavior makes supramolecular hydrogels highly attractive for many biomedical applications, e.g., ocular drug delivery. Zhang et al. were the first who reported thixotropic SMGel based on α-CD and a low-molecular-weight mPEG/PCL block copolymer for ocular drug encapsulation [69]. The SMGel containing diclofenac, known as an anti-inflammatory drug, showed relatively low cytotoxicity toward L-929 and HCEC cells. The hydrogel was nonirritant toward the rabbit eyes, what was confirmed by the Draize test, fluorescein staining, as well as histological observation. The application of Nile Red-labeled micellar supramolecular hydrogel proved that it significantly extends the retention time on the rabbit's corneal surface compared with a plain micellar formulation.

A variety of active substances were introduced to the PCL/PEG copolymer-based SMGels with pseudo-polyrotaxane structures formed by CD and PEG blocks. For example, by mixing PEG-b-PCL micelles that solubilize DOX, poly(ethylene glycol)-b-poly(acrylic acid) (PEG-b-PAA) micelles that host cisplatin, together with α-CD, results in the preparation of a dual-drug loaded pPR based hydrogel [88]. The erosion of the gels resulted in a discrete release of micelles from which the drugs were delivered. In vitro cytotoxicity studies proved that DOX-loaded hydrogel inhibited the growth of human bladder carcinoma EJ cells, whereas the dual drug-loaded SMGel showed significantly higher cytotoxicity against applied cells. The thixotropic and injectable SMGel based on PPRXs formation were described also by Xu et al. [66]. Authors confirmed that hydrophobic cores formed by self-assembly of amphiphilic polymer methoxy-poly(ethylene glycol)-b-poly(ε-caprolactone-co-1,4,8-trioxa[4.6]spiro-9-undecanone copolymer and the microcrystals of PRXs formed by α-CD and PEG blocks could serve as two-level cross-linking for the gel formation. The in vitro and in vivo degradation demonstrated the general release of NPs, which can be easily uptaken by cells and accumulate at the tumor site. Further studies on the paclitaxel controlled release and antitumor efficiency were also performed [89]. Most importantly, the obtained hydrogel was efficient in inhibiting tumor cells growth and prevented the diffusion of paclitaxel to other mice tissues.

Besides the small molecules, also proteins and genes were encapsulated and released from SMGels. The injectable and thixotropic SMGels of α-CD with methoxyPEG-poly(ε-caprolactone)-(dodecanedioic acid)-poly(ε-caprolactone)-methoxyPEG triblock polymer (α-CD/mPEG-b-PCL-b-mPEG) were proposed for sustained release of recombinant human erythropoietin (rhEPO) in an acute myocardial infarction rat model [90]. The rapid gelation of this system enabled effective encapsulation of rhEPO at the injection site, which improved cardiac function for 30 days after myocardial infarction and allows for avoidance of polycythaemia, a well-known collateral effect of rhEPO. Khodaverdi et al. reported γ-CD in preparation of an insulin-loaded supramolecular PCL-b-PEG-b-PCL based gel, with low hemolytic activity and superior biodegradability compared to those prepared with α-CD [86]. In this system, aggregations of γ-CD threading onto PEG blocks were supported by a small number of hydrophobic PCL blocks and a high number of hydrophilic blocks. The SMGels were obtained by mixing 10.5% (w/v) γ-CD and 2.5% (w/v) copolymer and revealed an excellent syringeability. Insulin was released up to 80% over 20 days, keeping its initial folding.

On the other hand, the methoxyPEG-poly(ε-caprolactone)-poly[2-(dimethylamino)ethyl methacrylate] triblock polymer (mPEG-b-PCL-b-PDMAEMA) and α-CD were used to form stable polyplexes with plasmid DNA (pDNA) [91]. The pDNA was electrostatically bonded to the cationic segment of the copolymer. The mPEG-PCL-PDMAEMA copolymers exhibit a good ability to condense pDNA into 275–405 nm polyplexes with hydrophilic mPEG in the outer corona. The multiple mPEG chains were used as cross-linking moieties to anchor the DNA nanoparticles within the α-CD/PEG supramolecular PPRXs hydrogel system. The obtained hydrogels revealed controlled release for several days without detrimental effects on protein expression level. In vitro gene transfection results showed that the supernatants containing pDNA released from the hydrogels at various time points had good bioactivity.

However, in vitro cytotoxicity of copolymers assay on COS7 cells confirmed that PDMAEMA chains length has a significant impact on the biocompatibility of the whole copolymers.

The SMGels revealing thermosensitivity, able to form an injectable solution at low temperatures and non-flowing gel at around physiological body temperature, were also extensively studied. The thermosensitive gels based on PCL/PEG copolymer and CD-ICs were presented in the literature as a promising systems for non-steroidal anti-inflammatory drug delivery [64,92]. A novel injectable, in situ gel-forming drug delivery system based on thermosensitive β-CD-modified PCL-b-PEG-b-PCL was studied by Wei et al. [64]. The applied copolymer can self-assemble in water to form a micelle solution, with a sol-gel transition occurring as the temperature increased, which was confirmed to be related to the polymer concentration. The linkage of β-CD to the hydrophobic macromolecule chain ends made the encapsulation of hydrophobic drug within the hydrogel networks more effective. Subsequently, the in vitro release behavior of indomethacin from the micelles was investigated. According to the cumulative release profile, the drug was sustainably released up to 50% over 9 days.

Indomethacin was released without remarkable initial burst release and the release behavior followed a linear course for 48 h, indicating that the obtained micelles could be applied as a depot for drug-controlled release. Additionally, two in vivo models, i.e., carrageenan-induced acute arthritis and Freund's complete adjuvant-induced arthritis were employed to evaluate the therapeutic effect of the drug after subcutaneous administration in the right-back paw of rats. A significant improvement in the anti-inflammatory effect of indomethacin in rats occurred after encapsulation in the obtained hydrogel network. Khodaverdi investigated thermosensitive PCL-b-PEG-b-PCL based SMGel obtained by IC with γ-CD as a carrier for sustained release of dexamethasone [92]. The SMGel with excellent syringeability was prepared by mixing 20 wt.% γ-CD and 10 wt.% of the copolymer in a few seconds. It is worth noting that the applied solution of the synthesized copolymer, with a PCL/PEG ratio of 1/5, could turn into a gel only in the presence of γ-CD due to the short PCL blocks and insufficient hydrophobic interactions between polymer chains. The rheological studies revealed the shear-thinning behavior of the obtained SMGel. The release profiles showed that formulation containing 0.1% dexamethasone released about 40–46% of the loaded drug after 23 days. It was shown that the release of dexamethasone from the supramolecular gel occurred slowly, with a slight initial burst release.

The pH-sensitive SMGels, formed within a few minutes in an aqueous medium, are investigated as potential "smart" drug delivery carriers. Hu et al. developed an injectable hydrogel based on inclusion complexes of the star-block copolymer from mPEG and PCL linked with acid-cleavable acetal groups ((mPEG-acetal-PCL-acetal)₃) (Figure 9a) [70]. The ICs aggregated into necklace-like crystalline PRXs and acted as physical crosslinking joints for the hydrogels, while the remaining uncovered hydrophilic PEG chains functioned as water-absorbing segments. The obtained SMGels revealed unique structure-related reversible gel-sol transition properties at a certain level of stress. Importantly, according to SEM observation, the lyophilized hydrogels exhibited a porous sponge-like structure and could be used as drug delivery depots (Figure 9b). In vitro drug release test showed that encapsulated DOX was released from the drug-loaded hydrogels in a controlled and pH-dependent manner (Figure 9c).

In this part of the review, the versatility of CD/PCL based supramolecular structures were presented, clearly demonstrating the suitability of IC assemblies in the form of nanocarriers or colloid-associated gels for diverse therapeutic demands. In all cases, the CD application resulted in improved delivery of hydrophobic drugs. The CD-ICs containing small bioactive molecules or even large molecules (e.g., enzymes) were designed and used for tailor/modulate release from multilayer nanofibrous structures or core-shell structures based on PCL and its copolymers. The necklace-like PRXs structures, stable due to hydrophobic and van der Waals interactions between the inner surface of the CDs and the PCL chains, were found as particularly useful for encapsulation/grafting of the biomolecules. However, both IC capability, as well as functionalization of hydroxyl groups present on external rims of CD and PCL macromolecules, are still not fully explored. The highly organized ICs based drug delivery systems capable of responding to external or internal stimuli, such as temperature, pH, light, or redox alterations, are presented as examples of useful carriers for several diseases treatment providing excellent extracellular stability and effective intracellular drug release. Nevertheless, it is anticipated that the studies of multi-stimuli responsive PCl/ICs structures will be continued and provide advanced carriers able to overcome both extra- and intracellular barriers. Although various successful in vitro and in vivo studies demonstrating pharmacologic, antimicrobial, and antioxidant effects are reported, the lack of comprehensive evaluation of systemic toxicology as well as biodegradability of CD-based supramolecular drug delivery systems opens a field for futher studies.

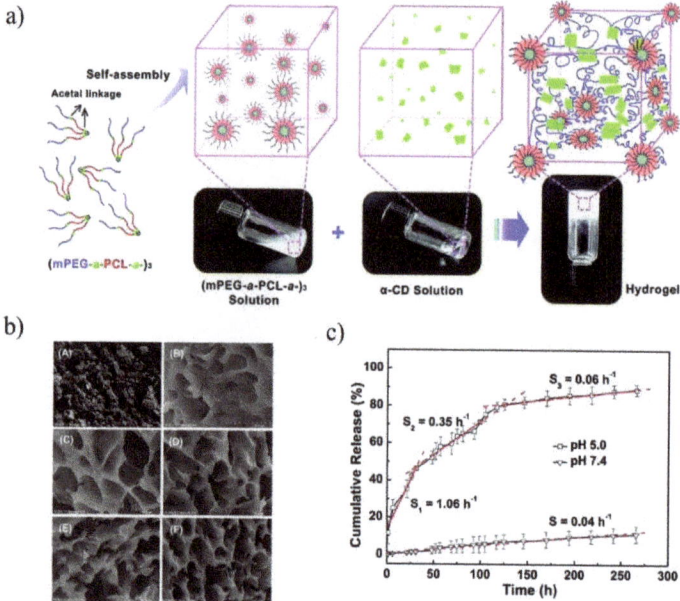

Figure 9. Illustration of acid-cleavable hydrogel networks based on inclusion complexes between (mPEG$_{45}$-acetal-PCL$_{27}$-acetal-)$_3$ and α-CD (**a**), SEM images of lyophilized hydrogels from copolymers varied with the content of hydrophilic and hydrophobic segments (**b**), (in vitro cumulative release of encapsulated DOX·HCl from (mPEG$_{45}$-acetal-PCL$_{27}$-acetal-)$_3$ hydrogel at 37 °C under different conditions: pH 5.0 buffer solution and pH 7.4 buffer solution (**c**) [70]. (Copyrights, 2016 The Royal Society of Chemistry).

4. Poly(Ethylene Glycol)/Cyclodextrin Systems for Biomedical Applications

PEG (Figure 10) is a gold standard typically used to modify biomacromolecules and synthetic macromolecules, that is subsequently applied to prepare the drug carriers with improved physicochemical properties and circulation in the body [120] The first example of this type modification was shown in pioneering work from the late 1970s, where PEGylation provided protection for proteins against destruction during their administration [121]. Currently, the term PEGylation is associated with the covalent coupling of PEG with biological and synthetic molecules [122]. Since CDs are typically used to host the drug molecules in their interior, the PEGylation was a rational strategy to additionally improve their biodistribution in vivo. The combination of PEG and CDs is frequently used for the preparation of NPs, micelles, gels and hydrogels (Table 3). In this respect, Rojas-Aguirre et al. [123] proposed using click chemistry to combine β-CD with different molar mass PEG chains (5000, 2000, and 550 g/mol) and prepared star-shaped PEGylated β-cyclodextrin. Copper(I)-catalysed azide–alkyne cycloaddition (CuAAC) [124] between alkyne modified PEG and azide-functionalized CDs, was also employed, however, the mixture of products with different level of substitution was obtained. Subsequently, the synthesized star-shaped PEGylated β-CD was tested against human monocytes,

Vero and HeLa cells, and after incubation no effect on the viability of most cells was observed with except, however, for β-CD-PEG550, which reduced the viability of HeLa cells and human monocytes. The observed effect was ascribed to the PEG molecular weight and architecture of PEGylated β-CD. Although, the copper was extracted during the synthesis, its content level in the final products was not investigated. Since copper may affect the toxicity [125] residual traces may influence the observed results. The most often described literature examples concerns the interactions between CDs and adamantyl (Ada)-functionalized, water-soluble polymers [126]. Using this approach, the dextran (DXT) was modified with β-CD, Ada, and PEG-Ada/PEG-CD, in which PEG macromolecules were used as a flexible spacer. After simple mixing, the nanoassemblies were spontaneously formed in water. The presence of PEG spacer leads to the formation of less compact and smaller nanoparticles due to the higher binding constant of guest polymer (DXT-gPEG-Ada) in relation to the DTX-Ada. However, the opposite relation was observed for DXT-gPEG-CD/DTX-CD for which the presence of spacer decreases decreased the binding constant. These results implicate that by the careful adjusting of the binding constant of host-guest interactions between macromolecules, the control over the structure of resulting nanoassemblies could be achieved [127].

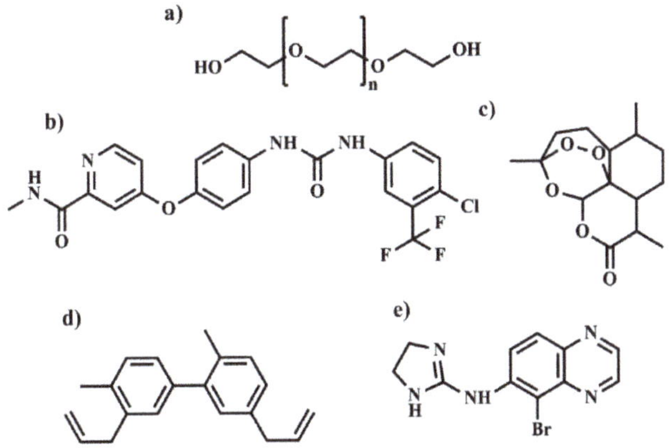

Figure 10. The chemical structure of (**a**) the carrier polymer-PEG, and chemical structure of (**b**) sorafenib, (**c**) artemisinin, (**d**) honokiol, (**e**) brimonidine.

Polyrotaxanes (PRXs) and polypseudorotaxanes (PPRXs) are different examples of host-guest interlocked complexes in which linear molecule or polymer are encircled by macrocyclic components (e.g., CDs) [128]. The influence of the formation of PPRXs during the microparticles (MPs) preparation was investigated for the system composed of CDs and PEG [129]. The emulsifying process, using polypropylene glycol (PPG) as an oil phase and CDs with and without PEG as a water phase, was used to prepare water-in-oil (W/O) emulsion. The solidification into desired MPs occurs during the lowering of the temperature.

The authors claim that the irregular MPs morphology was observed due to the formation of CD/PEG PPRX, whereas the regular structure of MPs was formed for the CD/PPG PPRX. Moreover, α-CD was essential for the formation of PPRXs, as the complex was not formed for β- and γ-CDs. The strategy based on the application of polyrotaxanes for the construction of drug delivery carriers was proposed also by Moon et al. [130]. The described preparation required a four-step process which consists of (1) inclusion complexation of β-CDs with amine-terminated PEG, (2) the blocking of PEG end groups with L-tyrosine (L-Tyr), and (3) modification of β-CDs by succinic anhydride, to formed PRXs for DOX delivery. The opened succinic anhydride were applied, to attach the DOX to the PRXs to induce control release of DOX from PRXs by cleavage the hydrolyzable ester bond. To prove this concept, the in vitro release experiment in the phosphate buffer saline was performed. For the first

48 h, the zero-order kinetics without significant burst release was observed. The control release was also observed for PPRXs micelles composed of β-CDs and PEG terminated with protoporphyrin (PpIX) [131]. The PpIX-functionalized PEG by a transformation of its hydroxyl end groups into the amine end groups, and, subsequently coupling carboxylic functionalities in PpIX were performed. The micelles (MCs) were prepared by simply mixing both components in water and CMC value was 12 µg/mL, as a consequence spherical nanoparticles with a diameter less than 100 nm in size were obtained. The formation of PPRXs in the MCs was verified by XRD where two types of the structure dominate: head to head or tail to tail tunnel structure. As a further step, three different DOX-loaded micelles were prepared in which PPRXs differs with the number of β-CDs (PPRX-2, PPRX-9, PPRX-13). It was observed that with the increasing number of β-CDs in PPRXs the size of MCs increased whereas the drug loading content decreased. Moreover, the stability and size of the MCs composed PPRX-2 and PPRX-9 was invariant after DOX encapsulation, however, the MCs sizes from PPRX-13 increased in time. As a result, the release of DOX from these nanoparticles depends on the ratio of PEG to PPRX in the resulting nanocarrier. Both in pH = 5.0 and pH = 7.0, the fastest DOX release for PPRX-13 MCs was observed, and it was attributed to the enhanced swelling of nanocarriers with the increase of the β-CDs number and lower π-π conjugation level of DOX with PpIX. Subsequently, HepG2 cell lines (human liver cancer) were chosen to test the anticancer activity, IC_{50} values were again the lowest for PPRX-13 MCs what reflects to the DOX release rate. In contrary, the cellular uptake was the highest for PPRX-2 MCs due to their smallest size ~45 nm in comparison to the PPRX-9 MCs (~75 nm) and PPRX-13 MCs (~89–150 nm).

Table 3. Summary of described hydrogels, micelles and nanoparticles for drug delivery based on combination of PEG and its copolymers with different CDs.

Type of Drug Delivery System	Platform	Type of CD	Drug	Release Medium	In Vitro/In Vivo Studies	Ref.
Micelles	PpIX-PEG	α-CDs	DOX	buffer (pH = 5.0 or pH = 7.4)	HepG2 cells	[131]
Nanoparticles	Star-shaped polymers CD-g-TPGS with different TPGS	β-CDs	DOX	PBS	MCF7 and ADR/MCF7 Cells/H22 sarcoma model	[132]
Nanoparticles	Folic acid–poly-ethylene glycol–β-cyclodextrin (FA–PEG–β-CD)	β-CDs	DOX	PBS (pH = 5.0 or pH = 7.4)	HepG2 cells	[133]
Nanoparticles	CDPF consisting of β-CD, PEG, and FA	β-CDs	DOX	PBS (pH 5.5, 6.8 and 7.4)	MCF7 cells/ Male NCRNU nude mice	[134]
Micelles	Ferrocene conjugated PEG (PEG-Fc) and β-CD-hydrazone-DOX	β-CDs	DOX	PBS (pH = 5.0 or pH = 7.4)	HeLa cells	[135]
Nanoparticles	PEG-HPG-BM and FA-CD	β-CDs	DOX	PBS (pH 5.3, 6.8 or 7.4)	HeLa, HepG2, L929 cells	[136]
Nanoparticles	PEG-CD/AD/SF	β-CDs	DOX sorafenib	PBS (pH = 5.0 or pH = 7.4)	HepG2 cells	[137]
Nanospheres	DMPE-mPEG2000/ γ-CD-C10	γ-CDs	artemisinin	-	Intravenously injection to Wistar rats	[138]
Micelles	F127-CD conjugate	β-CDs	honokiol	PBS (pH = 7.4)	Candida albicans as test strain	[139]
Nanoparticles	PEI-CD·PEG-AD·FA-AD	β-CDs	pDNA	-	FR-negative HEK293 and FR-positive KB cells	[140]
Nanoparticles	FA-PEG-GUG-β-CDE/ DOX/siPLK1	β-CDs	siRNA/DOX	PBS (pH 7.4) or citrate buffer (pH 5.5)	KB cells and BALB/c nu/nu mice	[141]
Micelles	P(Asp-co-AspGA)/ P(Asp-co-AspPBA)	α-CDs	vancomycin	PBS (pH = 7.4)	-	[142]

Table 3. Cont.

Type of Drug Delivery System	Platform	Type of CD	Drug	Release Medium	In Vitro/In Vivo Studies	Ref.
Nanoparticles	PNSC@APEG	β-CDs	5-FU	PBS (pH = 7.4)	NIH3T3 cell lines	[143]
Nanoparticles	PNS-SS-A CD-HPEG	β-CDs	DOX	PBS (pH = 7.4)	HeLa cells	[144]
Nanoparticles	β-CD-PEG capped ZnO	β-CDs	Cur	PBS (pH 7.4) or acetate buffer (pH = 4.8)	MCF7 cells	[145]
Gel	β-CD/PEG	β-CDs	diclofenac	PBS (pH = 7.4)/pig skins	-	[146]
Hydrogel	α-CD/PEO–PHB–PEO	α-CDs	dextran-FITC	PBS	-	[147]
Hydrogel	β-CD-NCO/ NH$_2$-PEG-NH$_2$	β-CDs	lysozyme, β-estradiol quinine	PBS (pH = 7.2)	-	[148]
Hydrogel	α-CD/ 4-arm-PEG	α-CDs	brimonidine	PBS	-	[149]
Hydrogel	α-CD/ A-PEG-A T-PEG-T	α-CDs	DOX	PBS (pH = 7.4)	L929 cells/Sprague Dawley (SD) rats/Chinese Kunming (KM) female mice	[150]
Hydrogel	PEG-β-CyD/PEG-Ad	β-CDs	Tf-AF647	PBS	HeLa cells	[151]
Hydrogel	β-CD/ Pluronic® 127	β-CDs	curcumin	PBS (pH = 7.4)/acidic buffer solution (pH 1.2)	HeLa, MCF-7 and L929 cells	[152]
Hydrogel	α-CD/ NPOD-PEG	α-CDs	DOX	PBS (pH 7.4) or acetate buffer (pH 5.0)	A549 cells	[153]

Nowadays, there is a growing interest in the application of host-guest interactions of CDs/drugs with a combination of hydrophilic PEG macromolecules in the preparation of polymeric nanoassemblies for DOX delivery. Both the covalent conjugation [132–134] and supramolecular complexation [135–137] were used. The first strategy was focused on the conjugation of targeting molecules along with hydrophilic PEG macromolecules applying two types of targeting molecules: folic acid (FA) and D-α-tocopheryl (α-TC). The FA is a vitamin which exhibits remarkable tumor targeting ability because it is overexpressed on the surfaces of a variety of human cancers such as breast, nasopharyngeal, cervical, ovarian, and colorectal cancers [154]. Therefore, folic acid–polyethylene glycol–β-cyclodextrin (FA–PEG–β-CD) was prepared to improve DOX delivery to targeted lines human liver cancer cells (HepG2) [133] and breast cancer (MCF-7) [134], as shown in Figure 11. The desired FA–PEG–β-CD was obtained by the reaction of carboxyl-functionalized FA-PEG-COOH [133] or amine-functionalized FA-PEG-NH$_2$ with β-CD [134].

The spherical DOX-loaded nanoparticles with the diameter ranging from 40 to 55 nm were obtained by simple mixing both components in water and purified by dialysis. The obtained NPs show the pH-dependent release of DOX and higher release in pH 5.5 in comparison to pH 7.4, 7.2, and 6.8, which was attributed to the protonation/deprotonation of β-CD and DOX in acidic conditions. For the HepG2 cells, it was shown that the DOX was delivered successfully to cells, however, there was no comparison to the free drug. In contrary, it was done for MCF-7 cells and their viability was significantly reduced after incubation with FA–PEG–β-CD (CDPF) and DOX as compared to free DOX. This indicates that the inclusion complex formation between β-CD and DOX enhanced the cytotoxicity rate towards cancer cells. To prove this concept, the in vivo experiments were performed and after intravenous injection of CDPF/DOX the tumour volume was lower in comparison to control, CDPF, and free DOX, as shown in Figure 11a–c. Moreover, CDPF/DOX treatment induced the overall necrosis cancer tissue, not only partial as for free DOX, as shown in Figure 11d. Thus, it could be concluded that the FA–PEG–β-CD/DOX drug delivery system can effectively deliver the DOX to the tumour tissue and decrease its side effects.

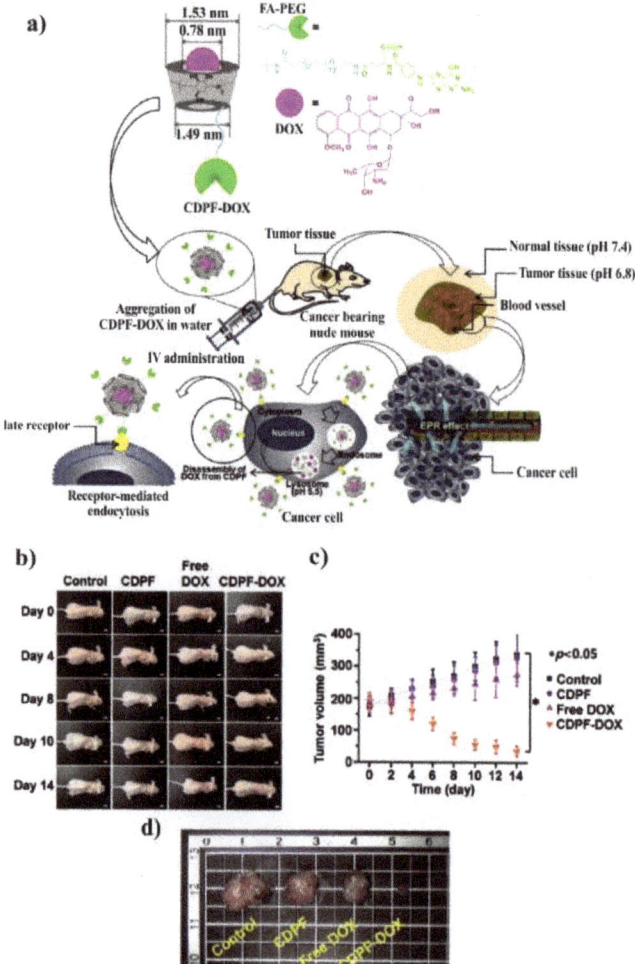

Figure 11. (a) Inclusion complex between FA–PEG–β-CD and DOX and schematic illustration of the endocytosis of FA–PEG–β-CD-DOX into cancer cells. (b) Gross appearances of tumor tissues in control, CDPF, free DOX _HCl and CDPF-DOX treated mice observed at 0, 4, 8, 10 and 14 days. (c) Tumor volume (mm3) and of mice treated with the samples at 0, 2, 4, 6, 8, 10, 12 and 14 days. Free DOX HCl and CDPF-DOX were intravenously injected via the lateral tail vein on days 1 and 8. Error bars represent mean SD (n = 5); the measurement of the cancer volume was repeated three times (* $p < 0.05$ compared with control). (d) Tumor images. (Copyright 2019 Elsevier).

In addition, D-α-tocopheryl polyethylene glycol succinate (TPGS) can be also used as targeting moiety against different tumours to enhance intracellular drug concentration which, as a result, increases the efficiency of the therapy [155,156]. It was found that one of the major mechanism of multidrug resistance in cancer cells is enhanced drug efflux by energy-dependent pump [157]. For instance, P-glycoprotein (P-gp) is responsible for such efflux, since it is overexpressed in many cancer cells. It was a rational strategy to combine CDs with TPGS because both are the P-gp inhibitors. Therefore, the multistep reactions were employed for the synthesis of CD-TPGS conjugates with a different number of arms (CD-2TPGS-2arms, CD-4TPGS-4arms, CD-6TPGS-6arms) in which the final step was the condensation of hydrazine-functionalized β-CD and aldehyde group of TPGS (prepared after its reactions with 4-formylbenzoic acid). Subsequently, the nanoparticles with the size ranging from 207.5 to 222.3 nm were obtained by a solvent evaporation method, however those with two TPGS were unstable, therefore the CD-4TPGS and CD-6TPGS were further investigated. The drug loading (41.3, 43.5%) and release (74.7, 76%) was similar for both formulations, DOX/CD-6TPGS and DOX/CD-4TPGS, respectively. The DOX-loaded nanoparticles composed of CD-4TPGS and CD-6TPGS showed similar cytotoxicity against MCF-7 cells to the free DOX, however great supremacy of cytotoxicity against MCF7/ADR drug-resistant cancer cell in comparison to free drug. Subsequently, in vivo antitumor effect was tested on the H22 sarcoma model, and the tumour inhibition was similar for CD-4TPGS and CD-6TPGS and slightly better than free DOX. This result was consistent with the cellular uptake and in vitro cytotoxicity. Moreover, NPs can be safely used as a drug delivery system, since the test of hepatotoxicity indicate that there was not a significant effect for different organs after their administration.

Stimuli-responsive DOX-loaded nanocarriers based on CD and PEG macromolecules were also proposed to improve the efficiency of antitumor therapeutics. In this regard, the pH-sensitive nanoparticles composed of benzimidazole end-capped poly(ethylene glycol)-hyperbranched polyglycerol (PEG-HPG-BM) and FA-modified β-CD (FA-CD) were designed for the preparation of targeted supramolecular nanoparticles (TSNs) [136]. The nanoparticles were formed by the supramolecular interactions between BM functionalities and CD due to its sensitivity to different pH values. The formed NPs (78–88 nm) showed pH-sensitive DOX-release in comparison to non-sensitive NPs (without CD-FA complexation) and this was attributed to the protonation of BM in acidic conditions. The cytotoxicity of NPs against the HeLa cells was highest for the TSNs due to the presence of FA targeting molecule. However, for the HepG2 cells for which FA is not targeting moiety, the effect of TSNs is less pronounced, therefore the combination of pH-sensitivity and targeting is required to efficiently enhance the therapy outcome. In addition, Song et al. proposed to use Ada and CD inclusion complexation for the preparation of a reduction-responsive drug delivery system for the delivery of both DOX and sorafenib (SF) [137]. To achieve this aim, the synthesis of both Ada-terminated doxorubicin prodrug with disulphide bonds and PEG-CD was performed. The NPs were prepared by inverse nanoprecipitation and after the addition of SF the size of NPs increase from 166 to 186 nm. The release of DOX and SF in pH 5 and pH 7.4 was much faster in the presence of dithiothreitol (DTT) which was a reducing agent able to cleave the disulphide bond. However, the released amounts of SF were significantly higher because it can only interact with the building blocks of NPs by hydrophobic interactions or p-p stacking, whereas DOX was connected by a chemical-reversible bond. Most importantly, the in vitro cytotoxicity essay demonstrate that PEG-CD/AD/SF NPs exhibit great efficiency against HepG2 cells in comparison to free SF and PEG-CD/AD+SF physical mixtures. Furthermore, dual responsive MCs based on β-CD-hydrazone-DOX and ferrocene-functionalized PEG (PEG-Fc) were designed. The conjugation of DOX to the CD by hydrazide groups was performed in the three-step procedure, whereas the PEG end-group modification was done by the reaction with ferrocenecarboxylic acid. The MCs were formed by spontaneous self-assembly in water and their size was ranging from 42 and 62 nm. The DOX-release profiles revealed that the DOX can be rapidly released at the acidic conditions (pH 5) and by reactive oxygen species (H_2O_2), typically present in cancer cells. This result was attributed to the hydrolysis of hydrazine bonds between DOX and CD, whereas the accelerated release with the addition of H_2O_2 was related to the oxidation of PEG-Fc and shedding of MCs shell. As a final step, it was shown that DOX was accumulated in the nucleus of HeLa cells slightly slower than the free DOX due to the fact that in the MCs the DOX should be cleaved in endosomal pH to subsequently reach the nucleus.

Apart from DOX, different kinds of drugs were encapsulated in the nanocarriers composed of CD and PEG molecules. For instance, artemisinin-loaded γ-CD NPs combined with polyethylene glycol (PEG) derivatives (polysorbate 80 and DMPE-mPEG2000) [138] and honokiol-loaded pluronic F127-cyclodextrin MCs [139] were tested in vivo. The first systems showed good activity against malaria, whereas the second one against fungal infections (*Candida albicans*). For both systems, the encapsulation of the drug into the nanocarrier could enhance the circulation time in the blood flow (longer elimination half-life) what increase the therapeutic effect of the drug. Moreover, the targeting by FA was used for the delivery of plasmid DNA (pDNA) with nanoparticles build by supramolecular complexation from Ada-functionalized FA, PEG-Ada, and β-CD-grafted with branched polyethylenimine with low molecular weight (PEI-CD). The two modes of complexation were employed in which (A) PEI-CD, FA-Ada, and PEG-Ada were complexed before the addition of pDNA, whereas in (B) firstly the pDNA was complexed with PEI-CD and subsequently the host-guest complexation was induced by the presence of FA-Ada and PEG-Ada. The aim was to obtain the polyplexes with the size in the range 50–200 nm that predispose their facile cellular uptake. The ability of each formulation to bind pDNA at various N/P ratios was examined by agarose gel electrophoresis. Method A gives rise to NPs of slightly lower size in comparison to method B, and the polyplex A was able to inhibit the DNA migration at nitrogen/phosphorus (N/P) ratios of 3, whereas for polyplex B inhibition occurs at N/P ratio of 10. However, for in vitro transfection, two polyplexes were prepared with N/P 20, naked pDNA (ND), and commercially available PEI-25KD, and tested by means of a luciferase activity assay. The activity was tested out in folate receptors (FR)-negative HEK293 and FR-positive KB cells. As expected, the effect of polyplexes was the highest for FR-positive cells, however, for both cellular models the efficiency followed the order Method A > Method B > PEI-CD. This could be attributed to the higher stability and lower size of polyplexes and as a result their better cellular internalization. It was concluded that especially polyplex A could be a good alternative for branched PEI-25KD due to lower toxicity and better transfection efficiency [140]. Furthermore, FA-targeting was used to selective, simultaneous delivery of siRNA and DOX by a ternary complex of FA-PEG-GUG-β-CDE/DOX/siPLK1. The main part of a carrier was FA-polyethylene glycol (PEG)-appended polyamidoamine (PAMAM) dendrimer (generation 3; G3) conjugated with glucuronylglucosyl-β-CyD (GUG-β-CyD). The positively charged nanoparticles with a size of 92 nm were prepared by mixing the carrier with DOX in a ratio of 1:3 and carrier to siRNA ratio of 50. It was shown that the ternary complex reaches the 94 % cytotoxicity activity in KB cells which was significantly higher than binary complex FA-PEG-GUG-β-CDE/siPLK1. Subsequently, an intravenous injection of the polyplex was administered twice a week to BALB/c nu/nu mice, whereby both the tumour size and weight were also significantly lower for FA-PEG-GUG-β-CDE/DOX/siPLK1 than the binary complex. Moreover, there were no side-effects of the therapy due to negligible mice body change after the administration. The obtained results showed great potential in a tumour selective-therapy [141]. In addition, stimuli-responsive glucose-responsive NPs were prepared by the complexation PEG-b-poly(aspartic acid) derivatives, such as: glucosamine (GA)-functionalized block copolymer PEG$_{45}$-b-P(Asp-co-AspGA), phenylboronic acid grafted-block copolymers of PEG$_{114}$-b-P(Asp-co-AspPBA) and α-CD. This sophisticated system self-assemble into core-shell (CS) micelles in PBS due to the formation of α-CD/PEG$_{45}$ inclusion complex and GA/PBA cycloborate. Subsequently, α-CD was removed by dialysis and the desired hollow vehicles with a diameter of 40–60 nm were obtained composed of cross-linked P(Asp-co-AspGA)/P(Asp-co-AspPBA). It was shown that the nanovesicles swell in the presence of glucose and their size increases depending on the glucose concentration. Moreover, the release of vancomycin (glycopeptide antibiotic) can be also controlled by varying the type of added sugar (fructose versus glucose). The higher release of vancomycin in the presence of fructose was attributed to its higher affinity for PBA in comparison to glucose [142].

Another strategy allowing nanocarriers formulation and drug encapsulation is based on silica [143,144] or zinc oxide [145] nanoparticles surface modification with CD and PEG macromolecules. For instance, the CD was conjugated on porous nanosilica (PNS), subsequently, the supramolecular complexation with Ada-PEG was performed along with the addition of 5-fluorouracil (5-FU). The resulting 5-FU loaded NPs size was 50 nm and the release of the drug lasted for 3 days. However, although the nanocarrier possesses acceptable cytotoxicity against NIH3T3 cell lines, its efficiency was lower than the free drug [143]. In contrary, Nguyen et al. proposed to functionalize PNS with adamantylamine (A) by disulfide bonds (PNS-SS-A) which were subsequently supramolecularly complexed with cyclodextrin-heparin-polyethylene glycol (CD-HPEG) [144]. The size of nanoparticles was approximately 40–50 nm and drug encapsulation efficiency of DOX was 56%. The release of DOX could be enhanced by the presence of DTT due to the redox-sensitive dissociation of disulfide bonds present on the surface of PNS. After incubation with HeLa cells, the dose-dependent cytotoxicity was observed for nanoparticles and free DOX. However, once again the free drug and DOX loaded PNS were more efficient than the supramolecular complexed NPs. The authors ascribed this phenomenon to free DOX aqueous solubility and membrane permeability. Moreover, luminescent zinc oxide nanoparticles were coated with PEG and β-CD by wet co-precipitation method and used for curcumin (Cur) delivery. The author describes the formed NPs as oval shape carriers with visible micellar core-shell morphology and NPs diameter was in the range of 20–23 nm. The release was tested in the physiological pH of 7.4 and tumor lysosomal pH environment of 4.8 and the higher amount of the drug was released in the acidic environment. The authors concluded that the drug release profiles are following fickian diffusion in which initially the first-order kinetics is observed followed by sustained release via zero order kinetics. The advantage of application of these NPs was the strong fluorescence of zinc oxide NPs which allows for their tracking in cells and brine shrimp. Moreover, the efficiency against the Staph bacteria of PEG-β-CD coated ZnO nanoparticles was moderate, in contrary good anticancer activity against MCF-7 cells was observed, however not better than drug-loaded zinc oxide NPs without supramolecular modification [145].

Gels and hydrogels can be also proposed as drug delivery depots for the administration of drugs and proteins [158]. The gels and hydrogels are a perfect environment that protects the drug and allow for controlled diffusion of their payloads by adjusting the crosslinking density [159]. There is only one example of gels composed of the PEG and CD for the diclofenac delivery [146]. The gels were prepared with a concentration of 0.1 M β-CD and 0.54 M K_2CO_3 in PEG400, respectively [160]. The occurrence of the gelation was ascribed to the formation of PPRXs supramolecular structure. Two types of thermo-reversible gels were formed at the gelation temperature close to the human body. It was observed that the release of diclofenac for the obtained gels was significantly higher in comparison to commercial products. Its release rate also reflects in the skin penetration properties, after the initial lag phase, the higher flux and K_p (permeability coefficient) for the PPRXs gels indicate that the gel was suitable for dermal formulation.

A different strategy was proposed by Li et al. [147] based on inclusion complexation between a biodegradable poly(ethylene oxide)–poly[(R)-3-hydroxybutyrate]–poly(ethylene oxide) (PEO–PHB–PEO) triblock copolymer and α-CD. Due to the formation of PPRX structure between α-CD and hydrophobic interactions between the middle PHB blocks, the strong network of supramolecular hydrogels was created. Subsequently, the release of fluorescein isothiocyanate labelled dextran (dextran-FITC) from the hydrogels with different composition was investigated. The best performance was obtained for the hydrogel composed of α-CD in a high concentration (9.7 wt.%) and PEO-PHB-PEO with a molecular weight of 5000-3140-5000 g/mol, respectively. For this formulation, the sustained release lasting for one month was achieved, whereas for other formulation faster release kinetics was observed. Therefore, by the delicate tuning, both the PHB block length and inclusion between the α-CD/PEG, the optimal strength of the hydrogel can be obtained. It was concluded that such drug-loaded hydrogel can be easily injected under pressure to desired tissue due to its thixotropic properties and subsequently slowly release its cargo. Moreover, CD/PEG hydrogels were prepared by conjugation of NH_2-PEG-NH_2 with β-CD with isocyanate groups in DMSO and incubation in distilled water [148]. The CD was functionalized with approximately five isocyanate groups and gelation occurred immediately after the addition of amino-functionalized PEG, however to slightly slow down this process, the acetic acid was added during the formulation. The release of the three different model drugs: lysozyme, β-estradiol and quinine release, was investigated in PBS. The drug physicochemical properties strongly influence the release from CD/PEG matrix since lysozyme is a hydrophilic, quinine is also hydrophilic able to form weak inclusion complexes with CD whereas hydrophobic β-estradiol create a strong inclusion complex with CD. Moreover, the release was additionally controlled by the CD content which was in an agreement with swelling/loading experiments, because a tighter network decreases the diffusion/penetration process. In conclusion, the lysozyme was rapidly released, whereas for quinine and β-estradiol the bi-modal profiles were observed due to drug-CD inclusion complexes. Additionally, the shear stress can be used to control the release of the drug from PEG/CD hydrogels [149]. For this purpose, 4-arm-PEG were mixed α-CD to form supramolecular hydrogel with polyrotaxane structure that exhibits reversible gel-sol transition. Its structure can be easily broken by shear forces and such thixotropic behaviour allows for hydrogel injection through a syringe needle. This shear-tinning behaviour was used to control the release of brimonidine by shaking a gel with PBS solution in different rates. As could be expected, the release of the model drug was much faster under external stress due to disassembly of hydrogel supramolecular structure.

Supramolecular hydrogel was also prepared by mixing adenine- and thymine-functionalized PEG (A-PEG-A/T-PEG-T) with α-CD for the delivery of DOX [150], as shown in Figure 12a,b. The presence of A and T end groups enhanced the elastic modulus, however, their influence on the drug release in different pH was notfigureinvestigated. Nevertheless, the sustained release of DOX in PBS (pH 7.4) was observed, depending on the drug loading content. Subsequently, the cytotoxicity test showed that freeze-dried powder of hydrogels is relatively safe at lower concentrations, whereas at higher concentrations the viability of L929 cells decreases due to the effect of α-CD which is toxic at the concentration of 12.5 mg/mL. Most importantly, the supramolecular hydrogel can be also formed in vivo after subcutaneous injection into SD rats by a needle, as shown in Figure 12d,e. In addition, the injected hydrogels preserved its porous structure after injection (Figure 12f) which allows for its good permeability to the biological medium. The in vivo intra-tumoral injection indicated that the A-T complexed gel (DOX-loaded G2 gel) exhibit better therapeutic effects due to restricted tumour growth in comparison to the controls and gel composed of α-CD/PEG without nucleobases (DOX-loaded PEG10k/α-CD (10 wt.%/10 wt.%) gel), as shown in Figure 12c. It was correlated to the elasticity of hydrogel containing nucleobases which lead to the longer release of DOX. Since, the application of supramolecular complexed hydrogel does not cause the reduction of the rats' body weight, it was concluded that it is a good candidate for anti-cancer therapy in which the sustained release of the drug is required.

Hennink et al. proposed [161] the use of star-shaped 8-arm PEG (PEG8) modified either by β-CD or cholesterol moiety with a hydrolyzable ester bonds. The hydrogels were prepared in ammonium acetate buffer (pH 4.7) due to the presence of reversible supramolecular interactions (inclusion complexes CD/cholesterol), the obtained material was thermo-reversible. In the following paper, this strategy was used to investigate a degradation and protein/peptides release from supramolecular complexed polymer networks [162]. Four different model payloads were used to test the hydrogel properties as drug delivery depot: lysozyme, bovine serum albumin (BSA), and immunoglobulin G (IgG). The sustained release of loaded proteins was observed regardless of their hydrodynamic diameter, however by an increase of solid content of hydrogel from 22.5% (w/w) to 35% (w/w) prolonged release could be achieved. It was concluded that the release is mainly governed by the surface erosion of hydrogel. Moreover, it was shown for a small peptide (bradykinin) that at high concentration of hydrogel components the release of a peptide depends also on the hydrogel surface erosion, however, it follows first-order kinetics. The alternative way for achieving the controlled release of proteins from CD/PEG hydrogels is to use the ultrasounds as a trigger for their release [151]. To acquire such ultrasound-responsive structure the host-guest between adamantine-PEG and β-CD-functionalized eight-branched PEG was used. It was shown that Tf-AF647 (transferrin) can be specifically released after the ultrasound exposure at the desired part of a hydrogel due to its structure degradation. Nevertheless, the hydrogel degraded after exposure, however, it could spontaneously be re-formed after removal of stimulus, and the Tf-AF647 clearly moved to the upper part of PBS indicating its successful release. As a model of a living body, the HeLa cells with a collagen layer were used to observe the intracellular accumulation of transferrin. To induce the transport through collagen layer to the cells, the hydrogel should be exposed to the ultrasound stimulus and it could be done by a repeating pulsation which causes cleavage of host-guest binding. Therefore, this system could release the proteins in an ultrasound-guided manner.

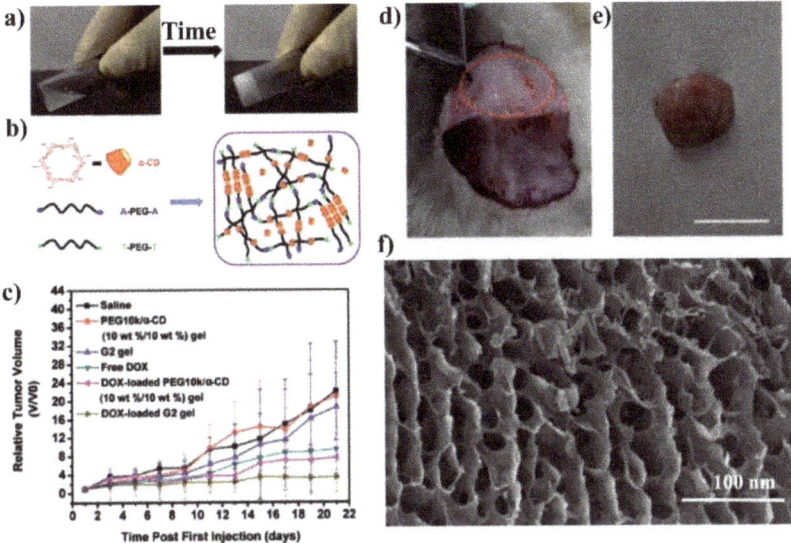

Figure 12. (**a**) Photographs of an A-PEG10k-A/T-PEG10k-T/α-CD aqueous solution (PEG/α-CD, 10 wt.%/ 10 wt.% and a G2 hydrogel. (**b**) Schematic illustration of the gelation mechanism of the supramolecular hydrogel. Changes in relative tumor volume (**c**) and in relative body weight. Each solution was injected into xenograft-bearing mice (U14) after the initial tumor volume had reached 150–250 mm³. In vivo formation of the G2 hydrogel in the subcutaneous tissue after 30 min (marked as red-dotted curves) (**d**), the G2 gel removed from the rat (bar = 1.5 cm) (**e**) and the corresponding SEM of the lyophilized hydrogel (bar = 100 mm) (**f**). (Copyright 2013, RSC).

Thermo-responsive hydrogels were prepared from β-CD/PEG mixed with poly (ethylene oxide)-poly (propylene oxide)-poly(ethylene oxide) tri-block copolymer (Pluronic® 127) for the delivery of curcumin (Cur) [152]. The micellar type hydrogel with LCST characteristics was formed in which PPO was encompassed by β-CD as a center block and PEO macromolecules as the micelle shell. By heating the supramolecular hydrogel above and below LCST, the reversible gel to sol transition can be achieved. Moreover, the hydrogel exhibit much higher swelling in PBS (pH 7.4) in comparison to acidic buffer (pH 1.2), therefore the release of Cur was tested by varying both the temperature and pH. The optimum conditions to obtain the highest Cur release were pH 7.4 (at 35 °C) due to the solvent uptake which causes the hydrogel pores opening and facilitates the diffusion of the drug. A slight decrease at 40 °C was attributed to the aggregation of hydrophobic domains. In contrary, the influence of temperature in acidic medium was negligible due to the deionization of functional groups and low degree of swelling. Additionally, with increasing the concertation of PEG the release of Cur in pH 7.4 (in 35 °C) was faster, whereas increasing content P127 slow down the release of Cur at both temperatures and in both dissolution media. Subsequently, the cytocompatibility of unloaded and Cur-loaded hydrogels was tested against mouse fibroblast cell line and the obtained material exhibited good safety in comparison to commercially available Triton-X 100. After the inhibition of hydrogel with HeLa and MCF7 cells, hydrogel exhibits good efficiency in killing cancer cells, however lower than the free drug. Therefore, it can be used for storage and the sustained delivery of Cur to cancer cells. Moreover, Shi et al. proposed to use α-CD/PEG hydrogels for the co-delivery of both hydrophobic (NPOD) and hydrophilic (DOX) anticancer drugs as a consequence of acidic environment present in the tumor tissues [153]. To achieve this aim PEG (2000 g/mol) was firstly reacted with 4-formylbenzoic to obtain aldehyde-functionalized PEG which was subsequently used for the reaction with 4-aminopodophyllotoxin (NPOD). As a result, NPOD-PEG conjugate was synthesized with pH-sensitive imine bond. The XRD studies confirm the

PPRX structure of the obtained hydrogel in which the α-CD rings are stacked along the NPOD-PEG chains and additionally this structure is stabilized by strong hydrogen-bond interaction between NPOD end groups. Such a structure composed of supra-cross-links exhibited the good shear-thinning behavior which is required for the hydrogel biomedical applications. Subsequently, the ability of hydrogels for the triggered, site-specific drug delivery was tested under two different pH values (pH 5.0, 7.4). The release of DOX and NPOD was relatively low (approximately 40%) at pH 7.4 while the accelerated release of DOX and steady release of NPOD was observed after decreasing the pH to 5.0. This was attributed to the fact that in pH 7.4 the release was mainly caused by diffusion whereas at pH 5.0 it was mainly controlled by a breakup of supramolecular structure. The lower release of NPOD in acidic conditions was correlated to the inclusion complex formation of PEG-NPOD with α-CD which slow down the hydrolysis. As a final step, cytotoxicity against the A549 human lung cancer cells was tested and the obtained hydrogel exhibited strong therapeutic effect in comparison to controls and free NPOD in all concentrations. This indicated the advantage of the combination of hydrophobic and hydrophilic drugs in cancer therapy with single-modality treatment.

The combination of PEG/CD in one material leads to the development of novel drug delivery systems both in the nanoscale (nanoparticles, micelles) and macroscale (gels, hydrogels). These systems were mainly used for the delivery of anticancer drugs and proteins. To enhance the therapeutic effect of encapsulated drugs targeting moieties were incorporated or a different methods of the stimulus (pH, temperature, ultrasound, shear) were used to control the release of loaded drug in the desired target. Moreover, a lot of effort has been made to test obtained systems both in vitro and in vivo. The investigated systems possess good biocompatibility and efficiency in therapy after drug encapsulation. Despite of the fact that many drug delivery systems based on CD have been reported by various research groups, those focused on the delivery of anti-cancer drugs are limited. For instance, CRLX101 polymeric nanoparticle-containing camptothecin (CPT) based on the CD/PEG conjugate was in the clinical trials [163]. Therefore, we believe that the effort of the scientist and the pharmaceutical industry should be focused on the translation of PEG/CD drug delivery systems from the laboratory to clinics.

5. Cyclodextrin Conjugate to Polysaccharides for Drug Delivery

Natural polysaccharides (Figure 13) are biocompatible, biodegradable, renewable linear or branched polymers consisting of different glucose units crosslinked by glycoside bonds [164,165]. These natural biomolecules have shown remarkable pharmaceutical activities such as anti-cancer activity [166], antiviral property [167] anti-inflammation activity [168] cardiovascular protection, and anti-mutation effects [169]. Moreover, the presence of many reactive groups (hydroxyl, amino, and carboxylic) allows for various chemical modification enabling, for instance, co-polymer grafting or atom transfer radical polymerization (ATRP) to improve the physical and chemical properties of polysaccharides [170].

β-CD is one of the most popular agents for modification of polysaccharides aimed for biomedical applications [171,172]. Therefore, the polysaccharides are excellent scaffolding material for cyclodextrin which enables for the preparation of different drug delivery systems (Table 4) with high encapsulation efficiency [173]. The crosslinking methods were used to combine β-CD with alginate gel (CCALs) for controlled drug delivery. The resulting hydrogels were obtained by simple mixing of sodium alginate (AL) with previously prepared crosslinker (β-CD activated by ethylene diamine). Next, the obtained hydrogels were immersed in $CaCl_2$ solution for the replacement of Na^+ ions for Ca^{2+}. The complexation of free carboxylic groups in AL chain by Ca^{2+} additionally improve the mechanical properties of CCAL. To elucidate the release profile under the mechanical stimulus (Figure 14a), the ondansetron (ODN) was employed as a model drug. The controlled release of ODN from CCAL hydrogel can be achieved by mechanical stimuli. The faster release of ODN was observed in the case of increasing compression above 30%, due to the regulation of the host-guest interaction. The β-CD moiety is distorted as a result of applied force, which probably causes the guest to be pushed out of the cavity and thus, the release of drug become faster however it is difficult to determine the exact effect of mechanical stimuli on the polymer structure. This method could show a novel approach for controlled drug release from hydrogels in comparison to other stimuli such as heat, light, and magnetic force. In addition, the delayed release was observed as a result of the formation of IC between ODN and β-CD [174].

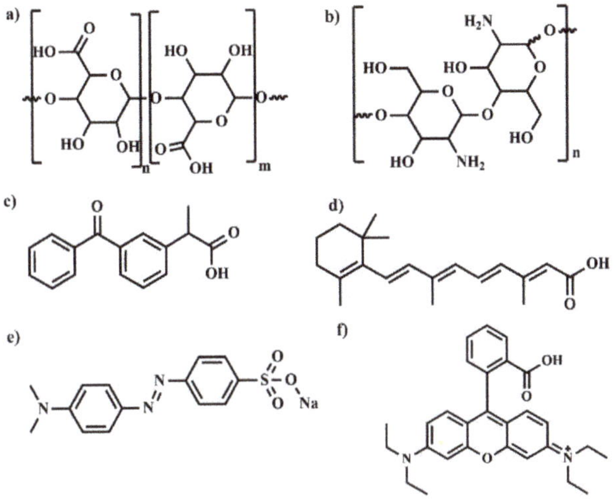

Figure 13. The chemical structure of the carrier polymer (a) alginate and (b) chitosan, and chemical structure of (c) ketprofen, (d) retinoic acid, (e) methyl orange, (f) rhodamine B.

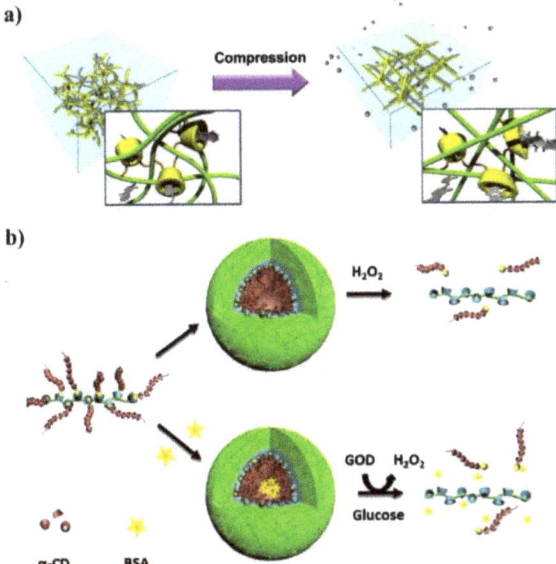

Figure 14. Two different approaches to controlled release: controlled release under mechanical compression (a) and oxidizing agents (b). (Copyrights 2019 Scientific report, Copyright 2018 Frontiers in Pharmacology).

Table 4. The summary of described hydrogels, microparticles and fibres for drug delivery based on combination of polysaccharides and its copolymers with different CDs.

Type of Drug Delivery System	Platform	Type of CD	Drug	Release Medium	In Vitro Studies	Ref.
Nanoparticle	Alginates	β-CD	BSA	PBS	CT26 cells	[174]
Hydrogel	Alginates	β-CD	Methyl orange	0.9% NaCl	nd	[175]
Hydrogel	Alginates	β-CD	Retinolic acid	PBS	Zebrafish embryos	[176]
Microcapsule	Dextran	α-CD	Rhodamine B	PBS or pH 5.5	HeLa cells	[177]
Microcapsule	Dextran	α-CD	Rhodamine B	PBS	nd	[178]
Nanoparticle	Chitosan	β-CD	Ketoprofen	PBS	A549 cells	[179]
Tablet	Chitosan	β-CD	Ketoprofen	pH 6.0	Mouse mucosa	[180]
Nanoparticle	Chitosan	β-CD	Insulin	PBS	NIH 3T3 cells	[181]
Hydrogel	Chitosan/PCL	α-CD	BSA	PBS	nd	[182]

A similar approach for the synthesis of alginate conjugated with β-CD was applied by Achamann and co-workers [175]. Oxidation of alginate, reductive amination, and azido-alkyne cycloaddition click-reaction were employed to covalently anchor β-CD to alginate chain. Subsequently, the water solution of $CaCl_2$ was mixed with a different type of alginates to form four different hydrogel beads: unmodified alginate, unmodified alginate in physical mixture with β-CD, and unmodified alginate with a mixture of alginate conjugated β-CD in two concentrations (25% and 50%). The release studies showed that hydrogel beads with β-CD chemically crosslinked to alginate are able to absorb higher amount of methyl orange and slow down its release than the other formulations. However, the obtain hydrogel beads are strongly unstable during release studies performed in 0.9% NaCl solution as a result of replacement Ca^{2+} ions for Na^+ [175].

As an alternative to AL, dextran is a biopolymer that can be used for hydrogel preparation. Simple modification of dextran by maleic anhydride leads to the obtaining of reactive scaffold suitable for the hydrogels production [176]. The injectable hydrogels were prepared by a multi thiol-ene click reaction. Dextran modified by maleic anhydride could be easily crosslinked by applying per-6-thio-β-CD (PSCD) or di-thiolated end-capping poly(ethylene glycol) (DSPEG). After mixing of components, the microscopic hydrogels (CD-gels or CD-PEG-gels) were created via Michael addition between carbon-carbon double bond of maleic anhydride and thiol groups of PSCD or DSPEG. The obtained hydrogels were composed of particles with a diameter less than 100 nm and the obtained materials were non-toxic against zebrafish embryos. The retinoic acid (RA) was used as a simple model thus CD-gels or CD-PEG-gels loaded with RA were prepared. The presence of IC between RA and β-CD lead to control release of RA in PBS what could be correlated with in vivo release studies [176].

Microcapsules are a promising carriers for drug delivery due to their simple administration, the effective protection of encapsulated drugs against degradation in their interior, and as a result, the possibility to control their release over time [183]. For the controlled release of rhodamine B (Rod B), biodegradable microcapsules consisting of dextran/poly(aspartic acid) (PASP) were prepared by Zhang et al. [177]. The multistep reaction included: a) functionalization of dextran by β-CD in Schiff's base reaction, b) preparation of PASP with adamantane backbones, c) deposition of both polymers on $CaCO_3$ modified by polycations surface, and finally d) removal of $CaCO_3$ by EDTA, which allows for obtaining pH-responsive hollow microcapsules for Rod B delivery. The stability of microcapsules is based on the IC formation between β-CD grafted to dextran and adamantane group anchored to PASP. Moreover, the Schiff base bond between β-CD and dextran (-C=N-) could be easily hydrolysed in a weakly acidic condition, therefore, the release behaviour of Rod B was investigated in physiological (PBS pH = 7.4) and acidic (5.5) environments (Figure 15a). The cleavage of -C=N- bond leads to separation between dextran and PASP and the controlled release of Rod B could be achieved. The fast hydrolysis of the Schiff base bond in acidic buffer accelerates the release of Rod B from microcapsule in comparison to release in physiological medium [177]. The same release behaviour of Rod B was achieved by Zhang and co-workers by mixing α-CD attached to dextran and poly(acrylic acid) with photo-switchable *p*-aminoazobenzene end groups. The controlled release has been reached by dissociation of microcapsules after cleavage of IC as a result of shifting the *cis* form of the azo bond to *trans* after UV light irradiation (365 nm). The release studies showed that irradiation dramatically accelerate the release of Rod B from microcapsules within 300 min as shown in Figure 15b,c [178].

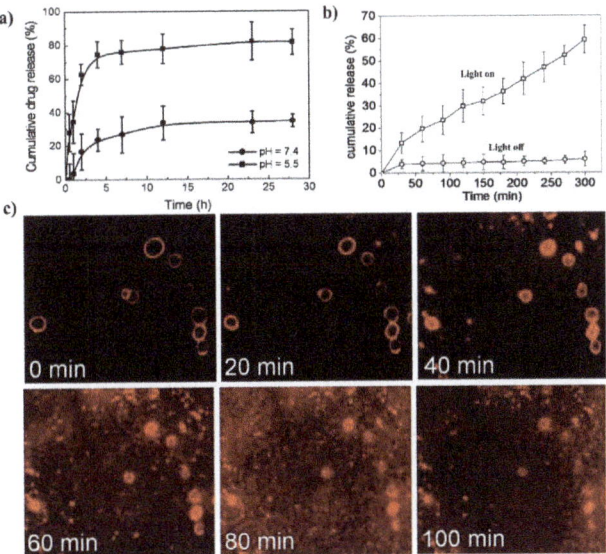

Figure 15. Drug release from (**a**) pH-sensitive dextran microcapsules, (**b**) from photo-switchable dextran microcapsules, and (**c**) snapshots of the photo-dissociation of photo-switchable dextran microcapsules-loaded with Rod A. (Copyrights 2011 ACS (1a), Copyrights 2011 ACS (1b)(1c)).

Moreover, the bio-based nanoparticles (NPs) composed of naturally occurring polysaccharides can be also used for controllable drug delivery. To prepare reversible, supramolecular systems self-assembled into NPs in water, the alginate modified by β-CD, metoxypolyetylene ferrocene end-capping (Fc-mPEG) and α-CD were mixed. The formation of NPs is based on two supramolecular host-guest interactions: a) creation of IC between β-CD attached to alginates chain and ferrocene end-group of metoxypolyetylene, and b) creation of IC between metoxypolyetylene chain and α-CD. However, the second interaction based on formation IC with α-CD is the determining process of NPs formation. Moreover, this strategy allows for obtaining NPs with a diameter of 100 nm for controlled delivery of the enzyme bovine serum albumin (BSA). The controlled release of BSA from NPs in glucose/glucose oxidase (GOD) environment can be achieved by gentle oxidation of Fc end-group (Figure 14b). The oxidation agent (H_2O_2) can be prepared as a by-product of glucose oxidation by GOD. After oxidation of Fc, NPs became unstable as a result of cleavage of the IC between β-CD and further BSA may be released in a controlled manner. Moreover, the presence of β-CD in the NPs structure enhance the encapsulation efficiency [184]. Also, the well-known interaction between β-CD and adamantane [185] was used to prepared self-assembled NPs from two series of dextran polymers a) dextran modified by β-CD and b) dextran modified by adamantyl groups. After simple mixing of these supramolecular polymers, the IC between β-CD and adamantyl end group is formed and the NPs with a diameter ranging from 100 to 250 nm could be obtained. The NPs were formed only when the degree of the β-CD substitution in dextran was higher than 4%, and, as the amount of β-CD increased, the NP size decreased whereas the stability of obtained NPs decreased with increasing degree of β-CD substitution [186].

Chitosan (CS) is a natural cationic polysaccharide used for the preparation drug delivery systems due to its biological activity and ability for effective entrapping of cargo molecules [187]. However, the poor solubility of CS in water or common organic solvent causes low reactivity and limits its application. To overcome this limitation some derivatives of CS for instance, maleoyl chitosan [179] or thiolated chitosan [180] were prepared. The most common reaction allowing the attachment β-CD to CS chain is the reaction with *p*-toluenosulfonyl derivatives [179,181,188,189]. This strategy leads to chitosan drug delivery systems functionalized with β-CD moieties for drug [179], DNA [190] or protein delivery [181]. To prepare hydrogels based on chitosan derivatives, the metoxypolyetylene glycol-*co*-polycaprolactone carboxyl-terminated (mPEG-PCL) was attached to chitosan backbone. It was observed that the amphiphilic nature of prepared copolymer cannot lead to the hydrogel formation in room or slightly elevated temperature. However, after mixing of 2.5% water solution of the copolymer with 12% aqueous solution of α-CD within a few minutes or seconds rapid formation of hydrogel occurred as a result of IC formation between mPEG-PCL chains and α-CD cavity. To estimate the release profile, the BSA loaded hydrogels was also prepared by adding BSA to the α-CD solution. The release rate of BSA from obtained hydrogel was strongly dependent on grafting degree, copolymer and α-CD concentration. The release profiles of BSA from obtained hydrogels are significantly faster as compared with reported before for triblock copolymer [poly(ethylene glycol-*co*-3-hydroxybutyrate-*co*-ethylene glycol)] with the α-CD moiety [147,182]. The preparation of supramolecular hydrogels based on PLA attached to the chitosan chain was also shown by Hu et al. In this work, the ability to the formation of the IC between PLA chain and β-CD was employed to prepared self-assembly hydrogel for simultaneous delivery of BSA and Heparin. Furthermore, the presence of weak, labile interaction makes the polymer sensitive for temperature. Due to the temperature-sensitiveness of prepared hydrogels, the release profile was investigated as a function of temperature. The increase in temperature from 20 °C to 30 °C causes an increase of release rate of about 27% for heparin and 13% for BSA [191]. The CD cavity can be also used for purification of water from micropollutants. Sillanpaa et al. insoluble EDTA-crosslinked chitosan bearing β-CD trifunctional material for water purification. The one-pot synthesis leads for preparation high efficient network for metal and drugs removal [188].

This section introduces the aspects of natural polysaccharides containing cyclodextrin moiety as a drug delivery systems. with This section introduces the aspects of natural polysaccharides containing cyclodextrin moiety as a drug delivery systems. Currently alginates, dextran, chitosan and its derivatives are widely used in the field of biomedical applications due to its beneficial properties. However, the polysaccharides require modification to be used as a drug delivery system. In this respect, the polymers or organic compounds could be successfully used to modify polysaccharides and thus enhance theirs low reactivity and solubility in water or organic solvents. The combination of CDs with different types of polysaccharides provides new hybrid materials which successfully slow down the release and reduce the side effects of the drug. Additionally, the CDs used as a crosslinking agent might enhance the unsatisfactory mechanical properties of bio-based materials. However, still existing limitations give opportunities to search for new solutions to improve the utility of polysaccharides as drug delivery systems.

6. Conclusions

In summary, by the construction of complex systems based on biocompatible polymers and CDs, a variety of multifunctional materials were developed. The development in macromolecular synthesis allows for the preparation of supramolecular assemblies with various architectures. Typically, nano- or micro-particle, hydrogels, and fibers are used as drug delivery carriers. It was demonstrated that these extraordinary materials, after efficient drug loading, show the controllable drug release by taking advantage of host-guest interactions and other noncovalent forces. The liberation of several important biologically active agents, including antibiotics, vitamins, hormones, enzymes, anticancer drugs, nonsteroidal anti-inflammatory drugs, and physiologically active lipid compounds has been examined in vitro. Although gradual progress has been achieved in the preparation of systems containing biocompatible polymers and cyclodextrins, some issues remain a challenge. Especially, when the synthetic strategy involves multiple steps procedure, the precise control on the composition and architecture of the final material appears to be difficult to accomplish. Moreover, the problem concerning the toxicity and pharmacokinetic study of these cyclodextrin-based carriers within the body should be investigated because most of the information is based on in vitro cell models. Besides, the comprehensive evaluation of the biodegradability of CD-based supramolecular drug delivery systems is still desired. Nevertheless, the advanced technology which exploits the advantages of cyclodextrin and biocompatible polymers represents a promising and innovative strategy allowing drug carriers design and preparation. Bearing in mind the extraordinary features of the system, further work should be intensified to obtain new biomedical materials for future therapies.

Author Contributions: B.K.—The author of Sections 2 and 5, responsible for the edition of the manuscript, M.B. (Marek Brzeziński)—The author of Sections 1 and 4, the correction of the manuscript, M.S.—The author of Section 3, M.B. (Małgorzata Baśko)—The author of Section 6, the language correction, T.B.—The correction of the manuscript. All authors have read and agreed to the published version of the manuscript.

Funding: Authors acknowledge support from the National Science Center Poland Grant: DEC-2016/23/D/ST5/02458. The funders had no role in the design of the study; in the collection, analyses, or interpretation of data; in the writing of the manuscript, or in the decision to publish the results.

Conflicts of Interest: The authors declare no conflict of interest.

References

1. Yang, L.; Tan, X.; Wang, Z.; Zhang, X. Supramolecular polymers: Historical development, preparation, characterization, and functions. *Chem. Rev.* **2015**, *115*, 7196–7239. [CrossRef] [PubMed]
2. Geng, W.C.; Sessler, J.L.; Guo, D.S. Supramolecular prodrugs based on host-guest interactions. *Chem. Soc. Rev.* **2020**, *49*, 2303–2315. [CrossRef] [PubMed]
3. Hu, Q.D.; Tang, G.P.; Chu, P.K. Cyclodextrin-based host-guest supramolecular nanoparticles for delivery: From design to applications. *Acc. Chem. Res.* **2014**, *47*, 2017–2025. [CrossRef] [PubMed]
4. Zhou, J.; Ritter, H. Cyclodextrin functionalized polymers as drug delivery systems. *Polym. Chem.* **2010**, *1*, 1552–1559. [CrossRef]
5. Rusa, C.C.; Bullions, T.A.; Fox, J.; Porbeni, F.E.; Wang, X.; Tonelli, A.E. Inclusion compound formation with a new columnar cyclodextrin host. *Langmuir* **2002**, *18*, 10016–10023. [CrossRef]
6. Connors, K.A. The stability of cyclodextrin complexes in solution. *Chem. Rev.* **1997**, *97*, 1325–1357. [CrossRef]
7. Takashima, Y.; Osaki, M.; Harada, A. Cyclodextrin-initiated polymerization of cyclic esters in bulk: Formation of polyester-tethered cyclodextrins. *J. Am. Chem. Soc.* **2004**, *126*, 13588–13589. [CrossRef]
8. Faugeras, P.A.; Boëns, B.; Elchinger, P.H.; Brouillette, F.; Montplaisir, D.; Zerrouki, R.; Lucas, R. When cyclodextrins meet click chemistry. *Eur. J. Org. Chem.* **2012**, 4087–4105. [CrossRef]
9. Concheiro, A.; Alvarez-Lorenzo, C. Chemically cross-linked and grafted cyclodextrin hydrogels: From nanostructures to drug-eluting medical devices. *Adv. Drug Deliv. Rev.* **2013**, *65*, 1188–1203. [CrossRef]
10. Yang, J.S.; Yang, L. Preparation and application of cyclodextrin immobilized polysaccharides. *J. Mater. Chem. B* **2013**, *1*, 909–918. [CrossRef]
11. Wang, L.; Li, L.L.; Fan, Y.S.; Wang, H. Host-guest supramolecular nanosystems for cancer diagnostics and therapeutics. *Adv. Mater.* **2013**, *25*, 3888–3898. [CrossRef] [PubMed]
12. Hettiarachchi, G.; Nguyen, D.; Wu, J.; Lucas, D.; Ma, D.; Isaacs, L.; Briken, V. Toxicology and drug delivery by cucurbit[n]uril type molecular containers. *PLoS ONE* **2010**, *5*, 2–11. [CrossRef] [PubMed]
13. Yan, X.; Wang, F.; Zheng, B.; Huang, F. Stimuli-responsive supramolecular polymeric materials. *Chem. Soc. Rev.* **2012**, *41*, 6042–6065. [CrossRef] [PubMed]
14. Cova, T.F.; Murtinho, D.; Pais, A.A.C.C.; Valente, A.J.M. Combining cellulose and cyclodextrins: Fascinating designs for materials and pharmaceutics. *Front. Chem.* **2018**, *6*, 1–19. [CrossRef]
15. Pawar, R.P.; Tekale, S.U.; Shisodia, S.U.; Totre, J.T.; Domb, A.J. Biomedical applications of poly(lactic acid). *Rec. Pat. Regen. Med.* **2014**, *4*, 40–51. [CrossRef]
16. Cheng, Y.; Deng, S.; Chen, P.; Ruan, R. Polylactic acid (PLA) synthesis and modifications: A review. *Front. Chem. China* **2009**, *4*, 259–264. [CrossRef]
17. Xiao, R.Z.; Zeng, Z.W.; Zhou, G.L.; Wang, J.J.; Li, F.Z.; Wang, A.M. Recent advances in PEG-PLA block copolymer nanoparticles. *Int. J. Nanomed.* **2010**, *5*, 1057–1065.

18. Basu, A.; Kunduru, K.R.; Doppalapudi, S.; Domb, A.J.; Khan, W. Poly(lactic acid) based hydrogels. *Adv. Drug Deliv. Rev.* **2016**, *107*, 192–205. [CrossRef]
19. Khodaverdi, E.; Tekie, F.S.M.; Hadizadeh, F.; Esmaeel, H.; Mohajeri, S.A.; Tabassi, S.A.S.; Zohuri, G. Hydrogels composed of cyclodextrin inclusion complexes with PLGA-PEG-PLGA triblock copolymers as drug delivery systems. *AAPS PharmSciTech* **2014**, *15*, 177–188. [CrossRef]
20. Poudel, A.J.; He, F.; Huang, L.; Xiao, L.; Yang, G. Supramolecular hydrogels based on poly (ethylene glycol)-poly (lactic acid) block copolymer micelles and α-cyclodextrin for potential injectable drug delivery system. *Carbohydr. Polym.* **2018**, *194*, 69–79. [CrossRef]
21. Aytac, Z.; Kusku, S.I.; Durgun, E.; Uyar, T. Encapsulation of gallic acid/cyclodextrin inclusion complex in electrospun polylactic acid nanofibers: Release behavior and antioxidant activity of gallic acid. *Mater. Sci. Eng. C* **2016**, *63*, 231–239. [CrossRef] [PubMed]
22. Kost, B.; Svyntkivska, M.; Brzeziński, M.; Makowski, T.; Piorkowska, E.; Rajkowska, K.; Kunicka-Styczyńska, A.; Biela, T. PLA/β-CD-based fibres loaded with quercetin as potential antibacterial dressing materials. *Colloids Surf. B* **2020**, 110949. [CrossRef] [PubMed]
23. Vermet, G.; Degoutin, S.; Chai, F.; Maton, M.; Flores, C.; Neut, C.; Danjou, P.E.; Martel, B.; Blanchemain, N. Cyclodextrin modified PLLA parietal reinforcement implant with prolonged antibacterial activity. *Acta Biomater.* **2017**, *53*, 222–232. [CrossRef] [PubMed]
24. Bucatariu, S.; Constantin, M.; Ascenzi, P.; Fundueanu, G. Poly(lactide-co-glycolide)/cyclodextrin (polyethyleneimine) microspheres for controlled delivery of dexamethasone. *React. Funct. Polym.* **2016**, *107*, 46–53. [CrossRef]
25. Cannavà, C.; Tommasini, S.; Stancanelli, R.; Cardile, V.; Cilurzo, F.; Giannone, I.; Puglisi, G.; Ventura, C.A. Celecoxib-loaded PLGA/cyclodextrin microspheres: Characterization and evaluation of anti-inflammatory activity on human chondrocyte cultures. *Colloids Surf. B* **2013**, *111*, 289–296. [CrossRef] [PubMed]
26. Gupta, V.; Davis, M.; Hope-Weeks, L.J.; Ahsan, F. PLGA microparticles encapsulating prostaglandin E 1-hydroxypropyl-β-cyclodextrin (PGE 1-HPβCD) complex for the treatment of pulmonary arterial hypertension (PAH). *Pharm. Res.* **2011**, *28*, 1733–1749. [CrossRef]
27. Li, F.; Wen, Y.; Zhang, Y.; Zheng, K.; Ban, J.; Xie, Q.; Wen, Y.; Liu, Q.; Chen, F.; Mo, Z.; et al. Characterisation of 2-HP-β-cyclodextrin-PLGA nanoparticle complexes for potential use as ocular drug delivery vehicles. *Artif. Cells, Nanomed. Biotechnol.* **2019**, *47*, 4097–4108. [CrossRef]
28. Li, S.; He, Q.; Chen, T.; Wu, W.; Lang, K.; Li, Z.M.; Li, J. Controlled co-delivery nanocarriers based on mixed micelles formed from cyclodextrin-conjugated and cross-linked copolymers. *Colloids Surf. B Biointerfaces* **2014**, *123*, 486–492. [CrossRef]
29. Zhang, Z.; Lv, Q.; Gao, X.; Chen, L.; Cao, Y.; Yu, S.; He, C.; Chen, X. pH-Responsive poly(ethylene glycol)/poly(L-lactide) supramolecular micelles based on host-guest interaction. *ACS Appl. Mater. Interfaces* **2015**, *7*, 8404–8411. [CrossRef]
30. Aytac, Z.; Keskin, N.O.S.; Tekinay, T.; Uyar, T. Antioxidant α-tocopherol/γ-cyclodextrin–inclusion complex encapsulated poly(lactic acid) electrospun nanofibrous web for food packaging. *J. Appl. Polym. Sci.* **2017**, *134*, 1–9. [CrossRef]
31. Kayaci, F.; Umu, O.C.O.; Tekinay, T.; Uyar, T. Antibacterial electrospun poly(lactic acid) (PLA) nanofibrous webs incorporating triclosan/cyclodextrin inclusion complexes. *J. Agric. Food Chem.* **2013**, *61*, 3901–3908. [CrossRef] [PubMed]
32. Sudakaran, S.V.; Venugopal, J.R.; Vijayakumar, G.P.; Abisegapriyan, S.; Grace, A.N.; Ramakrishna, S. Sequel of MgO nanoparticles in PLACL nanofibers for anti-cancer therapy in synergy with curcumin/β-cyclodextrin. *Mater. Sci. Eng. C* **2017**, *71*, 620–628. [CrossRef] [PubMed]
33. Nilkumhang, S.; Basit, A.W. The robustness and flexibility of an emulsion solvent evaporation method to prepare pH-responsive microparticles. *Int. J. Pharm.* **2009**, *377*, 135–141. [CrossRef] [PubMed]
34. Giri, T.K.; Choudhary, C.; Ajazuddin; Alexander, A.; Badwaik, H.; Tripathi, D.K. Prospects of pharmaceuticals and biopharmaceuticals loaded microparticles prepared by double emulsion technique for controlled delivery. *Saudi Pharm. J.* **2013**, *21*, 125–141. [CrossRef]
35. Malaekeh-nikouei, B.; Sajadi, S.A.; Jaafari, M.R. Preparation and characterization of PLGA microspheres loaded by cyclosporine-cyclodextrin complex. *Iran. J. Pharm. Sci.* **2005**, *1*, 195–201.
36. Brzeziński, M.; Kost, B.; Wedepohl, S.; Socka, M.; Biela, T.; Calderón, M. Stereocomplexed PLA microspheres: Control over morphology, drug encapsulation and anticancer activity. *Colloids Surf. B* **2019**, *184*, 110544. [CrossRef]

37. Uekama, K.; Hieda, Y.; Hirayama, F.; Arima, H.; Sudoh, M.; Yagi, A.; Terashima, H. Stabilizing and solubilizing effects of sulfobutyl ether β-cyclodextrin on prostaglandin E1 analogue. *Pharm. Res.* **2001**, *18*, 1578–1585. [CrossRef]
38. Çirpanli, Y.; Bilensoy, E.; Lale Doğan, A.; Çalış, S. Comparative evaluation of polymeric and amphiphilic cyclodextrin nanoparticles for effective camptothecin delivery. *Eur. J. Pharm. Biopharm.* **2009**, *73*, 82–89. [CrossRef]
39. He, Q.; Wu, W.; Xiu, K.; Zhang, Q.; Xu, F.; Li, J. Controlled drug release system based on cyclodextrin-conjugated poly(lactic acid)-b-poly(ethylene glycol) micelles. *Int. J. Pharm.* **2013**, *443*, 110–119. [CrossRef]
40. Gu, W.X.; Zhu, M.; Song, N.; Du, X.; Yang, Y.W.; Gao, H. Reverse micelles based on biocompatible β-cyclodextrin conjugated polyethylene glycol block polylactide for protein delivery. *J. Mater. Chem. B* **2015**, *3*, 316–322. [CrossRef]
41. Zhang, L.; Lu, J.; Jin, Y.; Qiu, L. Folate-conjugated beta-cyclodextrin-based polymeric micelles with enhanced doxorubicin antitumor efficacy. *Colloids Surf. B* **2014**, *122*, 260–269. [CrossRef]
42. Kost, B.; Brzeziński, M.; Cieślak, M.; Królewska-Golińska, K.; Makowski, T.; Socka, M.; Biela, T. Stereocomplexed micelles based on polylactides with β-cyclodextrin core as anti-cancer drug carriers. *Eur. Polym. J.* **2019**, *120*, 109271. [CrossRef]
43. Gao, H.; Wang, Y.N.; Fan, Y.G.; Ma, J.B. Synthesis of a biodegradable tadpole-shaped polymer via the coupling reaction of polylactide onto mono(6-(2-aminoethyl)amino-6-deoxy)-β- cyclodextrin and its properties as the new carrier of protein delivery system. *J. Control. Release* **2005**, *107*, 158–173. [CrossRef] [PubMed]
44. Gao, H.; Yang, Y.W.; Fan, Y.G.; Ma, J.B. Conjugates of poly(dl-lactic acid) with ethylenediamino or diethylenetriamino bridged bis(β-cyclodextrin)s and their nanoparticles as protein delivery systems. *J. Control. Release* **2006**, *112*, 301–311. [CrossRef] [PubMed]
45. Gao, H.; Wang, Y.-N.; Fan, Y.-G.; Ma, J.-B. Conjugates of poly(DL-lactide-co-glycolide) on amino cyclodextrins and their nanoparticles as protein delivery system. *J. Biomed. Mater. Res. Part A* **2007**, *80A*, 111–122. [CrossRef]
46. Li, W.; Fan, X.; Wang, X.; Shang, X.; Wang, Q.; Lin, J.; Hu, Z.; Li, Z. Stereocomplexed micelle formation through enantiomeric PLA-based Y-shaped copolymer for targeted drug delivery. *Mater. Sci. Eng. C* **2018**, *91*, 688–695. [CrossRef]
47. Jiang, Q.; Zhang, Y.; Zhuo, R.; Jiang, X. Supramolecular host-guest polycationic gene delivery system based on poly(cyclodextrin) and azobenzene-terminated polycations. *Colloids Surf. B* **2016**, *147*, 25–35. [CrossRef]
48. Feng, X.; Ding, J.; Gref, R.; Chen, X. Poly(β-cyclodextrin)-mediated polylactide-cholesterol stereocomplex micelles for controlled drug delivery. *Chin. J. Polym. Sci.* **2017**, *35*, 693–699. [CrossRef]
49. Yi, W.J.; Li, L.J.; He, H.; Hao, Z.; Liu, B.; Chao, Z.S.; Shen, Y. Synthesis of poly(l-lactide)/β-cyclodextrin/citrate network modified hydroxyapatite and its biomedical properties. *New J. Chem.* **2018**, *42*, 14729–14732. [CrossRef]
50. Woodruff, M.A.; Hutmacher, D.W. The return of a forgotten polymer-Polycaprolactone in the 21st century. *Prog. Polym. Sci.* **2010**, *35*, 1217–1256. [CrossRef]
51. Labet, M.; Thielemans, W. Synthesis of polycaprolactone: A review. *Chem. Soc. Rev.* **2009**, *38*, 3484–3504. [CrossRef] [PubMed]
52. Sinha, V.R.; Bansal, K.; Kaushik, R.; Kumria, R.; Trehan, A. Poly-ε-caprolactone microspheres and nanospheres: An overview. *Int. J. Pharm.* **2004**, *278*, 1–23. [CrossRef]
53. Dash, T.K.; Konkimalla, V.B. Poly-ε-caprolactone based formulations for drug delivery and tissue engineering: A review. *J. Control. Release* **2012**, *158*, 15–33. [CrossRef] [PubMed]
54. Lecomte, P.; Riva, R.; Schmeits, S.; Rieger, J.; Van Butsele, K.; Jérôme, C.; Jérôme, R. New prospects for the grafting of functional groups onto aliphatic polyesters. ring-opening polymerization of α- or γ-substituted ε-caprolactone followed by chemical derivatization of the substituents. *Macromol. Symp.* **2006**, *240*, 157–165. [CrossRef]
55. Seyednejad, H.; Ghassemi, A.H.; van Nostrum, C.F.; Vermonden, T.; Hennink, W.E. Functional aliphatic polyesters for biomedical and pharmaceutical applications. *J. Control. Release* **2011**, *152*, 168–176. [CrossRef]
56. Parrish, B.; Breitenkamp, R.B.; Emrick, T. PEG- and peptide-grafted aliphatic polyesters by click chemistry. *J. Am. Chem. Soc.* **2005**, *127*, 7404–7410. [CrossRef]
57. Hu, Y.; Zhang, L.; Cao, Y.; Ge, H.; Jiang, X.; Yang, C. Degradation behavior of poly(ε-caprolactone)-b-poly(ethylene glycol)-b-poly(ε-caprolactone) micelles in aqueous solution. *Biomacromolecules* **2004**, *5*, 1756–1762. [CrossRef]
58. Dorj, B.; Kim, M.-K.; Won, J.-E.; Kim, H.-W. Functionalization of poly(caprolactone) scaffolds by the surface mineralization for use in bone regeneration. *Mater. Lett.* **2011**, *65*, 3559–3562. [CrossRef]

59. Sousa, I.; Mendes, A.; Pereira, R.F.; Bártolo, P.J. Collagen surface modified poly(ε-caprolactone) scaffolds with improved hydrophilicity and cell adhesion properties. *Mater. Lett.* **2014**, *134*, 263–267. [CrossRef]
60. Kawaguchi, Y.; Nishiyama, T.; Okada, M.; Kamachi, M.; Harada, A. Complex formation of poly(ε-caprolactone) with cyclodextrins. *Macromolecules* **2000**, *33*, 4472–4477. [CrossRef]
61. Silva, C.V.N.S.; Barbosa, J.A.P.; Ferraz, M.S.; Silva, N.H.; Honda, N.K.; Rabello, M.M.; Hernandes, M.Z.; Bezerra, B.P.; Cavalcanti, I.M.F.; Ayala, A.P.; et al. Molecular modeling and cytotoxicity of diffractaic acid: HP-β-CD inclusion complex encapsulated in microspheres. *Int. J. Biol. Macromol.* **2016**, *92*, 494–503. [CrossRef] [PubMed]
62. Puglisi, A.; Bayir, E.; Timur, S.; Yagci, Y. pH-Responsive polymersome microparticles as smart cyclodextrin-releasing agents. *Biomacromolecules* **2019**, *20*, 4001–4007. [CrossRef] [PubMed]
63. Li, Y.; Chen, Y.; Dong, H.; Dong, C. Supramolecular, prodrug-based micelles with enzyme-regulated release behavior for controlled drug delivery. *Med. Chem. Comm.* **2015**, *6*, 1874–1881. [CrossRef]
64. Wei, X.; Lv, X.; Zhao, Q.; Qiu, L. Thermosensitive β-cyclodextrin modified poly(ε-caprolactone)- poly(ethylene glycol)-poly(ε-caprolactone) micelles prolong the anti-inflammatory effect of indomethacin following local injection. *Acta Biomater.* **2013**, *9*, 6953–6963. [CrossRef]
65. Li, J.; Li, X.; Liu, H.; Ren, T.; Huang, L.; Deng, Z.; Yang, Y.; Zhong, S. GSH and light dual stimuli-responsive supramolecular polymer drug carriers for cancer therapy. *Polym. Degrad. Stab.* **2019**, *168*, 108956. [CrossRef]
66. Xu, S.; Yin, L.; Xiang, Y.; Deng, H.; Deng, L.; Fan, H.; Tang, H.; Zhang, J.; Dong, A. Supramolecular hydrogel from nanoparticles and cyclodextrins for local and sustained nanoparticle delivery. *Macromol. Biosci.* **2016**, 1188–1199. [CrossRef] [PubMed]
67. Narayanan, G.; Shen, J.; Boy, R.; Gupta, B.S.; Tonelli, A.E. Aliphatic polyester nanofibers functionalized with cyclodextrins and cyclodextrin-guest inclusion complexes. *Polymers* **2018**, *10*, 428. [CrossRef]
68. Aytac, Z.; Sen, H.S.; Durgun, E.; Uyar, T. Sulfisoxazole/cyclodextrin inclusion complex incorporated in electrospun hydroxypropyl cellulose nanofibers as drug delivery system. *Colloids Surf. B* **2015**, *128*, 331–338. [CrossRef]
69. Zhang, Z.; He, Z.; Liang, R.; Ma, Y.; Huang, W.; Jiang, R.; Shi, S.; Chen, H.; Li, X. Fabrication of a micellar supramolecular hydrogel for ocular drug delivery. *Biomacromolecules* **2016**, *17*, 798–807. [CrossRef]
70. Hu, J.; Zhang, M.; He, J.; Ni, P. Injectable hydrogels by inclusion complexation between a three-armed star copolymer (mPEG-acetal-PCL-acetal-)3 and α-cyclodextrin for pH-triggered drug delivery. *RSC Adv.* **2016**, *6*, 40858–40868. [CrossRef]
71. Aytac, Z.; Uyar, T. Antioxidant activity and photostability of α-tocopherol/β-cyclodextrin inclusion complex encapsulated electrospun polycaprolactone nanofibers. *Eur. Polym. J.* **2016**, *79*, 140–149. [CrossRef]
72. Canbolat, M.F.; Celebioglu, A.; Uyar, T. Drug delivery system based on cyclodextrin-naproxen inclusion complex incorporated in electrospun polycaprolactone nanofibers. *Colloids Surf. B* **2014**, *115*, 15–21. [CrossRef] [PubMed]
73. Masoumi, S.; Amiri, S.; Bahrami, S.H. PCL-based nanofibers loaded with ciprofloxacin/cyclodextrin containers. *J. Text. Inst.* **2018**, *109*, 1044–1053. [CrossRef]
74. Canbolat, M.F.; Savas, H.B.; Gultekin, F. Improved catalytic activity by catalase immobilization using γ-cyclodextrin and electrospun PCL nanofibers. *J. Appl. Polym. Sci.* **2017**, *134*, 44404. [CrossRef]
75. Canbolat, M.F.; Savas, H.B.; Gultekin, F. Enzymatic behavior of laccase following interaction with γ-CD and immobilization into PCL nanofibers. *Anal. Biochem.* **2017**, *528*, 13–18. [CrossRef] [PubMed]
76. Souza, S.O.L.; Cotrim, M.A.P.; Oréfice, R.L.; Carvalho, S.G.; Dutra, J.A.P.; de Paula Careta, F.; Resende, J.A.; Villanova, J.C.O. Electrospun poly(ε-caprolactone) matrices containing silver sulfadiazine complexed with β-cyclodextrin as a new pharmaceutical dosage form to wound healing: Preliminary physicochemical and biological evaluation. *J. Mater. Sci. Mater. Med.* **2018**, *29*, 67. [CrossRef]
77. Monteiro, A.P.F.; Rocha, C.M.S.L.; Oliveira, M.F.; Gontijo, S.M.L.; Agudelo, R.R.; Sinisterra, R.D.; Cortés, M.E. Nanofibers containing tetracycline/β-cyclodextrin: Physico-chemical characterization and antimicrobial evaluation. *Carbohydr. Polym.* **2017**, *156*, 417–426. [CrossRef] [PubMed]
78. Varan, C.; Bilensoy, E. Development of implantable hydroxypropyl-β-cyclodextrin coated polycaprolactone nanoparticles for the controlled delivery of docetaxel to solid tumors. *J. Incl. Phenom. Macrocycl. Chem.* **2014**, *80*, 9–15. [CrossRef]

79. Varan, C.; Wickström, H.; Sandler, N.; Aktaş, Y.; Bilensoy, E. Inkjet printing of antiviral PCL nanoparticles and anticancer cyclodextrin inclusion complexes on bioadhesive film for cervical administration. *Int. J. Pharm.* **2017**, *531*, 701–713. [CrossRef]
80. Ahmed, A.; Wang, H.; Yu, H.; Zhou, Z.Y.; Ding, Y.; Hu, Y. Surface engineered cyclodextrin embedded polymeric nanoparticles through host-guest interaction used for drug delivery. *Chem. Eng. Sci.* **2015**, *125*, 121–128. [CrossRef]
81. Kuplennik, N.; Sosnik, A. Enhanced nanoencapsulation of sepiapterin within PEG-PCL nanoparticles by complexation with triacetyl-beta cyclodextrin. *Molecules* **2019**, *24*, 2715. [CrossRef] [PubMed]
82. Gao, Y.; Li, G.; Zhou, Z.; Guo, L.; Liu, X. Supramolecular assembly of poly(β-cyclodextrin) block copolymer and benzimidazole-poly(ε-caprolactone) based on host-guest recognition for drug delivery. *Colloids Surf. B* **2017**, *160*, 364–371. [CrossRef]
83. Li, X.; Liu, H.; Li, J.; Deng, Z.; Li, L.; Liu, J.; Yuan, J.; Gao, P.; Yang, Y.; Zhong, S. Micelles via self-assembly of amphiphilic beta-cyclodextrin block copolymers as drug carrier for cancer therapy. *Colloids Surf. B* **2019**, *183*, 110425. [CrossRef] [PubMed]
84. Peng, L.; Wang, Z.; Feng, A.; Huo, M.; Fang, T.; Wang, K.; Wei, Y.; Yuan, J. Star amphiphilic supramolecular copolymer based on host-guest interaction for electrochemical controlled drug delivery. *Polymer* **2016**, *88*, 112–122. [CrossRef]
85. Zuo, C.; Peng, J.; Cong, Y.; Dai, X.; Zhang, X.; Zhao, S.; Zhang, X.; Ma, L.; Wang, B.; Wei, H. Fabrication of supramolecular star-shaped amphiphilic copolymers for ROS-triggered drug release. *J. Colloid Interface Sci.* **2018**, *514*, 122–131. [CrossRef]
86. Khodaverdi, E.; Heidari, Z.; Tabassi, S.A.S.; Tafaghodi, M.; Alibolandi, M.; Tekie, F.S.M.; Khameneh, B.; Hadizadeh, F. Injectable supramolecular hydrogel from insulin-loaded triblock PCL-PEG-PCL copolymer and γ-cyclodextrin with sustained-release property. *AAPS PharmSciTech* **2014**, *16*, 140–149. [CrossRef]
87. Tabassi, S.A.S.; Tekie, F.S.M.; Hadizadeh, F.; Rashid, R.; Khodaverdi, E.; Mohajeri, S.A. Sustained release drug delivery using supramolecular hydrogels of the triblock copolymer PCL-PEG-PCL and α-cyclodextrin. *J. Sol-Gel Sci. Technol.* **2014**, *69*, 166–171. [CrossRef]
88. Zhu, W.; Li, Y.; Liu, L.; Chen, Y.; Xi, F. Supramolecular hydrogels as a universal scaffold for stepwise delivering Dox and Dox/cisplatin loaded block copolymer micelles. *Int. J. Pharm.* **2012**, *437*, 11–19. [CrossRef]
89. Yin, L.; Xu, S.; Feng, Z.; Deng, H.; Zhang, J.; Gao, H.; Deng, L.; Tang, H.; Dong, A. Supramolecular hydrogel based on high-solid-content mPECT nanoparticles and cyclodextrins for local and sustained drug delivery. *Biomater. Sci.* **2017**, *5*, 698–706. [CrossRef]
90. Wang, T.; Jiang, X.-J.; Lin, T.; Ren, S.; Li, X.-Y.; Zhang, X.-Z.; Tang, Q. The inhibition of postinfarct ventricle remodeling without polycythaemia following local sustained intramyocardial delivery of erythropoietin within a supramolecular hydrogel. *Biomaterials* **2009**, *30*, 4161–4167. [CrossRef]
91. Li, Z.; Yin, H.; Zhang, Z.; Li Liu, K.; Li, J. Supramolecular anchoring of DNA polyplexes in cyclodextrin-based polypseudorotaxane hydrogels for sustained gene delivery. *Biomacromolecules* **2012**, *13*, 3162–3172. [CrossRef] [PubMed]
92. Khodaverdi, E.; Gharechahi, M.; Alibolandi, M.; Tekie, F.M.; Khashyarmanesh, B.; Hadizadeh, F. Self-assembled supramolecular hydrogel based on PCL-PEG-PCL triblock copolymer and γ-cyclodextrin inclusion complex for sustained delivery of dexamethasone. *Int. J. Pharm. Investig.* **2016**, *6*, 78–85. [PubMed]
93. Conte, C.; Ungaro, F.; Maglio, G.; Tirino, P.; Siracusano, G.; Sciortino, M.T.; Leone, N.; Palma, G.; Barbieri, A.; Arra, C.; et al. Biodegradable core-shell nanoassemblies for the delivery of docetaxel and Zn(II)-phthalocyanine inspired by combination therapy for cancer. *J. Control. Release* **2013**, *167*, 40–52. [CrossRef] [PubMed]
94. Park, J.; Ye, M.; Park, K. Biodegradable polymers for microencapsulation of drugs. *Molecules* **2005**, *10*, 146–161. [CrossRef]
95. O'Donnell, P.B.; McGinity, J.W. Preparation of microspheres by the solvent evaporation technique. *Adv. Drug Deliv. Rev.* **1997**, *28*, 25–42. [CrossRef]
96. MacEwan, S.R.; Chilkoti, A. From composition to cure: A systems engineering approach to anticancer drug carriers. *Angew. Chemie Int. Ed.* **2017**, *56*, 6712–6733. [CrossRef]
97. Simões, S.M.N.; Rey-Rico, A.; Concheiro, A.; Alvarez-Lorenzo, C. Supramolecular cyclodextrin-based drug nanocarriers. *Chem. Commun.* **2015**, *51*, 6275–6289. [CrossRef]

98. Oster, M.; Schlatter, G.; Gallet, S.; Baati, R.; Pollet, E.; Gaillard, C.; Avérous, L.; Fajolles, C.; Hébraud, A. The study of the pseudo-polyrotaxane architecture as a route for mild surface functionalization by click chemistry of poly(ε-caprolactone)-based electrospun fibers. *J. Mater. Chem. B* **2017**, *5*, 2181–2189. [CrossRef]
99. Narayanan, G.; Gupta, B.S.; Tonelli, A.E. Poly(ε-caprolactone) Nanowebs Functionalized with α- and γ-Cyclodextrins. *Biomacromolecules* **2014**, *15*, 4122–4133. [CrossRef]
100. Narayanan, G.; Ormond, B.R.; Gupta, B.S.; Tonelli, A.E. Efficient wound odor removal by β-cyclodextrin functionalized poly (ε-caprolactone) nanofibers. *J. Appl. Polym. Sci.* **2015**, *132*, 42782. [CrossRef]
101. Narayanan, G.; Aguda, R.; Hartman, M.; Chung, C.-C.; Boy, R.; Gupta, B.S.; Tonelli, A.E. Fabrication and characterization of poly(ε-caprolactone)/α-cyclodextrin pseudorotaxane nanofibers. *Biomacromolecules* **2015**, *17*, 271–279. [CrossRef] [PubMed]
102. Zhan, J.; Singh, A.; Zhang, Z.; Huang, L.; Elisseeff, J.H. Multifunctional aliphatic polyester nanofibers for tissue engineering. *Biomatter* **2012**, *2*, 202–212. [CrossRef] [PubMed]
103. Panday, R.; Poudel, A.J.; Li, X.; Adhikari, M.; Ullah, M.W.; Yang, G. Amphiphilic core-shell nanoparticles: Synthesis, biophysical properties, and applications. *Colloids Surf. B* **2018**, *172*, 68–81. [CrossRef] [PubMed]
104. Boarca, B.; Lungu, I.I.; Holban, A.M. Core–shell nanomaterials for infection and cancer therapy. In *Materials for Biomedical Engineering*; Elsevier: Amsterdam, The Netherlands, 2019; pp. 197–211.
105. Chen, G.; Wang, Y.; Xie, R.; Gong, S. A review on core–shell structured unimolecular nanoparticles for biomedical applications. *Adv. Drug Deliv. Rev.* **2018**, *130*, 58–72. [CrossRef] [PubMed]
106. Zhao, J.; Weng, G.; Li, J.; Zhu, J.; Zhao, J. Polyester-based nanoparticles for nucleic acid delivery. *Mater. Sci. Eng. C* **2018**, *92*, 983–994. [CrossRef]
107. Yi, Y.; Lin, G.; Chen, S.; Liu, J.; Zhang, H.; Mi, P. Polyester micelles for drug delivery and cancer theranostics: Current achievements, progresses and future perspectives. *Mater. Sci. Eng. C* **2018**, *83*, 218–232. [CrossRef]
108. Alexander, A.; Dwivedi, S.; Ajazuddin; Giri, T.K.; Saraf, S.; Saraf, S.; Tripathi, D.K. Approaches for breaking the barriers of drug permeation through transdermal drug delivery. *J. Control. Release* **2012**, *164*, 26–40. [CrossRef]
109. Conte, C.; Costabile, G.; d'Angelo, I.; Pannico, M.; Musto, P.; Grassia, G.; Ialenti, A.; Tirino, P.; Miro, A.; Ungaro, F.; et al. Skin transport of PEGylated poly(ε-caprolactone) nanoparticles assisted by (2-hydroxypropyl)-β-cyclodextrin. *J. Colloid Interface Sci.* **2015**, *454*, 112–120. [CrossRef]
110. Lu, Y.; Zou, H.; Yuan, H.; Gu, S.; Yuan, W.; Li, M. Triple stimuli-responsive supramolecular assemblies based on host-guest inclusion complexation between β-cyclodextrin and azobenzene. *Eur. Polym. J.* **2017**, *91*, 396–407. [CrossRef]
111. Liu, X.; Chen, B.; Li, X.; Zhang, L.; Xu, Y.; Liu, Z.; Cheng, Z.; Zhu, X. Self-assembly of BODIPY based pH-sensitive near-infrared polymeric micelles for drug controlled delivery and fluorescence imaging applications. *Nanoscale* **2015**, *7*, 16399–16416. [CrossRef]
112. Cameron, D.J.A.; Shaver, M.P. Aliphatic polyester polymer stars: Synthesis, properties and applications in biomedicine and nanotechnology. *Chem. Soc. Rev.* **2011**, *40*, 1761–1776. [CrossRef] [PubMed]
113. Yin, H.; Kang, S.-W.; Han Bae, Y. Polymersome formation from AB2 type 3-miktoarm star copolymers. *Macromolecules* **2009**, *42*, 7456–7464. [CrossRef]
114. Wang, F.; Bronich, T.K.; Kabanov, A.V.; David Rauh, R.; Roovers, J. Synthesis and characterization of star poly(ε-caprolactone)-b-poly(ethylene glycol) and poly(l-lactide)-b-poly(ethylene glycol) copolymers: Evaluation as drug delivery carriers. *Bioconjug. Chem.* **2008**, *19*, 1423–1429. [CrossRef]
115. Gou, P.-F.; Zhu, W.-P.; Xu, N.; Shen, Z.-Q. Synthesis and self-assembly of well-defined cyclodextrin-centered amphiphilic A14B7 multimiktoarm star copolymers based on poly(ε-caprolactone) and poly(acrylic acid). *J. Polym. Sci. Part A Polym. Chem.* **2010**, *48*, 2961–2974. [CrossRef]
116. Gou, P.-F.; Zhu, W.-P.; Shen, Z.-Q. Synthesis, self-assembly, and drug-loading capacity of well-defined cyclodextrin-centered drug-conjugated amphiphilic A14B7 miktoarm star copolymers based on poly(ε-caprolactone) and poly(ethylene glycol). *Biomacromolecules* **2010**, *11*, 934–943. [CrossRef] [PubMed]
117. Yasen, W.; Dong, R.; Zhou, L.; Wu, J.; Cao, C.; Aini, A.; Zhu, X. Synthesis of a cationic supramolecular block copolymer with covalent and noncovalent polymer blocks for gene delivery. *ACS Appl. Mater. Interfaces* **2017**, *9*, 9006–9014. [CrossRef] [PubMed]
118. Fu, C.; Lin, X.; Wang, J.; Zheng, X.; Li, X.; Lin, Z.; Lin, G. Injectable micellar supramolecular hydrogel for delivery of hydrophobic anticancer drugs. *J. Mater. Sci. Mater. Med.* **2016**, *27*, 73. [CrossRef] [PubMed]

119. Wu, D.-Q.; Wang, T.; Lu, B.; Xu, X.-D.; Cheng, S.-X.; Jiang, X.-J.; Zhang, X.-Z.; Zhuo, R.-X. Fabrication of supramolecular hydrogels for drug delivery and stem cell encapsulation. *Langmuir* **2008**, *24*, 10306–10312. [CrossRef]
120. Veronese, F.M.; Mero, A. The impact of PEGylation on biological therapies. *BioDrugs* **2008**, *22*, 315–329. [CrossRef]
121. Milton Harris, J.; Chess, R.B. Effect of pegylation on pharmaceuticals. *Nat. Rev. Drug Discov.* **2003**, *2*, 214–221. [CrossRef]
122. Joralemon, M.J.; McRae, S.; Emrick, T. PEGylated polymers for medicine: From conjugation to self-assembled systems. *Chem. Commun.* **2010**, *46*, 1377–1393. [CrossRef] [PubMed]
123. Rojas-Aguirre, Y.; Torres-Mena, M.A.; López-Méndez, L.J.; Alcaraz-Estrada, S.L.; Guadarrama, P.; Urucha-Ortíz, J.M. PEGylated β-cyclodextrins: Click synthesis and in vitro biological insights. *Carbohydr. Polym.* **2019**, *223*, 115113. [CrossRef] [PubMed]
124. Liang, L.; Astruc, D. The copper(I)-catalyzed alkyne-azide cycloaddition (CuAAC) "click" reaction and its applications. An overview. *Coord. Chem. Rev.* **2011**, *255*, 2933–2945. [CrossRef]
125. Gaetke, L.M.; Chow-Johnson, H.S.; Chow, C.K. Copper: Toxicological relevance and mechanisms. *Arch. Toxicol.* **2014**, *88*, 1929–1938. [CrossRef] [PubMed]
126. Hashidzume, A.; Tomatsu, I.; Harada, A. Interaction of cyclodextrins with side chains of water soluble polymers: A simple model for biological molecular recognition and its utilization for stimuli-responsive systems. *Polymer* **2006**, *47*, 6011–6027. [CrossRef]
127. Antoniuk, I.; Plazzotta, B.; Wintgens, V.; Volet, G.; Nielsen, T.T.; Pedersen, J.S.; Amiel, C. Host–guest interaction and structural ordering in polymeric nanoassemblies: Influence of molecular design. *Int. J. Pharm.* **2017**, *531*, 433–443. [CrossRef]
128. Huang, F.; Gibson, H.W. Polypseudorotaxanes and polyrotaxanes. *Prog. Polym. Sci.* **2005**, *30*, 982–1018. [CrossRef]
129. Shinohara, K.; Yamashita, M.; Uchida, W.; Okabe, C.; Oshima, S.; Sugino, M.; Egawa, Y.; Miki, R.; Hosoya, O.; Fujihara, T.; et al. Preparation of polypseudorotaxanes composed of cyclodextrin and polymers in microspheres. *Chem. Pharm. Bull.* **2014**, *62*, 962–966. [CrossRef]
130. Moon, C.; Kwon, Y.M.; Lee, W.K.; Park, Y.J.; Yang, V.C. In vitro assessment of a novel polyrotaxane-based drug delivery system integrated with a cell-penetrating peptide. *J. Control. Release* **2007**, *124*, 43–50. [CrossRef]
131. Xu, T.; Li, J.; Cao, J.; Gao, W.; Li, L.; He, B. The effect of α-cyclodextrin on poly(pseudo)rotaxane nanoparticles self-assembled by protoporphyrin modified poly(ethylene glycol) for anticancer drug delivery. *Carbohydr. Polym.* **2017**, *174*, 789–797. [CrossRef]
132. Yang, C.; Qin, Y.; Tu, K.; Xu, C.; Li, Z.; Zhang, Z. Star-shaped polymer of β-cyclodextrin-g-vitamin E TPGS for doxorubicin delivery and multidrug resistance inhibition. *Colloids Surf. B* **2018**, *169*, 10–19. [CrossRef] [PubMed]
133. Fan, W.; Xu, Y.; Li, Z.; Li, Q. Folic acid-modified β-cyclodextrin nanoparticles as drug delivery to load DOX for liver cancer therapeutics. *Soft Mater.* **2019**, *17*, 437–447. [CrossRef]
134. Hyun, H.; Lee, S.; Lim, W.; Jo, D.; Jung, J.S.; Jo, G.; Kim, S.Y.; Lee, D.-W.; Um, S.; Yang, D.H.; et al. Engineered beta-cyclodextrin-based carrier for targeted doxorubicin delivery in breast cancer therapy in vivo. *J. Ind. Eng. Chem.* **2019**, *70*, 145–151. [CrossRef]
135. Wang, Y.; Wang, H.; Chen, Y.; Liu, X.; Jin, Q.; Ji, J. PH and hydrogen peroxide dual responsive supramolecular prodrug system for controlled release of bioactive molecules. *Colloids Surf. B* **2014**, *121*, 189–195. [CrossRef]
136. Chen, X.; Yao, X.; Wang, C.; Chen, L.; Chen, X. Hyperbranched PEG-based supramolecular nanoparticles for acid-responsive targeted drug delivery. *Biomater. Sci.* **2015**, *3*, 870–878. [CrossRef] [PubMed]
137. Xiong, Q.; Cui, M.; Yu, G.; Wang, J.; Song, T. Facile fabrication of reduction-responsive supramolecular nanoassemblies for co-delivery of doxorubicin and sorafenib toward hepatoma cells. *Front. Pharmacol.* **2018**, *9*, 1–11. [CrossRef]
138. Gérard Yaméogo, J.B.; Mazet, R.; Wouessidjewe, D.; Choisnard, L.; Godin-Ribuot, D.; Putaux, J.L.; Semdé, R.; Gèze, A. Pharmacokinetic study of intravenously administered artemisinin-loaded surface-decorated amphiphilic γ-cyclodextrin nanoparticles. *Mater. Sci. Eng. C* **2020**, *106*, 110281. [CrossRef]
139. Feng, R.; Deng, P.; Zhou, F.; Feng, S.; Song, Z. Pluronic F127-cyclodextrin conjugate micelles for encapsulation of honokiol. *J. Nanoparticle Res.* **2018**, *20*, 261. [CrossRef]

140. Liao, R.; Yi, S.; Liu, M.; Jin, W.; Yang, B. Folic-acid-targeted self-assembling supramolecular carrier for gene delivery. *ChemBioChem* **2015**, *16*, 1622–1628. [CrossRef]
141. Mohammed, A.F.A.; Higashi, T.; Motoyama, K.; Ohyama, A.; Onodera, R.; Khaled, K.A.; Sarhan, H.A.; Hussein, A.K.; Arima, H. In vitro and in vivo co-delivery of siRNA and doxorubicin by folate-PEG-appended dendrimer/glucuronylglucosyl-β-cyclodextrin conjugate. *AAPS J.* **2019**, *21*, 1–10. [CrossRef]
142. Yu, Y.; Chen, C.K.; Law, W.C.; Weinheimer, E.; Sengupta, S.; Prasad, P.N.; Cheng, C. Polylactide-graft-doxorubicin nanoparticles with precisely controlled drug loading for pH-triggered drug delivery. *Biomacromolecules* **2014**, *15*, 524–532. [CrossRef]
143. Tran, T.V.; Vo, U.V.; Pham, D.Y.; Tran, D.L.; Nguyen, T.H.; Tran, N.Q.; Nguyen, C.K.; Thu, L.V.; Nguyen, D.H. Supramolecular chemistry at interfaces: Host-guest interactions for attaching PEG and 5-fluorouracil to the surface of porous nanosilica. *Green Process. Synth.* **2016**, *5*, 521–528. [CrossRef]
144. Nguyen Thi, T.T.; Tran, T.V.; Tran, N.Q.; Nguyen, C.K.; Nguyen, D.H. Hierarchical self-assembly of heparin-PEG end-capped porous silica as a redox sensitive nanocarrier for doxorubicin delivery. *Mater. Sci. Eng. C* **2017**, *70*, 947–954. [CrossRef]
145. Sawant, V.J.; Bamane, S.R. PEG-beta-cyclodextrin functionalized zinc oxide nanoparticles show cell imaging with high drug payload and sustained pH responsive delivery of curcumin in to MCF-7 cells. *J. Drug Deliv. Sci. Technol.* **2018**, *43*, 397–408. [CrossRef]
146. Klaewklod, A.; Tantishaiyakul, V.; Hirun, N.; Sangfai, T.; Li, L. Characterization of supramolecular gels based on β-cyclodextrin and polyethyleneglycol and their potential use for topical drug delivery. *Mater. Sci. Eng. C* **2015**, *50*, 242–250. [CrossRef] [PubMed]
147. Li, J.; Li, X.; Ni, X.; Wang, X.; Li, H.; Leong, K.W. Self-assembled supramolecular hydrogels formed by biodegradable PEO-PHB-PEO triblock copolymers and α-cyclodextrin for controlled drug delivery. *Biomaterials* **2006**, *27*, 4132–4140. [CrossRef] [PubMed]
148. Salmaso, S.; Semenzato, A.; Bersani, S.; Matricardi, P.; Rossi, F.; Caliceti, P. Cyclodextrin/PEG based hydrogels for multi-drug delivery. *Int. J. Pharm.* **2007**, *345*, 42–50. [CrossRef]
149. Wang, J.; Williamson, G.S.; Yang, H. Branched polyrotaxane hydrogels consisting of alpha-cyclodextrin and low-molecular-weight four-arm polyethylene glycol and the utility of their thixotropic property for controlled drug release. *Colloids Surf. B* **2018**, *165*, 144–149. [CrossRef]
150. Kuang, H.; He, H.; Zhang, Z.; Qi, Y.; Xie, Z.; Jing, X.; Huang, Y. Injectable and biodegradable supramolecular hydrogels formed by nucleobase-terminated poly(ethylene oxide)s and α-cyclodextrin. *J. Mater. Chem. B* **2014**, *2*, 659–667. [CrossRef]
151. Yamaguchi, S.; Higashi, K.; Azuma, T.; Okamoto, A. Supramolecular polymeric hydrogels for ultrasound-guided protein release. *Biotechnol. J.* **2019**, *14*, 1–7. [CrossRef]
152. Khan, S.; Minhas, M.U.; Ahmad, M.; Sohail, M. Self-assembled supramolecular thermoreversible β-cyclodextrin/ethylene glycol injectable hydrogels with difunctional Pluronic®127 as controlled delivery depot of curcumin. Development, characterization and in vitro evaluation. *J. Biomater. Sci. Polym. Ed.* **2018**, *29*, 1–34. [CrossRef] [PubMed]
153. Yu, J.; Ha, W.; Chen, J.; Shi, Y.P. PH-Responsive supramolecular hydrogels for codelivery of hydrophobic and hydrophilic anticancer drugs. *RSC Adv.* **2014**, *4*, 58982–58989. [CrossRef]
154. Müller, C.; Schubiger, P.A.; Schibli, R. In vitro and in vivo targeting of different folate receptor-positive cancer cell lines with a novel 99mTc-radiofolate tracer. *Eur. J. Nucl. Med. Mol. Imaging* **2006**, *33*, 1162–1170. [CrossRef]
155. Bao, Y.; Yin, M.; Hu, X.; Zhuang, X.; Sun, Y.; Guo, Y.; Tan, S.; Zhang, Z. A safe, simple and efficient doxorubicin prodrug hybrid micelle for overcoming tumor multidrug resistance and targeting delivery. *J. Control. Release* **2016**, *235*, 182–194. [CrossRef]
156. Zhu, D.; Tao, W.; Zhang, H.; Liu, G.; Wang, T.; Zhang, L.; Zeng, X.; Mei, L. Docetaxel (DTX)-loaded polydopamine-modified TPGS-PLA nanoparticles as a targeted drug delivery system for the treatment of liver cancer. *Acta Biomater.* **2016**, *30*, 144–154. [CrossRef] [PubMed]
157. Brzeziński, M.; Wedepohl, S.; Kost, B.; Calderón, M. Nanoparticles from supramolecular polylactides overcome drug resistance of cancer cells. *Eur. Polym. J.* **2018**, *109*, 117–123. [CrossRef]
158. Caló, E.; Khutoryanskiy, V.V. Biomedical applications of hydrogels: A review of patents and commercial products. *Eur. Polym. J.* **2015**, *65*, 252–267. [CrossRef]

159. Wang, W.; Zhang, Y.; Liu, W. Bioinspired fabrication of high strength hydrogels from non-covalent interactions. *Prog. Polym. Sci.* **2017**, *71*, 1–25. [CrossRef]
160. Klaewklod, A.; Tantishaiyakul, V.; Sangfai, T.; Hirun, N.; Rugmai, S. Chemometric and experimental investigations of organogelation based on β-cyclodextrin. *Adv. Mater. Res.* **2014**, *1060*, 133–136. [CrossRef]
161. Van de Manakker, F.; Van der Pot, M.; Vermonden, T.; Van Nostrum, C.F.; Hennink, W.E. Self-assembling hydrogels based on β-cyclodextrin/cholesterol inclusion complexes. *Macromolecules* **2008**, *41*, 1766–1773. [CrossRef]
162. Van Manakker, F.D.; Braeckmans, K.; Morabit, N.E.; De Smedt, S.C.; Van Nostrum, C.F.; Hennink, W.E. Protein-release behavior of self-assembled PEG-ß-cyclodextrin/PEG- cholesterol hydrogels. *Adv. Funct. Mater.* **2009**, *19*, 2992–3001. [CrossRef]
163. Eliasof, S.; Lazarus, D.; Peters, C.G.; Case, R.I.; Cole, R.O.; Hwang, J.; Schluep, T.; Chao, J.; Lin, J.; Yen, Y.; et al. Correlating preclinical animal studies and human clinical trials of a multifunctional, polymeric nanoparticle. *Proc. Natl. Acad. Sci. USA* **2013**, *110*, 15127–15132. [CrossRef] [PubMed]
164. Ren, Y.; Bai, Y.; Zhang, Z.; Cai, W.; Del Rio Flores, A. The preparation and structure analysis methods of natural polysaccharides of plants and fungi: A review of recent development. *Molecules* **2019**, *24*, 3122. [CrossRef] [PubMed]
165. Gopinath, V.; Saravanan, S.; Al-Maleki, A.R.; Ramesh, M.; Vadivelu, J. A review of natural polysaccharides for drug delivery applications: Special focus on cellulose, starch and glycogen. *Biomed. Pharmacother.* **2018**, *107*, 96–108. [CrossRef] [PubMed]
166. Fan, S.; Zhang, J.; Nie, W.; Zhou, W.; Jin, L.; Chen, X.; Lu, J. Antitumor effects of polysaccharide from Sargassum fusiforme against human hepatocellular carcinoma HepG2 cells. *Food Chem. Toxicol.* **2017**, *102*, 53–62. [CrossRef]
167. Faccin-Galhardi, L.C.; Aimi Yamamoto, K.; Ray, S.; Ray, B.; Carvalho Linhares, R.E.; Nozawa, C. The in vitro antiviral property of Azadirachta indica polysaccharides for poliovirus. *J. Ethnopharmacol.* **2012**, *142*, 86–90. [CrossRef]
168. Olafsdottir, E.S.; Ingólfsdottir, K. Polysaccharides from lichens: Structural characteristics and biological activity. *Planta Med.* **2001**, *67*, 199–208. [CrossRef]
169. Yu, Y.; Shen, M.; Song, Q.; Xie, J. Biological activities and pharmaceutical applications of polysaccharide from natural resources: A review. *Carbohydr. Polym.* **2018**, *183*, 91–101. [CrossRef]
170. Cumpstey, I. Chemical modification of polysaccharides. *Int. Sch. Res. Not.* **2011**, *2013*, 383–406. [CrossRef]
171. Auzély-Velty, R. Self-assembling polysaccharide systems based on cyclodextrin complexation: Synthesis, properties and potential applications in the biomaterials field. *Comptes Rendus Chim.* **2011**, *14*, 167–177. [CrossRef]
172. Pushpamalar, J.; Veeramachineni, A.K.; Owh, C.; Loh, X.J. Biodegradable polysaccharides for controlled drug delivery. *ChemplusChem* **2016**, *81*, 504–514. [CrossRef] [PubMed]
173. Ramírez, H.L.; Valdivia, A.; Cao, R.; Torres-Labandeira, J.J.; Fragoso, A.; Villalonga, R. Cyclodextrin-grafted polysaccharides as supramolecular carrier systems for naproxen. *Bioorganic Med. Chem. Lett.* **2006**, *16*, 1499–1501. [CrossRef] [PubMed]
174. Izawa, H.; Kawakami, K.; Sumita, M.; Tateyama, Y.; Hill, J.P.; Ariga, K. β-Cyclodextrin-crosslinked alginate gel for patient-controlled drug delivery systems: Regulation of host-guest interactions with mechanical stimuli. *J. Mater. Chem. B* **2013**, *1*, 2155–2161. [CrossRef] [PubMed]
175. Omtvedt, L.A.; Dalheim, M.; Nielsen, T.T.; Larsen, K.L.; Strand, B.L.; Aachmann, F.L. Efficient grafting of cyclodextrin to alginate and performance of the hydrogel for release of model drug. *Sci. Rep.* **2019**, *9*, 1–11. [CrossRef] [PubMed]
176. Peng, K.; Cui, C.; Tomatsu, I.; Porta, F.; Meijer, A.H.; Spaink, H.P.; Kros, A. Cyclodextrin/dextran based drug carriers for a controlled release of hydrophobic drugs in zebrafish embryos. *Soft Matter* **2010**, *6*, 3778. [CrossRef]
177. Li, C.; Luo, G.F.; Wang, H.Y.; Zhang, J.; Gong, Y.H.; Cheng, S.X.; Zhuo, R.X.; Zhang, X.Z. Host-guest assembly of pH-responsive degradable microcapsules with controlled drug release behavior. *J. Phys. Chem. C* **2011**, *115*, 17651–17659. [CrossRef]
178. Xiao, W.; Chen, W.H.; Zhang, J.; Li, C.; Zhuo, R.X.; Zhang, X.Z. Design of a photoswitchable hollow microcapsular drug delivery system by using a supramolecular drug-loading approach. *J. Phys. Chem. B* **2011**, *115*, 13796–13802. [CrossRef] [PubMed]

179. Hou, X.; Zhang, W.; He, M.; Lu, Y.; Lou, K.; Gao, F. Preparation and characterization of β-cyclodextrin grafted N-maleoyl chitosan nanoparticles for drug delivery. *Asian J. Pharm. Sci.* **2017**, *12*, 558–568. [CrossRef]
180. Prabaharan, M.; Gong, S. Novel thiolated carboxymethyl chitosan-g-β-cyclodextrin as mucoadhesive hydrophobic drug delivery carriers. *Carbohydr. Polym.* **2008**, *73*, 117–125. [CrossRef]
181. Zhang, X.; Wu, Z.; Gao, X.; Shu, S.; Zhang, H.; Wang, Z.; Li, C. Chitosan bearing pendant cyclodextrin as a carrier for controlled protein release. *Carbohydr. Polym.* **2009**, *77*, 394–401. [CrossRef]
182. Zhao, S.; Lee, J.; Xu, W. Supramolecular hydrogels formed from biodegradable ternary COS-g-PCL-b-MPEG copolymer with α-cyclodextrin and their drug release. *Carbohydr. Res.* **2009**, *344*, 2201–2208. [CrossRef] [PubMed]
183. Singh, M.N.; Hemant, K.S.Y.; Ram, M.; Shivakumar, H.G. Microencapsulation: A promising technique for controlled drug delivery. *Res. Pharm. Sci.* **2010**, *5*, 65–77. [PubMed]
184. Dong, Z.; Kang, Y.; Yuan, Q.; Luo, M.; Gu, Z. H_2O_2-responsive nanoparticle based on the supramolecular self-assemble of cyclodextrin. *Front. Pharmacol.* **2018**, *9*, 1–10. [CrossRef] [PubMed]
185. Carrazana, J.; Jover, A.; Meijide, F.; Soto, V.H.; Tato, J.V. Complexation of adamantyl compounds by β-cyclodextrin and monoaminoderivatives. *J. Phys. Chem. B* **2005**, *109*, 9719–9726. [CrossRef]
186. Wintgens, V.; Nielsen, T.T.; Larsen, K.L.; Amiel, C. Size-controlled nanoassemblies based on cyclodextrin-modified dextrans. *Macromol. Biosci.* **2011**, *11*, 1254–1263. [CrossRef]
187. Elgadir, M.A.; Uddin, M.S.; Ferdosh, S.; Adam, A.; Chowdhury, A.J.K.; Sarker, M.Z.I. Impact of chitosan composites and chitosan nanoparticle composites on various drug delivery systems: A review. *J. Food Drug Anal.* **2015**, *23*, 619–629. [CrossRef]
188. Zhao, F.; Repo, E.; Yin, D.; Chen, L.; Kalliola, S.; Tang, J.; Iakovleva, E.; Tam, K.C.; Sillanpää, M. One-pot synthesis of trifunctional chitosan-EDTA-β-cyclodextrin polymer for simultaneous removal of metals and organic micropollutants. *Sci. Rep.* **2017**, *7*, 1–14. [CrossRef]
189. Chen, Y.; Ye, Y.; Wang, L.; Guo, Y.; Tan, H. Synthesis of chitosan C6-substituted cyclodextrin derivatives with tosyl-chitin as the intermediate precursor. *J. Appl. Polym. Sci.* **2012**, *125*, E378–E383. [CrossRef]
190. Liu, Y.; Yu, Z.L.; Zhang, Y.M.; Guo, D.S.; Liu, Y.P. Supramolecular architectures of β-cyclodextrin-modified chitosan and pyrene derivatives mediated by carbon nanotubes and their DNA condensation. *J. Am. Chem. Soc.* **2008**, *130*, 10431–10439. [CrossRef]
191. Hu, Y.; Wu, X.Y.; JinRui, X. Self-assembled supramolecular hydrogels formed by biodegradable PLA/CS diblock copolymers and β-cyclodextrin for controlled dual drug delivery. *Int. J. Biol. Macromol.* **2018**, *108*, 18–23. [CrossRef]

 © 2020 by the authors. Licensee MDPI, Basel, Switzerland. This article is an open access article distributed under the terms and conditions of the Creative Commons Attribution (CC BY) license (http://creativecommons.org/licenses/by/4.0/).

Review

Stereocomplex Polylactide for Drug Delivery and Biomedical Applications: A Review

Seung Hyuk Im [1,2,†], Dam Hyeok Im [3,†], Su Jeong Park [1,4], Justin Jihong Chung [4], Youngmee Jung [4,5] and Soo Hyun Kim [1,4,6,*]

1. NBIT, KU-KIST Graduate School of Converging Science and Technology, Korea University, 145 Anam-ro, Seongbuk-gu, Seoul 02841, Korea; ishidh@korea.ac.kr (S.H.I.); airplane96@kist.re.kr (S.J.P.)
2. enoughU Inc., 114 Goryeodae-ro, Seongbuk-gu, Seoul 02856, Korea
3. Department of Mechanical Engineering, Graduate School, Korea University, 145, Anam-ro, Seongbuk-gu, Seoul 02841, Korea; idhish1@naver.com
4. Center for Biomaterials, Biomedical Research Institute, Korea Institute of Science and Technology (KIST), Seoul 02792, Korea; chungjj@kist.re.kr (J.J.C.); winnie97@kist.re.kr (Y.J.)
5. School of Electrical and Electronic Engineering, Yonsei-KIST Convergence Research Institute, Yonsei University, Seoul 03722, Korea
6. Korea Institute of Science and Technology (KIST) Europe, Campus E 7.1, 66123 Saarbrueken, Germany
* Correspondence: soohkim@kist.re.kr
† These authors contributed equally to this work.

Abstract: Polylactide (PLA) is among the most common biodegradable polymers, with applications in various fields, such as renewable and biomedical industries. PLA features poly(D-lactic acid) (PDLA) and poly(L-lactic acid) (PLLA) enantiomers, which form stereocomplex crystals through racemic blending. PLA emerged as a promising material owing to its sustainable, eco-friendly, and fully biodegradable properties. Nevertheless, PLA still has a low applicability for drug delivery as a carrier and scaffold. Stereocomplex PLA (sc-PLA) exhibits substantially improved mechanical and physical strength compared to the homopolymer, overcoming these limitations. Recently, numerous studies have reported the use of sc-PLA as a drug carrier through encapsulation of various drugs, proteins, and secondary molecules by various processes including micelle formation, self-assembly, emulsion, and inkjet printing. However, concerns such as low loading capacity, weak stability of hydrophilic contents, and non-sustainable release behavior remain. This review focuses on various strategies to overcome the current challenges of sc-PLA in drug delivery systems and biomedical applications in three critical fields, namely anti-cancer therapy, tissue engineering, and anti-microbial activity. Furthermore, the excellent potential of sc-PLA as a next-generation polymeric material is discussed.

Keywords: polylactide; stereocomplex; biodegradable polymers; drug delivery system; biomedical applications

1. Introduction

Over the past few decades, polylactide (PLA) has emerged as a common biomaterial in biomedical applications owing to its favorable properties, such as complete biodegradability, mechanical properties, biocompatibility, processability, and transparency [1,2]. A PLA possesses two types of three-dimensional helical structures that twist in clockwise (D-configured) and counter-clockwise (L-configured) directions. Since Ikada et al. first reported stereocomplex formation between enantiomeric PLA in 1987 [3], stereocomplexation between poly(D-lactide) (PDLA) and poly(L-lactide) (PLLA) enantiomers has been the subject of continuous study. This research has accelerated with the rapid growth of practical use and the potential worth of PLA as a representative biodegradable polymer. Stereocomplex crystallites with a 3/1 helical structure in the PLA material can overcome inferior mechanical and thermal characteristics of homo-crystallites having a 10/3 helical structure. The combination of two enantiomeric polymers increases the melting point (T_m) and crystallinity through the compact orientation of crystals in the material. Ultimately, this change in a biodegradable polymer can result in an increase in thermal stability, mechanical strength, resistance against solvent penetration, and external forces. Stereocomplex crystals in PLA have commonly been formed by solution blending, melt blending, emulsion blending, precipitation into non-solvent, and supercritical fluid (SCF) techniques [4–9]. Each method has different advantages and disadvantages regarding the yield and stereocomplexation efficiency, processability, solubility, time, and cost.

Recently, numerous studies have reported that stereocomplex polylactide (sc-PLA) with improved physical characteristics can be used in drug delivery and as molecular carriers [10–12]. Nanoparticles, such as microspheres and micelles of sc-PLA, have the advantage of controlling drug uptake and release patterns through their synthesis and modification, as well as natural adsorption. Despite these advantages, many challenges remain for the application of sc-PLA as a drug carrier, including inferior encapsulation efficiency, low stability of hydrophilic drugs and proteins, and the burst release phenomenon. To resolve these issues, various strategies, such as polymerization, self-assembly, surface modification, and polymer grafting, have been studied [13–17]. This review discusses various synthesis and processing methods for the application of sc-PLA as a drug and molecular carrier, and it suggests future directions to stimulate the application of sc-PLA with therapeutic molecules in drug delivery systems and biomedical applications.

2. Drug Delivery

2.1. Stereocomplexed Micelle System

Micelles exhibiting specific core–shell structures are widely applied in drug delivery systems, because they can load a variety of drugs owing to their good loading capacity. Furthermore, micelles have a higher thermodynamic stability than colloids under physiological conditions owing to the lower critical micelle concentration. Therefore, polymeric micelles induced by the self-assembly of amphiphilic block copolymers have been vigorously researched for their biomedical roles, such as target-specific carriers, nano-bioreactors, and non-viral gene vectors. However, polymeric micelles have limited applications because of a short circulation time, due to their rapid excretion via urine after intravenous injection, and difficulty in accumulation and providing sustainable drug release at the target site [18]. Strategies for chemical cross-linking of hydrophilic poly(ethylene glycol) (PEG) segments have been proposed since the 1990s to improve the stability of polymeric micelles. Gref et al. (1994) fabricated nanospheres composed of a core and shell formed by biodegradable polymers, such as poly(lactic co-glycolic acid) (PLGA), polycaprolactone (PCL), and their copolymers, covalently bonded with PEG [19]. The PEG coating could significantly increase the blood circulation time of carriers by reducing their detection and opsonization by macrophages in the reticuloendothelial system and decrease their accumulation in the liver. In addition, this injectable nanoparticle carrier could encapsulate up to 45% of its weight in a one-step procedure. Micelles induced by biodegradable polymers enable the control of drug release kinetics and time, as the degradation period varies with the molecular weight (M_n) of the polymer. Furthermore, this feature allows higher kinetic and thermodynamic stabilities than those of a surfactant micelle with a lower molecular weight [20]. Kang et al. reported that monodisperse stereocomplexed micelles could be obtained by self-assembly of the PLA-PEG block copolymer [21]. This report was the first to verify that PLA-based micelles can form stereocomplex configurations in aqueous conditions. The micelles exhibited improved kinetic stability, as well as both physical and chemical stabilities. In particular, secondary aggregation, which is known to be the main problem of conventional polymeric micelles, was reduced. Figure 1a shows an atomic force microscopy (AFM) image of a stereocomplexed micelle with a spherical shape of approximately 46 nm diameter and a narrow size distribution. Figure 1b indicates that the micelle has a normal X-ray diffraction (XRD) pattern and a small crystalline domain of sc-PLA.

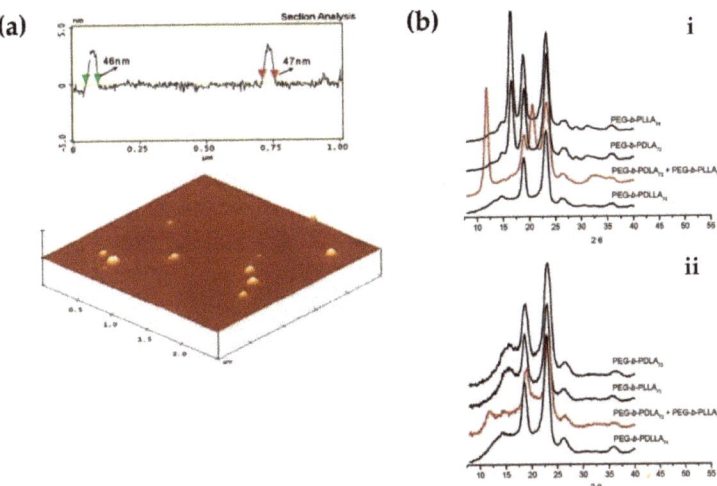

Figure 1. (a) AFM image of stereocomplexed micelle induced by mixing of PEG-b-PDLA$_{72}$ and PEG-b-PLLA$_{73}$ at equal proportions. (b) XRD patterns of PEG-b-PLA films (i) and stereocomplexed micelles composed of equimolar amounts of PEG-b-PDLA$_{72}$ and PEG-b-PLLA$_{73}$ (ii) [21], Copyright 2005. Reproduced with permission from the American Chemical Society.

2.2. Self-Assembled Nanoparticle

Stereocomplexes can build nanoparticles by self-assembly, apart from micelle formation. Non-covalent interactions, such as electrostatic interactions, hydrogen bonding, and hydrophobic-hydrophobic interactions are driving forces to induce self-assembly of amphiphilic block copolymers. The stability of the self-assembly can be influenced by several environmental factors, including the pH, temperature, polymer concentration, and ionic strength [22–25]. Bishara et al. synthesized stereocomplex particles composed of an enantiomeric triblock copolymer (PLA-PEG$_{2000}$-PLA) by blending in acetonitrile solutions [26]. Figure 2a shows that the stereocomplex nanoparticles have smooth surfaces resulting from PEG segments and sizes ranging from the nanometer to micrometer range. The size of stereocomplex particles increased with increasing concentration of PLA in the triblock copolymer (Figure 2b). This size of biodegradable particles affects the drug release rate from a particular carrier. Smaller-sized stereocomplex particles have a faster degradation rate owing to their larger surface area, as shown in Figure 2c. In the in vitro release profile test, stereocomplex particles with triblock copolymer could encapsulate 80% of the water-soluble drug dexamethasone, which was completely released for 30 days (Figure 2d). Similar to the previous degradation results, particles with a higher PLA concentration could release drugs at a slower rate. Biodegradable particles were fully degraded during the two months after complete release of dexamethasone phosphate. The hydrophilic drug was assumed to accelerate the infiltration of water into the polymer material compared to the original polymer.

Liu et al. produced pH-sensitive stereocomplex nanoparticles consisting of methoxy poly(ethylene glycol)-poly(L-histidine)-polylactide (mPEG$_{45}$-PH$_{30}$-PLA$_{82}$) tri-block copolymer by self-assembly [27] (Figure 3a). In this study, the mPEG$_{45}$-PH$_{30}$-PLLA$_{82}$/mPEG$_{45}$-PH$_{30}$-PDLA$_{82}$ stereocomplex stably maintained a mean diameter of 90 nm at pH 6.8, whereas the diameter of mPEG$_{45}$-PH$_{30}$-PLA$_{82}$ increased to the micrometer scale under the same conditions. The mean diameter of the stereocomplex nanoparticles slightly decreased when the pH changed from 5.0 to 7.9, as shown in Figure 3b,c. It was considered that lower pH conditions caused swelling of the nanoparticles with protonation of poly(L-histidine) in the tri-block copolymer [28]. Transmission electron microscopy (TEM) revealed that the stereocomplex particles retained their spherical shape, ranging from pH 5.0 to 7.4, and the TEM image confirmed a reduction in the particle size at pH 7.9 (Figure 3d,e). Furthermore, the cell viability of the stereocomplex nanoparticles was 90 % higher than that of homopolymer nanoparticles in co-culture with mouse 3T3 fibroblasts. This was attributed to the reduction of cytotoxic PDLA segments through stereocomplexation. Numerous studies have demonstrated that stereocomplex particles are capable of loading not only proteins, but also drugs.

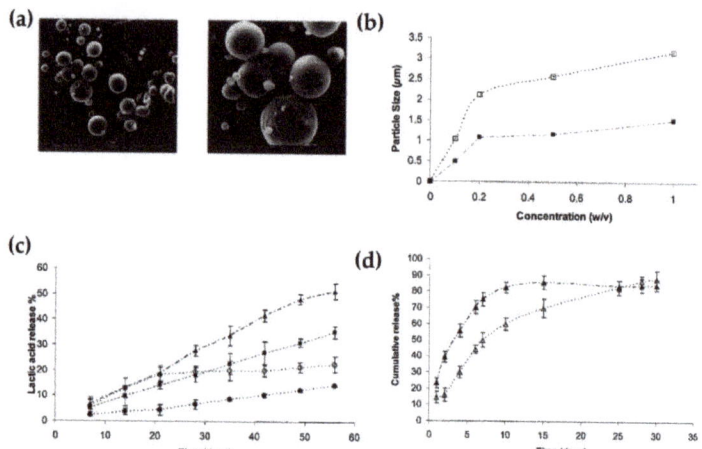

Figure 2. (a) Scanning electron microscopy (SEM) images of stereocomplex nanoparticles composed of PLLAx-PEG2000-PLLAy with D-PLA ($x + y = 25$, left) and PLLAx-PEG2000-PLLAy with 10% w/w dexamethasone phosphate with D-PLA, respectively (right). (b) Stereocomplex particle size at different concentrations. Closed and open squares depict PLLAx-PEG2000-PLLAy with D-PLA ($x + y = 25$) and PLLAx-PEG2000-PLLAy with D-PLA ($x + y = 50$) specimens, respectively. (c) HPLC analysis for lactic acid release from copolymers and stereocomplexes. Open and closed circles depict PLLAx-PEG2000-PLLAy ($x + y = 25$) and PLLAx-PEG2000-PLLAy ($x + y = 50$), respectively. Closed squares and triangles depict PLLAx-PEG2000-PLLAy ($x + y = 25$) with D-PLA specimen and same material containing 10% w/w dexamethasone phosphate, respectively. (d) HPLA analysis for in vitro release of dexamethasone from stereocomplexes. Closed and open triangles depict specimens composed of PLLAx-PEG2000-PLLAy ($x + y = 25$) with D-PLA and PLLAx-PEG2000-PLLAy ($x + y = 50$) with D-PLA, respectively. This experiment was conducted in phosphate buffer (pH 7.4) at 37 °C [26], Copyright 2005. Reproduced with permission from WILEY-VCH Verlag GmbH & Co.

Lim and Park (2000) synthesized stereocomplex microspheres based on the solvent-casting method, after polymerization of PLLA-PEG-PLLA and PDLA-PEG-PDLA of a tri-block ABA copolymer [29]. Then, a bovine serum albumin (BSA) protein could be encapsulated in the stereocomplex by the double emulsion solvent evaporation method. Their study reported that stereocomplex microspheres showed a more sustainable and predictable release pattern of the protein than that of the homopolymer. This can be attributed to the hydrophilic PEG unit in the tri-block microspheres preventing aggregation and non-specific adsorption of the protein. In the in vitro release profiles of BSA, PEG stereocomplex microspheres and PEG tri-block copolymer microspheres exhibited a higher initial burst effect than that of PLLA microspheres, but the microspheres based on PEG showed a larger cumulative release (Figure 4). This is attributed to the water uptake capacity of the microspheres that increased owing to the PEG of the hydrophilic unit. After the burst release in the initial stage, the microspheres showed a relatively sustained release by a diffusion-controlled mechanism over 50 days.

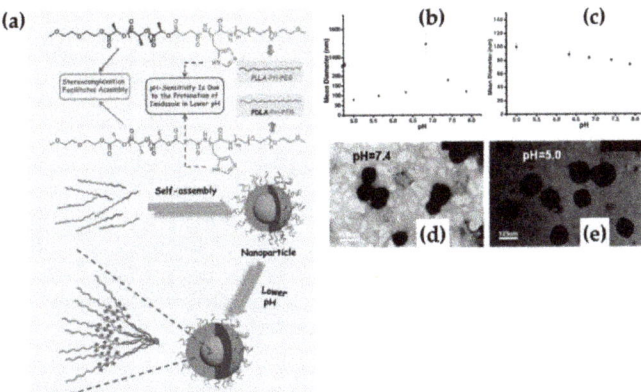

Figure 3. (a) Formation process schematic of mPEG$_{45}$-PH$_{30}$-PLLA$_{82}$/mPEG$_{45}$-PH$_{30}$-PDLA$_{82}$ stereocomplex nanoparticles. Mean diameters of mPEG$_{45}$-PH$_{30}$-PLLA$_{82}$ (b) and mPEG$_{45}$-PH$_{30}$-PLLA$_{82}$/mPEG$_{45}$-PH$_{30}$-PDLA$_{82}$ stereocomplex nanoparticles (c) with various pH conditions. TEM images of the stereocomplex nanoparticles at pH 7.4 (d) and pH 5.0 (e) [27], Copyright 2012. Reproduced with permission from WILEY-VCH Verlag GmbH & Co.

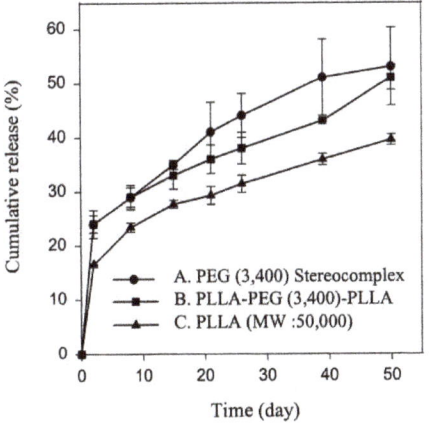

Figure 4. In vitro BSA release profile from three specimens [29], Copyright 2000. Reproduced with permission from John Wiley & Sons, Inc.

2.3. Emulsion Blending

Generally, sc-PLA is prepared by mixing two enantiomeric polymers in solution or in a melted state. However, these methods result in low stereocomplexation efficiency, loss of molecular weights, and original properties. The layer-by-layer (LbL) technique based on a Pickering emulsion template has been commonly used to fabricate nano- and microcapsules [30–34]. Despite its facilitated application to various fields, this approach has several concerns, including the requirement of inorganic solid multi-layer precursors, linkers, and templates, and inferior integrity and loading efficiency of a capsule [35–37]. To overcome these limitations, emulsion blending based on the droplet-in-droplet method has emerged as an alternative method to fabricate polymeric micro-carriers. Brzeziński proposed a microfluidic approach based on a water-in-oil-in-water (W/O/W) double emulsion for the synthesis of hollow stereocomplex microcapsules [38]. In this study, 2-ureido-4[1H]-pyrimidinone (UPy)-functionalized PLA enantiomers formed stereocomplex microcapsules at the water-chloroform interface via a one-step microfluidic self-assembly (Figure 5a). The capsule could reversibly control the assembly and disassembly to the supramolecular functionality of the interfacial assembly, unlike other microcarriers. This enables the capsule to freely adjust the stiffness and permeability of its shell and drug release. Figure 5b shows the morphological and structural reorganization of the sc-PLA microparticles induced by the W/O/W double emulsion. In this observation, microdroplets with a mean diameter of approximately 260 µm appear to have a high monodispersity and narrow size distribution. In particular, microcapsules were divided from the water phase by spontaneous dewetting after the oil droplets shrank. The inner oil droplet could maintain hollow stereocomplexed microcapsules induced at the interface between water and chloroform [39–41]. Finally, the sc-PLA-UPy microcapsules were precipitated with a mean diameter of approximately 160 µm. Stereocomplex crystallites were assumed to act as efficient nucleating agents and interfacial enhancers during this reaction [42].

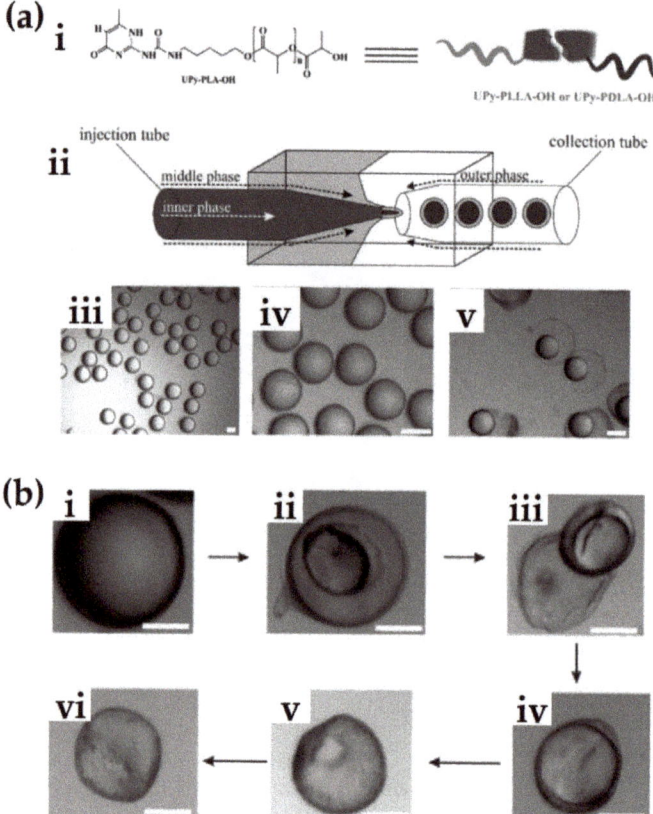

Figure 5. (a) Schematic illustrations of UPy-PLA-OH chemical structure (i) and microfluidic device for induction of W/O/W emulsion droplets (ii). Optical microscope observations of monodisperse stereocomplexed drops (iii,iv), and UPy-PLA-OH unstable drops (v). Scale bars: 200 µm. (b) Change of structural arrangement of W/O/W double emulsion droplets. (i) sc-PLA-UPy double emulsion droplet, (ii,iii) separation of microcapsule in water phase, (iv–vi) shell solidification of the microcapsule induced by evaporation. Scale bars: 100 µm [38], Copyright 2017. Reproduced with permission from WILEY-VCH Verlag GmbH & Co.

Im et al. developed a novel strategy for blending two homopolymers of PLA in an oil-in-water (O/W) emulsion state [43]. This O/W emulsion blending method facilitated the rapid combination of solutions of PLLA and PDLA enantiomers with the addition of an emulsifier and mechanical stirring in a one-pot reactor (Figure 6a) [43]. During blending, stereocomplex crystals could be formed simultaneously with the diffusion of oil-phased PLLA and PDLA into water, emulsification induced by an emulsifier, and mechanical mixing. Unlike other phases, the emulsion phase could induce significantly rapid stereocomplexation by promoting supramolecular interactions derived from lower interfacial tension. Moreover, this method can significantly improve the time and cost, availability, and efficiency compared to conventional methods. As a result, sc-PLA particles fabricated by emulsion blending showed spherical morphology with an improved stereocomplexation efficiency of up to 99% (Figure 6b,c). Furthermore, fluorouracil (5-FU), a cancer drug, can be encapsulated into sc-PLA particles for stereocomplexation during O/W emulsion blending, as shown in Figure 7a. In the drug release profiling test for 5-FU, sc-PLA particles induced by emulsion blending loaded 13 wt% of 5-FU, and the drug was slowly released for eight days after the initial burst of drug release (Figure 7b). To the best of our knowledge, this is the first report on an O/W emulsion blending method for simultaneously inducing stereocomplexation and drug encapsulation in an emulsion state.

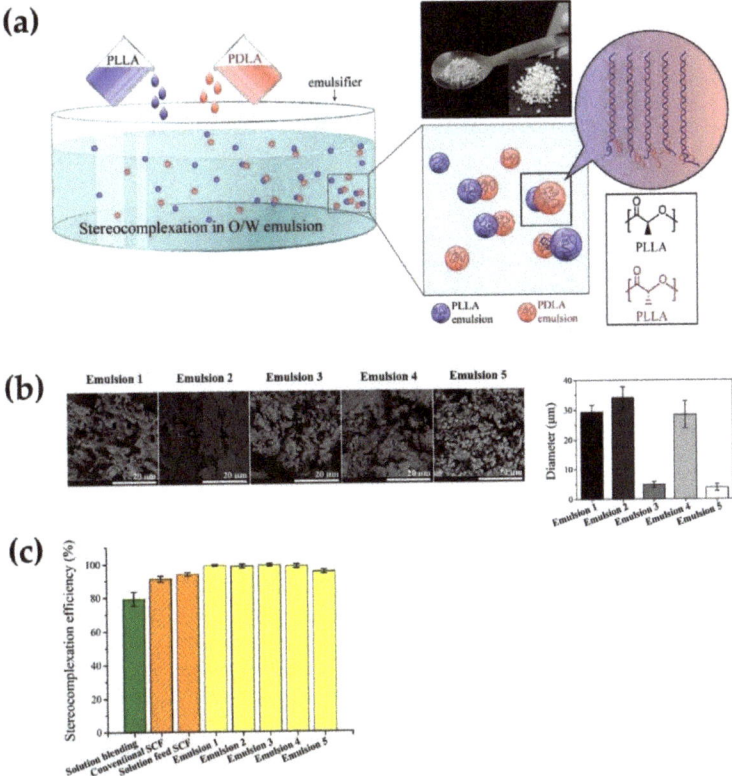

Figure 6. (a) Scheme for O/W emulsion blending approach for inducing sc-PLA. Inserted image exhibits sc-PLA particles prepared by emulsion blending. (b) SEM observations (X10000, left) and diameters (right) of sc-PLA particles prepared by various O/W emulsion blending named Emulsion 1-5. (c) Comparison of stereocomplexation efficiency (%) of sc-PLA particles prepared Emulsion 1-5. SCF: supercritical fluid technology [43], Copyright 2020. Reproduced with permission from the American Chemical Society.

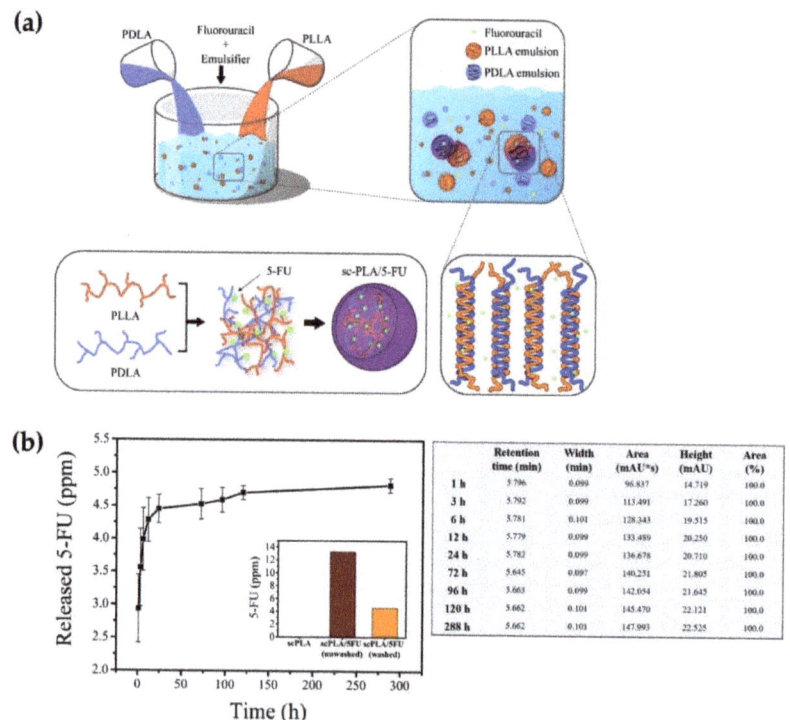

Figure 7. (a) Schematic illustration of O/W emulsion blending for simultaneously inducing stereocomplexation and infiltration of 5-FU drugs. (b) Release profile of 5-FU from washed sc-PLA particles prepared by O/W emulsion blending over 12 days. Insert presents graph of released 5-FU concentration from neat, washed, and unwashed sc-PLA/5-FU specimens. Data are plotted as mean values ± standard deviation (SD) (n = 3) [43], Copyright 2020. Reproduced with permission from the American Chemical Society.

2.4. Inkjet Printing

Recently, inkjet printing technology has been used in various industrial areas, including tissue engineering and organic or inorganic electronics. This technique can rapidly and exquisitely deposit printing materials, such as polymers, metals, nano- and micro-particles, and proteins and cells with highly controlled volume and patterning of link-droplets. Consequently, polymeric biomaterials can be LbL-assembled on substrates. Akagi et al. (2012) devised a fast printing method for sc-PLA based on LbL deposition using an inkjet printer without a redundant rinsing step [44]. Each PLLA and PDLA solution was separately printed at the same point, and stereocomplex crystals were then formed with enantiomeric homocrystallites during solvent evaporation. Figure 8 shows the schematics of the inkjet printing process for the LbL formation of sc-PLA. The first LbL-assembly method, employed by the authors for formation of the stereocomplex, involved the dissolution of PLLA and PDLA in chloroform, followed by the PLLA solution being sprayed onto a glass substrate. Then, the PDLA solution was printed over the substrate after drying the PLLA droplets (Figure 8a). One cycle indicates that PLLA and PDLA solutions were printed once on the substrate in turn. In the second LbL-assembly method, PLLA and PDLA were dissolved in chloroform, and the blended solution was printed on the substrate in the first step. Subsequently, the blended solution was printed over dried mixed droplets once again in the second step (Figure 8b). This method holds potential to provide rapid fabrication, as 1×10^5 droplets were sprayed, and the processing time was approximately

100 s in each step. As shown in Figure 9a, the X-ray diffraction (XRD) patterns of PLA composites fabricated by the second method showed peaks at 12°, 21°, and 24° of 2θ degrees. This indicates that the specimens had an orthorhombic or pseudo-orthorhombic unit cell with a 10_3 helical conformation, indicating an α-form crystal. The peak intensities of the sc-PLA for both XRD and FTIR results were amplified by increasing the number of printing steps with no observation of any peak indicating homopolymers. Furthermore, the thicknesses of the PLA composites increased with the increasing number of steps (220, 600, and 980 nm for 2, 5, and 10 steps, respectively). As shown in Figure 9b, the XRD patterns analyzed the effect of the number of printing cycles on the structural sc-PLA fabricated by inkjet printing. The L-D LbL-1 specimen, which was fabricated with one cycle of printing, showed two peaks at 12° and 17°, indicating stereocomplex crystals and α-form crystals, respectively. In contrast, homocrystallite formation was no longer observed with increasing cycle numbers for the printed specimens. The intensity of the peaks corresponding to β-form stereocomplex crystals was higher at 2θ = 12°, 21°, and 24° with an increasing cycle number, instead of the appearance of a peak at 2θ = 17°. Consequently, the higher cycle of inkjet printing could render sc-PLA of higher purity and increase its crystallinity. The right graph in Figure 9b shows the XRD patterns of the PLA specimens prepared using PLLA/PDLA mixed solutions. All specimens prepared by inkjet printing using a PLLA and PDLA mixed solution showed no significant increase in the peak intensity despite increasing the number of printing cycles. Furthermore, it only exhibited peaks with no homocrystallites at 12°, 21°, and 24°.

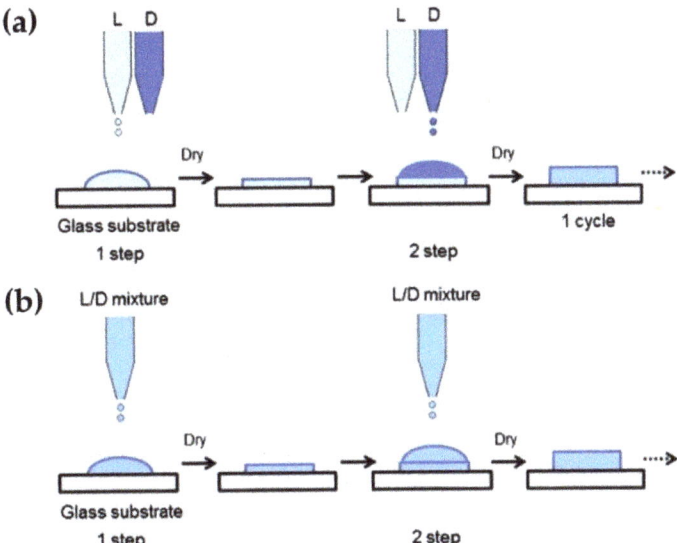

Figure 8. Schematic illustration for sc-PLA formation using the inkjet printing based on LbL deposition. (a) First LbL-assembly method; alternately printing of PLLA and PDLA solution onto a substrate in sequence (2 steps = 1 cycle), (b) Second LbL-assembly method; simultaneous printing of PLLA/PDLA mixed solution onto a substrate at one time [44], Copyright 2012. Reproduced with permission from WILEY-VCH Verlag GmbH & Co.

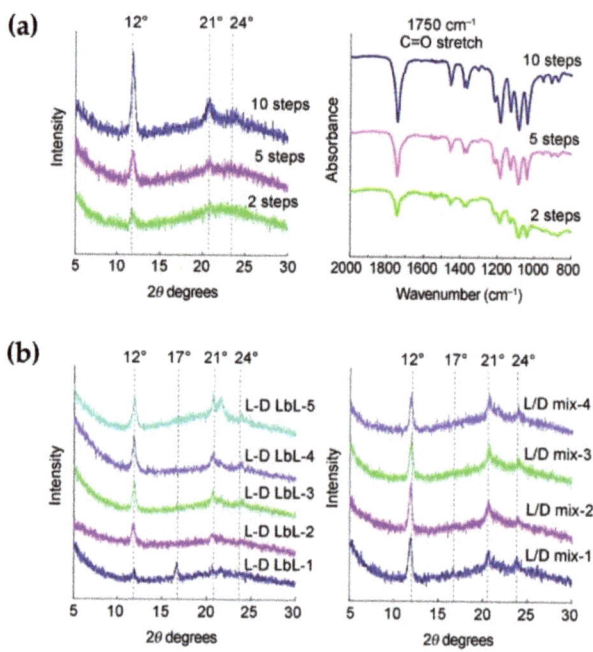

Figure 9. (a) XRD patterns (left) and FTIR spectra (right) of sc-PLA fabricated by inkjet printing of the second LbL-assembly method. The steps of three specimens indicate the number of repeated times of the printing. (b) Influence of the number of printing cycle on crystal formation of inkjet-printed sc-PLA. XRD patterns of sc-PLA product prepared by first (left) and second (right) LbL-assembly method [44], Copyright 2012. Reproduced with permission from WILEY-VCH Verlag GmbH & Co.

Akashi et al. (2014) reported an advanced method that models drugs such as an 8-mer peptide, ovalbumin (OVA), where proteins could be loaded onto the sc-PLA substrate based on inkjet printing technology [45]. The sc-PLA composites could encapsulate drugs through alternate overprinting of PLLA, PDLA, and drugs on one substrate (Figure 10a). PLLA and PDLA were dissolved in chloroform at a concentration of 0.5 mg/mL. Each polymer solution was alternately printed onto the same substrate. First, the PLLA solution was printed onto a glass substrate and dried at room temperature, and then the PDLA solution was printed onto the substrate. Finally, 0.1 mg/mL of the drug, like OVA-protein, OVA-NPs, or peptide was printed onto the printed sc-PLA composite. A single cycle for the entire inkjet-printing process comprised these three steps. This process was conducted for up to a maximum of 10 cycles. As shown in Figure 10b, 25% of the peptide in the sc-PLA carrier was released in the initial 1 h, and 50% of the peptide was released after 1 day. Then, 90% of the peptide was released for 5 days via diffusion. In contrast, 40% of ovalbumin in the sc-PLA carrier was released for 1 day, and the remainder of the loaded protein failed to be released. This was assumed to be due to the aggregation of ovalbumin proteins in the sc-PLA matrix. In contrast, the sc-PLA with OVA-NPs group, which was printed by ovalbumin encapsulated nanoparticles (OVA-NPs), released 30% of the protein for the initial 1 day, and subsequently the remainder of the contents could be sustainably released during the following 30 days. It was considered that denaturation and aggregation of protein could occur to a lesser extent in the sc-PLA composite, because encapsulation of ovalbumin in the nanoparticle could offer resistance to protein denaturation and stability to the carrier. As shown in Figure 10c, the in vitro release profile of sc-PLA with ovalbumin encapsulated nanoparticles and homopolymer (PLLA and PDLA) with ovalbumin encapsulated nanoparticles was analyzed. All groups exhibited burst release kinetics of the protein for 24 h, and their release rates gradually decreased. The sc-PLA composite group showed a lower cumulative release of ovalbumin than the homopolymers after 24 h (28 vs. 35 vs. 41% for sc-PLA-OVA-NPs, PLLA-OVA-NPs, and PDLA-OVA-NPs, respectively). Furthermore, the sc-PLA-OVA-NPs had a lower release rate with sustainable release compared to the other groups. It was supposed that sc-PLA had superior hydrolysis resistance and carrier stability owing to its higher crystallinity compared to homopolymers.

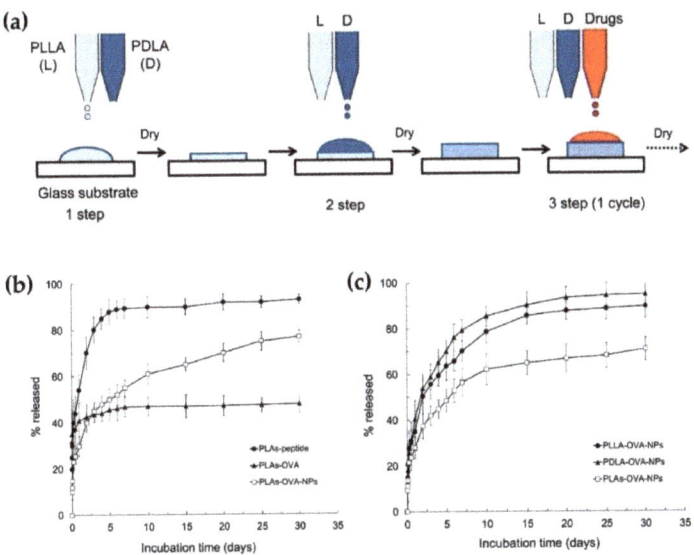

Figure 10. (a) Schematic illustration for sc-PLA with drugs composites fabricated by inkjet printing based on LbL-assembly. (b) Drug release profiles of the sc-PLA with drug (peptide or OVA) composites in phosphate-buffered saline (PBS). The data were plotted as mean values ± SD (n = 3). (c) OVA release behaviors from sc-PLA-OVA-NPs (named PLAs-OVA-NPs) and PLLA or PDLA-OVA-NPs in PBS. Results were plotted as mean values ± SD (n = 3) [45], Copyright 2014. Reproduced with permission from the American Chemical Society.

Recently, Akashi et al. (2017) reported that sc-PLA composites could be conjugated to benzyl alcohol and 3,4-diacetoxycinnamic acid (DACA) of bio-based aromatic compounds at the hydroxyl groups of both terminals using inkjet printing [46]. This DACA conjugation enhanced the thermal stability of sc-PLA by increasing the thermal decomposition temperature by 10% to above 90 °C. Consequently, inkjet printing can be considered as an innovative technology for fabricating sc-PLA scaffolds rapidly and easily based on LbL deposition and assembly. Furthermore, it has the advantage of versatile control of the shape, thickness, and amount of printed scaffolds composed of sc-PLA. Therefore, this technology has the potential to bring innovation to some fields, such as tissue engineering and biomaterials, if it can be converged with 3D printing for freely customized fabrication of scaffolds and substrates.

2.5. Stereocomplex Hydrogel

A hydrogel is a network of cross-linked polymer chains. Hydrogels have been frequently used as scaffolds in tissue engineering and drug carriers and are known as the first biomaterials to be used in the human body. S.J. de Jong et al. reported that a stereocomplex hydrogel could be synthesized by mixing dextran-grafted L-lactate and D-lactate in an aqueous solution [47]. The stereocomplex hydrogel could encapsulate the IgG and lysozyme of the model protein, and the loaded contents were released by Fickian diffusion for six days. Even after the release of sensitive proteins from the gel, the stereocomplex hydrogel played the role of a stable protein carrier as well as the maintenance of enzymatic activity. Subsequently, Hennink et al. fabricated a stereocomplex hydrogel by mixing dextran-L or D-lactate without organic solvents or crosslinking agents in an aqueous environment, as shown in Figure 11a [48]. Enantiomeric PLA oligomers grafted to dextran did not require artificial agents, because they were already crosslinked by stereocomplexation. Moreover, this stereocomplex hydrogel has the advantage of full biodegradability and clinical safety, because dextran is a non-toxic water-soluble polymer. Figure 11b shows the rheological properties of dex-(L)lactate and a mixture of dex-(L)lactate and dex-(D)lactate. A mixture of enantiomers exhibited a growth in the storage modulus (G') and a reduction of tan δ with time, whereas dex-(L)lactate showed no change in G' and tan δ with time. These results indicate that the hydrogel network was formed, and that the polymer had a more elastic property. This was presumed to be due to self-assembly between the chains of L-lactate and D-lactate via stereocomplexation. As shown in Figure 11c, the lysozyme was released from the dex-lactate hydrogel faster than IgG in the same carrier during the initial stage. The hydrogel with higher polydispersity (PDI) of the lactate graft showed faster release of the two model proteins compared to those with lower PDI. The results showed that all groups exhibited complete release of the loaded contents from the hydrogel after eight days and retained enzymatic activity. Hence, the sc-PLA hydrogel could encapsulate proteins and control release. It is potentially applicable to drug carriers with good biocompatibility and gelation behavior.

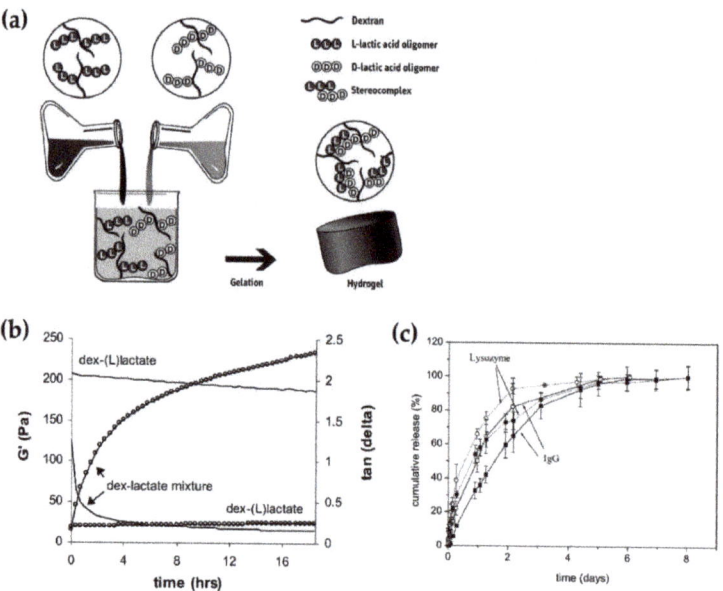

Figure 11. (a) Formation of stereocomplex hydrogel schematic via blending dextran-L or D-lactate without organic solvents or crosslinking agents. (b) Measurements of rheological properties of dex-(L)lactate and a mixture of dex-(L)lactate and dex-(D)lactate. Open circles and lines in the graph depict G' and tan δ, respectively [45], Copyright 2003. Reproduced with permission from Elsevier B.V. (c) Protein release profiles of lysozyme (indicated by dotted line in the graph) and IgG (indicated by solid line in the graph) from dex-lactate hydrogel exhibiting high (open symbols) and low (filled symbols) PDI. The measurement environment was at pH 7.0 and 37 °C. The data were plotted as mean values ±SD (n = 4) [47], Copyright 2001. Reproduced with permission from Elsevier Science B.V.

A strategy using in situ gelling systems has been used for the transformation of drug/polymer precursor complexes from solution after injection into the human body to gel form by physiological conditions of target tissues or artificial stimuli, such as pH or temperature change, UV irradiation, solvent exchange, catalytic ions, or molecules [49–51]. Generally, an in situ gelling hydrogel can be synthesized by various chemical reactions, including enzyme-catalyzed cross-linking, Schiff-base reaction, photo-induced polymerization, and Michael-type addition [52–55]. This drug delivery system can prevent adverse events in non-target tissues with improved availability of administration. Based on this strategy, sc-PLA-based hydrogels have the potential to improve the mechanical strength and durability and delay the degradation rate of the carrier induced by stereocomplexation in the future.

3. Biomedical Applications

3.1. Anti-Cancer Therapy

Recently, numerous studies have reported that biodegradable polymeric nanocarriers possess a superior stealth function for the detection of the reticuloendothelial system in the human body and exhibit an enhanced permeability and retention effect [56–61]. In particular, these polymeric nanoparticles are capable of having several beneficial properties, such as biomimetics, stimuli-sensitivity, easy modification, and exquisite target specificity compared to carriers composed of other materials. These advantages support the possibility that stereocomplex nanoparticles can be used as a promising approach in the field of anti-cancer therapy.

Goldberg proposed that oligomers of PDLA were used for the formation of stereocomplex crystals with L-lactate in the human body to induce lactate deficiency in cancer cells [62]. This study suggested that tumor growth could be inhibited and terminated by stereocomplexation through lactate deficiency to retain the electrical neutrality of tumors. In preliminary experiments, it was demonstrated that high concentrations of PDLA could induce stereocomplexation with lactate in the body, which exhibited cytotoxic effects on the tumor. Li et al. (2016) fabricated sc-PLA-coated nanoparticles with multiple functionalities for a highly tunable drug delivery system via simple LbL self-assembly [14]. TEM images show that the nanoparticles had a spherical shape with a core–shell structure and a diameter of approximately 190 nm (Figure 12b). The in vitro drug release profiles of doxorubicin (DOX)-loaded nanoparticles were analyzed at different pH and temperature conditions to demonstrate the ability to adjust the drug delivery of the sc-PLA-coated nanoparticles. As shown in Figure 12c, the drug release rate was significantly reduced with increasing pH from 3.5 to 7.4. The cumulative release of DOX was 73.1 vs. 55.6% at pH 3.5 and 7.4 after 12 h, respectively. This is because acidic conditions could stimulate protonation of the tertiary amine groups in the outermost layer of the sc-PLA-coated nanoparticles, resulting in the swelling of nanoparticles. Furthermore, a lower pH yields the nanoparticles pH-responsive properties through an increase in the solubility of DOX, which retains phenols and amines. In addition to pH, the cumulative release of DOX was 39.8 vs. 60% at 37 and 20 °C after 12 h, respectively. A lower temperature condition could facilitate a more rapid release via swelling of the outer layer of sc-PLA-coated nanoparticles. Consequently, the nanoparticles based on sc-PLA could tune the rate and amount of drug release, depending on the physiological conditions. To identify the cytotoxicity of DOX-loaded sc-PLA-coated nanoparticles to breast cancer cells, MCF-7 cells were incubated with free DOX and two types of DOX-loaded sc-PLA nanoparticles (Figure 12d). During the incubation of cancer cells with free DOX, most of the DOXs were localized in the cell nuclei, instead of the cytoplasm. In contrast, the DOXs released from all sc-PLA particles infiltrated the cells, and their accumulation was significantly increased. These results indicated that sc-PLA-coated nanoparticles could effectively act as anti-cancer drug carriers by improving the cell uptake efficiency.

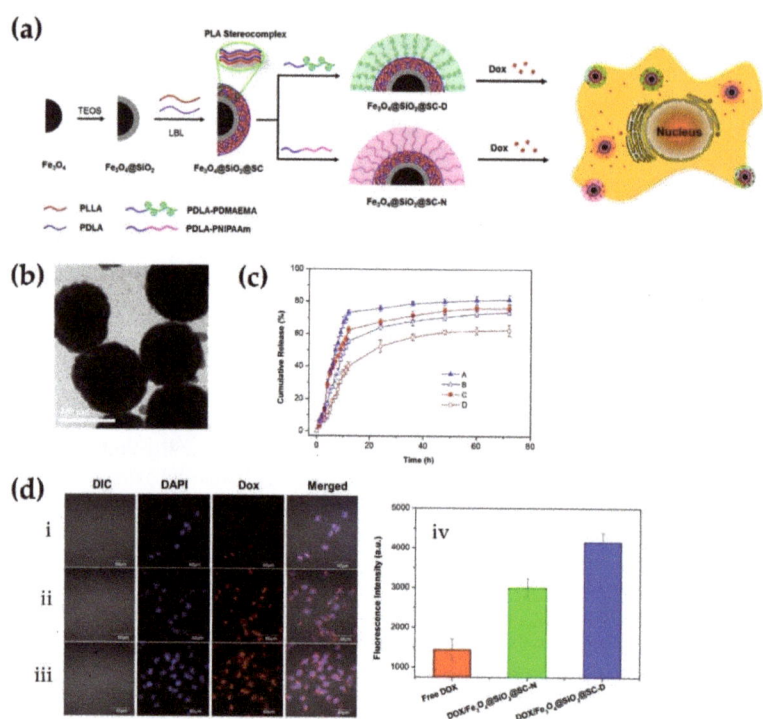

Figure 12. (a) Schematic representation for highly tunable sc-PLA coated nanoparticles and drug delivery. (b) TEM image of Fe$_3$O$_4$@SiO$_2$@SC-N nanoparticles. Scale bar: 200 nm. (c) In vitro drug release profiles of DOX-loaded Fe$_3$O$_4$@SiO$_2$@SC-D nanoparticles in acidic condition of pH 3.5 (A), physiological condition of pH 7.4 (B), and Fe$_3$O$_4$@SiO$_2$@SC-N nanoparticles at 20 °C (C) and 37 °C (D). (d) Confocal laser scanning microscope observations of MCF-7 cells incubated with free DOX (i), DOX-loaded Fe$_3$O$_4$@SiO$_2$@SC-N (ii), and Fe$_3$O$_4$@SiO$_2$@SC-D (iii) nanoparticles. Each panel named DIC, DAPI, Dox, and Merged depicts a differential interference contrast (DIC) image, cell nuclei staining by DAPI, DOX fluorescence in cells, and overlay of the all images, respectively. (iv) Comparison of fluorescence intensity in three specimens calculated by ImageJ [14], Copyright 2015. Reproduced with permission from the American Chemical Society.

Brzeziński et al. synthesized DOX-loaded stereocomplexed microspheres using spontaneous precipitation after the polymerization of L-proline-functionalized PLLA and PDLA via coordination polymerization (Figure 13a) [63]. Therein, the size of the microparticles was dependent on whether the L-proline end groups were blocked or unblocked in the microspheres. Based on this correlation, they obtained spherical microspheres with various sizes ranging from 0.5 to 10 μm through adjusting the solvent and functionalization. In the in vitro release profiles of DOX, stereocomplexed microspheres with Boc-protected L-proline exhibited cumulative release within 10% of loaded contents for 100 h, while microspheres with unblocked L-proline showed a faster cumulative release of 41–81% of loaded contents for the same duration (Figure 13b). This can be attributed to the low surface area of the Boc-protected group, which delayed hydrolysis. In contrast, the unblocked group could easily release DOX due to the localization of DOX on the surface of the microsphere, resulting in an initial burst release. To evaluate the cytotoxicity of the DOX-loaded stereocomplexed microspheres, A549 lung cancer cells were incubated with medium extracts of the two types of microspheres. Figure 13b shows a very slow reduction in the cell viability of cancer cells cultured with Boc-protected microspheres after 24 h, whereas the cell viability of the cancer cells cultured with unblocked microspheres was dramatically decreased after incubation for 2 h and decreased to below 25% after 24 h. This can be attributed to the difference in drug release rate between the blocked and unblocked microspheres. In particular, the microspheres prepared in tetrahydrofuran (THF) showed a significantly high anti-proliferative effect.

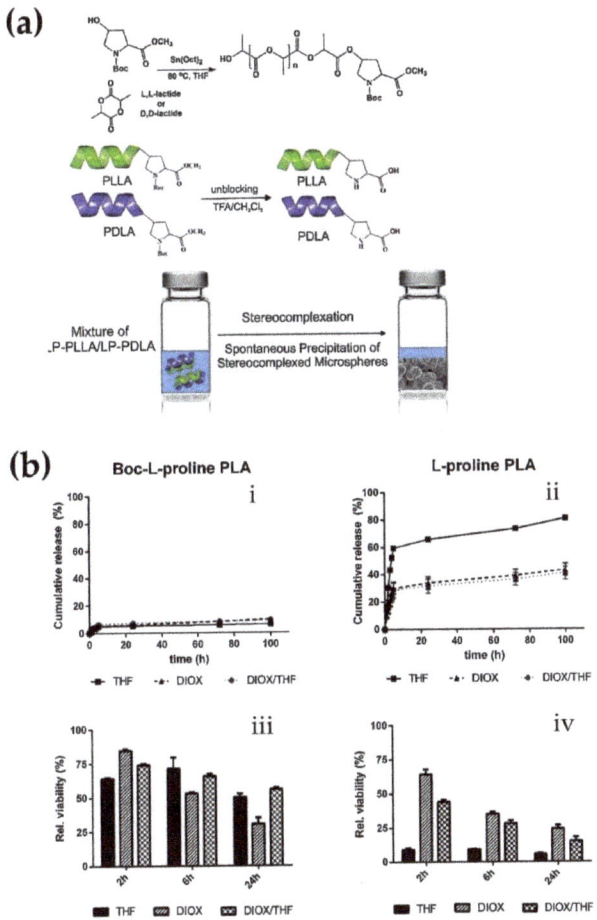

Figure 13. (a) Synthesis process of Boc-L-proline functionalized PLLA and PDLA, unblocking of L-proline end groups (top). Schematic illustration for fabrication of sc-PLA microspheres using spontaneous precipitation from mixture of enantiomeric PLAs (bottom). (b) In vitro release profiles of DOX from sc-PLA microspheres functionalized with (i) Boc-L-proline and (ii) L-proline. In vitro cytotoxicity test on A459 cells incubated with medium extracts of the sc-PLA microspheres functionalized with (iii) Boc-L-proline and (iv) L-proline at incubation times of 2, 6, and 24 h [63], Copyright 2019. Reproduced with permission from Elsevier B.V.

In addition, Brzeziński et al. successfully fabricated stereocomplexed micelles with β-cyclodextrin (β-CD) core as an intracellular drug carrier [64]. Hence, micelles stability can be improved, and in vitro release rate of DOX from supramolecular nanocarriers can be controlled. The stereocomplexed micelles with DOX efficiently inhibited the proliferation of HeLa (cervical cancer) and K562 (chronic myelogenous leukemia). This study suggested that the supramolecular interactions facilitate effective establishment of drug delivery system as an anti-cancer therapy.

3.2. Tissue Engineering

Tissue engineering is an increasingly popular next-generation biomedical technology to treat defects and malfunctions in human organs. This technology has the potential to expand medical coverage and resolve problems, such as a lack of organ donation and transplant rejection [65–69]. Scaffolds that are suitable for tissue engineering require high biocompatibility, biodegradability, good processability, mechanical properties similar to those of native tissue, and proper flexibility [70–77]. sc-PLA has been extensively applied in tissue engineering owing to its excellent biocompatibility, full biodegradability, improved mechanical properties, and thermal stability. In particular, it is suitable for scaffolds that require robust properties for bone, cartilage, and orthopedic implants. V. Katiyar et al. (2017) fabricated orthopedic implants based on nano-hydroxyapatite (n-HAP)-grafted sc-PLA composites using 3D printing [78]. As shown in Figure 14a, n-HAP-grafted PDLA was polymerized by in situ ring-opening polymerization, and the sc-PLA/n-HAP filament with a diameter of approximately 1.6 mm for 3D printing was fabricated by melt mixing with PLLA in a twin-screw extruder. A middle phalanx bone composed of filaments was successfully manufactured using 3D printing. As shown in FE-SEM images, the fractured Sc-PLA/n-HAP nanocomposites exhibited a smooth surface and uniform dispersion of n-HAP of 60 nm size (Figure 14b). The n-HAP provided a reinforcement effect and expansion of the surface area as a filler in the sc-PLA matrix. As shown in Figure 14c, sc-PLA/n-HAP of 2.5% increased the ultimate tensile strength up to a maximum of 16% above that of neat sc-PLA (40.2 vs. 33.8%, respectively). This is because an increase in intermolecular bonding and cross-linking in stereocomplex crystals, together with strong interfacial bonding between the polymer matrix and the filler, could increase its crystallinities. Furthermore, the ductility of sc-PLA increased by the addition of n-HAP fillers, which increased the elongation at the break of sc-PLA/n-HAP up to a maximum of 131.6 %. Improving ductility could enhance the durability of biocomposites resulting from the prevention of fracture and abruption, thus expanding its application for implants that require robust resistance for high loads.

Subsequently, V. Katiyar et al. (2019) focused on the synthesis of linear block copolymers composed of hard and soft segments of PLLA/PDLA and PCL [79]. Diblock and stereotriblock copolymers were successfully polymerized with PCL as a macroinitiator using sequential ring-opening polymerization, as shown in Figure 15a. Values of the tensile strength and elongation at the break for diblock copolymers were improved from 14.8 to 28.9 MPa and from 6.4 to 17.8%, respectively, with an increase of the block length of PDLA (Figure 15b). The synthesized materials were thermally processed based on injection molding to manufacture cancellous and cortical bone screws, which are considered as orthopedic fixation devices, as shown in Figure 16a. In a study on thermo-mechanical stability, cancellous bone screws consisting of sc-PLA/PCL blends could stably maintain their shape and structure at 121 °C for 60 min, more than those of commercial homo PLA (Figure 16b). This is because the stereocomplex crystallites of the hard segment in the blend copolymer improved the thermal resistance of the scaffold. Consequently, the sc-PLA and PCL in the sc-PLA block copolymer enabled the scaffold to increase mechanical and thermal stability, and to reduce brittleness of PLA by its plasticization effect; thus, scaffolds composed of these biomaterials are considered suitable for biomedical implants with good clinical outcomes.

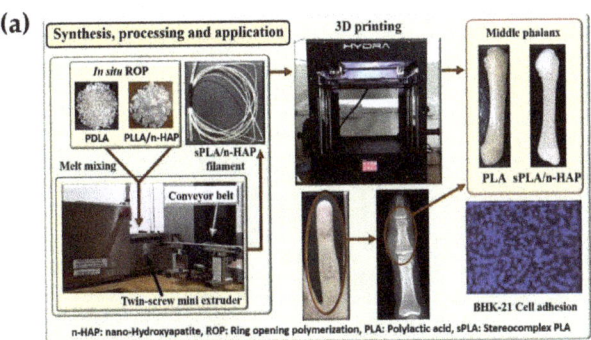

Figure 14. (a) A schematic representation for synthesis, processing, and application of sc-PLA/n-HAP. (b) Field emission scanning electron microscopy (FESEM) images of fractured surface of (i) a neat sc-PLA and (ii) a sc-PLA/n-HAP biocomposite. Pointed arrows depict n-HAP particles of approximately 60 nm diameter. (c) (i) Load–elongation curves of sc-PLA and sc-PLA/n-HAP. (ii) Comparison of ultimate tensile strength and elongation at break of sc-PLA and sc-PLA/n-HAP with diverse HAP contents [78], Copyright 2017. Reproduced with permission from the American Chemical Society.

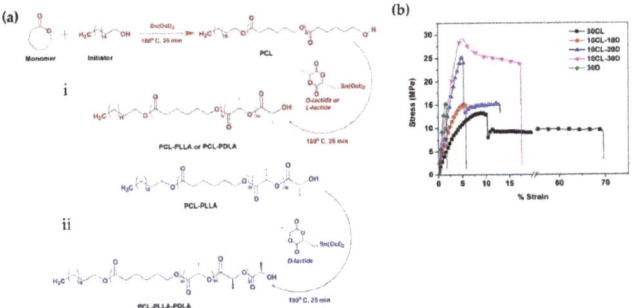

Figure 15. (a) Polymerization process of (i) diblock and (ii) stereotriblock copolymers. (b) Stress–strain (SS) curves of homopolymers and diblock copolymers [79], Copyright 2019. Reproduced with permission from the American Chemical Society.

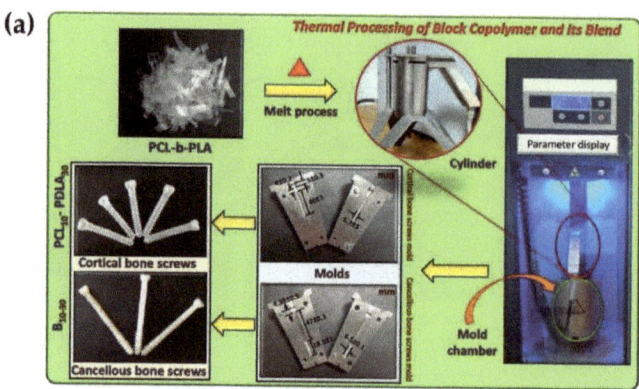
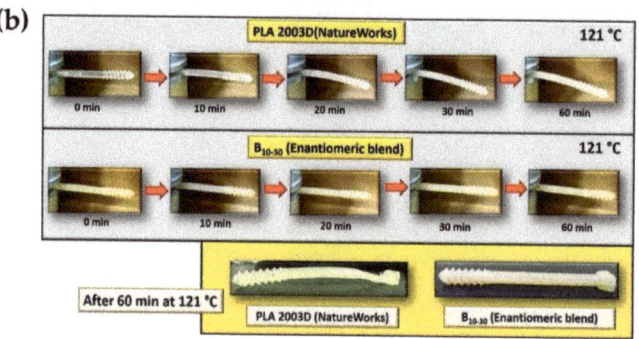

Figure 16. (a) Schematic illustration for thermal processing of diblock copolymer and enantiomeric diblock blend for fabrication of cortical and cancellous bone screws as orthopedic fixation devices. (b) Comparison of thermo-mechanical stability of cancellous bone screw comprising commercial PLA 2003D (Natureworks) and enantiomeric blend at 121 °C at intervals up to 60 min [79], Copyright 2019. Reproduced with permission from the American Chemical Society.

Enhancing mechanical properties of biodegradable polymers is critical to biomedical fields, such as bone fixation. Numerous processing methods have been developed to improve the strength of the polymers. Many studies have demonstrated that solid-state drawing (SSD) can induce self-reinforcement through the maximization of macromolecular chain orientation in polymeric materials [80–82]. Im et al. (2016, 2017) determined that the tensile strengths of PLLA monofilaments and films could be increased up to two- and nine-fold, respectively, by increasing the draw ratio using a directly designed processing machine for the SSD method (Figure 17a) [83,84]. Furthermore, this study showed that solid-state drawn PLLA enhanced blood compatibility and cell adhesion. Recently, Li et al. (2021) successfully oriented shish-kebab crystals in sc-PLA using the SSD method, as shown in Figure 17b [85]. The oriented sc-PLA scaffold had a tensile strength of 373 MPa and an elongation of 9%. This processing could lead to fibrous crystals of shish and kebabs with parallel lamellar microstructures along the direction of the drawing. This shish-kebab microstructure with a specific topography could provide a self-reinforcing effect and prevent cracking and collapse of aligned kebabs into biomaterials based on sc-PLA.

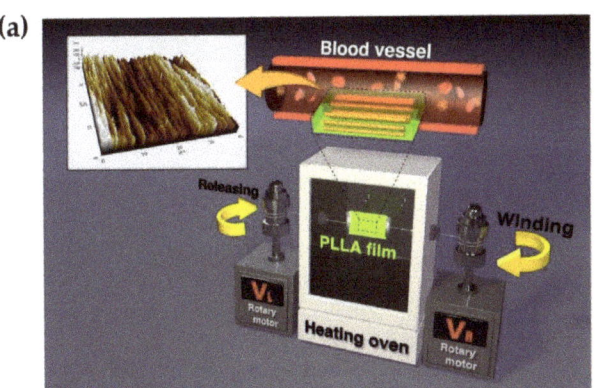

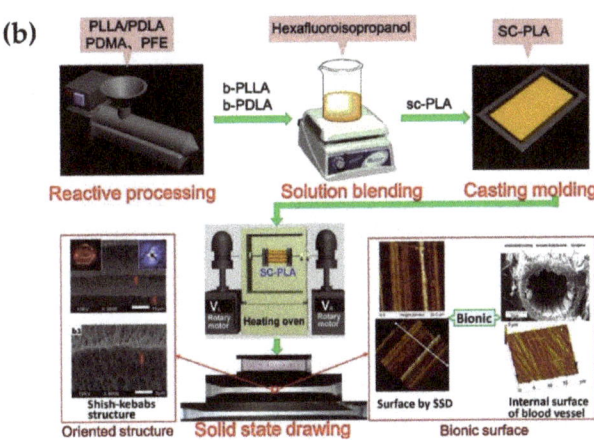

Figure 17. (**a**) Schematic of processing machine for SSD method with PLLA films or filaments [83], Copyright 2016. Reproduced with permission from IOP Publishing, Ltd. (**b**) Schematic representation for preparation process of oriented sc-PLA using SSD method [85], Copyright 2021. Reproduced with permission from the American Chemical Society.

C. Wang et al. (2019) synthesized injectable thermogels based on the stereocomplex 4-arm poly(ethylene glycol)-polylactide (PEG-PLA) and cholesterol-modified 4-arm PEG-PLA for optimized cartilage regeneration (Figure 18) [86]. The cholesterol-modified sc-PLA gels exhibited improved mechanical strength, lower critical gelation temperature, higher chondrocyte proliferation, and slower degradation than unmodified specimens. Moreover, cholesterol-modified sc-PLA gel-loaded chondrocytes showed considerably more cartilage-like tissues than fibrous- and bone-like tissues. This is attributed to the improved mechanical properties and microstructure induced by cholesterol modification of the sc-PLA gel.

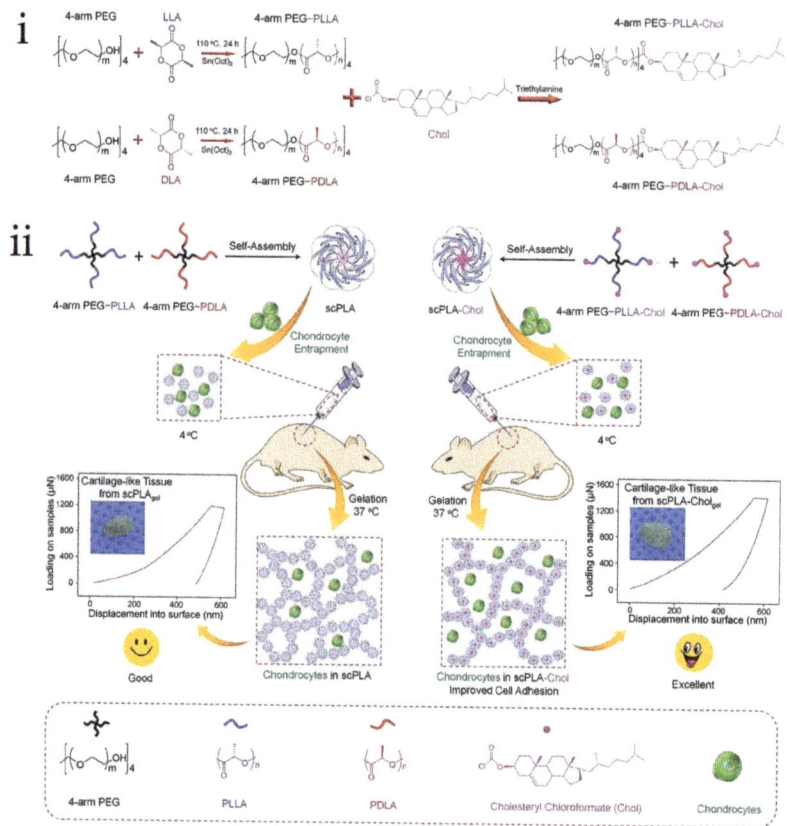

Figure 18. Schematic illustrations for (**i**) copolymer synthesis of 4-armPEG-PLL(D)A-cholesterol and (**ii**) in vivo cartilage regeneration of thermogels entrapped chondrocytes via subcutaneous injection into nude mice [86], Copyright 2019. Reproduced with permission from WILEY-VCH Verlag GmbH & Co.

3.3. Anti-Microbial Effect

Biomaterials and surgical implants based on the sc-PLA material are increasingly being applied for the above-mentioned biomedical applications, including tumor treatment and tissue engineering. However, preventing contamination from foreign microorganisms, such as bacteria and viruses is essential for the application of sc-PLA as a biomaterial in the human body [87–89]. Sterilization is necessary to prevent contamination before medical surgery, albeit it has been shown that contamination from bacteria such as *Staphylococcus aureus* (*S. aureus*) most frequently occurs during surgery [90–92]. Therefore, biomaterials based on sc-PLA are critical for securing anti-microbial effects to prevent bacterial proliferation to decrease adverse events and maximize clinical efficacy. Normally, biomaterials are sterilized by ethylene oxide gas, gamma ray irradiation, dry-heat sterilizer, microwave, autoclave, and supercritical carbon dioxide (CO_2) [93–97]. Unfortunately, these methods have remaining concerns not only regarding changes in the inherent properties of the material during the sterilization process, but also the difficulty in preventing secondary contamination. Thus, biomaterials must possess sustained anti-microbial effects to inhibit external microorganisms before and after implantation in the body. Spasova et al. (2010) fabricated electrospun sc-PLA fibers with antibacterial and hemostatic effects using diblock copolymers composed of poly(N,N-dimethylamino-2-ehtylmethacrylate) (PDMAEMA) [98]. After the incubation of mats composed of these fibers with *S. aureus* and *Escherichia coli*, the adhesion of these bacteria was observed. Consequently, sc-PLA mats containing PDMAEMA significantly inhibited bacterial adhesion and proliferation on the surface and exhibited effective antibacterial effects, while the control group showed bacterial adhesion and biofilm formation on the surface. This was attributed to the surface of tertiary amino groups from PDMAEMA blocks [99–101]. This surface modification strategy for antibacterial effects could remove concerns about contamination in surgical procedures and sustain its efficacy in the human body. Ajiro et al. (2016) polymerized PLLA and PDLA using catechin (CT) as an initiator precursor, which is an antibacterial compound [102]. Figure 19a shows the polymerization of CT-conjugated PLLA and PDLA at the chain end groups. Lactide was polymerized with benzyl catechin (BnCT), and then CT was chemically combined with PLAs to protect the phenolic hydroxyl groups. To assess antibacterial properties of the polymerized products, the ratio of total viable bacteria was calculated after 24 h according to the JIS Z 2801 test, as shown in Figure 19b. The CT-absorbed substrate reduced the ratio of total viable counts by 20% compared to the control. In contrast, both PLLA-CT and PLA-CT-SC substrates had approximately 50% significantly lower values than those of the control group. As shown in Figure 19c, the counts of killed bacteria per CT unit were compared with those of the CT-absorbed substrate, PLLA-CT, and PLA-CT-SC. Both PLLA-CT and PLA-CT-SC groups showed antibacterial properties mainly induced by the phenolic hydroxyl groups of CT. The antibacterial effect was dependent on the amount of CT, and it could be maintained for long-term use.

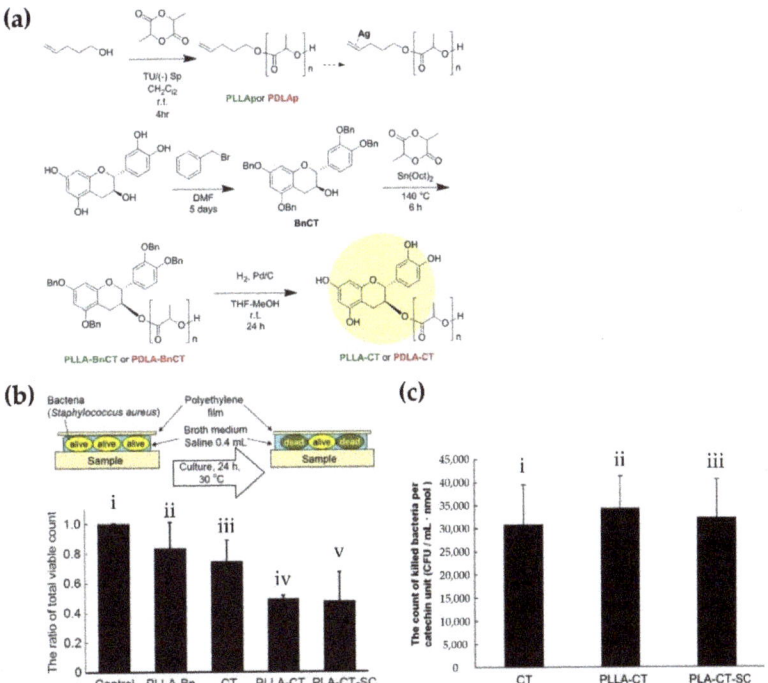

Figure 19. (a) Synthesis process schematic of chain-end modified PLLA and PDLA. (b) Comparison of antibacterial properties through ratio of live bacteria counts of (i) control, (ii) PLLA-Bn, (iii) CT, (iv) PLLA-CT, and (v) PLA-CT-SC against the control group. Top scheme indicates a method for measurement of antibacterial properties. The graphs were represented as mean values ± SD (n = 3). (c) Comparison of the counts of killed bacteria per CT unit of (i) CT, (ii) PLLA-CT, and (iii) PLA-CT-SC. The graphs were represented as mean values ± SD (n = 3) [102], Copyright 2016. Reproduced with permission from WILEY-VCH Verlag GmbH & Co.

Y. Li et al. (2013) suggested a novel approach for synthesizing charged hydrogels based on non-covalent interactions for disrupting biofilms and microbes [103]. They fabricated the antimicrobial gel by stereocomplexation between PLLA-PEG-PLLA and a charged PDLA-polycarbonate-PDLA (PDLA-CPC-PDLA) triblock polymer in aqueous solution. At physiological temperature (37 °C), the physical properties of the stereocomplex hydrogel were transformed into shear thinning behavior with supramolecular fiber and ribbon-like structures. This improved antimicrobial activity of the cationic hydrogel against diverse pathogenic microorganisms, such as fungi and both Gram-positive and Gram-negative bacteria. Up to 60% film biomass of *S. aureus*, *E. coli*, *Candida albicans* (*C. albicans*), and Methicillin-resistant *S. aureus* (MRSA) were eliminated after hydrogel treatment. In the safety test for skin sensitization, acute dermal toxicity, and skin irritation, the stereocomplex hydrogels appeared to provoke no adverse events in animal models of rats, guinea pigs, and rabbits. L. Mei et al. (2018) fabricated hybrid nanofibers by electrospinning in addition to PLA stereoisomers for inducing stereocomplexation between the stereoisomer chains and addition of chlorogenic acid (CA) of an antibacterial agent [104]. To prevent damage of the agent, the stereocomplex nanofibers were electrospun at relatively lower temperature (65 °C). The antibacterial fibers could effectively remove both gram-positive and gram-negative bacteria by quickly released CA within a few hours. The fibers based on sc-PLA have a potential to be applied in various fields such as filter, masks, and packages, owing to enhanced mechanical and thermal properties as well as full biodegradability. Recently, Y. Ren et al. (2019) reported successful fabrication of a novel eco-friendly sc-PLA nanofiber by electrospinning, with both functions of adsorption of heavy metal ions and inhibition of bacterial growth [105]. In the fabrication process, an antibacterial agent called HTA, which could be synthesized from tannic acid and hexamethylenediamine, was loaded into the products to provide an antibacterial effect. Furthermore, the electrospun nanofibers had improved tensile strength and Young's modulus, as well as thermal resistance owing to the formation of stereocomplex crystallites. The nanofiber mats based on sc-PLA exhibited excellent abilities for adsorption of Cr(VI) and capture of *E. coli* and *S. aureus*. Cr(VI) was converted to less toxic Cr(III) after its adsorption. This indicates that the heavy metal pollutant can be changed into eco-friendly and stable compounds in nature.

4. Conclusions

This review focuses on the research on sc-PLA with biodegradability and superior mechanical and thermal properties. sc-PLA can be used to produce therapeutic carriers in various forms, as it can be processed by micelles, self-assembly, emulsion, and 3D printing. In particular, 3D printing technology has commonly used PLA as filament material; however, sc-PLA has not been used in industrial practice to date. Because of these application fields and the extensive potential of 3D printing, inkjet printing using sc-PLA with improved characteristics must be a versatile technology for simultaneously inducing both stereocomplexation and fabrication in the future, if inkjet printing can be suitably converged with 3D printing techniques. In biomedical applications, sc-PLA nanoparticles are potentially promising carriers for anti-cancer therapy and scaffolds for tissue engineering, as numerous studies have demonstrated that they could be improved by diverse methods, including surface modification and co-polymerization, and attain additional functions, such as anti-microbial effects and immune stealth. sc-PLA nanoparticles effectively encapsulate therapeutic anti-tumor agents, such as DOX, and specifically transfer the agents to target lesions. This review suggests the fields for potential sc-PLA material applications by presenting current research trends. sc-PLA, as an advanced material from commonly used PLA, is expected to become a next-generation polymeric material owing to its excellent biocompatibility, biodegradability, and mechanical and thermal properties.

Author Contributions: Conceptualization, investigation, original draft preparation and writing, visualization, S.H.I.; resources, investigation, methodology, data curation, original draft preparation and writing, D.H.I.; resources, visualization, S.J.P.; review and editing, supervision, J.J.C.; review and editing, supervision, Y.J.; review and editing, supervision, project administration, funding acquisition, S.H.K. All authors have read and agreed to the published version of the manuscript.

Funding: This research was funded by the Ministry of Trade, Industry and Energy, Republic of Korea (project number 20008686) and by the KIST Institutional Program, and the KU-KIST Graduate School of Converging Science and Technology Program.

Institutional Review Board Statement: Not applicable.

Informed Consent Statement: Not applicable.

Data Availability Statement: Not applicable.

Conflicts of Interest: The authors declare no conflict of interest.

References

1. Im, S.H.; Park, S.J.; Chung, J.J.; Jung, Y.; Kim, S.H. Creation of polylactide vascular scaffolds with high compressive strength using a novel melt-tube drawing method. *Polymer* **2019**, *166*, 130–137. [CrossRef]
2. Im, S.H.; Jung, Y.; Kim, S.H. Current status and future direction of biodegradable metallic and polymeric vascular scaffolds for next-generation stents. *Acta Biomater.* **2017**, *60*, 3–22. [CrossRef] [PubMed]
3. Ikada, Y.; Jamshidi, K.; Tsuji, H.; Hyon, S.H. Stereocomplex formation between enantiomeric poly (lactides). *Macromolecules* **1987**, *20*, 904–906. [CrossRef]
4. Tsuji, H.; Hyon, S.H.; Ikada, Y. Stereocomplex formation between enantiomeric poly (lactic acid) s. 4. Differential scanning calorimetric studies on precipitates from mixed solutions of poly (D-lactic acid) and poly (L-lactic acid). *Macromolecules* **1991**, *24*, 5657–5662. [CrossRef]
5. Im, S.H.; Lee, C.W.; Bibi, G.; Jung, Y.; Kim, S.H. Supercritical fluid technology parameters affecting size and behavior of stereocomplex polylactide particles and their composites. *Polym Eng. Sci.* **2018**, *58*, 1193–1200. [CrossRef]
6. Tsuji, H.; Ikada, Y. Stereocomplex formation between enantiomeric poly (lactic acid) s. XI. Mechanical properties and morphology of solution-cast films. *Polymer* **1999**, *40*, 6699–6708.
7. Deng, S.; Bai, H.; Liu, Z.; Zhang, Q.; Fu, Q. Toward supertough and heat-resistant stereocomplex-type polylactide/elastomer blends with impressive melt stability via in situ formation of graft copolymer during one-pot reactive melt blending. *Macromolecules* **2019**, *52*, 1718–1730. [CrossRef]
8. Bibi, G.; Jung, Y.; Lim, J.-C.; Kim, S.H. Novel strategy of lactide polymerization leading to stereocomplex polylactide nanoparticles using supercritical fluid technology. *ACS Sustain. Chem. Eng.* **2016**, *4*, 4521–4528. [CrossRef]
9. Im, S.H.; Jung, Y.; Kim, S.H. In situ homologous polymerization of l-lactide Having a stereocomplex crystal. *Macromolecules* **2018**, *51*, 6303–6311.
10. Luo, F.; Fortenberry, A.; Ren, J.; Qiang, Z.J. Recent progress in enhancing poly (lactic acid) stereocomplex formation for material property improvement. *Front. Chem.* **2020**, *8*, 688.
11. Sánchez, A.; Mejía, S.P.; Orozco, J. Recent Advances in Polymeric Nanoparticle-Encapsulated Drugs against Intracellular Infections. *Molecules* **2020**, *25*, 3760. [CrossRef]
12. Tsuji, H. Poly (lactic acid) stereocomplexes: A decade of progress. *Adv. Drug Deliv. Rev.* **2016**, *107*, 97–135. [CrossRef]
13. Scheuer, K.; Bandelli, D.; Helbing, C.; Weber, C.; Alex, J.; Max, J.B.; Hocken, A.; Stranik, O.; Seiler, L.; Gladigau, F. Self-assembly of copolyesters into stereocomplex crystallites tunes the properties of polyester nanoparticles. *Macromolecules* **2020**, *53*, 8340–8351. [CrossRef]
14. Li, Z.; Yuan, D.; Jin, G.; Tan, B.H.; He, C. Facile layer-by-layer self-assembly toward enantiomeric poly (lactide) stereocomplex coated magnetite nanocarrier for highly tunable drug deliveries. *ACS Appl. Mater. Interfaces* **2016**, *8*, 1842–1853. [CrossRef]
15. Boi, S.; Dellacasa, E.; Bianchini, P.; Petrini, P.; Pastorino, L.; Monticelli, O.J.C. Encapsulated functionalized stereocomplex PLA particles: An effective system to support mucolytic enzymes. *Colloids Surf. B Colloid Surface B* **2019**, *179*, 190–198. [CrossRef]
16. Niu, K.; Yao, Y.; Xiu, M.; Guo, C.; Ge, Y.; Wang, J. Controlled drug delivery by polylactide stereocomplex micelle for cervical cancer chemotherapy. *Front. Pharmacol.* **2018**, *9*, 930. [CrossRef]
17. Li, W.; Fan, X.; Wang, X.; Shang, X.; Wang, Q.; Lin, J.; Hu, Z.; Li, Z. Stereocomplexed micelle formation through enantiomeric PLA-based Y-shaped copolymer for targeted drug delivery. *Mater. Sci. Eng. C* **2018**, *91*, 688–695. [CrossRef]
18. Burt, H.M.; Zhang, X.; Toleikis, P.; Embree, L.; Hunter, W.L. Development of copolymers of poly (D, L-lactide) and methoxypolyethylene glycol as micellar carriers of paclitaxel. *Colloids Surf. B Colloid Surface B* **1999**, *16*, 161–171. [CrossRef]
19. Gref, R.; Minamitake, Y.; Peracchia, M.T.; Trubetskoy, V.; Torchilin, V.; Langer, R. Biodegradable long-circulating polymeric nanospheres. *Science* **1994**, *263*, 1600–1603. [CrossRef]
20. Jones, M.-C.; Leroux, J.-C. Polymeric micelles–a new generation of colloidal drug carriers. *Eur. J. Pharm. Biopharm.* **1999**, *48*, 101–111. [CrossRef]

21. Kang, N.; Perron, M.-È.; Prud'Homme, R.E.; Zhang, Y.; Gaucher, G.; Leroux, J.-C. Stereocomplex block copolymer micelles: Core− shell nanostructures with enhanced stability. *Nano Lett.* **2005**, *5*, 315–319. [CrossRef] [PubMed]
22. An, L.; Cao, M.; Zhang, X.; Lin, J.; Tian, Q.; Yang, S. PH and glutathione synergistically triggered release and self-assembly of Au nanospheres for tumor theranostics. *ACS Appl. Mater. Interfaces* **2020**, *12*, 8050–8061. [CrossRef] [PubMed]
23. Dai, W.; Zhu, X.; Zhang, J.; Zhao, Y. Temperature and solvent isotope dependent hierarchical self-assembly of a heterografted block copolymer. *Chem. Commun.* **2019**, *55*, 5709–5712. [CrossRef] [PubMed]
24. Zhou, D.; Dong, S.; Kuchel, R.P.; Perrier, S.; Zetterlund, P.B. Polymerization induced self-assembly: Tuning of morphology using ionic strength and pH. *Polym. Chem.* **2017**, *8*, 3082–3089.
25. Gerbelli, B.B.; Vassiliades, S.V.; Rojas, J.E.; Pelin, J.N.; Mancini, R.S.; Pereira, W.S.; Aguilar, A.M.; Venanzi, M.; Cavalieri, F.; Giuntini, F. Hierarchical self-assembly of peptides and its applications in bionanotechnology. *Macromol. Chem. Phys.* **2019**, *220*, 1900085. [CrossRef]
26. Bishara, A.; Kricheldorf, H.R.; Domb, A.J. Stereocomplexes of Triblock Poly (lactide-PEG2000-lactide) as Carrier of Drugs. *Macromol. Symp.* **2005**, *225*, 17–30. [CrossRef]
27. Liu, R.; He, B.; Li, D.; Lai, Y.; Tang, J.Z.; Gu, Z. Stabilization of pH-Sensitive mPEG–PH–PLA nanoparticles by stereocomplexation between enantiomeric polylactides. *Macromol. Rapid Commun.* **2012**, *33*, 1061–1066. [CrossRef]
28. Lee, E.S.; Na, K.; Bae, Y.H. Super pH-sensitive multifunctional polymeric micelle. *Nano Lett.* **2005**, *5*, 325–329. [CrossRef]
29. Lim, D.W.; Park, T.G. Stereocomplex formation between enantiomeric PLA–PEG–PLA triblock copolymers: Characterization and use as protein-delivery microparticulate carriers. *J. App. Polym. Sci.* **2000**, *75*, 1615–1623. [CrossRef]
30. Kurapati, R.; Groth, T.W.; Raichur, A.M. Recent developments in layer-by-layer technique for drug delivery applications. *ACS Appl. Bio Mater.* **2019**, *2*, 5512–5527. [CrossRef]
31. Hellwig, J.; Strebe, J.; Klitzing, R.V. Effect of environmental parameters on the nano mechanical properties of hyaluronic acid/poly (l-lysine) multilayers. *Polymers* **2018**, *20*, 19082–19086. [CrossRef]
32. Apte, G.; Repanas, A.; Willems, C.; Mujtaba, A.; Schmelzer, C.E.; Raichur, A.; Syrowatka, F.; Groth, T. Effect of different crosslinking strategies on physical properties and biocompatibility of freestanding multilayer films made of alginate and chitosan. *Macromol. Biosci.* **2019**, *19*, 1900181. [CrossRef]
33. Luo, X.; Song, H.; Yang, J.; Han, B.; Feng, Y.; Leng, Y.; Chen, Z. Encapsulation of Escherichia coli strain Nissle 1917 in a chitosan—alginate matrix by combining layer-by-layer assembly with CaCl2 cross-linking for an effective treatment of inflammatory bowel diseases. *Colloids Surf. B Colloid Surf. B* **2020**, *189*, 110818. [CrossRef]
34. Ariga, K.; Ahn, E.; Park, M.; Kim, B.S. Layer-by-layer assembly: Recent progress from layered assemblies to layered nanoarchitectonics. *Chem. Asian J.* **2019**, *14*, 2553–2566. [CrossRef]
35. Kondo, K.; Kida, T.; Ogawa, Y.; Arikawa, Y.; Akashi, M.J. Nanotube formation through the continuous one-dimensional fusion of hollow nanocapsules composed of layer-by-layer poly (lactic acid) stereocomplex films. *J. Am. Chem. Soc.* **2010**, *132*, 8236–8237. [CrossRef]
36. Ahvenniemi, E.; Karppinen, M. In Situ atomic/molecular layer-by-layer deposition of inorganic–organic coordination network thin films from gaseous precursors. *Chem. Mater.* **2016**, *28*, 6260–6265. [CrossRef]
37. Dellacasa, E.; Zhao, L.; Yang, G.; Pastorino, L.; Sukhorukov, G.B. Fabrication and characterization of novel multilayered structures by stereocomplexion of poly (D-lactic acid)/poly (L-lactic acid) and self-assembly of polyelectrolytes. *Beilstein J. Nanotechnol.* **2016**, *7*, 81–90. [CrossRef]
38. Brzeziński, M. Hollow microcapsules with enhanced stability via stereocomplex assemblies. *Macromol. Chem. Phys.* **2017**, *218*, 1700018. [CrossRef]
39. Donath, E.; Sukhorukov, G.B.; Caruso, F.; Davis, S.A.; Möhwald, H. Novel hollow polymer shells by colloid-templated assembly of polyelectrolytes. *Angew. Chem.* **1998**, *37*, 2201–2205. [CrossRef]
40. Yow, H.N.; Routh, A.F. Formation of liquid core–polymer shell microcapsules. *Soft Matter* **2006**, *2*, 940–949. [CrossRef]
41. Ma, Q.; Song, Y.; Kim, J.W.; Choi, H.S.; Shum, H.C. Affinity partitioning-induced self-assembly in aqueous two-phase systems: Templating for polyelectrolyte microcapsules. *ACS Macro Lett.* **2016**, *5*, 666–670. [CrossRef]
42. Liu, H.; Bai, D.; Bai, H.; Zhang, Q.; Fu, Q. Constructing stereocomplex structures at the interface for remarkably accelerating matrix crystallization and enhancing the mechanical properties of poly (L-lactide)/multi-walled carbon nanotube nanocomposites. *J. Mater. Chem. A* **2015**, *3*, 13835–13847. [CrossRef]
43. Im, S.H.; Park, S.J.; Jung, Y.; Chung, J.J.; Kim, S.H. Strategy for stereocomplexation of polylactide using O/W emulsion blending and applications as composite fillers, drug carriers, and self-nucleating agents. *ACS Sustain. Chem. Eng.* **2020**, *8*, 8752–8761. [CrossRef]
44. Akagi, T.; Fujiwara, T.; Akashi, M. Rapid Fabrication of Polylactide Stereocomplex Using Layer-by-Layer Deposition by Inkjet Printing. *Angew. Chem.* **2012**, *124*, 5589–5592. [CrossRef]
45. Akagi, T.; Fujiwara, T.; Akashi, M. Inkjet printing of layer-by-layer assembled poly (lactide) stereocomplex with encapsulated proteins. *Langmuir* **2014**, *30*, 1669–1676. [CrossRef]
46. Tran, H.T.; Ajiro, H.; Hsiao, Y.-J.; Akashi, M. Thermally resistant polylactide layer-by-layer film prepared using an inkjet approach. *Polym. J.* **2017**, *49*, 327–334. [CrossRef]

47. De Jong, S.; Van Eerdenbrugh, B.; van Nostrum, C.V.; Kettenes-Van Den Bosch, J.; Hennink, W. Physically crosslinked dextran hydrogels by stereocomplex formation of lactic acid oligomers: Degradation and protein release behavior. *J. Control. Release* **2001**, *71*, 261–275. [CrossRef]
48. Hennink, W.; De Jong, S.; Bos, G.; Veldhuis, T.; Van Nostrum, C. Biodegradable dextran hydrogels crosslinked by stereocomplex formation for the controlled release of pharmaceutical proteins. *Int. J. Pharm.* **2004**, *277*, 99–104. [CrossRef]
49. Cao, L.; Cao, B.; Lu, C.; Wang, G.; Yu, L.; Ding, J. An injectable hydrogel formed by in situ cross-linking of glycol chitosan and multi-benzaldehyde functionalized PEG analogues for cartilage tissue engineering. *J. Mater. Chem. B* **2015**, *3*, 1268–1280. [CrossRef]
50. Kretlow, J.D.; Young, S.; Klouda, L.; Wong, M.; Mikos, A.G. Injectable biomaterials for regenerating complex craniofacial tissues. *Adv. Mater.* **2009**, *21*, 3368–3393. [CrossRef]
51. Purcell, B.P.; Lobb, D.; Charati, M.B.; Dorsey, S.M.; Wade, R.J.; Zellars, K.N.; Doviak, H.; Pettaway, S.; Logdon, C.B.; Shuman, J.A. Injectable and bioresponsive hydrogels for on-demand matrix metalloproteinase inhibition. *Nat. Mater.* **2014**, *13*, 653–661. [CrossRef] [PubMed]
52. Zhang, Y.; Cao, Y.; Zhao, H.; Zhang, L.; Ni, T.; Liu, Y.; An, Z.; Liu, M.; Pei, R.J. An injectable BMSC-laden enzyme-catalyzed crosslinking collagen-hyaluronic acid hydrogel for cartilage repair and regeneration. *J. Mater. Chem. B* **2020**, *8*, 4237–4244. [CrossRef] [PubMed]
53. Liu, J.; Li, J.; Yu, F.; Zhao, Y.-x.; Mo, X.-m.; Pan, J.-F. In situ forming hydrogel of natural polysaccharides through Schiff base reaction for soft tissue adhesive and hemostasis. *Int. J. Biol. Macromol.* **2020**, *147*, 653–666. [CrossRef] [PubMed]
54. Zhang, J.; Wang, S.; Zhao, Z.; Si, D.; Zhou, H.; Yang, M.; Wang, X. An In situ forming hydrogel based on photo-induced hydrogen bonding. *Macromol. Res.* **2020**, *28*, 1127–1133. [CrossRef]
55. Dong, Y.; Cui, M.; Qu, J.; Wang, X.; Kwon, S.H.; Barrera, J.; Elvassore, N.; Gurtner, G.C. Conformable hyaluronic acid hydrogel delivers adipose-derived stem cells and promotes regeneration of burn injury. *Acta Biomater.* **2020**, *108*, 56–66. [CrossRef]
56. Kang, H.; Rho, S.; Stiles, W.R.; Hu, S.; Baek, Y.; Hwang, D.W.; Kashiwagi, S.; Kim, M.S.; Choi, H.S. Size-dependent EPR effect of polymeric nanoparticles on tumor targeting. *Adv. Healthc. Mater.* **2020**, *9*, 1901223. [CrossRef]
57. Goos, J.A.; Cho, A.; Carter, L.M.; Dilling, T.R.; Davydova, M.; Mandleywala, K.; Puttick, S.; Gupta, A.; Price, W.S.; Quinn, J.F. Delivery of polymeric nanostars for molecular imaging and endoradiotherapy through the enhanced permeability and retention (EPR) effect. *Theranostics* **2020**, *10*, 567. [CrossRef]
58. Fang, J.; Islam, W.; Maeda, H.J. Exploiting the dynamics of the EPR effect and strategies to improve the therapeutic effects of nanomedicines by using EPR effect enhancers. *Adv. Drug Deliv. Rev.* **2020**, *157*, 142–160. [CrossRef]
59. Ding, Y.; Xu, Y.; Yang, W.; Niu, P.; Li, X.; Chen, Y.; Li, Z.; Liu, Y.; An, Y.; Liu, Y. Investigating the EPR effect of nanomedicines in human renal tumors via ex vivo perfusion strategy. *Nano Today* **2020**, *35*, 100970. [CrossRef]
60. Izci, M.; Maksoudian, C.; Manshian, B.B.; Soenen, S.J. The use of alternative strategies for enhanced nanoparticle delivery to solid tumors. *Chem. Rev.* **2021**, *121*, 1746–1803. [CrossRef]
61. Guo, Z.; Sui, J.; Ma, M.; Hu, J.; Sun, Y.; Yang, L.; Fan, Y.; Zhang, X.J. pH-Responsive charge switchable PEGylated ε-poly-l-lysine polymeric nanoparticles-assisted combination therapy for improving breast cancer treatment. *J. Control. Release* **2020**, *326*, 350–364. [CrossRef]
62. Goldberg, J.S. Stereocomplexes formed from select oligomers of polymer d-lactic acid (pDLA) and l-lactate may inhibit growth of cancer cells and help diagnose aggressive cancers—Applications of the Warburg effect. *Perspect. Med. Chem.* **2011**, *5*, 1–10. [CrossRef]
63. Brzeziński, M.; Kost, B.; Wedepohl, S.; Socka, M.; Biela, T.; Calderón, M. Stereocomplexed PLA microspheres: Control over morphology, drug encapsulation and anticancer activity. *Colloids Surf. B Colloid Surf. B* **2019**, *184*, 110544. [CrossRef]
64. Kost, B.; Brzeziński, M.; Cieślak, M.; Królewska-Golińska, K.; Makowski, T.; Socka, M.; Biela, T.J. Stereocomplexed micelles based on polylactides with β-cyclodextrin core as anti-cancer drug carriers. *Eur. Polym. J.* **2019**, *120*, 109271. [CrossRef]
65. Koons, G.L.; Diba, M.; Mikos, A.G. Materials design for bone-tissue engineering. *Nat. Rev. Mater.* **2020**, *5*, 584–603. [CrossRef]
66. Dong, R.; Ma, P.X.; Guo, B. Conductive biomaterials for muscle tissue engineering. *Biomaterials* **2020**, *229*, 119584. [CrossRef]
67. Qu, H.; Fu, H.; Han, Z.; Sun, Y. Biomaterials for bone tissue engineering scaffolds: A review. *RSC Adv.* **2019**, *9*, 26252–26262. [CrossRef]
68. Williams, D.F. Challenges with the development of biomaterials for sustainable tissue engineering. *Front. Bioeng. Biotechnol.* **2019**, *7*, 127. [CrossRef]
69. Pina, S.; Ribeiro, V.P.; Marques, C.F.; Maia, F.R.; Silva, T.H.; Reis, R.L.; Oliveira, J.M. Scaffolding strategies for tissue engineering and regenerative medicine applications. *Polymers* **2019**, *12*, 1824. [CrossRef]
70. Cheng, A.; Schwartz, Z.; Kahn, A.; Li, X.; Shao, Z.; Sun, M.; Ao, Y.; Boyan, B.D.; Chen, H. Advances in porous scaffold design for bone and cartilage tissue engineering and regeneration. *Tissue Eng. Part B Rev.* **2019**, *25*, 14–29. [CrossRef]
71. Hassanajili, S.; Karami-Pour, A.; Oryan, A.; Talaei-Khozani, T. Preparation and characterization of PLA/PCL/HA composite scaffolds using indirect 3D printing for bone tissue engineering. *Mater. Sci. Eng. C* **2019**, *104*, 109960. [CrossRef] [PubMed]
72. Sartore, L.; Inverardi, N.; Pandini, S.; Bignotti, F.; Chiellini, F. PLA/PCL-based foams as scaffolds for tissue engineering applications. *Mater. Today Proc.* **2019**, *7*, 410–417. [CrossRef]
73. Mader, M.; Jérôme, V.R.; Freitag, R.; Agarwal, S.; Greiner, A. Ultraporous, compressible, wettable polylactide/polycaprolactone sponges for tissue engineering. *Biomacromolecules* **2018**, *19*, 1663–1673. [CrossRef] [PubMed]

74. Gayer, C.; Ritter, J.; Bullemer, M.; Grom, S.; Jauer, L.; Meiners, W.; Pfister, A.; Reinauer, F.; Vučak, M.; Wissenbach, K. Development of a solvent-free polylactide/calcium carbonate composite for selective laser sintering of bone tissue engineering scaffolds. *Mater. Sci. Eng. C* **2019**, *101*, 660–673. [CrossRef]
75. Gritsch, L.; Conoscenti, G.; La Carrubba, V.; Nooeaid, P.; Boccaccini, A.R. Polylactide-based materials science strategies to improve tissue-material interface without the use of growth factors or other biological molecules. *Mater. Sci. Eng. C* **2019**, *94*, 1083–1101. [CrossRef]
76. Kang, Y.; Chen, P.; Shi, X.; Zhang, G.; Wang, C. Multilevel structural stereocomplex polylactic acid/collagen membranes by pattern electrospinning for tissue engineering. *Polymer* **2018**, *156*, 250–260. [CrossRef]
77. Im, S.H.; Kim, C.Y.; Lee, C.W.; Jung, Y.; Kim, S.H. Strategy for securing key patents in the field of biomaterials. *Macromol. Res.* **2020**, *28*, 87–98. [CrossRef]
78. Gupta, A.; Prasad, A.; Mulchandani, N.; Shah, M.; Ravi Sankar, M.; Kumar, S.; Katiyar, V. Multifunctional nanohydroxyapatite-promoted toughened high-molecular-weight stereocomplex poly (lactic acid)-based bionanocomposite for both 3D-printed orthopedic implants and high-temperature engineering applications. *ACS Omega* **2017**, *2*, 4039–4052. [CrossRef]
79. Mulchandani, N.; Gupta, A.; Masutani, K.; Kumar, S.; Sakurai, S.; Kimura, Y.; Katiyar, V. Effect of block length and stereocomplex-ation on the thermally processable poly (ε-caprolactone) and poly (lactic acid) block copolymers for biomedical applications. *ACS Appl. Polym. Mater.* **2019**, *1*, 3354–3365. [CrossRef]
80. Singh, A.A.; Wei, J.; Herrera, N.; Geng, S.; Oksman, K. Synergistic effect of chitin nanocrystals and orientations induced by solid-state drawing on PLA-based nanocomposite tapes. *Compos. Sci. Technol.* **2018**, *162*, 140–145. [CrossRef]
81. Lin, Y.; Tu, W.; Verpaalen, R.C.; Zhang, H.; Bastiaansen, C.W.; Peijs, T. Transparent, lightweight, and high strength polyethylene films by a scalable continuous extrusion and solid-state drawing process. *Macromol. Mater. Eng.* **2019**, *304*, 1900138. [CrossRef]
82. Walker, J.; Melaj, M.; Giménez, R.; Pérez, E.; Bernal, C. Solid-state drawing of commercial poly (lactic acid)(PLA) based filaments. *Front. Mater.* **2019**, *6*, 280. [CrossRef]
83. Im, S.H.; Jung, Y.; Jang, Y.; Kim, S.H. Poly (L-lactic acid) scaffold with oriented micro-valley surface and superior properties fabricated by solid-state drawing for blood-contact biomaterials. *Biofabrication* **2016**, *8*, 045010. [CrossRef]
84. Im, S.H.; Kim, C.Y.; Jung, Y.; Jang, Y.; Kim, S.H. Biodegradable vascular stents with high tensile and compressive strength: A novel strategy for applying monofilaments via solid-state drawing and shaped-annealing processes. *Biomater. Sci.* **2017**, *5*, 422–431. [CrossRef]
85. Li, J.; Ye, W.; Fan, Z.; Cao, L. A novel stereocomplex poly (lactic acid) with shish-kebab crystals and bionic surface structures as bioimplant materials for tissue engineering applications. *ACS Appl. Mater. Interfaces* **2021**, *13*, 5469–5477. [CrossRef]
86. Wang, C.; Feng, N.; Chang, F.; Wang, J.; Yuan, B.; Cheng, Y.; Liu, H.; Yu, J.; Zou, J.; Ding, J. Injectable cholesterol-enhanced stereocomplex polylactide thermogel loading chondrocytes for optimized cartilage regeneration. *Adv. Healthc. Mater.* **2019**, *8*, 1900312. [CrossRef]
87. Xue, Q.; Liu, X.-B.; Lao, Y.-H.; Wu, L.-P.; Wang, D.; Zuo, Z.-Q.; Chen, J.-Y.; Hou, J.; Bei, Y.-Y.; Wu, X.-F. Anti-infective biomaterials with surface-decorated tachyplesin I. *Biomaterials* **2018**, *178*, 351–362. [CrossRef]
88. Reigada, I.; Pérez-Tanoira, R.; Patel, J.Z.; Savijoki, K.; Yli-Kauhaluoma, J.; Kinnari, T.J.; Fallarero, A. Strategies to prevent biofilm infections on biomaterials: Effect of novel naturally-derived biofilm inhibitors on a competitive colonization model of titanium by Staphylococcus aureus and SaOS-2 cells. *Microorganisms* **2020**, *8*, 345. [CrossRef]
89. Spriano, S.; Yamaguchi, S.; Baino, F.; Ferraris, S. A critical review of multifunctional titanium surfaces: New frontiers for improving osseointegration and host response, avoiding bacteria contamination. *Acta Biomater.* **2018**, *79*, 1–22. [CrossRef]
90. Danese, P.N. Antibiofilm approaches: Prevention of catheter colonization. *Chem. Biol.* **2002**, *9*, 873–880. [CrossRef]
91. Campoccia, D.; Montanaro, L.; Arciola, C.R. The significance of infection related to orthopedic devices and issues of antibiotic resistance. *Biomaterials* **2006**, *27*, 2331–2339. [CrossRef]
92. Oliveira, W.; Silva, P.; Silva, R.; Silva, G.; Machado, G.; Coelho, L.; Correia, M.J. Staphylococcus aureus and Staphylococcus epidermidis infections on implants. *J. Hosp. Infect.* **2018**, *98*, 111–117. [CrossRef]
93. Kim, S.; Jeong, J.-O.; Lee, S.; Park, J.-S.; Gwon, H.-J.; Jeong, S.I.; Hardy, J.G.; Lim, Y.-M.; Lee, J.Y. Effective gamma-ray sterilization and characterization of conductive polypyrrole biomaterials. *Sci. Rep.* **2018**, *8*, 1–10. [CrossRef]
94. Harrington, R.E.; Guda, T.; Lambert, B.; Martin, J. Sterilization and disinfection of biomaterials for medical devices. *Biomater. Sci.* **2020**, 1431–1446.
95. Rodriguez Chanfrau, J.E. Evaluation of the influence of microwaves radiation on a biomaterial composed of three phases of calcium phosphates. *Biointerface Res. Appl. Chem.* **2020**, 5141–5144.
96. Henise, J.; Yao, B.; Ashley, G.W.; Santi, D.V. Autoclave sterilization of tetra-polyethylene glycol hydrogel biomaterials with β-eliminative crosslinks. *Eng. Rep.* **2020**, *2*, e12091. [CrossRef]
97. Soares, G.C.; Learmonth, D.A.; Vallejo, M.C.; Davila, S.P.; González, P.; Sousa, R.A.; Oliveira, A.L. Supercritical CO_2 technology: The next standard sterilization technique? *Mater. Sci. Eng. C* **2019**, *99*, 520–540. [CrossRef]
98. Spasova, M.; Manolova, N.; Paneva, D.; Mincheva, R.; Dubois, P.; Rashkov, I.; Maximova, V.; Danchev, D. Polylactide stereocomplex-based electrospun materials possessing surface with antibacterial and hemostatic properties. *Biomacromolecules* **2010**, *11*, 151–159. [CrossRef]

99. Yancheva, E.; Paneva, D.; Maximova, V.; Mespouille, L.; Dubois, P.; Manolova, N.; Rashkov, I. Polyelectrolyte complexes between (cross-linked) N-carboxyethylchitosan and (quaternized) poly [2-(dimethylamino) ethyl methacrylate]: Preparation, characterization, and antibacterial properties. *Biomacromolecules* **2007**, *8*, 976–984. [CrossRef]
100. Yancheva, E.; Paneva, D.; Danchev, D.; Mespouille, L.; Dubois, P.; Manolova, N.; Rashkov, I. Polyelectrolyte complexes based on (quaternized) poly [(2-dimethylamino) ethyl methacrylate]: Behavior in contact with blood. *Macromol. Biosci.* **2007**, *7*, 940–954. [CrossRef]
101. Huang, J.; Murata, H.; Koepsel, R.R.; Russell, A.J.; Matyjaszewski, K.J.B. Antibacterial polypropylene via surface-initiated atom transfer radical polymerization. *Biomacromolecules* **2007**, *8*, 1396–1399. [CrossRef] [PubMed]
102. Ajiro, H.; Ito, S.; Kan, K.; Akashi, M. Catechin-modified polylactide stereocomplex at chain End improved antibiobacterial property. *Macromol. Biosci.* **2016**, *16*, 694–704. [CrossRef] [PubMed]
103. Li, Y.; Fukushima, K.; Coady, D.J.; Engler, A.C.; Liu, S.; Huang, Y.; Cho, J.S.; Guo, Y.; Miller, L.S.; Tan, J.P. Broad-spectrum antimicrobial and biofilm-disrupting hydrogels: Stereocomplex-driven supramolecular assemblies. *Angew. Chem.* **2013**, *125*, 702–706. [CrossRef]
104. Mei, L.; Ren, Y.; Gu, Y.; Li, X.; Wang, C.; Du, Y.; Fan, R.; Gao, X.; Chen, H.; Tong, A. Strengthened and thermally resistant poly (lactic acid)-based composite nanofibers prepared via easy stereocomplexation with antibacterial effects. *ACS Appl. Mater. Interfaces* **2018**, *10*, 42992–43002. [CrossRef]
105. Ren, Y.; Mei, L.; Gu, Y.; Zhao, N.; Wang, Y.; Fan, R.; Tong, A.; Chen, H.; Yang, H.; Han, B. Stereocomplex crystallite-based eco-friendly nanofiber membranes for removal of Cr (VI) and antibacterial effects. *ACS Sustain. Chem. Eng.* **2019**, *7*, 16072–16083. [CrossRef]

Review

Crosslinking of Polylactide by High Energy Irradiation and Photo-Curing

Melania Bednarek [1,*], Katarina Borska [1,2] and Przemysław Kubisa [1]

[1] Centre of Molecular and Macromolecular Studies, Polish Academy of Sciences, Sienkiewicza 112, 90-362 Lodz, Poland; katarina.borska@savba.sk (K.B.); pkubisa@cbmm.lodz.pl (P.K.)
[2] Polymer Institute, Slovak Academy of Sciences, Dubravska Cesta 9, 845 41 Bratislava, Slovakia
* Correspondence: bednarek@cbmm.lodz.pl

Academic Editor: Dimitrios Bikiaris
Received: 29 September 2020; Accepted: 20 October 2020; Published: 23 October 2020

Abstract: Polylactide (PLA) is presently the most studied bioderived polymer because, in addition to its established position as a material for biomedical applications, it can replace mass production plastics from petroleum. However, some drawbacks of polylactide such as insufficient mechanical properties at a higher temperature and poor shape stability have to be overcome. One of the methods of mechanical and thermal properties modification is crosslinking which can be achieved by different approaches, both at the stage of PLA-based materials synthesis and by physical modification of neat polylactide. This review covers PLA crosslinking by applying different types of irradiation, i.e., high energy electron beam or gamma irradiation and UV light which enables curing at mild conditions. In the last section, selected examples of biomedical applications as well as applications for packaging and daily-use items are presented in order to visualize how a variety of materials can be obtained using specific methods.

Keywords: polylactide; poly(lactic acid); crosslinking; photo-crosslinking; irradiation; electron-beam; gamma rays

---

1. Introduction

Polylactide/poly(lactic acid) (PLA) is a biodegradable and biobased aliphatic polyester derived from renewable sources such as corn, potato, and sugar cane. Due to biodegradability and biocompatibility (PLA is approved by US Food and Drug Administration for contact with human cells) the early applications of polylactide (or its copolymers with polyglycolides) involved surgical sutures, implants or drug formulations [1–5]. The use of PLA was initially limited to these biomedical applications due to its high cost and low availability.

Biodegradability is considered as one of the major advantages of polylactide, thus, PLA materials (apart from biomedical applications) have been also used for the production of short-use items and packaging [1]. More recently, however, there have been some concerns related to the environmental impact of the utilization of biodegradable polymers as the dumping of biobased waste in landfills contributes to global warming and leachate [6,7]. Thus, in recent years the shift from degradability/"compostability" to "renewability" and an increasing interest in using PLA based products for long-term usage applications, even at the expense of reducing the biodegradability of the polymer may be observed [1]. It is expected that PLA will have broader applications in the medical and food industries, however, much has to be overcome to ensure actual sustainability, including enhancement of mechanical and thermal properties [8]. There are many efforts to improve the performance of PLA via different modification methods including copolymerization, blending, or crosslinking [9–12]. The design of novel structures such as networks based on polyesters offers a possibility for enhancing mechanical and thermal properties. The improvement of mechanical strength is needed for both medical devices and parts of daily-use items. However, when designing materials for these applications, an awareness that the crosslinking process leads not only to the enhancement of polymer toughness (usually with the improvement of thermal properties) but also to the modification of other properties as degradability, solubility, gas permeability, and so on must be maintained. Basically, crosslinking leads to a decrease in degradability and this may be a negative feature for materials designed for biomedical applications (implants and drug delivery systems). However, in some cases, longer times of implant destruction or longer delivery of the pharmaceutical agent could be required. Thus, very detailed studies of crosslinking of biobased and biocompatible materials include investigation of their degradation in different environments and their biocompatibility. The balance between mechanical strength and degradability of materials should be established at the stage of their design in order to fulfill the requirements of the target application.

Networks can be obtained through different approaches. Applied methods are usually classified into several main groups, that is chemical crosslinking and crosslinking by exposure to low- energy light or ionizing radiation. Chemical crosslinking is one of the largest groups because, among others, this group includes all methods based on radical crosslinking induced by peroxides [13].

Both high-energy radiation crosslinking and photo-crosslinking are relatively easy methods in comparison with chemical crosslinking, however, some advantages and weakness of each of them are known. Photo-crosslinking requires the presence of reactive -unsaturated groups in polymer chains (functionalization of synthesized polymers) and the presence of photo-initiators but may be accomplished under mild conditions (room temperature) for different materials (solid, liquid, containing different encapsulated compounds including sensitive proteins, etc.) [14]. High energy radiation, such as gamma and electron beam radiation has been applied for various treatments of polymers, and also crosslinking [15,16]. Crosslinking by irradiation may be performed on pristine polymer without the necessity of functionalization because radicals may be generated directly on polymer chains. However, it is accompanied or even dominated by chain scission. Thus, even in this case using additives (crosslinking monomers) is preferred.

This review does not include chemical crosslinking which has been broadly covered by other articles [13,17]. Instead, the present state of the art concerning the formation of crosslinked PLA-based materials by photo-crosslinking and high energy irradiation is presented. Totally different materials are obtained depending mainly on crosslinking density which, in turn, depends on the radiation type, its dose, conditions, and on PLA-based polymer (a composition and an architecture of this (co)polymer have an independent contribution to the overall properties).

2. PLA Crosslinking by Electron Beam or Gamma Irradiation

Ionizing radiation has been well known as a very convenient tool for the modification of polymeric materials through crosslinking, grafting, and degradation. PLA however, undergoes predominantly degradation under the influence of ionizing radiation. Thus, the mechanical and physical properties of polymers exposed to gamma rays or an electron beam, decrease due to the reduction in molecular weight.

Beginning with the advent of research related to the development of polyester materials for the production of implants, surgical sutures, and drug delivery systems, high energy irradiation, mainly gamma but also electron irradiation has been used for sterilization [18–21]. It has been found that, depending on the chemical structure of the polymers, the absorbed dose, the dose rate, and the temperature of irradiation, various reactions involving radicals generated along the polymer chain proceeded such as chain scission and crosslinking reactions which were often accompanied by the evolution of gaseous products [15,22–27]. Radical processes which may proceed during high energy irradiation [27] (see Figure 1) have been proposed.

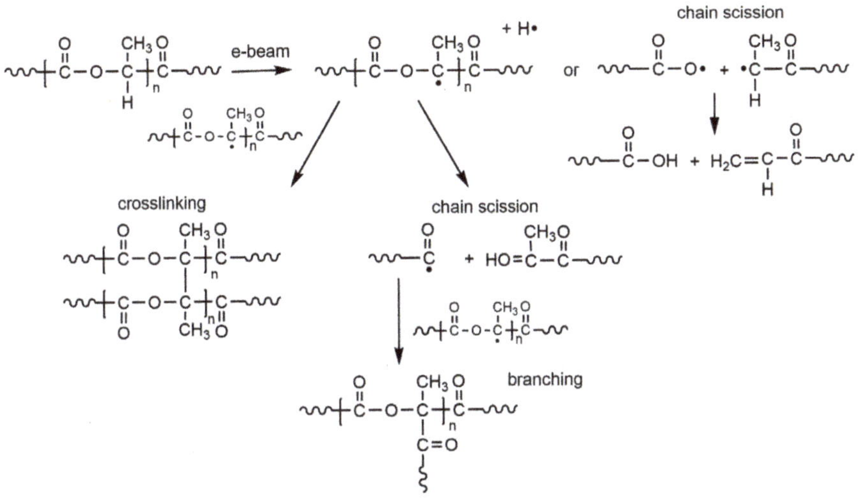

Figure 1. Possible processes involving radicals on polylactide (PLA) chains formed by irradiation with an electron beam (on the basis of Reference [27]).

Reactions under high energy radiation leading to the destruction of the PLA chain were useful for the controlled PLA degradation [19,28,29].

Many authors studied gamma radiation-induced changes (decreases) in the enthalpy of melting and cold crystallization, the degree of crystallinity, the glass transition temperature, and the thermal stability of polylactides [24,27,29]. It is obvious that the extent of the drop of mentioned parameters increases with the increase of radiation dose. The chain scission of polymer chains under irradiation is accompanied by the crosslinking process especially when higher doses are applied [20]. The polymer crosslinking may dominate at appropriate conditions (below doses of 250 kGy mainly chain scission proceeds [19]) and can be done on purpose. Biodegradable polymers that could be crosslinked by irradiation would be valuable not only in the medical field but for other industrial applications as well. Introducing crosslinking into biodegradable polymers, should result in an enhancement of mechanical properties and delayed hydrolysis of the polymer. To overcome the effect of significant molecular weight decrease, the curing of PLA is frequently performed in the presence of polyfunctional monomers. In irradiation-induced crosslinking such compounds as triallyl isocyanurate (TAIC), trimethallyl isocyanurate (TMAIC) trimethylolpropane triacrylate (TMPTA), trimethylolpropane trimethacrylate (TMPTMA), 1,6-hexanediol diacrylate (HDDA) and ethylene glycol bis (pentakis (glycidyl allyl ether)) ether were applied [30]. The structures of these monomers are shown in Figure 2.

Polyfunctional monomers have been applied predominantly in electron beam-induced crosslinking [27,30–35] although they have been also used for curing by γ-irradiation [36–38].

Mitomo and his group studied the effects of the type and the concentration of polyfunctional monomer as well as parameters of the irradiation with electron beam (the irradiation dose, temperature) on the crosslinking of poly(L-lactide) (PLLA) or mixture of poly(L-lactide)/poly(D-lactide) (PLLA/PDLA), the thermal properties, and the biodegradation of obtained crosslinked polymers [30,31]. It was found that the most optimal conditions to introduce crosslinking were around 3% of TAIC and the irradiation dose of 30–50 kGy [30,33]. The crosslinked PLA films had much improved heat stability and mechanical properties. The resultant properties of PLA samples were governed by crosslinking density which depended on the structure and length of PLA chains and on the radiation dose. The crosslinked PLA became harder and more brittle at low temperatures, but was rubbery, soft, and stable at higher temperatures, even over T_m. The degradation of irradiated – crosslinked PLA samples was considerably retarded.

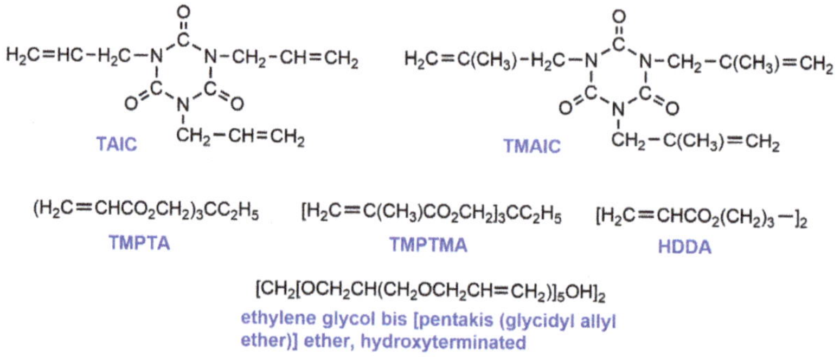

$H_2C=HC-H_2C-N\overset{\overset{O}{\|}}{\underset{\underset{\underset{|}{N}}{\overset{|}{C}_{\searrow_N}\nearrow C}}{C}}N-CH_2-CH=CH_2$ $H_2C=C(CH_3)-H_2C-N\overset{\overset{O}{\|}}{\underset{\underset{\underset{|}{N}}{\overset{|}{C}_{\searrow_N}\nearrow C}}{C}}N-CH_2-C(CH_3)=CH_2$

TAIC $\quad CH_2-CH=CH_2$ TMAIC $\quad CH_2-C(CH_3)=CH_2$

$(H_2C=CHCO_2CH_2)_3CC_2H_5$ $[H_2C=C(CH_3)CO_2CH_2]_3CC_2H_5$ $[H_2C=CHCO_2(CH_2)_3-]_2$

TMPTA   TMPTMA   HDDA

$[CH_2[OCH_2CH(CH_2OCH_2CH=CH_2)]_5OH]_2$
ethylene glycol bis [pentakis (glycidyl allyl ether)] ether, hydroxyterminated

Figure 2. Multifunctional crosslinking agents applied in radiation-induced crosslinking [30]. (Adapted with permission from Elsevier, 2005).

Some authors studied the results of the crosslinking of PLA blends and composites with other materials (PCL, poly(butylene adipate-co-terephthalate) (PLA/PBAT), flax fibers, montmorillonite, and others) under electron beam irradiation [34,39–43]. The physical properties, apart from crosslinking conditions, depended strongly on the blend composition. It was stated that TAIC was an efficient agent that hindered the phase separation and linked macromolecules of both the same and the different polymers. On the other hand, the addition of crosslinking monomers (TAIC) was claimed to hamper polyester degradation [41].

Various additives have been also added for PLA crosslinking by γ-rays. For example, blends of PLA with flax fibers were subjected to γ-irradiation [37], octavinyl-POSS (octavinyl polyhedral oligomeric silsesquioxane) was used as an additional crosslinking agent [44], or PLA was blended with epoxy-functional acrylic oligomer as a chain extender in order to receive higher M_n and improved properties of PLA-based material [38].

These composite materials have been prepared with the support of ionizing radiation in order to modify the PLA properties, mainly with the aim of their use in packaging and in the production of consumer goods. The radiation-induced radicals on different components of blend/composite are able to react forming linkages between separate phases leading to the increase of the compatibility between components or to the formation of a multicomponent network. Some examples of the preparation of composite materials for daily use applications are shown in Section 4.

The attempted crosslinking of different polylactide-based polymers by using high-energy irradiation and different crosslinking agents is presented in Table 1.

Table 1. Ionizing radiation curing of polylactide.

Polylactide-Based Polymer (M_n, g·mol^{-1})	Radiation Type, Dose	Curing Co-Agent	Gel Content, %	Achieved Results	Reference
PLLA (99,000)	Electron beam, 0–100 kGy	TAIC, TMAIC, TMPTA, TMPTMA, HDDA, derivative of EG	0.1–88	Together with annealing improved heat stability above T_g until T_m; lower solubility in any solvents; retarded enzymatic degradation.	[30]
PLLA	Electron beam, 0–50 kGy	TAIC, TMAIC, TMPTA, TMPTMA, HDDA, derivative of EG	10–83	Stability at higher melting temperature; application of the crosslinked PLLA on heat-shrinkable tubes, cups and plates.	[31]
PLLA (115,100), PDLLA (197,000)	Electron beam, 0–50 kGy	TAIC	~40–100	Shifts of T_{cc} to higher and T_m to lower temperatures; increase in tensile strength, young's modulus and decrease in elongation at break; the crosslinked PLA samples were harder and more brittle at low temperature, but rubbery and soft, then stable at higher temperature (over T_m); decreased rate of enzymatic hydrolysis.	[32]
Equimolar blend of PLLA and PDLA	Electron beam, 0–50 kGy	TAIC + supercritical CO_2	~30–90	Shift of T_m of homo crystals to lower temp.; improved toughness and tensile strength.	[33]
PLLA (155,500)	Electron beam, 0–90 kGy	TAIC	NA	Pristine PLA: Only degradation was observed; PLA/TAIC: Increase of T_g (69–75 °C), decrease of melt flow and water vapor permeability.	[35]
PLA (155,500)	Electron beam, 200–1000 kGy	TAIC	68.2–89.4	For neat PLA: only degradation; PLA/TAIC: decrease of gel content with increasing radiation dose; optimum crosslinking obtained at radiation dose of 40–200 kGy and 3–5 wt % of TAIC.	[25]
PLA (155,500)/PCL (82,500) blend	Electron beam, 0–90 kGy	TAIC	NA	PLA: Increase of flexural modulus, tensile strength, flexural strength, decrease of elongation at break; PLA/PCL blend: partial degradation of PLA phase, mechanical properties depending on ratios of the polymeric components.	[34]
PLA (91,000)/PBAT (35,000) blend	Electron beam, 0–90 kGy	TAIC	40–90	Crosslinking and degradation after irradiation mostly in PLA phase, PBAT less susceptible to radiation influence.	[39,40]

Table 1. Cont.

Polylactide-Based Polymer (M_n, g·mol^{-1})	Radiation Type, Dose	Curing Co-Agent	Gel Content, %	Achieved Results	Reference
PLLA + Reinforced by flax fiber (20 wt %)	Electron beam, 0–40 kGy	TAIC	7.6–62.5	Increase of tensile strength of about 20% in the presence of TAIC at 40kGy of irradiation dose; irradiation in the presence of TAIC led to reduced enzymatic degradation; decrease of interfacial adhesion of flax fibers and PLA matrix in the presence of TAIC.	[41]
PLA (210,000) + MMT (1,3,5 wt %)	Electron beam, 1 and 10 kGy	–	NA	Increase of T_g, crystallinity and young modulus, decrease of elongation at brake and oxygen permeability.	[42]
PLA/PEGM/HBN blend composite	Electron beam, 0–100 kGy	–	NA	At low doses: partial branching and crosslinking for neat PLA and PLA/PEGM; at higher doses: chain scission dominates. increase of T_g, notched impact strength and heat deflection temp. with radiation of blend-composites with higher amount of HBN; accelerated hydrolytic degradation of irradiated blend and blend-composites.	[43]
PLLA	γ-rays 2.5–50 kGy	TAIC	10–100	Decrease of swelling with increasing gel content, decrease in elongation (75%), maintenance of tensile strength, decrease of crystallinity (from 36 to 10%) and T_m (from 182 to 165 °C).	[36]
PLA + Flax fiber (5 wt %)	γ-rays 0–20 kGy	TAIC	70–90	Increase of the gel fraction in PLA/flax composite with the radiation dose, degradation at higher doses; improvement of tensile strength and toughness with the increase in the radiation dose, decrease of elongation at break.	[37]
PLA (106,000)	γ-rays 0–100 kGy	TAIC, Ov-POSS	Up to 80	Higher degree of crosslinking for PLA/OvPOSS in comparison to PLA/TAIC; irradiated composites exhibited decrease of crystallinity, lower elongation at break and higher E-modulus, higher thermal stability and heat deflection temp. than that of neat PLA	[44]

Table 1. Cont.

Polylactide-Based Polymer (M_n, g·mol^{-1})	Radiation Type, Dose	Curing Co-Agent	Gel Content, %	Achieved Results	Reference
PLA (72,000)	γ-rays 0–20 kGy	TAIC as crosslinking agent (CA), Epoxy functional acrylic oligomer (Joncryl® ADR 4368) as chain extender (CE)	1.2–46.2	Considerable gel formation was observed for PLA/CA at high irradiation dose; addition of CA or CE increased the shear viscosity of neat and irradiated PLA; addition of CA and CE enhanced T_c and decreased crystallinity; improvement of tensile properties was higher for CA.	[38]

TAIC—tiallyl isocyanurate; TMAIC—trimethylallyl isocyanurate; TMPTA—trimethylolpropane triacrylate; TMPTMA—trimethylolpropane trimethacrylate; HDDA—1,6-hexanediol diacrylate; EG—ethylene glycol bis(pentakis(glycidyl allyl ether)ether, hydroxy terminated; PDLLA—polylactide prepared from racemic mixture of D-LA and L-LA; PBAT—poly(butylene adipate-co-terepthalate); PEGM—poly(ethylene-co-glycidyl methacrylate); HBN—hexagonal boron nitride; Ov-POSS—octavinyl polyhedral oligomeric silsesquioxane; MMT—montmorillonite; NA—not available.

3. Photo-Crosslinked PLA

Photo-initiated crosslinking has many advantages for biomedical applications because it allows fast crosslinking under mild reaction conditions without solvents. Radical crosslinking using peroxides is not appropriate for biomedical applications because of the toxicity of the decomposition products of peroxides and not defined degradation products. Electron beam and γ-irradiation crosslinking require a large amount of radiation energy and the presence of a crosslinking agent to get the advantage of crosslinking over chain scission. Photo-crosslinking provides significant advantages over these two approaches, such as ease of use, safety, especially in connection to living systems, and low cost [45], although it may also be accompanied by other processes leading to polymer degradation. UV was also intentionally used to induce polylactide degradation (through, e.g., radiolysis or photo-oxidation) [46].

Polylactide intended for crosslinking is first functionalized at the chain ends with double bonds and then subjected to UV or visible light/laser irradiation which induces radical polymerization. To initiate radical polymerization photoactive additives are added such as substituted phenylacetophenones (irgacures) or camphorquinone. In comparison with peroxide-induced crosslinking, the photo-crosslinking can be accomplished at low temperatures [47].

As it was mentioned PLA-based materials which were photo-crosslinked were designed for medical applications (tissue scaffolds or drug carriers), thus, this crosslinking method concerns mostly lactide copolymers i.e., PLA/ polyethylene glycol (PEG), poly(tetramethylene oxide) (PTMO), poly(ε-caprolactone) (PCL), polyglycolide (PGA), poly(trimethylene carbonate) (TMC) copolymers [46,47]. These were ABA type copolymers or statistical copolymers obtained by tin octanoate catalyzed ring-opening (co)polymerization of lactide initiated by polyether diol (ABA block copolymers), alternatively, by low molecular diols or multifunctional alcohols in the case of the statistical or star-shape copolymer. Obtained difunctional or multifunctional –OH terminated polylactides were functionalized by esterification, usually with (meth)acryloyl chloride. Low molecular weight polymers (oligomers) were mixed with photoinitiator and, often in molds, were exposed to UV lamp irradiation.

Based on acrylated PDLLA-PEG-PDLLA copolymers or functionalized with fumarate groups, upon crosslinking, either hydrolyzable gels [48,49] or tissue scaffolds with controlled macroscopic architecture, potentially for bone regeneration [50,51] were prepared. Water-soluble PEG/PLA copolymers consisted of PEG fragment with M_n = 1000–10,000 and the attached 2–40 LA units [48]. The authors claim that obtained nontoxic macromers could be photo-polymerized in vivo in direct contact with tissue. Prepared by another group, hydrogels (with about 2–4 LA units attached to PEG with M_n = 4000) were applied for encapsulation of model proteins [49].

Mechanical properties of networks for scaffold were highly dependent on the number of lactic acid and ethylene glycol units in the oligomer backbone ranging from 2–8 EG units and 6-10 LA units [52]. Also, hydrophobicity/hydrophilicity balance varied with copolymer compositions what was important with regard to polymer degradation and cell attachment. In both studies, the complete degradation of networks to water-soluble products was performed in physiological conditions.

The authors of another work [53] prepared copolymers consisting of polyethers such as PEG, poly(propylene glycol) (PPG) or poly(tetramethylene glycol) (PTMG) and 7–65 wt % of D,L-lactide units which, after functionalization introducing acrylate end groups, were subjected to UV irradiation. Photo-polymerization resulted in the network with gel content equal to 78% for copolymers with long PEG chain (M_n = 10,000) and over 97% for PEG and PPG with M_n around 400. Hydrophilic PEG-based networks rapidly degraded into completely water-soluble products within 1 day, while the degradation times of the more hydrophobic PPG and PTMG-based networks varied from 1 to 7 days. Obtained materials can potentially be used as biodegradable lubricants for coating various medical products.

In several articles, the syntheses of PLA-based copolymers by copolymerization of D,L-lactide with ε-caprolactone (CL), or L-lactide with ε-caprolactone and glycolide (GL) initiated with diethylene glycol or tetra(ethylene glycol), followed by the end-(meth)acrylation and crosslinking were described [51,54]. In the case of the first cited work PLA/PCL copolymers with M_n in the range 1500–2400 resulted in a rather dense network after crosslinking while in a second work, M_n of PLA/PCL/PGA terpolymers was in the range 1800–10,200 giving networks with varied mechanical properties depending on copolymer composition and molecular weight. Degradation of all networks has been studied as well as their biocompatibility, both in respect to their potential application in tissue engineering. Copolymerization of D,L-lactide with glycolide was also initiated by PEG (M_n = 1500) leading to PLGA–PEG–PLGA copolymers which were further functionalized with itaconic anhydride [55]. Crosslinking by UV irradiation resulted in hydrogels which, according to the authors, could be used in moist wound healing or as carriers for controlled drug release.

For tissue elastic implants, materials from PDLLA/1,3-trimethylenecarbonate (TMC) copolymers have been synthesized by UV coupling of linear macromers with methacrylate groups and relatively high molecular weights (~30,000) [56]. Obtained networks with tunable thermal and mechanical properties depending on DLLA to TMC ratio could be used as implantable devices having different geometries as well as porous scaffolds with shape-memory properties.

Many authors used star-shaped polylactides and lactide copolymers instead of linear ones for the photo-crosslinking. Starting their works with homopolymers, Grijpma group prepared star-shaped poly(D,L-lactide) oligomers with 3 and 6 arms, with arm molecular weight in the range 200–5700 [57], which, after functionalization with methacryloyl chloride, were diluted with ethyl lactate and subjected to photo-crosslinking. Networks prepared from macromers of which the molecular weight per arm was 600 or higher had good mechanical properties, similar to linear high molecular weight poly(D,L-lactide). Films and porous scaffolds with gyroid architecture have been prepared by stereolithography, using a liquid resin based on a 2-arm PDLLA macromer and ethyl lactate. It appeared that pre-osteoblasts showed good adherence to these photo-crosslinked networks. The same group prepared 3-arm copolymers of D,L-lactide with ε-caprolactone and 1,3-trimethylenecarbonate with M_n = 3100–4000 which subsequently were functionalized with fumarate groups [58]. UV-initiated polymerization proved the sufficient reactivity of these groups and resulted in networks with high gel content (up to 96%) which physical properties varied depending on the composition, and molecular weight of the oligomeric precursors.

Other authors prepared 4-arm PDLLA/PCL copolymers functionalized with (meth)acrylate groups [59,60]. One of these works concerned the investigation of thermal properties of prepared thermoresponsive membranes from prepolymers with M_n in the range 3200–12,000, designed for drug delivery [60]. The other was focused on different techniques of resistor preparation to achieve shape accuracy and edge sharpness of samples prepared from crosslinked PLLA/PCL stars with short arm length, i.e., ~2500 [59].

Significant achievements in the field of synthesis of photo-crosslinked materials based on polylactide, designed for biomedical applications has a group of Amsden [61]. They worked on bioelastomers which could be used for the production of tissue scaffolds and implantable devices for drug delivery. For this reason, copolymers of lactide with such comonomers as ε-caprolactone and trimethylene carbonate (or substituted carbonate) introducing flexibility were prepared. Similarly as it was in the study by other researches, star 3-arm D,L-LA (co)polymers, or occasionally linear oligomers with M_n usually in the range 1000–5000 were functionalized with acrylate groups. Alternatively, acrylate groups were introduced as side groups by copolymerization of lactide with cyclic carbonate substituted with these groups [62,63]. Functionalized prepolymers were crosslinked in the presence of photo-active compounds and sometimes together with co-crosslinkers as, e.g., poly(ethylene glycol) diacrylate [64–67]. Obtained networks were studied concerning for their mechanical properties, degradation, and biocompatibility. A large part of the works concerned the study of the encapsulation of biologically active compounds and their release. Specific properties achieved in particular studies described by group of Amsden in numerous articles are shown in Table 2 (at the end of this section).

A similar approach as described above has been also used by other authors. Thus, 4-arm star PDLLA oligomers (containing ethylene glycol units in the initiator fragment), of different M_n (1500–9500) with either methacrylated or urethane methacrylated end groups have been synthesized and photochemically crosslinked [68]. High gel content networks (90–99%) had T_g strongly dependent on prepolymer molar mass. Mechanical properties depended on both the type of introduced end groups of prepolymer (methacrylate or urethane methacrylate) and molar mass [68].

Crosslinking of methacrylate-terminated linear D,L-lactide oligomers with M_n around 1300 has been also applied for the preparation of potential composite resin for stereolithography [69]. To enhance crosslinking, triethylene glycol dimethacrylate (TEGDMA) as reactive diluent has been added in the amount of 30% and 50%. PLAs together with TEGDMA was blended with hydroxyapatite (HA) in the amount of 20%, 30%, 40% and 60% to prepare composites that were next photopolymerized in the presence of photoinitiator giving products with gel content up to ~100 %. Analysis of the thermal properties of crosslinked composites showed that T_g significantly shifted to a higher temperature when HA was incorporated. It indicated the interaction between HA particles and PLA matrix, leading to a mobility restriction of the polymeric chains. The addition of HA also affected the thermal stability, as known from the thermogravimetric analysis—the shift to higher temperature was observed for crosslinked PLA containing HA. Degradation of composites has been investigated as well as changes in thermal and mechanical properties during degradation. Additionally, the cytocompatibility of cells in contact with composites with different HA contents during degradation has been studied. Lower cytotoxicity of degradation products was observed for a sample with a higher content of HA. As a conclusion, the authors claim, that materials showed their potential in a stereolithographic fabrication of bone implants.

Functionalized with (meth)acrylate groups star low molecular weight polylactides have been used for stereolithography also by other groups where two-photon polymerization (2PP) technique was applied for crosslinking [70–75]. Star-shaped methacrylate-terminated oligo(D,L-lactide)s with $M_n = 2800$ were prepared, and it was demonstrated that oligomer synthesis and their functionalization can be carried out in the same reactor [71]. Subsequently, 2PP technique was used to prepare hexagonal porous scaffold with 3D structures in the presence of photoinitiator. These fabricated scaffolds were shown as a beneficial microenvironment for osteogenesis and bone regeneration in vitro and in vivo. Similarly, fabricated scaffolds (2PP technique) were also used for supporting of Schwann cells growth and thus, as neural scaffolds in nerve repair [70]. Laser-induced crosslinked star-shaped methacrylate-terminated oligo(D,L-lactide)s ($M_n = 2400$) were used as a reinforcement of collagen materials [74,75]. The material exhibited improved resistance to biodegradation, while the direct multipotent stromal cell growth during their culture was observed. Reinforcement of collagen sponges resulted in near one order of magnitude increase of Young's modulus without affecting of cytotoxicity and developed matrix provided cell adhesion and proliferation. Based on the results, the authors suggested this material for tissue engineering applications.

All previous studies (above-mentioned works) were focused on crosslinking of PLA or PLA copolymers where curable groups were sited at the ends of polymer chains (end-functionalized polymers). In an alternative approach, poly(lactide-co-glycidyl methacrylate) (P(LA-co-GMA)) copolymer has been synthesized by ring-opening polymerization where curable C=C groups were placed in side-chains of the copolymer (pendant unsaturated groups) [76]. The copolymer was irradiated in the presence of an initiator and the influence of irradiation time, initiator concentration, as well as GMA content in polymer chain on crosslinking efficiency were followed by gel content measurement. Crosslinking led to the enhancement of mechanical and thermal properties and was dependent on the content of GMA units. In another study, P(LA-co-GMA) copolymer and its partly UV crosslinked counterpart were grafted with a pH-responsive polyacrylamide (PAAm), by UV-assisted reactions using acrylamide (AAm) and N,N'-methylene bisacrylamide monomers, and various photoinitiator systems [77]. These materials have the potential for use in biomedical and environmental applications due to their amphiphilic and pH-responsible properties.

A different example of crosslinking is the application of high molecular weight/commercial PLLA for UV-induced crosslinking [78]. PLLA powders containing different concentrations of benzophenone (2–3.6 mol% per LA repeating unit) were hot-pressed at 190 °C and obtained films were continuously

UV irradiated from both sides using different energy. Networks with gel content up to 98.5% have been prepared. By ^{1}H and solid state ^{13}C analyses of pristine PLA, the gel, and soluble fractions of the products, the authors suggested the mechanism of crosslinking which is presented in Figure 3. According to them, the photo-crosslinking may result from the recombination between primary and tertiary carbon radicals generated by the hydrogen abstraction from the PLA chain by the excited benzophenone.

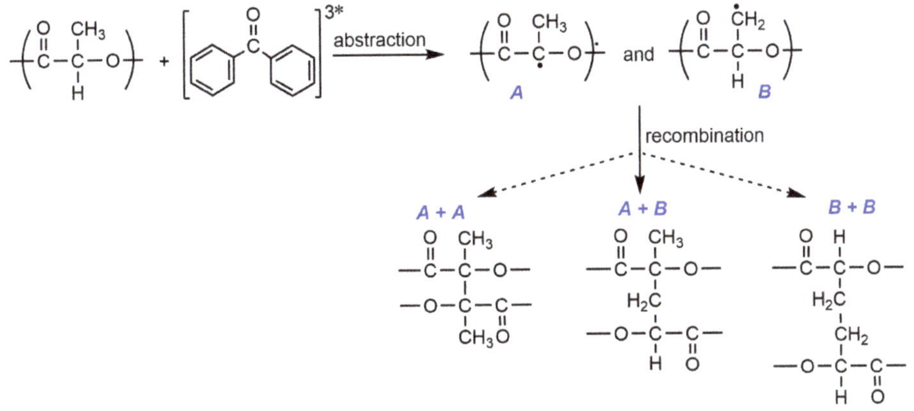

Figure 3. The proposed mechanism of photo-crosslinking of not functionalized PLA in the presence of benzophenone [78]. (Reproduced with permission from Wiley, 2013).

DSC and XRD analyses indicated that prepared networks were partially crystalline up to 93% of gel content. T_g slightly increased because of the introduction of crosslinked structure in PLA, both T_c and T_m shifted to higher temperatures and finally disappeared with increasing gel fraction. The authors found that the crosslinks have been formed not only in the amorphous region but also in the crystalline region incorporating into the crosslinked network. The photoinitiator may penetrate into the crystalline region by the sublimation during the film formation. The photo-crosslinking improved mechanical properties by increasing both tensile strength and modulus by 70% with a little less decrease in elongation at break. Unexpectedly the toughness of the crosslinked PLA also increased by 22.5%. The authors named this type of crosslinking "crystal crosslinking" and claimed that described by them photo-crosslinking was more efficient compared with conventional amorphous crosslinking (much more significant improvements in thermal and mechanical properties).

Photo-crosslinking has also been applied by several authors for the curing of PLA-based prepolymers using quite an alternative approach. In this approach, photochemically active groups were introduced not as photosensitive additives but directly into PLA chains. As photosensitive sites cinnamoyl groups were used which are able to dimerize upon UV light of appropriate wavelength according to the scheme shown in Figure 4, forming cyclobutane rings [79] and so bridges between PLA chains.

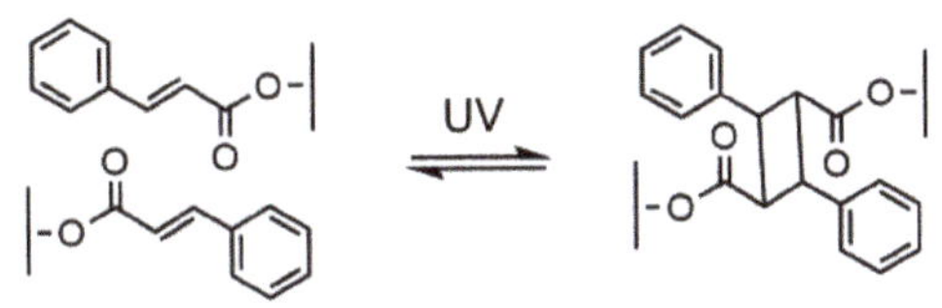

Figure 4. Cyclodimerization of cinnamoyl groups.

The dimerization of cinnamoyl groups has been mainly used for the preparation of reversible networks (dimerization of cinnamoyl groups is reversible and cyclobutane ring undergoes cleavage under UV irradiation with another wavelength) [80,81], however, some authors didn't study the mentioned reversibility. Thus, cinnamoyl groups were introduced into PLA chain by polycondensation [82,83] of PLLA diols (M_n in the range 1260–3010 or 2300–8900) with diacyl dichlorides containing these groups, i.e., with 5-cinnamoyloxyisophthalic acid (ICA) [82] or with diacyl chloride of 4,4'-(adipoyldioxy)dicinnamic acid (CAC) [83] (see Figure 5).

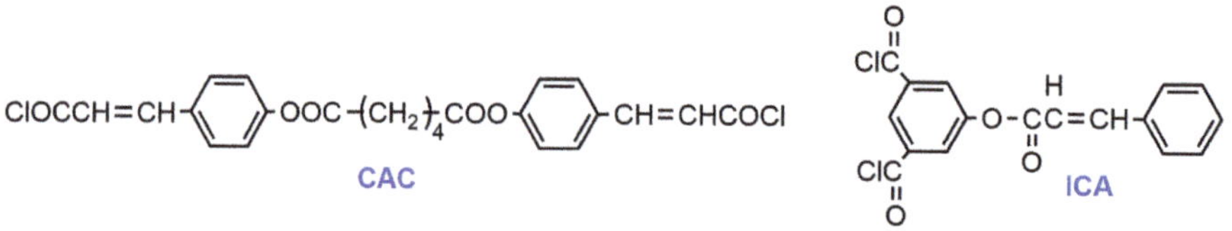

Figure 5. Compounds used for polycondensation with PLA diols.

Polycondensates were subsequently crosslinked with the light of λ = 282 nm. Dimerization of cinnamoyl groups appeared effective and after 2 h of irradiation, approximately 90% of these groups disappeared [82]. The authors observed a decrease of crosslinking rate and the amount of formed gel with increasing M_n of ICA/PLLA copolymer, which they assigned to lower concentration of a photosensitive component in the sample. While for PLLA with M_n ~ 4000 the amount of gel content was 100%, for PLLA with M_n ~ 9000 it was only 50 % [82]. From a comparison of the crosslinking rate of copolymers ICA/PLLA and CAC/PLLA with the same M_n of PLLA-diols, the authors concluded that cinnamoyl moiety in the side-chain was more photoreactive than that in the main-chain [82]. In both cases, decrease in degradation rate was observed after crosslinking in comparison with un-crosslinked functionalized PLLA and neat telechelic PLLA.

In another work, cinnamoyl moiety has been introduced into the PLLA side chain by copolymerization of lactide with cyclic carbonate monomer, i.e., 5-methyl-5-cinnamoyloxymethyl-1,3-dioxan-2-one (MC). Polymers of different ratios of MC/LA and M_n ranging from 12,900 to 65,100 were prepared. The crosslinking of the copolymer was followed by FTIR but no further properties of crosslinked material were discussed [84].

An original approach to PLA crosslinking has been recently presented by authors who applied multi(aryl azide) crosslinker for UV curing of PLA-Pluronic® copolymer not containing unsaturated groups [85]. They adopted an earlier reported strategy of UV-induced polyester crosslinking [86] relying on the UV-activation of the aryl azide group to generate highly reactive nitrene species that can insert into carbon-hydrogen bonds of the polymer backbone, thereby leading to crosslinking via amine groups (see scheme in Figure 6).

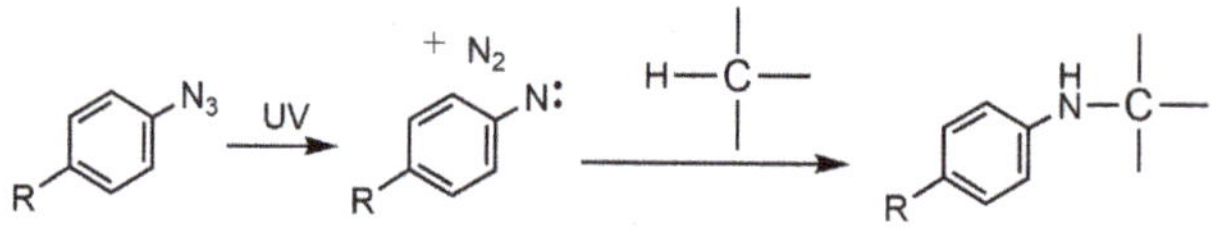

Figure 6. Mechanism of the formation of covalent bond between species bearing azide group and the compound with reactive hydrogen.

Applying this elegant and straightforward strategy, using polymeric multi-azide crosslinker, which can be also used for the crosslinking of other not pre-functionalized polymers, the authors prepared degradable elastomers for soft tissue engineering. Interesting elastic scaffold prepared by electrospinning from above described materials will be shown in the last section of the review.

Table 2. Conditions of photo-crosslinking of linear and star PLA low molecular weight (co)polymers and observed results. UV light was used to induce crosslinking (visible or laser light/2PP, when indicated).

PLA Structure (M_n, g·mol^{-1})	Crosslinking Group	Photoinitiator	Gel Content [a], %	Achieved Results	Ref.
PDLLA-b-PEG-b-PDLLA (1000–20,000)	Acrylate end group	2,2-dimethoxy-2-phenylacetophenone (Irgacure 651)	65–74	Degradation rate increased with increasing M_n of precursor; materials used in the sustained release of proteins.	[48]
PDLLA-b-PEG-b-PDLLA Or: P(DLLA-co-TMC)-b-PEG-b-P (DLLA-co-TMC (4500–5500)	Fumarate end group	2,2-dimethoxy-2-phenylacetophenone	>90	Hydrogels prepared in N-vinylpyrrolidone were used for the study of model protein release; the degradation behavior could be controlled by changing the composition of the hydrophobic segments.	[49]
PDLLA-b-PEG-b-PDLLA (~1600)	Acrylate end group	2,2-dimethoxy-2-phenyl acetophenone		Preparation of porous scaffolds for the study of the growth factor encapsulation and release and implantation in the case of cranial defect.	[50]
PDLLA-b-PEG-b-PDLLA (990–1240)	Acrylate end group	camphorquinone/ethyl-4-N,N-dimethylaminobenzoate	89–100	Modification of hydrophobicity (contact angle 123°–142°); T_g = 1.8–26 °C depending on the composition and crosslinking density; tensile modulus in the range 0.92–3.67 MPa and strain at break 0.19–0.65; preparation of scaffolds with various pore sizes by salt-leaching method.	[52]
PDLLA-b-PEG-b-PDLLA (1120–10,720)	Acrylate end group	2,2-dimethoxy-2-phenylacetophenone	78–100	Both lower crosslinking density (higher M_n of macromer) and the lower crystallinity (lower M_n) increased the degradation rate of the networks; the maximum improvement in penetration force, lubricant property, over control was 41% in the needle coated with PPG-based polymer network.	[53]
PDLLA-b-PPG-b-PDLLA (1150–4720)	Acrylate end group	2,2-dimethoxy-2-phenylacetophenone	93–99		
PDLLA-b-PTMG-b-PDLLA (1370–3620)	Acrylate end group	2,2-dimethoxy-2-phenylacetophenone	95–97		
PDLLA-b-PCL (1570–2390)	Methacrylate end group	camphorquinone/ethyl-4-dimethylaminobenzoate	Highly crosslinked	Decrease of T_g with increasing CL content (T_g in the range −30 to 60 °C). Storage moduli in the glassy regime were similar, in the rubbery regime dependent on crosslinking density; highly cross-linked scaffolds were cellularly compatible and promoted osteoblast attachment.	[51]
P(CL-co-LLA-co-GA) (1870–10,190)	Acrylate end group	2,2-dimethoxy-2-phenylacetophenone	>95	Increase of T_g of 2.8–14.9 °C, similar ultimate strength (σ = 2.39–3.76 MPa); Young's modulus (E = 1.66–12.29 MPa and maximum strain (ε = 21–176%); Excellent biocompatibility of films with smooth muscle cells.	[54]

Table 2. Cont.

PLA Structure (M_n, g·mol^{-1})	Crosslinking Group	Photoinitiator	Gel Content[a], %	Achieved Results	Ref.
P(LDLA-co-GA)-b-PEG-b-P(LDLA-co-GA) (~5300)	Itaconic end groups	camphorquinone	94–98[b]	Swelling properties depended on crosslinking time, thus crosslinking density; with longer UV exposure better hydrolytic stability of hydrogel was observed.	[55]
P(DLLA-co-TMC) (27,000–29,000)	Methacrylate end group	Irgacure 2959	74–90	Depending on the DLLA/TMC ratio, amorphous networks with T_g of 13 to 51 °C and elastic modulus from 3.6 MPa to 2.7 GPa were obtained; networks of more than 40 mol% of TMC are tough, flexible and elastomeric at r.t. with elongations at break of up to 800%. When DLLA:TMC = 60:40, T_g is between 25 and 37 °C, thus elastic medical devices with SM properties could be implanted in a temporary shape.	[56]
2,3- and 6-arm PDLLA (6600–34,200)	Methacrylate end group	2-hydroxy-1-[4-(hydroxyethoxy)phenyl]-2-methyl-1-propanone (Irgacure 2959)	96	T_g (55–76 °C) dependent on macromer chain length; mechanical properties similar to HMW PDLLA- suitable for stereolithography; mouse pre-osteoblasts readily adhered and proliferated well on networks.	[57]
3-arm P(TMC-co-DLLA) (3100–4000)	Fumaric acid monoethyl ester	2,2-dimethoxy-2-phenylacetophenone	67–81[c]	The E modulus decrease with TMC content, tensile strength and elongation at break unaffected. Relative low values of tensile strength (1–2 MPa), and E modulus (1–10 MPa) in comparison with HMW PDLLA and PTMC.	[58]
4-arm PDLLA-co-PCL (5000–10,000)	Acrylate end group	1-hydroxycyclohexylphenylketone (irgacure 184)	NA	Fabrication of microstructures by soft lithography. Possibility of using studied materials to culture mammalian cells.	[59]
4-arm P(LLA-b-CL) (Mn ~3200–12,000)	Methacrylate end group	Camphorquinone[d]	NA	Transition temperatures depended on the length of poly-CL segments. Decrease of T_m and crystallinity with increasing M_n. Thermo-responsive properties as permeability of a drug.	[60]
3-arm P(CL-co-DLLA) (1250–7800)	Acrylate end group	2,2-dimethoxy-2-phenylacetophenone	>95	T_g of elastomers below physiological temperature (even below 0 °C). The Young's modulus and stress at break inversely proportional but strain at break-proportional to the prepolymer M_n. The ability of elastomeric devices to encapsulate (glyco)proteins and release them according to an osmotic pressure delivery mechanism; confirmed ability to degradation in vitro and in vivo. Preparation of porous scaffolds capable to degradation with mechanical properties dependent on prepolymers M_n. Ability to adsorb proteins and to cell proliferation; dependence of adsorbed protein layer on the material stiffness.	[87–92]

Table 2. Cont.

PLA Structure (M_n, g·mol^{-1})	Crosslinking Group	Photoinitiator	Gel Content [a], %	Achieved Results	Ref.
3-arm poly(CL-co-DLLA) (1250, 2700 and 3900)	Acrylate end group and co-photo-crosslinker poly(ethylene glycol)diacrylate (PEGDA) (4000 and 24,000)	2,2-dimethoxy-2-phenylacetophenone	95–98	T_g, T_m and ΔH_f varied with prepolymer M_n, co-photo-crosslinker amount and M_n. Networks without PEGDA were amorphous, with PEGDA indicated melting; preparation of cylindrical elastomeric devices able to encapsulate Vitamin B_{12}.	[64]
3-arm poly(TMC-co-DLLA) (7800–8500)	Acrylate end group	2,2-dimethoxy-2-phenylacetophenone	79–88	With increasing amount of DLLA increase of Young's, stress at break, T_g and decrease of elongation at break. The possibility of osmotic pressure driven release of proteins. Study of the behavior of elastomers implanted into rats.	[93,94]
3-arm poly(TMC-DLLA-CL) (2300–7800)	Acrylate end group and co-photo-crosslinker poly(ethylene glycol)diacrylate (PEGDA)	2,2-dimethoxy-2-phenylacetophenone	86–99	T_g (−18 to 2 °C) varied with the monomer composition and the M_n of PEGDA. Preparation of cylindrical elastomeric devices able to swell and to encapsulate corticosteroid and growth factors utilizing the osmotic pressure mechanism.	[65–67]
Poly(LLA-co- CL-acryolyl carbonate) (17,900–22,600)	Pendant acrylate group	2,2-dimethoxy-2-phenylacetophenone	90	Preparation of fibrous scaffolds by melt electrospinning writing; Stiffness of the scaffolds increased significantly (up to ~10-fold) after crosslinking with UV compared with un-crosslinked scaffolds; the preservation of stiffness upon repetitive loading.	[62]
Poly(L-lactide-co-acryolyl carbonate) (55,900–72,100)	Pendant acrylate group	NA	84–94	Increase of T_g, decrease of T_m and degree of crystallinity after crosslinking; Electrospun and photo-crosslinked polymer resulted in scaffolds with increased tensile modulus in comparison with uncrosslinked fibrous scaffolds; good cytocompatibility toward fibroblasts of crimp-stabilized scaffolds.	[63]
3-arm Poly(DLLA-co-CL) (M_w 4800–10,900)	Acrylate end group and co-photo-crosslinker N-methacrylated glycol chitosan (MGC)	Irgacure 2959	98–100	Preparation of bi-continuous two-phase (elastomer/hydrogel) cell delivery device for the repair and/or replacement of load-bearing soft tissues. Decrease of elastic modulus with increasing content of MGC; using electrospinning for scaffold preparation.	[95–97]
3-arm Poly(DLLA-co-CL) (2700 and 5000)	Acrylate end group and co-photo-crosslinker diacrylate oligo (D,L-lactide)-b-poly (ethylene glycol)-b-oligo (D,L-lactide)	2,2-dimethoxy-2-phenylacetophenone	>95	Enhancing the degradation rate by introducing PEG fragment; regulation of the degradation rate and peptide release by M_n of PEG and M_n of prepolymer.	[98]

Table 2. Cont.

PLA Structure (M_n, g·mol^{-1})	Crosslinking Group	Photoinitiator	Gel Content [a], %	Achieved Results	Ref.
4-arm PDLLA (1500–9500)	Methacrylate end group (methacrylic anhydride or 2-isocyanatoethyl methacrylate)	2,2-dimethoxy-2-phenylacetophenone	90–99	Increasing of T_g with decreasing M_n of precursors; networks based on low M_n oligomers were generally more rigid, those based on high M_n exhibited higher elongation; mechanical properties differ with type of precursors methacrylate end group.	[68]
PDLLA (1310) + TEGDMA as reactive diluent + Hydroxyapatite (HA) as bioactive filler	Methacrylate end group	Camphorquinone/ N,N'-dimethylaminoethyl Methacrylate	77–100	T_g (38–55 °C), flexural strength (3.5–94 MPa) and flexural modulus (75–3980 MPa) were dependent on composition of polymer resin and an amount of HA; increasing thermal stability with increasing amount of filler. Higher gel content and higher concentration of HA led to decreased rate of degradation; higher HA content resulted in the less cytotoxic sample.	[69]
4-arm Poly(D,L-lactide) (2600 or 2400 or 450–820)	Methacrylate end group	4,4'-bis(dimethylamino) benzophenone [e]	NA	Preparation of scaffolds with Young's modulus even bigger than 4 GPa for the mesenchymal stem cells osteogenic differentiation; Independently—collagen reinforcement: about one order of magnitude increased Young's modulus for the hybrid matrix without affecting its cytotoxicity;	[72–75]
4-arm Poly(L-lactide) (M_w 1250)	Methacrylate end group	Irgacure 369 [e]	NA	Preparation of scaffolds for supporting Schwann cell growth—neural scaffolds in nerve repair.	[70]
Poly(LLA-co-GMA) (1650–3260)	Pendant methacrylate group	Camphorquinone/ N,N'-dimethylaminoethyl methacrylate	72–95	With increasing content of GMA (9.5–19.2 mol%) the increase of gel content, compressive stress (3–25.5 MPa) and the decrease of degree of swelling was observed; Increase of T_g by 15–20 °C in comparison with original copolymer.	[76,77]
PLLA (M_v 276,500)	-	Benzophenone	38–98.5	Slight increase of T_m in comparison with pristine PLA, decrease of T_m and crystallinity; improvement of thermal stability; with increase of gel fraction—increase of storage modulus (from 5.4 to 9.6 GPa at 0 °C), tensile strength (from 48 to 81 MPa), modulus (from 1.8 to 3.1 GPa), toughness (from 67 to 82 MPa) and decrease of strain (from 3.9 to 1.6%).	[78]
PLLA - diacyl of 5-cinnamoyloxyisophthalic acid (ICA) (10,000–34,500)	Pendant 3-phenylprop-2-ene group	-	50–100	Slight increase of T_g (from 50 to 53 °C), decrease of crystallinity (from 10 to 3 %), slight decreases of T_m, thermal decomposition T_d, increase of ultimate tensile strength (from 13 to 23 MPa), decrease in elongation (from 12 to 5.2%), increase of Young's modulus E (from 483 to 830 MPa); Decrease of degradation rate.	[82]

Table 2. Cont.

PLA Structure (M_n, g·mol^{-1})	Crosslinking Group	Photoinitiator	Gel Content [a], %	Achieved Results	Ref.
PLLA- diacyl of 4,4'-(adipoyldioxy)dicinnamic acid (8700–43,500) [f]	Main-chain 3-phenylprop-2-ene group	-	9–86	Increase of T_g (from 51 to 53 °C); decrease of ΔH_m (from 4.8 to 0.1 J·g^{-1}), small decrease of T_m (from 150 to 147 °C); increase of thermal decomposition T_d; increase of tensile strength and tensile modulus and decrease of elongation at break with increasing photocuring time and gel content; decrease of degradation rate.	[83]
P(LLA-co-MC) (12,900–65,100) [f]	Pendant phenylprop-2-ene group	-	NA	The kinetic of UV crosslinking was studied by FT IR spectroscopy.	[84]
PLA$_{50}$-Pluronic®-PLA$_{50}$ (50,000–200,000)	-C-H- bond in polymer chain	Aryl-azide group	Up to 55	Preparation of elastic microfibers (elastic limit-ε_y up to 182 %) for soft tissues by electrospinning.	[85]

[a] Gel content was dependent on M_n and/or copolymer composition; [b] determined by conversion of double bonds; [c] gel content was dependent on increasing UV energy and photoinitiator concentration; [d] visible light; [e] two-photon polymerization technique (2PP); [f] M_n of the polycondensation product; PPG—poly(propylene glycol); PTMG—poly(tetramethylene glycol); TMC – 1,3-methylene carbonate (1,3-dioxan-2-one); TEGDMA—triethylene glycol dimethacrylate; MC—5-methyl-5-cinnamoyloxymethyl-1,3-dioxan-2-one; GMA—glycidyl methacrylate; NA—data not available.

4. PLA-Based Materials by Photo- and High Energy Radiation Crosslinking

As mentioned in the previous sections, electron beam and gamma irradiations as well as UV light were predominantly used for crosslinking of PLA-based materials designed for biomedicine. However, in most articles concerning crosslinking by ionizing radiation only changes in properties of polylactide (co)polymers upon exposure to high energy radiation are discussed without indication of their specific medical application. On the other hand, some authors presented examples of daily use items prepared from gamma or electron beam irradiated PLAs.

Mitomo, who studied PLA crosslinking by electron beam irradiation for many years (the results are presented in Table 1 in Section 2) [29–33] demonstrated how irradiation improved the thermal stability of cups and plates prepared from poly(L-lactide) and how it enabled the preparation of heat-shrinkable tube which can be used as a cover for electric wire [31]. The tube was prepared by extrusion of PLLA blended with 3 wt % of TAIC at 180 °C and then irradiated at 50 kGy. The irradiated tube expanded two times at 180 °C and its shape was kept at room temperature. After that, the expanded tube shrunk up to the original size by re-heating, thus could bundle wire by heat shrinking. The result is shown in Figure 7A. In another attempt, PLLA with 3 wt % of TAIC was molded to cup and plate by the extruder and irradiated to form a crosslinked structure at 50 kGy. Boiling water was poured into unirradiated and irradiated cups. The unirradiated cup deformed and changed to milky-like transparency, but the crosslinked cup kept its original shape and transparency due to protection from the crystallization of crosslinked structure (see Figure 7B).

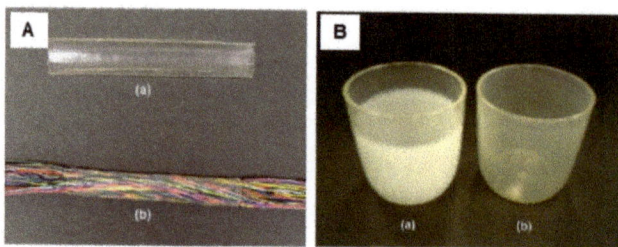

Figure 7. PLLA crosslinked by electron beam irradiation (50 kGy). (**A**) shrinkable tube (a); possible use (b). (**B**) Appearance of cups after using for hot water: (a) the unirradiated product, (b) the product crosslinked by irradiation [31]. (Adapted with permission from Elsevier, 2005).

As polylactide is also considered as an insulating material for different electronic applications, the influence of electron beam irradiation on electrical insulating properties of PLA has been studied [99]. Thus, to have better material for electric wire sheaths, a soft-resin as a plasticizer was added to polylactide but then the electric breakdown strength (E_B) decreased. To keep the E_B at the same level as that of neat PLA, the composite was irradiated by the electron beam at the dose 100 kGy what resulted in the PLA crosslinking.

Examples of the use of crosslinked by irradiation polylactide in the production of packaging can be found in both the scientific literature and patents [42,100–103].

Thus, new films based on PLA and montmorillonite with improved barrier and mechanical properties have been developed [42]. These were designed for use with foods being processed with electron beam technology for a shelf-life extension, phytosanitary treatment, and pathogen elimination. Only low radiation doses were applied, i.e., nanocomposite films were prepared at 1, 3, and 5 wt % of clay and exposed to target electron beam doses of 1 and 10 kGy. It was observed that PLA properties were influenced by the addition of clay and by electron beam irradiation treatment: the samples showed some surface irregularities, increases of T_g and Young modulus, and a decrease of oxygen permeability. This limited permeability was attributed to the presence of clay and crosslinks in PLA material.

The possible application of polylactide for the production of packaging made from PLA combined with another material, e.g., cardboard is reflected in some cited in this review patents [100–103]. PLA crosslinked by gamma rays or electron beam may also find the application as adhesives in conventional glue guns [104].

Contrary to attempts of PLA crosslinking by high energy radiation, PLA photo-crosslinking has been performed on the purpose of obtaining materials for biomedicine. Among them, elastic and stiff scaffolds were prepared as well as gels which could be applied as drug delivery systems.

Amsden and his group worked for a long time on biodegradable elastomers for biomedical applications, among them polylactide-based materials make up the majority. These materials were prepared predominantly from acrylated star PLLA-PCL but also from star PLLA-poly(trimethylene carbonate) copolymers which were crosslinked by UV light. Because the main purpose of the biodegradable elastomers synthesis was their application as implants capable of releasing biologically active compounds, most studies included the preparation of elastomeric devices containing a variety of proteins (growth factors, interferons), corticosteroids, peptides [61,64–66,88–90,93,98,105]. These devices were prepared by embedding the drug (together with accompanying/solubilizing compounds) in prepolymers in bulk or solution (containing photoinitiator) in a form and irradiation with UV light. Studies included the investigation of active compounds release.

Another group of works concerned the preparation of elastic porous scaffolds which could be applied for the cell culture but also for the release of biological compounds [62,63,91,95–97]. Initially fabricated porous scaffolds (including those by Amsden group) were prepared using porogens, e.g., paraffin beads or water, for the pores generation. Among these works, the preparation of porous scaffold from 3-arm acrylated PLA/ε-CL copolymer with dual porosity (due to the introduction of two porogens, i.e., paraffin beads and water) and pore interconnectivity is situated [91] (see Figure 8).

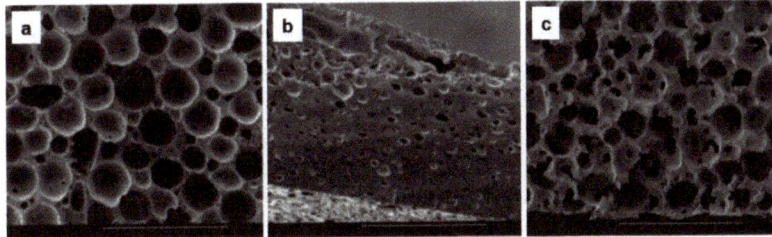

Figure 8. SEM images of preparation methods on elastomer scaffold structure: (**a**) scaffold made using only paraffin microbeads, (**b**) scaffold prepared using only water emulsified in the polymer solution, (**c**) scaffold prepared with combined emulsion and paraffin microbeads [91]. (Adapted with permission from Elsevier, 2009).

As another interesting example, the preparation of the combined system, i.e., elastomeric scaffold with the mechanical strength and a hydrophilic cell encapsulating hydrogel which formed a bi-continuous two-phase cell delivery device for the repair and/or replacement of load-bearing soft tissues can be presented [95]. Thus, an elastomer from a star-poly(ε-CL-co-D,L-lactide) triacrylate (CL:DL-LA = 0.5:0.5, M_w = 4000 and 8000) and an N-methacrylate glycol chitosan (MGC) hydrogel to distribute the cells from bovine articular cartilage and enable cells growth has been prepared. Functionalized chitosan containing cell culture was mechanically mixed with functionalized PCL-PDLA copolymer and after the addition of photoinitiator, this material was cured in the appropriate form, using UV light. The obtained scaffolds of bi-continuous morphology had mechanical properties resembling those of soft tissues. Cell culture experiments conducted with the selected scaffold demonstrated that the chondrocytes remained viable throughout the entire manufacturing process and were able to proliferate. The authors claimed the feasibility of the scaffolds as an injectable and in situ crosslinkable cell delivery system.

Many articles concern precisely designed porous tissue scaffolds prepared using stereolithography. Stereolithography is an additive fabrication process that uses a liquid light-curable photopolymer and a laser to create three-dimensional (3-D) structures [106]. Thus, porous PLA scaffolds with gyroid morphology have been fabricated using stereolithography, by visible light crosslinking of PLA macromers [57]. Complex structures could be built by illuminating sequential layers of a polymerizable resin using digital pixel masks or arrays of mirrors. In stereolithography, the thickness of a solidified layer is controlled by the light irradiation dose. It was possible to form relatively large structures (up to 42 × 33 × 200 mm) at high resolutions. The size of the smallest features that can be built was determined by the size of the light pixels (32 × 32 µm in the x and y directions), the layer thickness (25 µm), and the over cure. Although cell seeding of porous structures prepared from hydrophobic polymers, such as PDLLA is difficult, the very open structure of the gyroid architecture facilitated the penetration of water into PDLLA scaffolds prepared by stereolithography and enabled the cell seeding of mouse pre-osteoblasts. The achieved results are shown in Figure 9.

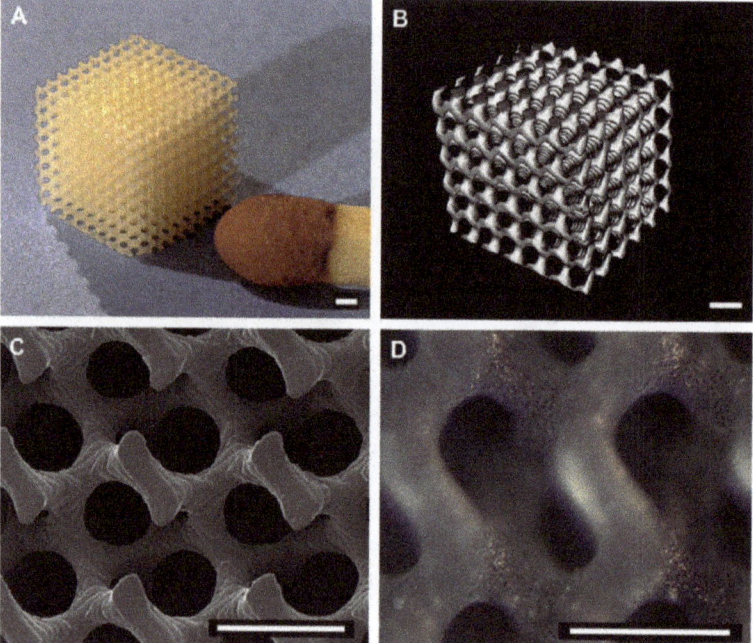

Figure 9. Images of PDLLA network scaffolds with a gyroid architecture prepared by stereolithography: (**A**) photograph, (**B**) microcomputed tomography (µCT) visualization and (**C**) SEM image. In (**D**) a light microscopy image is shown for a scaffold seeded with mouse pre-osteoblasts after 1 d of culturing. Scale bars represent 500 µm [57]. (With permission from Elsevier, 2009).

Photolithography was also applied for the network synthesized by a different approach [107]. PDLA network was prepared by a thiol-yne photo click reaction where alkyne functionalized star-shaped and linear PLAs were coupled with tetrafunctional thiols. Crosslinking was performed by UV irradiation of prepared polymer films. Amorphous crosslinked polymers were stable when hydrolyzed—no significant weight loss was observed during the first 10 weeks (around 4%). Films prepared by the casting of solutions containing functionalized PLA, tetrathiol and photoinitiator were also crosslinked using the direct laser writing (DLW) technique which enabled the preparation of structured patterns. Patterned samples were prepared by moving the thin film of the photopolymerizable material within the focal plane using a computer-controlled XY translation stage. Photopolymerization selectively took place in the exposed areas leaving the non-exposed material unreacted. This unexposed material was subsequently etched away using acetone as a solvent. The results are shown in Figure 10. Patterned films were used for the cell culture and independent experiments concerning cell viability indicated that studied materials based on crosslinked PLA were not toxic. The presented study showed that advanced photolithographic techniques allowed the microfabrication of well-defined micrometer-scale structures for cell patterning.

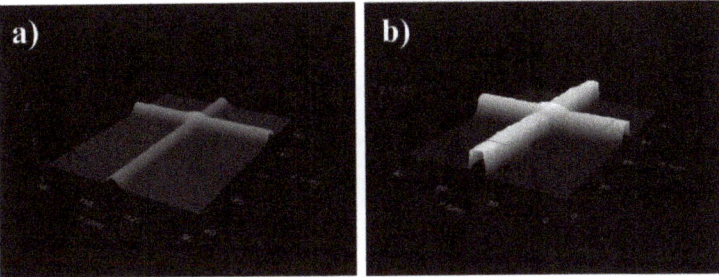

Figure 10. Topography images of crossing lines generated by direct laser writing using a formulation comprising the macromonomer: (**a**) linear-YNE and (**b**) star-YNE PLAs, both with a stoichiometric amount of the thiol (stoichiometry alkyne/thiol 1:2) and 3 wt % of photoinitiator. Images were obtained using a confocal microscope [107]. (Adapted with permission from Elsevier, 2017).

Micro-stereolithography has been also used for PLA composites. For example, a well-defined three-dimensional 3D pore network has been prepared starting from composite PDLLA/nanosized hydroxyapatite (HAP) [108]. The authors dispersed nano-HAP powder in a photo-curable PDLLA macromer in N-methyl pyrrolidone (as not reactive diluent) and after the addition of photoinitiator and some additives (inhibitor and dye improving the depth of light penetration), the composition was used to fabricate porous structure in a standard stereolithography apparatus. Subsequent layers were cured a dozen times by UV irradiation. As a result, a Schwarz pore network containing 5 wt % of nano-HAP has been fabricated what is illustrated in Figure 11. The ceramic component remained well dispersed in the polymeric matrix and HAP particles on the pore surface could allow the interaction between the bone-forming nano-HAP and cells. Investigation of mechanical properties showed that with increasing nano-HAP content the elasticity modulus of the composite PDLLA/nano-HAP network materials increased.

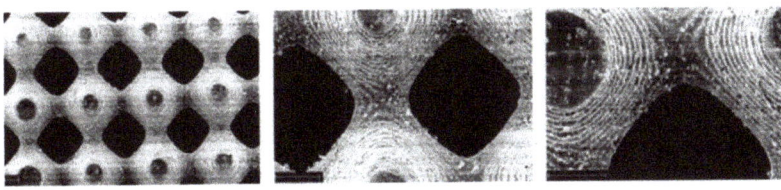

Figure 11. SEM images of porous structures with Schwarz primitive pore network architecture prepared by stereolithography from PDLLA and nano-HAP composite resins containing 5 wt % nano-HAP. Scale bars 200 μm [108]. (Adapted with permission from Elsevier, 2013).

Some authors applied a two-photon polymerization technique (2PP) as the type of stereolithography [109] for the preparation of UV crosslinked PLA-based materials designed for tissue scaffolds [70,71,73–75]. 2PP is a computer-aided microfabrication method by which it is possible to produce biomimetic synthetic scaffolds with high precision and reproducibility. This process uses simultaneous absorption of two photons of near-infrared (780 nm) or green (515 nm) laser light. For example, photoactive material was prepared by dissolving star-shaped methacrylate-functionalized poly(D,L-lactide) (M_n = 2600) in dichloromethane and mixing it with photoinitiator [71]. This material was next used for the fabrication of 3D structures (shown in Figure 12a) by the 2PP technique. It was demonstrated that the fabricated PLA-based scaffolds were a beneficial microenvironment for the osteogenic differentiation of mesenchymal stem cells in vitro and the potential of prepared scaffolds as implants in cranial defects was proved by tests in vivo upon their implantation into the cranial bone defect in mice. Figure 12 illustrates prepared PLA scaffolds and their behavior as implants in mice.

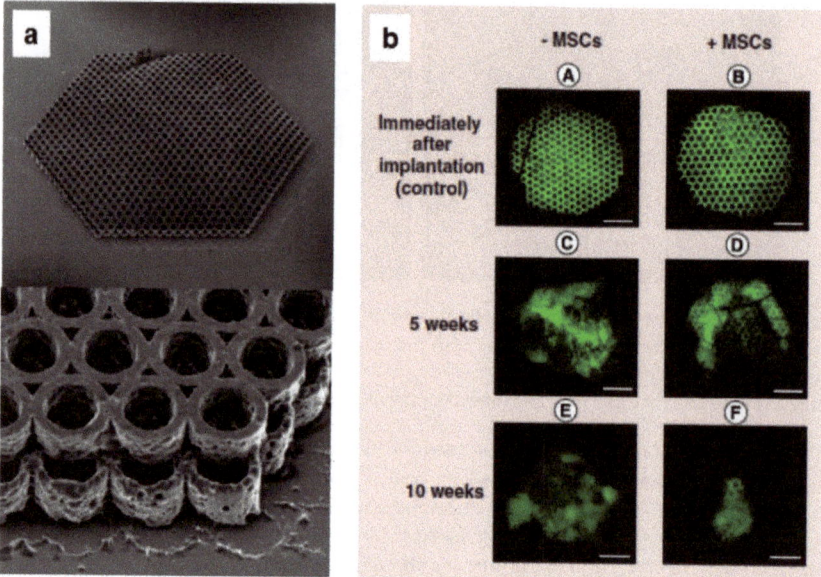

Figure 12. (**a**) Micrographs of a 2PP-fabricated PLA scaffold, (**b**) fluorescence of PLA scaffolds after implantation into mice; MSC—mesenchymal stem cell [71]. (Adapted from Future Medicine, 2016).

The above-presented selected examples of the formation of complicated 3D structures by different stereolithography techniques concern mainly the fabrication of porous scaffolds which, as authors claim, could be used as implants for bone regeneration. However, it seems that these methods may be also useful in the production of precise elements for some other applications, as, e.g., electronic devices.

5. Conclusions

Different types of radiation were successfully used to crosslink polylactide which appeared very sensitive to irradiation mainly due to the presence of methine hydrogen atoms along the polymer chain as well as the possibility of introducing functional end-groups. However, not only crosslinking but also other reaction as branching and chain scission proceeds upon irradiation. The proportions of these processes depend on many factors and the specified goal can be achieved by manipulating the irradiation conditions, by PLA functionalization and by the application of reactive additives. The choice of the irradiation type and its parameters depends on these specified goals.

Polylactide-based networks with high gel fraction were synthesized from both high molecular weight PLA and its copolymers as well as from low molecular weight PLAs functionalized at the chain ends by chain linking (coupling) methods.

The first of these approaches was mainly used in crosslinking by irradiation with high energy rays (electron beam and γ-rays). However, to have any control over crosslinking under irradiation, which can generate many radicals, multifunctional crosslinking agents containing unsaturated groups have to be added. As crosslinking by electron beam and gamma irradiation is always accompanied by chain scission, this method cannot be applied for all applications. Chain scission may be limited by adjusting the irradiation dose. The evident advantage of ionizing irradiation is the possibility to perform crosslinking at low temperatures and excellent penetration. One should, however, have in mind that the radiation affects many characteristics of the material, i.e., causes the decrease of glass transition, cold crystallization, and melting temperatures.

Crosslinking by exposure to less energetic UV (or visible) light undergoes in a more controlled manner, although undesired radical reactions may also proceed. Photo-curing is considered as the best method for crosslinking of PLA-based polymers designed for tissue engineering. For crosslinking by UV light, medium molecular weight polylactides (linear and star) functionalized with unsaturated end groups should be prepared. Polylactide macromers are mixed with photosensitive compounds being a source of radicals that initiate polymerization of unsaturated groups. Among different photo-initiators, the compounds which are reported to be not harmful in the context of biomedical applications can be chosen [110]. Because prepolymers used for crosslinking are characterized by low viscosity, it is often possible to mix polymer with the photoactive compound in bulk without using a solvent. This, in turn, enables in-situ crosslinking at low temperature, e.g., after placing liquid oligomers in a body. The weakness of this crosslinking method is the necessity of efficient PLA (meth)acrylation or another functionalization introducing reactive end groups. Moreover, radical chain-growth polymerization generates non-biodegradable high molecular weight acrylic chains. These acrylic chains become the major drawback as they are difficult to eliminate from the body. To avoid acrylic chains some authors prepared polylactide networks from prepolymers which were linked via different "click" type reactions; for biomedical application photo-induced, "thiol-ene" addition is well suited. Although photo-crosslinking seems to be a very suitable method for biomedical PLA applications, it also has a limitation, namely limited depth of penetration.

The methods presented here of radiation and photo-induced crosslinking of PLA-based materials could find an application at present (and possibly in the future) in different areas. The still developing sophisticated techniques in tissue engineering (e.g., photolithographic techniques) and in complicated drug delivery systems but also predictions for PLA mass production of durable bioplastics suggest also development of crosslinking methods for this biocompatible and bioderived polyester.

Author Contributions: Conceptualization, original draft preparation, supervision of the manuscript, M.B.; participation in writing, classification and preparation of the material for tables, K.B.; preparation of references, revisions at the stages of manuscript preparation, P.K. All authors have read and agreed to the published version of the manuscript.

Funding: This work was supported by the National Science Centre in Poland under Grant no. 2018/31/B/ST8/01969. The funder has no role in the design of the study; in the collection, analyses, or interpretation of data; in the writing of the manuscript, nor in the decision to publish the results.

Conflicts of Interest: There are no conflicts to declare.

References

1. Castro-Aguirre, E.; Iñiguez-Franco, F.; Samsudin, H.; Fang, X.; Auras, R. Poly(lactic acid)—Mass production, processing, industrial applications, and end of life. *Adv. Drug Deliv. Rev.* **2016**, *107*, 333–366. [CrossRef] [PubMed]
2. Bawa, K.K.; Oh, J.K. Stimulus-Responsive Degradable Polylactide-Based Block Copolymer Nanoassemblies for Controlled/Enhanced Drug Delivery. *Mol. Pharm.* **2017**, *14*, 2460–2474. [CrossRef] [PubMed]
3. Vacaras, S.; Baciut, M.; Lucaciu, O.; Dinu, C.; Baciut, G.; Crisan, L.; Hedesiu, M.; Crisan, B.; Onisor, F.; Armencea, G.; et al. Understanding the basis of medical use of poly-lactide-based resorbable polymers and composites – a review of the clinical and metabolic impact. *Drug Metab. Rev.* **2019**, *51*, 570–588. [CrossRef]
4. Chatterjee, S.; Saxena, M.; Padmanabhan, D.; Jayachandra, M.; Pandya, H.J. Futuristic medical implants using bioresorbable materials and devices. *Biosens. Bioelectron.* **2019**, *142*, 111489. [CrossRef] [PubMed]
5. Bano, K.; Pandey, R.; Jamal-e-Fatima and Roohi. New advancements of bioplastics in medical applications. *Int. J. Pharm. Sci. Res.* **2018**, *9*, 402–416. [CrossRef]
6. Thakur, S.; Chaudhary, J.; Sharma, B.; Verma, A.; Tamulevicius, S.; Thakur, V.K. Sustainability of bioplastics: Opportunities and challenges. *Curr. Opin. Green Sustain. Chem.* **2018**, *13*, 68–75. [CrossRef]

7. Emadian, S.M.; Onay, T.T.; Demirel, B. Biodegradation of bioplastics in natural environments. *Waste Manag.* **2017**, *59*, 526–536. [CrossRef] [PubMed]
8. Gutierrez, R.J. PLA Plastic/Material: All You Need to Know in 2020. Available online: https://all3dp.com/1/pla-plastic-material-polylactic-acid/ (accessed on 11 January 2020).
9. Murariu, M.; Dubois, P. PLA composites: From production to properties. *Adv. Drug Deliv. Rev.* **2016**, *107*, 17–46. [CrossRef] [PubMed]
10. Nagarajan, V.; Mohanty, A.K.; Misra, M. Perspective on Polylactic Acid (PLA) based Sustainable Materials for Durable Applications: Focus on Toughness and Heat Resistance. *ACS Sustain. Chem. Eng.* **2016**, *4*, 2899–2916. [CrossRef]
11. Nofar, M.; Sacligil, D.; Carreau, P.J.; Kamal, M.R.; Heuzey, M.-C. Poly (lactic acid) blends: Processing, properties and applications. *Int. J. Biol. Macromol.* **2019**, *125*, 307–360. [CrossRef]
12. Jem, K.J.; Tan, B. The development and challenges of poly (lactic acid) and poly (glycolic acid). *Adv. Ind. Eng. Polym. Res.* **2020**, *3*, 60–70. [CrossRef]
13. Mangeon, C.; Renard, E.; Thevenieau, F.; Langlois, V. Networks based on biodegradable polyesters: An overview of the chemical ways of crosslinking. *Mater. Sci. Eng. C* **2017**, *80*, 760–770. [CrossRef] [PubMed]
14. Parhi, R. Cross-Linked Hydrogel for Pharmaceutical Applications: A Review. *Adv. Pharm. Bull.* **2017**, *7*, 515–530. [CrossRef] [PubMed]
15. Gupta, M.C.; Deshmukh, V.G. Radiation effects on poly(lactic acid). *Polymer (Guildf)* **1983**, *24*, 827–830. [CrossRef]
16. Manas, D.; Ovsik, M.; Mizera, A.; Manas, M.; Hylova, L.; Bednarik, M.; Stanek, M. The Effect of Irradiation on Mechanical and Thermal Properties of Selected Types of Polymers. *Polymers (Basel)* **2018**, *10*, 158. [CrossRef]
17. Bednarek, M.; Borská, K.; Kubisa, P. New Polylactide -Based Materials by Chemical Crosslinking of Polylactide. *Polymer Rev.* **2020**. submitted.
18. Gilding, D.K.; Reed, A.M. Biodegradable polymers for use in surgery—poly(ethylene oxide) poly(ethylene terephthalate) (PEO/PET) copolymers: 1. *Polymer (Guildf)* **1979**, *20*, 1454–1458. [CrossRef]
19. Birkinshaw, C.; Buggy, M.; Henn, G.G.; Jones, E. Irradiation of poly-D,L-lactide. *Polym. Degrad. Stab.* **1992**, *38*, 249–253. [CrossRef]
20. Sintzel, M.B.; Merkli, A.; Tabatabay, C.; Gurny, R. Influence of Irradiation Sterilization on Polymers Used as Drug Carriers—A Review. *Drug Dev. Ind. Pharm.* **1997**, *23*, 857–878. [CrossRef]
21. Yoshioka, S.; Aso, Y.; Otsuka, T.; Kojima, S. The effect of γ-irradiation on drug release from poly(lactide) microspheres. *Radiat. Phys. Chem.* **1995**, *46*, 281–285. [CrossRef]
22. Babanalbandi, A.; Hill, D.J.T.; Whittaker, A.K. Volatile products and new polymer structures formed on 60Co γ-radiolysis of poly(lactic acid) and poly(glycolic acid). *Polym. Degrad. Stab.* **1997**, *58*, 203–214. [CrossRef]
23. Babanalbandi, A.; Hill, D.J.T.; O'Donnell, J.H.; Pomery, P.J.; Whittaker, A. An electron spin resonance study on γ-irradiated poly(l-lactic acid) and poly(D,L-lactic acid). *Polym. Degrad. Stab.* **1995**, *50*, 297–304. [CrossRef]
24. Milicevic, D.; Trifunovic, S.; Galovic, S.; Suljovrujic, E. Thermal and crystallization behaviour of gamma irradiated PLLA. *Radiat. Phys. Chem.* **2007**, *76*, 1376–1380. [CrossRef]
25. Malinowski, R. Effect of high energy β-radiation and addition of triallyl isocyanurate on the selected properties of polylactide. *Nucl. Instrum. Methods Phys. Res. Sect. B Beam Interact. Mater. Atoms* **2016**, *377*, 59–66. [CrossRef]
26. Adamus-Wlodarczyk, A.; Wach, R.; Ulanski, P.; Rosiak, J.; Socka, M.; Tsinas, Z.; Al-Sheikhly, M. On the Mechanisms of the Effects of Ionizing Radiation on Diblock and Random Copolymers of Poly(Lactic Acid) and Poly(Trimethylene Carbonate). *Polymers (Basel)* **2018**, *10*, 672. [CrossRef] [PubMed]
27. Shin, B.Y.; Han, D.H.; Narayan, R. Rheological and Thermal Properties of the PLA Modified by Electron Beam Irradiation in the Presence of Functional Monomer. *J. Polym. Environ.* **2010**, *18*, 558–566. [CrossRef]
28. Chu, C.C. Degradation phenomena of two linear aliphatic polyester fibres used in medicine and surgery. *Polymer (Guildf)* **1985**, *26*, 591–594. [CrossRef]
29. Nugroho, P.; Mitomo, H.; Yoshii, F.; Kume, T. Degradation of poly(l-lactic acid) by γ-irradiation. *Polym. Degrad. Stab.* **2001**, *72*, 337–343. [CrossRef]
30. Mitomo, H.; Kaneda, A.; Quynh, T.M.; Nagasawa, N.; Yoshii, F. Improvement of heat stability of poly(l-lactic acid) by radiation-induced crosslinking. *Polymer (Guildf)* **2005**, *46*, 4695–4703. [CrossRef]

31. Nagasawa, N.; Kaneda, A.; Kanazawa, S.; Yagi, T.; Mitomo, H.; Yoshii, F.; Tamada, M. Application of poly(lactic acid) modified by radiation crosslinking. *Nucl. Instruments Methods Phys. Res. Sect. B Beam Interact. Mater. Atoms* **2005**, *236*, 611–616. [CrossRef]
32. Quynh, T.M.; Mitomo, H.; Nagasawa, N.; Wada, Y.; Yoshii, F.; Tamada, M. Properties of crosslinked polylactides (PLLA & PDLA) by radiation and its biodegradability. *Eur. Polym. J.* **2007**, *43*, 1779–1785. [CrossRef]
33. Quynh, T.M.; Mitomo, H.; Zhao, L.; Asai, S. The radiation crosslinked films based on PLLA/PDLA stereocomplex after TAIC absorption in supercritical carbon dioxide. *Carbohydr. Polym.* **2008**, *72*, 673–681. [CrossRef]
34. Malinowski, R. Mechanical properties of PLA/PCL blends crosslinked by electron beam and TAIC additive. *Chem. Phys. Lett.* **2016**, *662*, 91–96. [CrossRef]
35. Malinowski, R.; Rytlewski, P.; Żenkiewicz, M. Effects of electron radiation on properties of PLA. *Arch. Mater. Sci. Eng.* **2011**, *49*, 25–32.
36. Jin, F.; Hyon, S.-H.; Iwata, H.; Tsutsumi, S. Crosslinking of Poly(L-lactide) by γ-Irradiation. *Macromol. Rapid Commun.* **2002**, *23*, 909–912. [CrossRef]
37. Xia, X.; Shi, X.; Liu, W.; He, S.; Zhu, C.; Liu, H. Effects of gamma irradiation on properties of PLA/flax composites. *Iran. Polym. J.* **2020**, *29*, 581–590. [CrossRef]
38. Hachana, N.; Wongwanchai, T.; Chaochanchaikul, K.; Harnnarongchai, W. Influence of Crosslinking Agent and Chain Extender on Properties of Gamma-Irradiated PLA. *J. Polym. Environ.* **2017**, *25*, 323–333. [CrossRef]
39. Malinowski, R.; Janczak, K.; Moraczewski, K.; Raszkowska-Kaczor, A. Analysis of swelling degree and gel fraction of polylactide/poly(butylene adipate-co-terephthalate) blends crosslinked by radiation. *Polimery* **2018**, *63*, 25–30. [CrossRef]
40. Malinowski, R.; Moraczewski, K.; Raszkowska-Kaczor, A. Studies on the Uncrosslinked Fraction of PLA/PBAT Blends Modified by Electron Radiation. *Materials (Basel)* **2020**, *13*, 1068. [CrossRef]
41. Rytlewski, P.; Stepczyńska, M.; Gohs, U.; Malinowski, R.; Budner, B.; Żenkiewicz, M. Flax fibres reinforced polylactide modified by ionizing radiation. *Ind. Crops Prod.* **2018**, *112*, 716–723. [CrossRef]
42. Salvatore, M.; Marra, A.; Duraccio, D.; Shayanfar, S.; Pillai, S.D.; Cimmino, S.; Silvestre, C. Effect of electron beam irradiation on the properties of polylactic acid/montmorillonite nanocomposites for food packaging applications. *J. Appl. Polym. Sci.* **2016**, *133*, 42219. [CrossRef]
43. Kumar, A.; Venkatappa Rao, T.; Ray Chowdhury, S.; Ramana Reddy, S.V.S. Optimization of mechanical, thermal and hydrolytic degradation properties of Poly (lactic acid)/Poly (ethylene-co-glycidyl methacrylate)/Hexagonal boron nitride blend-composites through electron-beam irradiation. *Nucl. Instrum. Methods Phys. Res. Sect. B Beam Interact. Mater. Atoms* **2018**, *428*, 38–46. [CrossRef]
44. Kodal, M.; Wis, A.A.; Ozkoc, G. The mechanical, thermal and morphological properties of γ-irradiated PLA/TAIC and PLA/OvPOSS. *Radiat. Phys. Chem.* **2018**, *153*, 214–225. [CrossRef]
45. Seppälä, J.; Korhonen, H.; Hakala, R.; Malin, M. Photocrosslinkable Polyesters and Poly(ester anhydride)s for Biomedical Applications. *Macromol. Biosci.* **2011**, *11*, 1647–1652. [CrossRef]
46. Borská, K.; Danko, M.; Mosnácek, J. Photodegradation and Photochemical Crosslinking of Polylactide. *Chem. List.* **2014**, *108*, 1030–1039.
47. van Bochove, B.; Grijpma, D.W. Photo-crosslinked synthetic biodegradable polymer networks for biomedical applications. *J. Biomater. Sci. Polym. Ed.* **2019**, *30*, 77–106. [CrossRef]
48. Sawhney, A.S.; Pathak, C.P.; Hubbell, J.A. Bioerodible hydrogels based on photopolymerized poly(ethylene glycol)-co-poly(.alpha.-hydroxy acid) diacrylate macromers. *Macromolecules* **1993**, *26*, 581–587. [CrossRef]
49. Jansen, J.; Mihov, G.; Feijen, J.; Grijpma, D.W. Photo-Crosslinked Biodegradable Hydrogels Prepared from Fumaric Acid Monoethyl Ester-Functionalized Oligomers for Protein Delivery. *Macromol. Biosci.* **2012**, *12*, 692–702. [CrossRef]
50. Burdick, J.A.; Frankel, D.; Dernell, W.S.; Anseth, K.S. An initial investigation of photocurable three-dimensional lactic acid based scaffolds in a critical-sized cranial defect. *Biomaterials* **2003**, *24*, 1613–1620. [CrossRef]
51. Davis, K.A.; Burdick, J.A.; Anseth, K.S. Photoinitiated crosslinked degradable copolymer networks for tissue engineering applications. *Biomaterials* **2003**, *24*, 2485–2495. [CrossRef]

52. Burdick, J.A.; Philpott, L.M.; Anseth, K.S. Synthesis and characterization of tetrafunctional lactic acid oligomers: A potential in situ forming degradable orthopaedic biomaterial. *J. Polym. Sci. Part A Polym. Chem.* **2001**, *39*, 683–692. [CrossRef]
53. Kim, B.S.; Hrkach, J.S.; Langer, R. Biodegradable photo-crosslinked poly(ether-ester) networks for lubricious coatings. *Biomaterials* **2000**, *21*, 259–265. [CrossRef]
54. Shen, J.Y.; Pan, X.Y.; Lim, C.H.; Chan-Park, M.B.; Zhu, X.; Beuerman, R.W. Synthesis, Characterization, and In Vitro Degradation of a Biodegradable Photo-Cross-Linked Film from Liquid Poly(ε-caprolactone- co -lactide- co -glycolide) Diacrylate. *Biomacromolecules* **2007**, *8*, 376–385. [CrossRef] [PubMed]
55. Michlovská, L.; Vojtová, L.; Humpa, O.; Kučerík, J.; Žídek, J.; Jančář, J. Hydrolytic stability of end-linked hydrogels from PLGA–PEG–PLGA macromonomers terminated by α,ω-itaconyl groups. *RSC Adv.* **2016**, *6*, 16808–16816. [CrossRef]
56. Sharifi, S.; Grijpma, D.W. Resilient Amorphous Networks Prepared by Photo-Crosslinking High-Molecular-Weight D,L-Lactide and Trimethylene Carbonate Macromers: Mechanical Properties and Shape-Memory Behavior. *Macromol. Biosci.* **2012**, *12*, 1423–1435. [CrossRef]
57. Melchels, F.P.W.; Feijen, J.; Grijpma, D.W. A poly(D,L-lactide) resin for the preparation of tissue engineering scaffolds by stereolithography. *Biomaterials* **2009**, *30*, 3801–3809. [CrossRef]
58. Grijpma, D.W.; Hou, Q.; Feijen, J. Preparation of biodegradable networks by photo-crosslinking lactide, ε-caprolactone and trimethylene carbonate-based oligomers functionalized with fumaric acid monoethyl ester. *Biomaterials* **2005**, *26*, 2795–2802. [CrossRef]
59. Leclerc, E.; Furukawa, K.; Miyata, F.; Sakai, Y.; Ushida, T.; Fujii, T. Fabrication of microstructures in photosensitive biodegradable polymers for tissue engineering applications. *Biomaterials* **2004**, *25*, 4683–4690. [CrossRef]
60. Aoyagi, T.; Miyata, F.; Nagase, Y. Preparation of cross-linked aliphatic polyester and application to thermo-responsive material. *J. Control. Release* **1994**, *32*, 87–96. [CrossRef]
61. Amsden, B.G. Biodegradable elastomers in drug delivery. *Expert Opin. Drug Deliv.* **2008**, *5*, 175–187. [CrossRef]
62. Chen, F.; Hochleitner, G.; Woodfield, T.; Groll, J.; Dalton, P.D.; Amsden, B.G. Additive Manufacturing of a Photo-Cross-Linkable Polymer via Direct Melt Electrospinning Writing for Producing High Strength Structures. *Biomacromolecules* **2016**, *17*, 208–214. [CrossRef] [PubMed]
63. Chen, F.; Hayami, J.W.S.; Amsden, B.G. Electrospun Poly(D,L-lactide- co -acryloyl carbonate) Fiber Scaffolds With a Mechanically Stable Crimp Structure For Ligament Tissue Engineering. *Biomacromolecules* **2014**, *15*, 1593–1601. [CrossRef]
64. Amsden, B.; Misra, G.; Marshall, M.; Turner, N. Synthesis and Characterization of Biodegradable Networks Providing Saturated-Solution Prolonged Delivery. *J. Pharm. Sci.* **2008**, *97*, 860–874. [CrossRef] [PubMed]
65. Chapanian, R.; Amsden, B.G. Combined and sequential delivery of bioactive VEGF165 and HGF from poly(trimethylene carbonate) based photo-cross-linked elastomers. *J. Control. Release* **2010**, *143*, 53–63. [CrossRef]
66. Amsden, B.G.; Marecak, D. Long-Term Sustained Release from a Biodegradable Photo-Cross-Linked Network for Intraocular Corticosteroid Delivery. *Mol. Pharm.* **2016**, *13*, 3004–3012. [CrossRef] [PubMed]
67. Chapanian, R.; Tse, M.Y.; Pang, S.C.; Amsden, B.G. Osmotic Release of Bioactive VEGF from Biodegradable Elastomer Monoliths is the Same In Vivo As In Vitro. *J. Pharm. Sci.* **2012**, *101*, 588–597. [CrossRef] [PubMed]
68. Karikari, A.S.; Edwards, W.F.; Mecham, J.B.; Long, T.E. Influence of Peripheral Hydrogen Bonding on the Mechanical Properties of Photo-Cross-Linked Star-Shaped Poly(D,L-lactide) Networks. *Biomacromolecules* **2005**, *6*, 2866–2874. [CrossRef] [PubMed]
69. Tanodekaew, S.; Channasanon, S.; Uppanan, P. Preparation and degradation study of photocurable oligolactide-HA composite: A potential resin for stereolithography application. *J. Biomed. Mater. Res. Part B Appl. Biomater.* **2014**, *102*, 604–611. [CrossRef]
70. Koroleva, A.; Gill, A.A.; Ortega, I.; Haycock, J.W.; Schlie, S.; Gittard, S.D.; Chichkov, B.N.; Claeyssens, F. Two-photon polymerization-generated and micromolding-replicated 3D scaffolds for peripheral neural tissue engineering applications. *Biofabrication* **2012**, *4*, 025005. [CrossRef]
71. Timashev, P.; Kuznetsova, D.; Koroleva, A.; Prodanets, N.; Deiwick, A.; Piskun, Y.; Bardakova, K.; Dzhoyashvili, N.; Kostjuk, S.; Zagaynova, E.; et al. Novel biodegradable star-shaped polylactide scaffolds for bone regeneration fabricated by two-photon polymerization. *Nanomedicine* **2016**, *11*, 1041–1053. [CrossRef]

72. Shashkova, V.T.; Matveeva, I.A.; Glagolev, N.N.; Zarkhina, T.S.; Cherkasova, A.V.; Kotova, S.L.; Timashev, P.S.; Solovieva, A.B. Synthesis of polylactide acrylate derivatives for the preparation of 3D structures by photo-curing. *Mendeleev Commun.* **2016**, *26*, 418–420. [CrossRef]
73. Kuznetsova, D.; Ageykin, A.; Koroleva, A.; Deiwick, A.; Shpichka, A.; Solovieva, A.; Kostjuk, S.; Meleshina, A.; Rodimova, S.; Akovanceva, A.; et al. Surface micromorphology of cross-linked tetrafunctional polylactide scaffolds inducing vessel growth and bone formation. *Biofabrication* **2017**, *9*, 025009. [CrossRef] [PubMed]
74. Bardakova, K.N.; Grebenik, E.A.; Istranova, E.V.; Istranov, L.P.; Gerasimov, Y.V.; Grosheva, A.G.; Zharikova, T.M.; Minaev, N.V.; Shavkuta, B.S.; Dudova, D.S.; et al. Reinforced Hybrid Collagen Sponges for Tissue Engineering. *Bull. Exp. Biol. Med.* **2018**, *165*, 142–147. [CrossRef] [PubMed]
75. Bardakova, K.N.; Grebenik, E.A.; Minaev, N.V.; Churbanov, S.N.; Moldagazyeva, Z.; Krupinov, G.E.; Kostjuk, S.V.; Timashev, P.S. Tailoring the collagen film structural properties via direct laser crosslinking of star-shaped polylactide for robust scaffold formation. *Mater. Sci. Eng. C* **2020**, *107*, 110300. [CrossRef]
76. Petchsuk, A.; Submark, W.; Opaprakasit, P. Development of crosslinkable poly(lactic acid-co-glycidyl methacrylate) copolymers and their curing behaviors. *Polym. J.* **2013**, *45*, 406–412. [CrossRef]
77. Rahman, M.; Thananukul, K.; Supmak, W.; Petchsuk, A.; Opaprakasit, P. Synthesis and quantitative analyses of acrylamide-grafted poly(lactide-co-glycidyl methacrylate) amphiphilic copolymers for environmental and biomedical applications. *Spectrochim. Acta Part A Mol. Biomol. Spectrosc.* **2020**, *225*, 117447. [CrossRef]
78. Koo, G.-H.; Jang, J. Preparation of melting-free poly(lactic acid) by amorphous and crystal crosslinking under UV irradiation. *J. Appl. Polym. Sci.* **2013**, *127*, 4515–4523. [CrossRef]
79. Poplata, S.; Tröster, A.; Zou, Y.-Q.; Bach, T. Recent Advances in the Synthesis of Cyclobutanes by Olefin [2 + 2] Photocycloaddition Reactions. *Chem. Rev.* **2016**, *116*, 9748–9815. [CrossRef]
80. Habault, D.; Zhang, H.; Zhao, Y. Light-triggered self-healing and shape-memory polymers. *Chem. Soc. Rev.* **2013**, *42*, 7244–7256. [CrossRef]
81. Kaur, G.; Johnston, P.; Saito, K. Photo-reversible dimerisation reactions and their applications in polymeric systems. *Polym. Chem.* **2014**, *5*, 2171–2186. [CrossRef]
82. Nagata, M.; Inaki, K. Synthesis and characterization of photocrosslinkable poly(l-lactide)s with a pendent cinnamate group. *Eur. Polym. J.* **2009**, *45*, 1111–1117. [CrossRef]
83. Nagata, M.; Sato, Y. Photocurable biodegradable polyesters from poly(L-lactide) diols. *Polym. Int.* **2005**, *54*, 386–391. [CrossRef]
84. Hu, X.; Chen, X.; Cheng, H.; Jing, X. Cinnamate-functionalized poly(ester-carbonate): Synthesis and its UV irradiation-induced photo-crosslinking. *J. Polym. Sci. Part A Polym. Chem.* **2009**, *47*, 161–169. [CrossRef]
85. Gangolphe, L.; Déjean, S.; Bethry, A.; Hunger, S.; Pinese, C.; Garric, X.; Bossard, F.; Nottelet, B. Degradable multi(aryl azide) star copolymer as universal photo-crosslinker for elastomeric scaffolds. *Mater. Today Chem.* **2019**, *12*, 209–221. [CrossRef]
86. Rupp, B.; Ebner, C.; Rossegger, E.; Slugovc, C.; Stelzer, F.; Wiesbrock, F. UV-induced crosslinking of the biopolyester poly(3-hydroxybutyrate)-co-(3-hydroxyvalerate). *Green Chem.* **2010**, *12*, 1796–1802. [CrossRef]
87. Chapanian, R.; Tse, M.Y.; Pang, S.C.; Amsden, B.G. Long term in vivo degradation and tissue response to photo-cross-linked elastomers prepared from star-shaped prepolymers of poly(ε-caprolactone- co -D,L-lactide). *J. Biomed. Mater. Res. Part A* **2009**, *92A*, 830–842. [CrossRef]
88. Gu, F.; Neufeld, R.; Amsden, B. Maintenance of vascular endothelial growth factor and potentially other therapeutic proteins bioactivity during a photo-initiated free radical cross-linking reaction forming biodegradable elastomers. *Eur. J. Pharm. Biopharm.* **2007**, *66*, 21–27. [CrossRef]
89. Gu, F.; Neufeld, R.; Amsden, B. Sustained release of bioactive therapeutic proteins from a biodegradable elastomeric device. *J. Control. Release* **2007**, *117*, 80–89. [CrossRef]
90. Gu, F.; Younes, H.M.; El-Kadi, A.O.S.; Neufeld, R.J.; Amsden, B.G. Sustained interferon-γ delivery from a photocrosslinked biodegradable elastomer. *J. Control. Release* **2005**, *102*, 607–617. [CrossRef]
91. Ilagan, B.G.; Amsden, B.G. Macroporous photocrosslinked elastomer scaffolds containing microposity: Preparation and in vitro degradation properties. *J. Biomed. Mater. Res. Part A* **2009**, *93A*, 211–218. [CrossRef]
92. Vyner, M.C.; Liu, L.; Sheardown, H.D.; Amsden, B.G. The effect of elastomer chain flexibility on protein adsorption. *Biomaterials* **2013**, *34*, 9287–9294. [CrossRef] [PubMed]
93. Chapanian, R.; Amsden, B.G. Osmotically driven protein release from photo-cross-linked elastomers of poly(trimethylene carbonate) and poly(trimethylene carbonate-co-D,L-lactide). *Eur. J. Pharm. Biopharm.* **2010**, *74*, 172–183. [CrossRef] [PubMed]

94. Timbart, L.; Tse, M.Y.; Pang, S.; Amsden, B.G. Tissue Response to, and Degradation Rate of, Photocrosslinked Trimethylene Carbonate-Based Elastomers Following Intramuscular Implantation. *Materials (Basel)* **2010**, *3*, 1156–1171. [CrossRef]
95. Hayami, J.W.S.; Waldman, S.D.; Amsden, B.G. A Photocurable Hydrogel/Elastomer Composite Scaffold with Bi-Continuous Morphology for Cell Encapsulation. *Macromol. Biosci.* **2011**, *11*, 1672–1683. [CrossRef]
96. Hayami, J.W.S.; Waldman, S.D.; Amsden, B.G. Injectable, High Modulus, And Fatigue Resistant Composite Scaffold for Load-Bearing Soft Tissue Regeneration. *Biomacromolecules* **2013**, *14*, 4236–4247. [CrossRef] [PubMed]
97. Hayami, J.W.S.; Surrao, D.C.; Waldman, S.D.; Amsden, B.G. Design and characterization of a biodegradable composite scaffold for ligament tissue engineering. *J. Biomed. Mater. Res. Part A* **2009**, *92A*, 1407–1420. [CrossRef]
98. Amsden, B.; Qi, B. Anti-atherosclerotic peptide delivery from a photocrosslinkable biodegradable network. *Int. J. Pharm.* **2010**, *388*, 32–39. [CrossRef]
99. Shinyama, K. Influence of Electron Beam Irradiation on Electrical Insulating Properties of PLA with Soft Resin Added †. *Polymers (Basel)* **2018**, *10*, 898. [CrossRef]
100. Oyj, S.E. Method of Use of Polylactide and Manufacturing a Heat-Sealed Paper or Board Container or Package. U.S. Patent 10 414 105B2, 24 January 2014. Available online: https://patentswarm.com/patents/US10414105B2. (accessed on 22 October 2020).
101. Oyj, S.E. Heat-Sealable Biodegradable Packaging Material, a Method for Its Manufacture, and a Product Package Made from the Material. U.S. Patent 9181010B2, 11 May 2011. Available online: https://patents.google.com/patent/EP2544957A1. (accessed on 22 October 2020).
102. Oyj, S.E. Method for Improving the Heat Sealibility of Packaging Material and Method for Manufacturing Heat-Sealed Container or Package. International Application No. PCT/FI2011/050381; U.S. Patent Application No. 13695496, U.S. Patent 20130137562 3 November 2011. Available online: https://patentscope.wipo.int/search/en/detail.jsf?docId=WO2011135182. (accessed on 22 October 2020).
103. Lai, W.-J.; Huang, C.-H. Wei Mon Ind Co Polylactide-Coated Paperboard. U.S. Patent 20100209636A1, 9 April 2009. Available online: https://patents.google.com/patent/US20100209636A1. (accessed on 22 October 2020).
104. Taleyarkhan, R.; Bakken, A.C.; Fisher, K.F.; Hagen, A.R.; Kostry, N.P. Polylactic Acid Adhesive Compositions and Methods for Their Preparation and Use. U.S. Patent 10442966B2, 15 November 2013. Available online: https://patents.google.com/patent/WO2014078720A1 (accessed on 22 October 2020).
105. Amsden, B.G.; Misra, G.; Gu, F.; Younes, H.M. Synthesis and Characterization of a Photo-Cross-Linked Biodegradable Elastomer. *Biomacromolecules* **2004**, *5*, 2479–2486. [CrossRef]
106. Kim, K.; Yeatts, A.; Dean, D.; Fisher, J.P. Stereolithographic Bone Scaffold Design Parameters: Osteogenic Differentiation and Signal Expression. *Tissue Eng. Part B Rev.* **2010**, *16*, 523–539. [CrossRef]
107. Concellón, A.; Asín, L.; González-Lana, S.; de la Fuente, J.M.; Sánchez-Somolinos, C.; Piñol, M.; Oriol, L. Photopolymers based on ethynyl-functionalized degradable polylactides by thiol-yne 'Click Chemistry'. *Polymer (Guildf)* **2017**, *117*, 259–267. [CrossRef]
108. Ronca, A.; Ambrosio, L.; Grijpma, D.W. Preparation of designed poly(D,L-lactide)/nanosized hydroxyapatite composite structures by stereolithography. *Acta Biomater.* **2013**, *9*, 5989–5996. [CrossRef] [PubMed]
109. Akopova, T.A.; Timashev, P.S.; Demina, T.S.; Bardakova, K.N.; Minaev, N.V.; Burdukovskii, V.F.; Cherkaev, G.V.; Vladimirov, L.V.; Istomin, A.V.; Svidchenko, E.A.; et al. Solid-state synthesis of unsaturated chitosan derivatives to design 3D structures through two-photon-induced polymerization. *Mendeleev Commun.* **2015**, *25*, 280–282. [CrossRef]
110. Bryant, S.J.; Nuttelman, C.R.; Anseth, K.S. Cytocompatibility of UV and visible light photoinitiating systems on cultured NIH/3T3 fibroblasts in vitro. *J. Biomater. Sci. Polym. Ed.* **2000**, *11*, 439–457. [CrossRef]

Publisher's Note: MDPI stays neutral with regard to jurisdictional claims in published maps and institutional affiliations.

© 2020 by the authors. Licensee MDPI, Basel, Switzerland. This article is an open access article distributed under the terms and conditions of the Creative Commons Attribution (CC BY) license (http://creativecommons.org/licenses/by/4.0/).

Review

Supramolecular Interactions in Hybrid Polylactide Blends—The Structures, Mechanisms and Properties

Anna Kowalewska * and Maria Nowacka

Centre of Molecular and Macromolecular Studies, Polish Academy of Sciences, Sienkiewicza 112, 90-363 Łódź, Poland; mnowacka@cbmm.lodz.pl
* Correspondence: anko@cbmm.lodz.pl; Tel.: +48-42-680-3350

Academic Editor: Sylvain Caillol
Received: 19 June 2020; Accepted: 20 July 2020; Published: 23 July 2020

Abstract: The conformation of polylactide (PLA) chains can be adjusted by supramolecular interactions (the formation of hydrogen bonds or host-guest complexes) with appropriate organic molecules. The structures formed due to those intermolecular interactions may act as crystal nuclei in the PLA matrix ("soft templating"). In this review, the properties of several supramolecular nucleating systems based on synthetic organic nucleators (arylamides, hydrazides, and 1,3:2,4-dibenzylidene-D-sorbitol) are compared to those achieved with biobased nucleating agents (orotic acid, humic acids, fulvic acids, nanocellulose, and cyclodextrins) that can also improve the mechanical properties of PLA. The PLA nanocomposites containing both types of nucleating agents/additives are discussed and evaluated in the context of their biomedical applicability.

Keywords: polylactide; composites; supramolecular interactions; crystallization

1. Introduction

Polylactide (PLA) is a biodegradable semi-crystalline polymer that has attracted enormous attention over recent years as a biocompatible and environment-friendly material [1–3]. It has been approved by the U.S. Food and Drug Administration for biomedical applications and contact with body fluids, e.g., as bioresorbable artificial ligaments or drug delivery systems [4]. PLA is also one of the most important thermoplastic materials for 3D printing [5]. It is easily processable, although its brittleness prevents tensile drawing. The tensile modulus and strength of neat PLA can be increased and the strain at break can be reduced when the polylactide matrix contains a significant amount of crystalline fraction [6]. The crystallinity degree also governs the barrier properties of polylactide [6]. The crystallization behaviour of polylactide has been extensively investigated (details can be found in comprehensive reviews, e.g., [6–8]). In general, the type of crystal structure depends on the crystallization conditions. The most common α'- and α-crystals that have similar chain conformations and belong to the same crystal system are formed in melt crystallizations. The less ordered α'-form is obtained exclusively, or as an admixture coexisting with the α-form, when PLA is crystallized isothermally at temperatures below 120 °C. A less thermally stable β-form of a frustrated structure is obtained by stretching the α-form at high draw ratios in the hot-drawing of melt- or solution-spun fibres. The γ-form was obtained by the epitaxial crystallization of PLA on hexamethylbenzene.

Considerable attention was given to the improvement of crystallization kinetics (nucleation and crystal growth) that can be enhanced by nucleators and/or plasticizers. Numerous potential nucleating systems have been examined in the literature, including "green" nucleating agents. PLA-based polymer composites and nanocomposites containing nanoclays [9–11], nanosilicas [12,13], carbon nanotubes [14,15], or graphene [16] are well known. Another class of additives for polylactide that can improve its crystallization rate are species that exploit supramolecular phenomena, e.g., hydrogen bonding or host-guest effects, to interact with PLA macromolecules ("soft templating").

Hydrogen bonding is also important for the formation of poly(L-lactide)/poly(D-lactide) (PLLA/PDLA) stereocomplex structures (SC) [17–20]. The specific C-H···O=C interactions between the paired stereoisomeric PLLA and PDLA chains play a very important role in the formation of SC crystals. It was found that the racemic ($3_2/3_1$) helical conformation of the pair of macromolecules starts to emerge in the melt of a racemic blend, and the formed structures subsequently act as nucleating sites upon cooling [21]. Thus, the conformation of the polylactide chains may be changed by intermolecular interactions with neighbouring macromolecules. It means that the crystallization of the PLA matrix can also be adjusted by hydrogen bonding species, even before the true crystal nuclei emerge. Hydrogen bonding between polylactide and nucleating agents has been postulated, e.g., for PLA blends with amino acids or poly(amino acids) [22], carbon nanotubes [23], phtalimide [24], bisurea derivatives [25], or D-gluconic acid derivatives [26]. PLA was also modified with macromolecular nucleators, such as linear polysilsesquioxanes [27–29] or their cyclosiloxane analogues [30] with side substituents acting as donors/acceptors of hydrogen bonds (-OH···O=C-; -COOH···O=C-; and -C-F···H$_3$C-) to/from the polyester backbone.

Extremely efficient nucleating systems are based on arylamides and arylhydrazides that may self-organize in the polymer melt. Unfortunately, not all supramolecular nucleators are biodegradable, bioabsorbable, or nontoxic. The potential problems with some of those organic compounds can be a significant concern in the biomedical field. Therefore, polylactide matrices have been also blended with harmless biobased nucleating agents (e.g., orotic acid, humic acids, fulvic acids, nanocellulose, and cyclodextrins). In this review, the research progress on PLA nanocomposites containing both groups of nucleating agents has been evaluated in the context of their biomedical applicability.

2. Organic Nucleators

Several groups of organic compounds, such as aryl amide and hydrazide derivatives and relative compounds, can dissolve into PLA melt and then separate and self-organize into supramolecular frameworks upon cooling [31–34]. The molecular mechanism of such interactions involves hydrogen bonding between C=O in the polylactide macromolecules and the hydrogen bond donating groups (-OH, -NH, and -NH$_2$) in the nucleating agents. Those compounds should also have relatively high thermal decomposition temperatures. Self-assembly nucleators have been increasingly used to guide the crystallization of PLA. The heterogeneous nucleation of the polymer matrix is based on crystallization of those organic compounds soluble in PLA melt. The process of organic crystal formation involves nucleation and growth. It makes the crystallite morphology highly dependent on the applied cooling rates, concentrations, and supersaturation. The nucleating efficiency of self-assembling nucleators is largely determined by their surface area-to-volume ratio after crystallization. In the following sections, the PLA nucleation ability of arylamides and hydrazides, orotic acid, and 1,3:2,4-dibenzylidene-D-sorbitol is discussed in comparison to humic acids and fulvic acids.

2.1. Arylamide Nucleators

The addition of N,N',N"-tricyclohexyl-1,3,5-benzenetricarboxylamide (TCB, Scheme 1) to PLA can promote the nucleation of the polymer matrix and accelerate the overall crystallization rate [35,36]. The half-time and rate constants of non-isothermal crystallizations carried out at different cooling rates showed that TCB significantly accelerated the process at a very low content of 0.3 wt%. The crystallinity of neat PLA decreased from 47% to 6% when the cooling rate was increased from 1 to 5 °C/min (crystallization did not occur at 10 °C/min), while for PLA nucleated by TCB, it remained almost constant at 45–51% irrespectively of the cooling rates (2.5–10 °C/min).

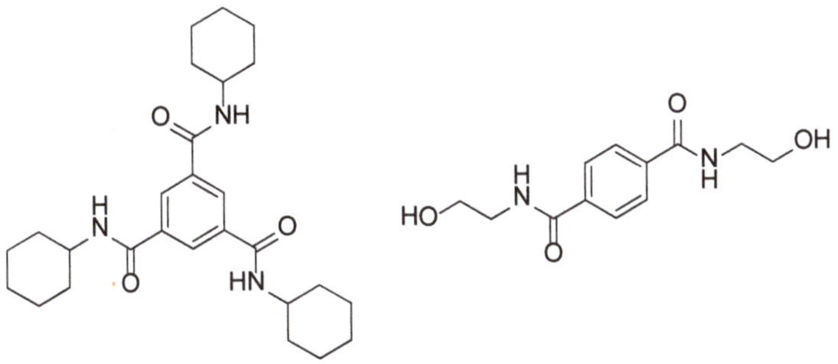

Scheme 1. Chemical structures of N,N′,N″-tricyclohexyl-1,3,5-benzenetricarboxylamide (TCB) and N,N′-bis(2-hydroxyethyl)-terephthalamide (BHET).

Three unique crystal superstructures, including cone-like, shish-kebab, and needle-like structures, were obtained by the melt crystallization of PLLA nucleated by TCB [37]. N,N′,N″-tricyclohexyl-1,3,5-benzenetricarboxylamide dissolved in the polymer melt self-organizes upon cooling into fine fibrils prior to PLLA crystallization. The fibrils serve, subsequently, as a "shish" to induce the epitaxial growth ("soft templating") of "kebab-like" structures approximately orthogonal to the long axis (Figure 1).

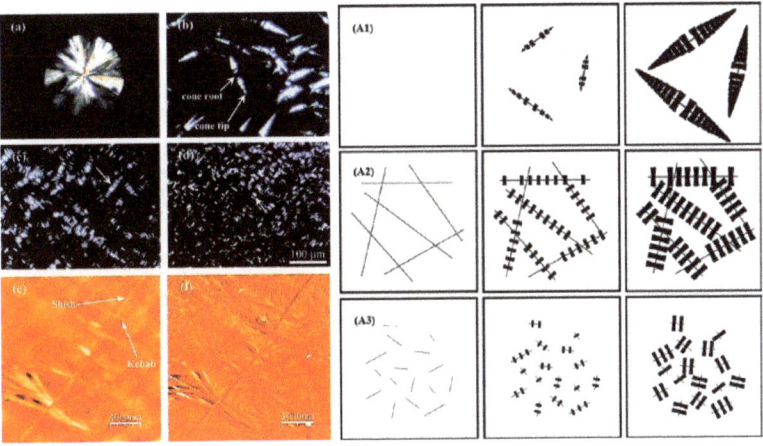

Figure 1. Left panel: Polarized Optical Microscopy (POM) micrographs of the crystal morphology for poly(L-lactide) (PLLA) containing different amounts of N,N′,N″-tricyclohexyl-1,3,5-benzenetricarboxylamide (TCB): (**a**) neat PLLA, (**b**) 0.2 wt%, (**c**) 0.3 wt% and (**d**) 0.5 wt% prepared by isothermal crystallization at 130 °C for 55, 7, 10 and 10 min, respectively, and Atomic Force Microscope (AFM) height (**e**) and phase (**f**) images of PLLA containing 0.2 wt% TCB of a typical shish-kebab-like superstructure. Right panel: schematic representation of the evolution of crystal morphologies during the crystallization of PLLA containing TCB: (**A1**) 0.2 wt%, (**A2**) 0.3 wt% and (**A3**) 0.5 wt% [37]. Reprinted with permission from Macromolecules. Copyright (2011), American Chemical Society.

The effect of TCB on the crystallization behaviour of PLA was studied on a molecular level by time-resolved FTIR and wide angle X-ray diffraction [38]. The observed vibrational changes indicate that the arylamide molecules can accelerate the formation of a skeletal conformational-ordered structure but their presence is more favourable to the formation of a 10_3 helix structure that is characteristic of α and α' crystals. The value of the Avrami exponent was lower for PLA nucleated by TCB than that of neat PLA, indicating changes in the crystallization mechanism, although they had no impact on the crystal form. It was also shown that 0.3 wt% of TCB effectively increased the crystallization rate and yield upon cooling from melt at 10 °C/min (Figure 2).

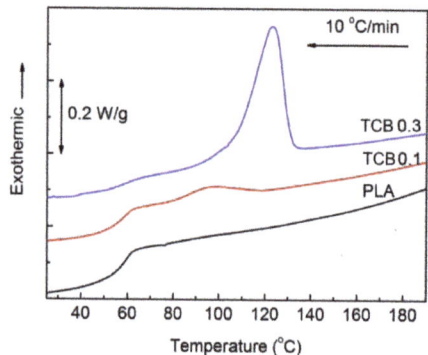

Figure 2. Differential scanning calorimetry (DSC) cooling traces recorded for polylactide (PLA) and its compositions with TCB (0.1 and 0.3 wt%) [38]. Adapted from Polymer Testing. Copyright (2017), with permission from Elsevier.

The effect of the melting temperatures (ranging from 190 to 240 °C), concentrations of self-assembling TCB (0–0.5 wt%), and cooling rates (2.5–20 °C/min) was investigated for PLA/TCB mixtures [39]. The solubility of TCB was largely dependent on the processing temperature and the concentration of the compound in the polyester matrix. At 240 °C, TCB could dissolve completely and then self-assemble into supramolecular frameworks upon cooling. The crystallization peak temperature of PLA showed a bell-shaped dependence on the concentration of TCB and the cooling rate applied. TCB was also used as a self-assembly nucleating agent for melt-blended PLA/poly(ethylene oxide) (PEO) [40]. Both PEO and TCB exhibited a synergistic effect on promoting PLA crystallization as well as a toughening effect on the blended material (Figure 3). Moreover, TCB prominently reinforced both neat PLA and PLA/PEO blends in the glass transition region and at $T > T_g$, indicating an improvement of their heat resistance. The cooperative effect on promoting PLA crystallization was explained by nucleation with TCB and plasticization with PEO chains.

N,N'-Bis(2-hydroxyethyl)-terephthalamide (BHET, Scheme 1) can be a versatile additive for PLLA, operating both as a plasticizer for processing purposes and as a nucleating agent [41]. BHET crystallizes from the PLA melt during cooling, and the formed crystals facilitate heterogeneous nucleation in the PLA matrix. The formation of BHET crystals with a high surface area-to-volume ratio is favoured at high undercooling/supersaturation. Very importantly, it was proved that the hydroxyl groups of BHET were not involved in transesterification with polylactide chains during extrusion at 200 °C, and the molecular weight of the polyester was not changed. When allowed to crystallize during processing, BHET induced the formation of PLA crystals oriented along the flow direction, enhancing the tensile modulus of the blend. Interestingly, a rapid cooling to $T < T_g$ prevented the crystallization of BHET and only a plasticizing effect was indicated by a decrease in both the melt viscosity and glass-transition temperature. A characteristic suppression in the yield point of the amorphous PLA/BHET blend was also observed during mechanical tests with increasing BHET concentration.

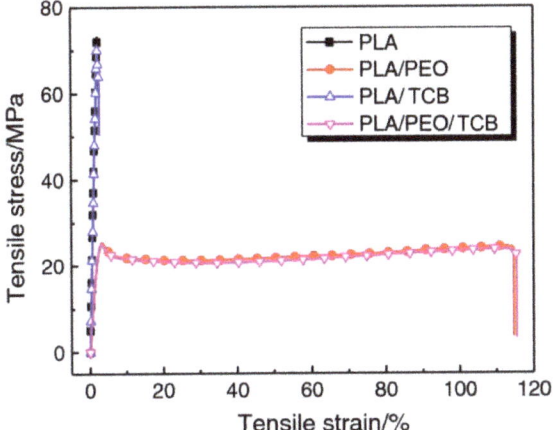

Figure 3. Comparison of tensile stress–strain curves of injection-moulded neat PLA and PLA/poly(ethylene oxide) (PEO) blends with or without TCB [40]. Adapted with permission from Springer Nature: Journal of Thermal Analysis and Calorimetry. Copyright (2018).

2.2. Hydrazide Nucleators

Aryl hydrazides used as nucleators for the enhancement of polylactide crystallization have two characteristic structural features—a carbohydrazide linkage (-C(O)NH-NHC(O)-) that can take part as an acceptor/donor during the formation of hydrogen bonds, and at least one aryl group providing a system of π-electrons for π-π interactions (Scheme 2). The combination of these two functions allows for an effective supramolecular organization of hydrazide derivatives within polymer melts. Even the small molecule of phthalhydrazide (PH, Scheme 2) can be an efficient nucleating agent that enhances the crystallization of polylactide. It was reported that the non-isothermal melt crystallization of PLA started much earlier in the presence of even a low content of PH (0.1 wt%) [42]. The analysis of Avrami plots indicated that the crystallization mechanism was not changed. The overall effect was ascribed to the increased number of nucleation sites.

However, larger hydrazide derivatives built of at least two aryl groups linked by two carbohydrazide systems separated by an alkyl or aryl spacer are much more effective. They can crystallize within the polymer melt and form structures that can serve as templates for growing polymer crystals. Tetramethylenedicarboxylic dibenzoylhydrazide (TMC, $n = 2$, Scheme 2)—the shortest dihydrazide—exhibited an excellent nucleating effect on PLA. With the addition of 0.05 wt% of TMC, the crystallization half-time of PLA decreased from 26.06 to 6.13 min at 130 °C [43]. TMC did not change the crystallization mechanism in the PLA matrix, as indicated by comparison of the Avrami values. However, the density of nuclei in the PLA was increased, while no discernible effect on the crystalline structure was noted. Non-isothermal crystallization studies indicated that N,N'-bis(benzoyl)suberic acid dihydrazide (BSDH, $n = 4$, Scheme 2) can accelerate the overall PLLA crystallization rate due to a heterogeneous nucleation effect. With the incorporation of BSDH, the PLLA crystallization peak became sharper and was shifted to a higher temperature range as the sample was cooled down from the melt at a rate of 1 °C/min [44]. The presence of BSDH also affected the isothermal crystalline behaviours (a shorter crystallization time and a faster overall crystallization due to an increased number of small spherulites).

Octamethylenedicarboxylic dibenzoylhydrazide (OMBH, $n = 6$, Scheme 2) was found to be very effective for the acceleration of PLA crystallization under a high cooling rate (50 °C/min with 1 wt% of OMBH) [33]. Very short isothermal crystallization half-times were recorded within 90–130 °C. The moulding cycle time of PLA/OMBH was <3 min. The physical and mechanical properties of the blend were improved as well (a heat distortion temperature of 124 °C, flexural modulus of

4.1 GPa, and Izod impact strength of 7.9 kJ/m^2). Decamethylenedicarboxylic dibenzoylhydrazide (DMBH, n = 8, Scheme 2) also increased the crystallization temperature (T$_c$) of PLA [33]. The overall effect of larger dihydrazide molecules is rather complicated and dependent on the crystallization temperature and the crystalline structure. The nucleation efficiency of OMBH significantly depended on its solubility during the thermal annealing (generally increasing with temperature and decreasing with a concentration increase) [45]. However, under certain conditions, the crystallization temperature and nucleation efficiency of OMBH increased with the concentration of the additive, resulting in higher crystallization enthalpy.

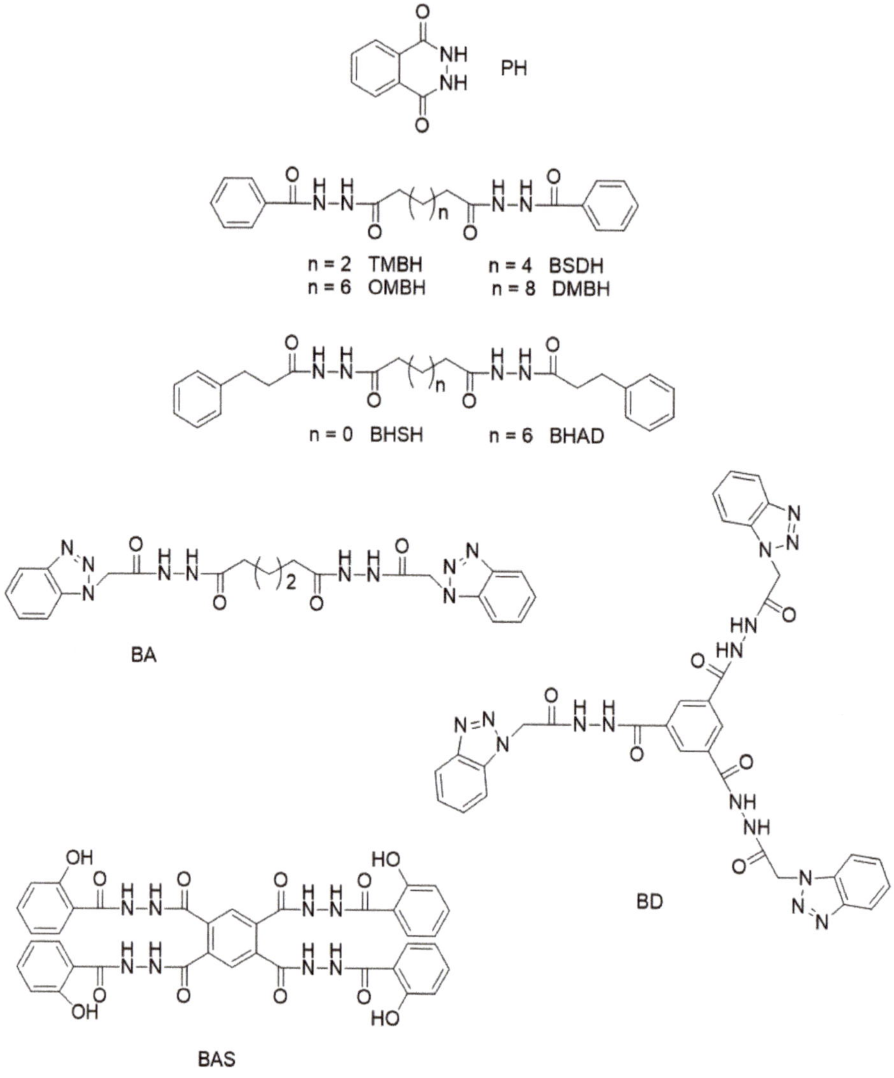

Scheme 2. Chemical structures of phthalhydrazide and symmetrical or branched di-, tri-, and tetra-hydrazides.

Time-resolved spectroscopic studies on the interactions between OMBH and polylactide elucidated the nucleation process at the molecular level [46]. The results showed that the crystallization of PLLA in the presence of the dihydrazide nucleator involves not only heterogeneous nucleation with OMBH but also the conformational regulation of polyester chains by hydrogen bonding between the two components of the blend (Scheme 3). The results indicated that due to NH···O=C interactions between the dissolved nucleator and PLLA, the building blocks of the PLLA chains were transformed into trans-gauche conformers before the self-assembling of OMBH into nanocrystals and their phase-separation from the PLLA melt. It resulted in a decrease in the energy barrier to the formation of α-crystals of polylactide. Once the dissolved molecules of OMBH start to self-assemble into nanostructures upon cooling, the polylactide chains with an increased population of trans-gauche conformers begin to form primary nuclei on the surface of OMBH nanofibrils and crystals. Therefore, conformational regulation was proposed for the crystalline manipulation of PLLA by hydrazide nucleators.

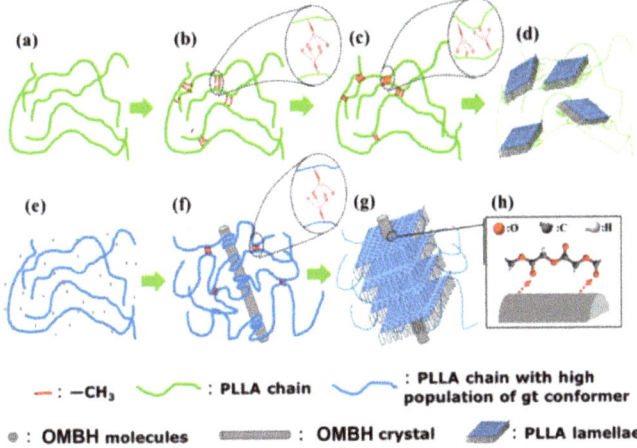

Scheme 3. Crystallization of neat PLLA (**a**–**d**) compared to the formation of α-crystals due to "soft templating", and conformational regulation of PLLA chains through supramolecular interactions with octamethylenedicarboxylic dibenzoylhydrazide (OMBH) (**e**–**h**) [46]. Adapted with permission from Biomacromolecules. Copyright (2017), American Chemical Society.

It was also shown that OMBH can self-assemble into diverse frameworks that induce the formation of various crystalline superstructures of polylactide, depending on their content in the polymer melt and the processing conditions (Figure 4). Larger amounts of OMBH (1 wt%) first self-assembled into star multiarm frameworks (at 170 °C), and then each of the resultant arms served as a "shish" to induce the growth of PLLA lamellae after cooling down to 150 °C. The nucleating sites on the surface of such structures are scarce, and thus, the lamellae grew as branched "calabash" structures. A slightly lower amount of OMBH (0.5 wt%) self-assembled at 125.3 °C into short fibril-like frameworks with a sufficient number of available nucleating sites. It resulted in a transcrystalline superstructure with PLLA crystals growing epitaxially orthogonally to the long axis of the OMBH fibrils. An interesting sunflower-like superstructure with big PLLA spherulite centres, surrounded by hybrid fibril-like trans-crystalline superstructures at the boundary between the spherulite and the amorphous region, was obtained in the presence of a very small amount of OMBH (0.3 wt%).

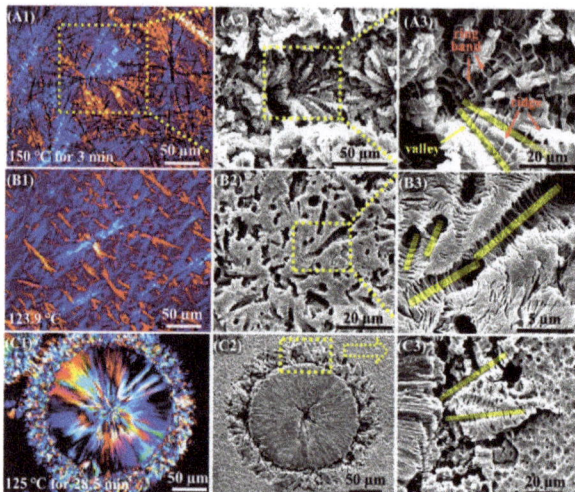

Figure 4. POM (**A1**–**C1**) and SEM micrographs (**A2**–**C2** and **A3**–**C3**) of the crystalline superstructures formed by PLLA in the presence of OMBH (A: 1 wt%; B: 0.5 wt%; C: 0.3 wt%) after different thermal treatments [47]. Reprinted with permission from Biomacromolecules. Copyright (2017), American Chemical Society.

The crystallization of polylactive was also studied in the presence of more structurally complicated dihydrazides. The studies revealed that both the cooling rate and the melting temperature affected the non-isothermal crystallization behaviour of PLLA in the presence of N,N'-succinic bis(hydrocinnamic acid) dihydrazide (BHSH, Scheme 2) [47] and N,N'-sebacic bis(hydrocinnamic acid) dihydrazide (HAD, Scheme 2) [48]. BHSH accelerated both the melt crystallization and the non-isothermal crystallization of PLLA. At a content of 2 wt%, the half time of crystallization at 100 °C decreased by twelve times, compared to that of neat PLLA (48.6 vs. 575.7 s). HAD (1 wt%) increased both the crystallization temperature and non-isothermal crystallization enthalpy from 94.5 °C and 0.1 J/g (neat PLLA) to 131.6 °C and 48.5 J/g, respectively. It was also found that the cold crystallization behaviour of PLLA/HAD was almost independent of the HAD concentration, when it was larger than 2 wt%. Although those compounds vastly improved the crystallization of PLLA, their presence decreased the thermal stability and light transmittance of the PLLA films (Figure 5). This effect was ascribed to the increased crystallinity of the blends as well as the colour of the additives.

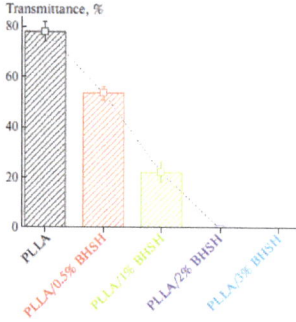

Figure 5. Light transmittance of PLLA and its composites with bis(hydrocinnamic acid) dihydrazide (BHSH) [47]. Reprinted with permission from Springer Nature: Polymer Science, Series A. Copyright (2019).

The non-isothermal crystallization of PLLA was also significantly promoted by the presence of 1 wt% of N,N'-bis(1H-benzotriazole) adipic acid acethydrazide (BA, Scheme 2) [49], 0.5 wt% of N,N,N'-tris(1H-benzotriazole) trimesinic acid acethydrazide (BD, Scheme 2) [50], and 3 wt% of N,N,N,N'-salicylic tetra(1,2,4,5-benzenetetracarboxylic acid) hydrazide (BAS, Scheme 2) [51]. Upon cooling at a rate of 1 °C/min, the onset of crystallization temperature was shifted from 101.4 to 111.3 °C upon the addition of 0.5 wt% BD to the PLLA. The enthalpy of non-isothermal crystallization increased from 0.1 to 38.6 J/g. The isothermal crystallization half-time decreased as well. An enhanced nucleation density was indicated by double-melting peaks of PLLA/0.5% BD, assigned to melting-recrystallization. The equilibrium melting point of the PLLA/0.5% BD blend was set at lower temperatures than those of neat PLLA. A similar effect was observed for blends of PLLA and BAS. This additive is a highly efficient nucleating agent that can significantly promote the crystallization of PLLA upon cooling at a rate of 20 °C/min. The best acceleration of the melt crystallization of PLLA carried out upon cooling at 1 °C/min was achieved with 3 wt% of BAS. The final melting temperature was another important factor. The best temperature for the melt-crystallization of the blend was 200 °C. Despite the improvement of the thermal stability and fluidity compared with the neat PLLA, a decrease in the light transmittance through the PLLA/BAS blends was observed.

Hydrazide nucleating species were also obtained by an in situ reaction between 4,4'-diphenylmethane diisocyanate (MDI) and benzohydrazine (P) in the polylactide matrix (Scheme 4) [52]. The rate of PLLA crystallization was enhanced by blending those reagents with the polyester upon melting. The reaction between the components was confirmed by NMR spectroscopy. This procedure increased the compatibility between the nucleating agents and the polylactide matrix. The crystallinity of the PLLA increased from 10.3% to 42.1% upon adding 0.25% (MDI + P) and melting them for 8 min. The PLLA crystallization half-time at 130 °C decreased from 42.0 to 1.1 min. Moreover, the heat resistance of the PLA was enhanced and good mechanical properties were achieved.

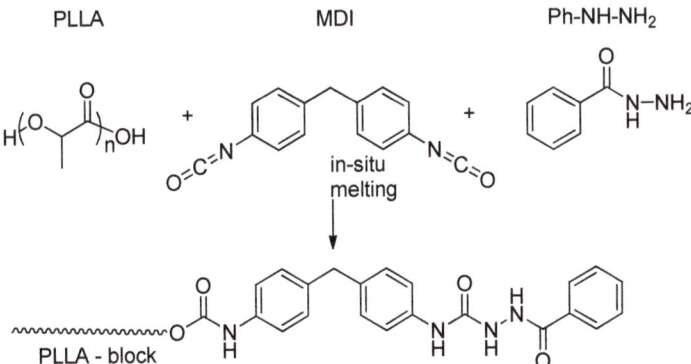

Scheme 4. Postulated pathway of the reaction between 4,4'-diphenylmethane diisocyanate, benzohydrazine, and PLLA.

2.3. PLA/Orotic Acid Blends

Orotic acid (2,6-dioxo-1,2,3,6-tetrahydropyrimidine-4-carboxylic acid, OA) (Scheme 5) is a compound synthesized by living organisms as a key substance in the biosynthesis of naturally occurring pyrimidines [53,54]. Owing to its hydrogen-bond donating/accepting properties, it may also serve as an excellent biobased nucleating agent for polylactide. The temperature of the melting of OA (345–346 °C) is high enough to allow melt compounding with PLA.

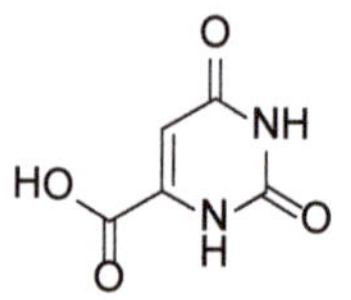

Scheme 5. Chemical structure of 2,6-dioxo-1,2,3,6-tetrahydropyrimidine-4-carboxylic acid (orotic acid).

Both the non-isothermal and isothermal melt crystallization kinetics of PLLA were enhanced significantly by orotic acid, even at a very low content (0.3 wt%) [55]. The presence of OA effectively induced the crystallization of PLLA upon fast cooling (10 °C/min) (Figure 6). The crystallization peak temperature was 123.9 °C and the crystallization enthalpy was equal to 34.1 J/g when the sample was crystallized non-isothermally from melt at a cooling rate of 10 °C/min. The half-time of isothermal crystallization at 120 °C was 0.64 min.

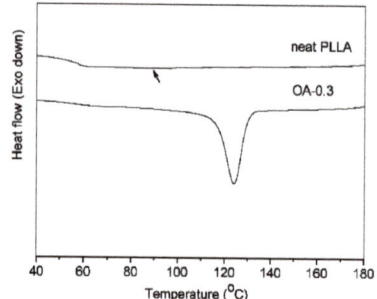

Figure 6. A comparison of DSC cooling (10 °C/min) traces of neat PLLA and the composite containing 0.3 wt% of OA (OA-0.3) [55]. Reprinted with permission from Industrial & Engineering Chemistry Research. Copyright (2011), American Chemical Society.

Tests with anhydrous (OA-a) and monohydrated (OA-m) orotic acid were carried out to gain an insight into the role that water molecules play in the activity of orotic acid as a nucleating agent in PLA crystallization [56]. The compounds were mixed with PLLA melt, and their nucleation effectiveness in the non-isothermal and isothermal melt crystallization of PLLA was investigated. Although OA-a showed more prominent nucleation efficiency than OA-m, both forms of orotic acid improved the nucleation density, the degree of crystallinity of the PLLA, and the overall crystallization rate. The data derived from Avrami plots (Figure 7) suggest a simultaneous nucleation and a three-dimensional crystal growth. The number of spherulite nucleation sites increases with increasing OA content, while their size decreases. The Avrami exponent n, indicating the nature of nucleation and the dimensionality of crystal growth, is very close to 3 regardless of the degree of orotic acid hydration. Nevertheless, the molecules of water bound to OA-m and its dehydration transition seem to decrease the nucleation effect. The nucleation density is higher and the spherulite size distribution is more uniform in PLLA/OA-a blends. This can be linked to the better dispersion and more uniform distribution of active nucleating agents in the blends.

The isothermal crystallization behaviour of PLLA/OA blends was investigated with time-resolved FTIR spectroscopy [56]. The spectral evolution of conformational changes and chain packing during thermal annealing at 140 °C showed that for both PLLA admixed with OA-a and OA-m, the crystallization proceeded with changes in the interchain interactions that were followed by the formation of the 10_3 helix. However, the induction time and crystallization half-times were much shorter for the sample containing anhydrous OA. The formation of crystal structure is similar to that observed for pure PLLA at, for example, 120 °C, despite the fact that without OA, PLLA hardly crystallized at 140 °C. The photodegradation and biodegradation of PLA in the presence of orotic acid were also studied [57]. Rheological measurements showed that OA slightly enhanced the UV-induced cleavage of polyester chains. Photodegradation was shown to accelerate the subsequent biodegradation, as macromolecules of smaller molecular weight are more prone to decomposition. Orotic acid itself promoted biodegradation both in native and photodegraded samples, by shortening the lag phase and increasing the rate of the process.

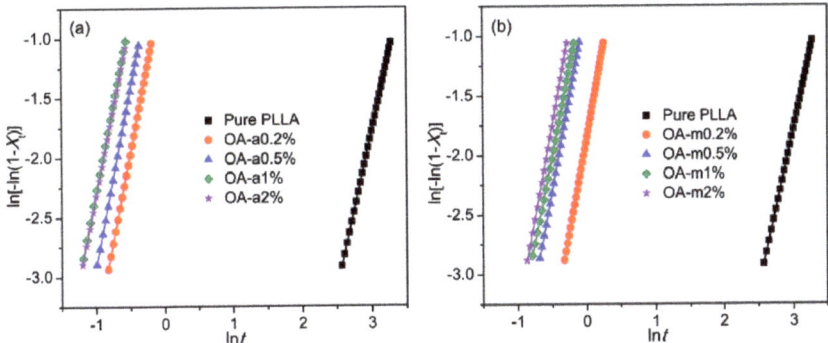

Figure 7. Avrami plots for blends of PLLA and (**a**) anhydrous or (**b**) monohydrate orotic acid isothermally melt-crystallized at 130 °C [56]. Published by The Royal Society of Chemistry (2017).

2.4. PLA/Sorbitol Blends

1,3:2,4-Bibenzylidene-D-sorbitol (DBS, Scheme 6), derived from the natural sugar alcohol D-glucitol, is considered an environmentally friendly material of acceptably low toxicity. It is a well-known low-molecular-weight supramolecular gelator of "butterfly-like" structure that has considerable potential in applications such as personal care products, adhesives, dental composites, gel electrolytes, or liquid crystalline materials and as a polymer nucleator [58].

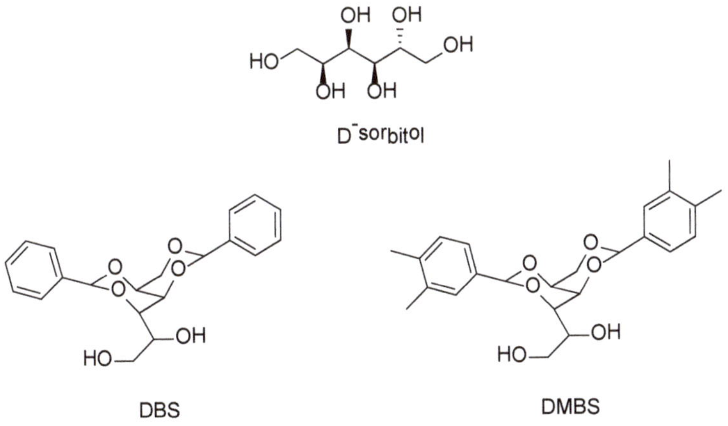

Scheme 6. Chemical structure of D-sorbitol, 1,3:2,4-dibenzylidene-D-sorbitol (DBS), and 1,3,2,4-bis(3,4-dimethylobenzylideno)sorbitol (DMBS).

The benzylidene groups and hydroxyl moieties on the sorbitol backbone are two molecular recognition motifs underpinning the role of DBS in the formation of a nanoscale network and the gelation mechanism. It involves interactions between 5-OH/6-OH hydrogen-bond-donating groups of one molecule of DBS and the hydrogen-bond-accepting 5-OH/6-OH or the cyclic acetals of another molecule. The π-π stacking/solvophobic interactions between the aromatic substituents are other important factor in the molecular recognition pathway. In non-polar media, the intermolecular hydrogen bonding plays a key role in the self-assembly, whereas in polar, protic solvents, the π-π stacking and/or solvophobic interactions between the aromatic groups become more important. The primary one-dimensional nanostructures (nanofibrils) formed by DBS molecules bundle together into 3D nanofibers (10 nm to 0.8 µm in diameter) of helical or non-helical structure (depending on the environment's polarity), and a three-dimensional network is subsequently formed.

DBS can direct the crystallization of polylactide. It influenced the crystal structure of PLLA and affected its crystallization rate and melting behaviour. FTIR spectroscopy showed that the C=O stretching band in the PLLA sample at 1755 cm^{-1} became much broader as DBS was added, presumably due to hydrogen bonding between DBS and PLLA [52]. In the presence of DBS, the formation of the perfect α-form was favoured over the less ordered α'-structures [59]. The α-crystals were obtained more easily, and the disorder-to-order (α'-to-α) phase-transition was shifted to lower temperatures with an increasing concentration of DBS. This was explained in terms of supramolecular templating exerted by DBS nanofibrils. DBS molecules stacked together through π-π interactions to form a strand that was linked to PLLA macromolecules by hydrogen bonding (Scheme 7). The equilibrium melting point and glass-transition temperature of PLLA admixed with DBS were not significantly changed.

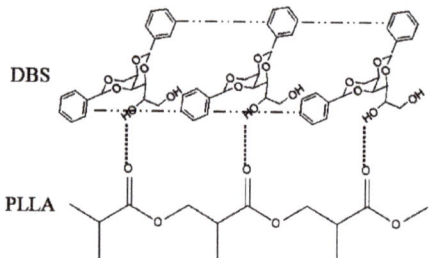

Scheme 7. Postulated interactions between the polylactide backbone and nanofibrils of self-assembled DBS [59]. Reprinted with permission from The Journal of Physical Chemistry B. Copyright (2011), American Chemical Society.

The X-ray diffraction peaks at 2θ = 12.5°, 14.7°, 16.7°, 19.1° and 22.4° correspond to the (103), (010), (200)/(110), (203) and (015) reflections characteristic of the α-crystals of PLLA, whereas the one at 2θ = 24.5° is representative of the less perfect α'-crystals. The latter are typically formed as a single form or coexist with α-crystals at crystallization temperatures <120 °C. It was found that with an increase in DBS content, the reflections (200)/(110) moved forward to the higher 2θ range and, contrary to the behaviour of neat PLLA, they were not significantly changed during isothermal crystallization at 120, 130 and 140 °C (Figure 8) [59]. The α-crystals predominate even at T < 120 °C. The temperature of the α'-to-α phase-transition of PLLA was lowered with an increase in the DBS concentration, owing to the interactions between the polyester chains and the self-organized nanofibrils.

The DBS architectures could be tuned by varying the DBS contents and PLLA crystallization temperatures [52]. Micron-sized fibrillar rings or disks and "concentric-circled" PLLA spherulites were formed due to the aggregation of nanofibrils in samples with DBS contents >3 wt% crystallized above 120 °C (Figure 9). The dispersed nanofibrils affected the orientation of PLLA lamellae and caused a change in birefringence, not significantly affecting the spherulitic growth rate. If the local concentration of DBS was low then PLLA crystallized in a typical manner and DBS molecules were excluded outside the crystals. Their concentration in the amorphous region increased. When the PLLA crystallization was complete, the concentration of the DBS molecules ejected outside the PLLA crystals became high enough to let them self-assemble at the PLLA growth front and form nanofibrils between the spherulites. For higher DBS contents (3–4 wt%) the formation of nanofibrils was much easier. Therefore, they were formed during the process of PLLA crystallization and aggregated inside the spherulites. More regular PLLA crystals were formed at lower temperatures when larger amounts of

DBS were added [60]. Moreover, the spherulitic growth rate of PLLA depended inversely on the fold surface free energy, which increased with the amount of the additive. DBS also enhanced the hardness and stiffness of the PLLA on cooling [61]. Furthermore, the hydrophilicity of the PLLA was significantly improved by an increase in the DBS concentration, which is important for biomedical applications of PLA. The crystallization of the PLA melt was also enhanced in a more complex blend, via the synergistic effect of DBS nanofibrils as the nucleating agent, combined with a plasticizer (poly(ethylene glycol), PEG) and a multifunctional monomer (pentaerythritol triacrylate, PETA) [62]. It was crucial to prepare the DBS/PEG gel before mixing those components with PLA. The mixing temperature was also critical. The acceleration of crystallization was ascribed to the increase in the nucleation density as well as the faster growth rate of spherulites in the presence of the plasticizer.

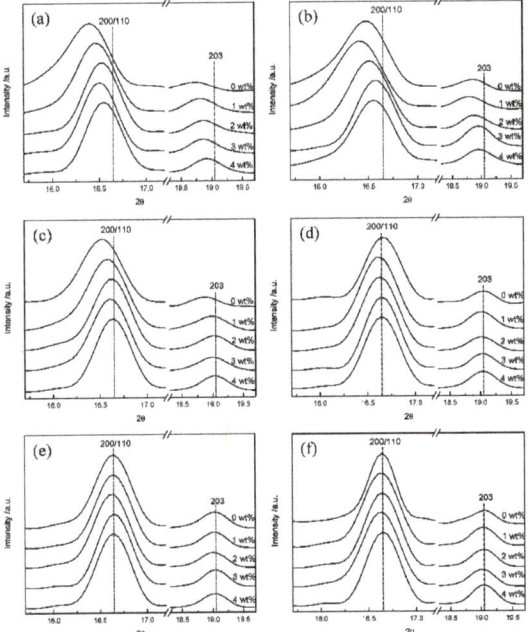

Figure 8. Wide-angle X-ray Scattering (WAXS) patterns of PLLA containing 0–4 wt% of DBS isothermally crystallized at (**a**) 90, (**b**) 100, (**c**) 110, (**d**) 120, (**e**) 130, and (**f**) 140 °C [59]. Reprinted with permission from The Journal of Physical Chemistry B. Copyright (2011), American Chemical Society.

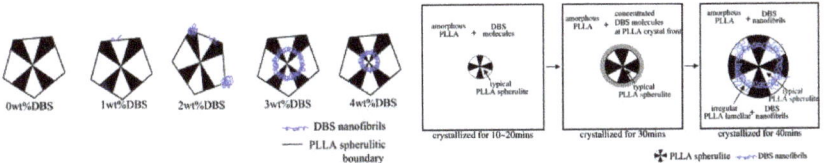

Figure 9. Concentration-dependent formation of nanofibrils of DBS during crystallization of PLLA [52]. Republished with permission of Royal Society of Chemistry. Permission conveyed through Copyright Clearance Center, Inc.

An analogue of DBS—1,3,2,4-bis(3,4-dimethylobenzylideno)sorbitol (DMBS) (Scheme 6)—was applied to poly(lactic acid) as a nucleating and clarifying agent [63]. The nucleated PLA crystallized earlier, and a reduction in the crystallization temperature was observed at DMBS concentrations of 0.25–10 wt%, while the glass transition temperature was decreased by ~10 °C. The composite maintained similar clarity at all DMBS concentrations. The tensile modulus and tensile strength increased slightly with the content of DMBS up to 1.5 wt% but dropped at higher concentrations of the additive. Biodegradable PLA composites were prepared using a combination of wood fiber (WF) and DMBS [64]. 1,3,2,4-bis(3,4-dimethylobenzylideno)sorbitol acted as an effective nucleating agent for those composites, improved their thermal stability and mechanical properties, and slowed down their enzymatic degradation.

2.5. PLA/Humic Substances

Humic substances (HS) are organic matter distributed in the natural environment. They comprise a complex mixture of humic acids (HA), fulvic acid (FA), and humins (Scheme 8) [65]. Humic substances can be considered as environmentally and biologically safe natural antioxidants. Blends of HA/FA and PLA are interesting because of their potential applicability as biomedical scaffolds or packaging materials for oxidation-sensitive food. Owing to its biological origin, a black mixture of HA has an undefined composition that may vary depending on their origin and the process of extraction. The components generally contain aromatic and aliphatic structures with attached carbonyl, carboxylic, and phenolic groups, as well as a minor amount of amine and amide residues that may contribute as donors/acceptors for the formation of hydrogen bonds. Thus, HA may display interfacial activity, hydrophilicity, and cation exchange and complexation capacities. HA have an amphiphilic character and are soluble in aqueous alkaline media. They are not soluble in neutral to acidic media and form micelle-like structures under such conditions. Fulvic acids are the yellowish brown, low molecular weight fraction of the amorphous polymeric mixture of humic acids. FA are soluble in aqueous media of different pH as well as in some organic solvents (acetone and ethanol).

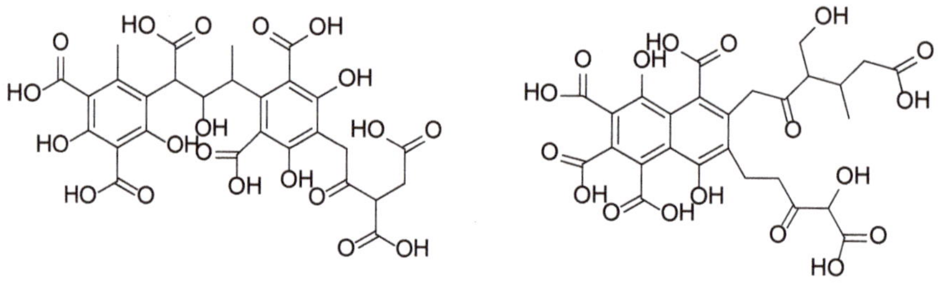

Scheme 8. Exemplary components of Humic Acids (HA) and Fulvic Acids (FA).

The polycondensation of L-lactic acid was performed at 150 °C in the presence of HA (0.01% w/v), resulting in a 93% yield of a hybrid polymer of molecular weight 6.4×10^5 g/mol [66]. The incorporation of HA slightly enhanced the thermal stability of the polyester matrix. The glass transition temperature and the temperature of melting were reduced, although the elongation at break and ductility were enhanced. The presence of HA in the blend significantly improved the hydrophilicity of the polyester film, total content of phenolics (0.075 µmol GAE/g film), and water absorption capacity (90.65%). An improvement in the antioxidant activities and radical scavenging properties was also observed.

Humic acids may be also derivatized to increase their compatibility with the PLA matrix in the hybrid blends. For example, the amount of amide groups was increased by the amidation of HA using dodecylamine in the presence of carbonyl diimidazole as the coupling agent (HA-amide-1) [67] or aniline in the presence of a phosphorus trichloride catalyst (HA-amide-2) [68]. The non-isothermal crystallization kinetics showed that HA amides incorporated during the thermoplastic processing may serve as nucleating agents that enhance the crystallization rate of PLA. Inclusions of HA amide provided a large number of heterogeneous nucleation sites thus increasing the degree of the crystallinity of the PLA/HA-amide composites (Figure 10). The crystallization behaviour varied under different cooling conditions. The mechanical properties of the blends were also improved after the introduction of HA amides that increased the rigidity of the network structure in the PLA matrix (Figure 11).

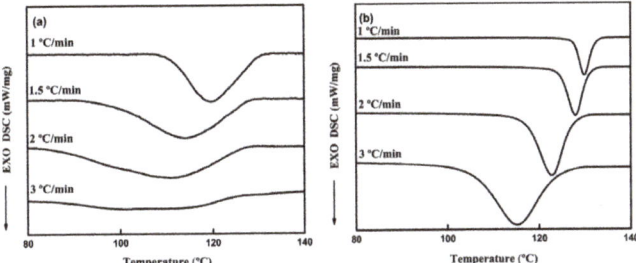

Figure 10. DSC thermograms of non-isothermal crystallization for PLA (**a**) and PLA containing 0.3 wt% of HA-amide-1 (**b**) [68]. Reprinted with permission from Springer Nature. Copyright (2017).

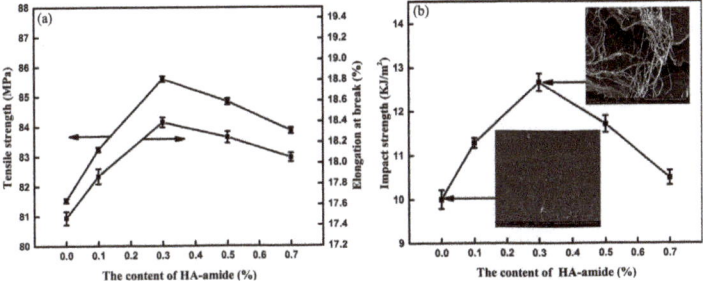

Figure 11. Effects of HA-amide-1 on mechanical performance of PLA blends: (**a**) tensile strength and elongation at break; (**b**) impact strength (insets: impact fracture SEM images) [67]. Reprinted with the permission of the publisher (Taylor & Francis Ltd., http://www.tandfonline.com).

Analogously to HA, FA structural units consist of polycyclic aromatic structures and functional groups (e.g., -COOH, -OH, R-CH=CH-OH, and C=O). They can thus be modified through sulfonation, nitration, etc. reactions in order to broaden the application scope. Fulvic acids have also been transformed into amide derivatives in order to augment their interactions with polylactide chains. Fulvic acid amide (FAA) was synthesized with fulvic acids and urea [69]. FA derivatives were also prepared by coupling FA with benzhydrazide (FA-BH) [70] or p-phenylenediamine (MFA) [71].

Polylactide blends containing the derivatives of fulvic acids were obtained upon melt blending. FAA accelerated the PLA crystallization rate and improved toughness of the PLA/FAA composites [69]. FAA acted as a heterogeneous nucleation agent and enhanced the three-dimensional spherulitic crystal growth in PLA. The nucleating mechanism of PLA/FAA proposed to explain the observed phenomena was based on the hydrogen bonding between C=O in the polyester backbone and N-H residues in FA derivatives. The rheological behaviour of the PLA/FAA blends showed an increase in the storage modulus induced by FAA. The apparent viscosities and thermal stability of the composites were much higher after the blending. The rheological behaviour indicated good interfacial compatibility between PLA and FA grafted with p-phenylenediamine (MFA) [71]. Mechanical tests showed that the impact strength of PLA admixed with 0.5 wt% MFA was improved by 97.2% and the impact destruction resulted in the characteristic ductile fracture. Moreover, the rate of nucleation was improved and the crystallinity of the PLA increased from 4.9% to 36.9% in the presence of MFA. Interestingly, MFA also influenced the mechanism of enzymatic degradation, increasing the Km of proteinase K and somewhat inhibiting the degradation. The mechanical performance of composites containing 0.1 wt% FA-BH was improved [70]. The network structure of the PLA/FA-BH blend was more rigid. The tensile strength, tensile modulus, elongation at break, and impact strength increased, respectively, by 6.38%, 27.47%, 28.75% and 74.56% (Figure 12). However, larger amounts of FA-BH deteriorated the properties of the PLA. FA-BH acted efficiently as nucleating agents (Figure 13). The crystallization rate of the PLA matrix was improved by 0.1 wt% of FA-BH. The degree of crystallinity increased to 41.88% (Figure 14) and the temperature of crystallization increased from 97.2 to 116.4 °C upon cooling from melt.

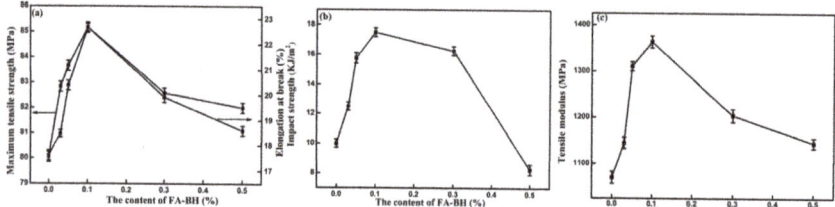

Figure 12. Effects of fulvic acid–benzhydrazide (FA-BH) on mechanical properties of PLA-based composites: (**a**) maximum tensile strength and elongation at break; (**b**) impact strength; (**c**) tensile modulus [70]. Reproduced with permission of John Wiley & Sons Ltd. from Liu P., Zhen W., Bian S., Wang X., Advances in Polymer Technology, 2018; 37, 2788–2798.

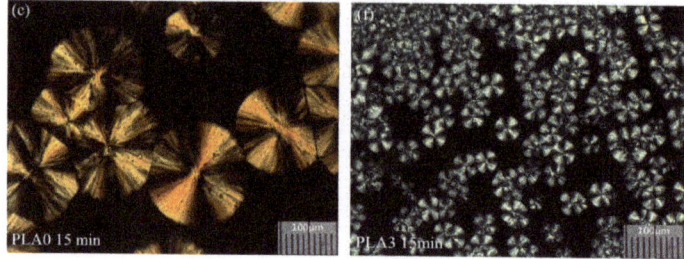

Figure 13. Evolution of spherulitic structures in PLA (**c**) and PLA3 (0.1 wt%) (**f**) melted at 200 °C and then crystallized at 130 °C for 15 min, observed with polarized optical microscopy [70]. Reproduced with permission of John Wiley & Sons Ltd.

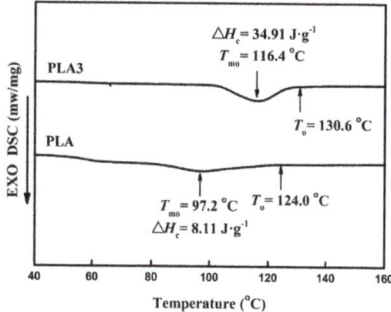

Figure 14. Thermal characteristics of PLA and PLA/FA-BH hybrid blend (PLA3) containing 0.1 wt% of the additive upon cooling at 2 °C/min [70]. Reproduced with permission of John Wiley & Sons Ltd. Copyright (2018).

Polylactide has been also modified with an FA-based hybrid macromolecular nucleator—a poly(lactic acid)-fulvic acid graft copolymer (PLA-FA) [72]. PLA-FA was synthesized with lactic acid monomer and FA as shown in Scheme 9. The obtained hybrid polymers had molecular weight (Mw) of about 14,300 g/mol and polydispersity index (PDI) = 1.3. The narrow molecular weight distribution implies that the grafted PLA chains were rather short and the PLA-FA macromolecules had a compact quasi-spherical structure. The presence of polyester components improved the compatibility of FA with the PLA matrix. The structure characterization and tests demonstrated that the PLA-FA used as a hybrid filler effectively enhanced the performance of the PLA composites prepared by melt blending. A plasticization effect of PLA-FA was indicated by the results of rheological analysis. Various plasticizers of low or high molecular weight are well known for the improvement of ductility, flexibility, and processability of PLA due to the increased number of independently moving segments [6]. Plasticisation can also be beneficial for the growth of crystals in amorphous matrices. Consequently, the PLA-FA additive promoted not only the nucleation (Figure 15) but also the rate of the non-isothermal crystallization of the PLA composites and improved their thermal stability and toughness.

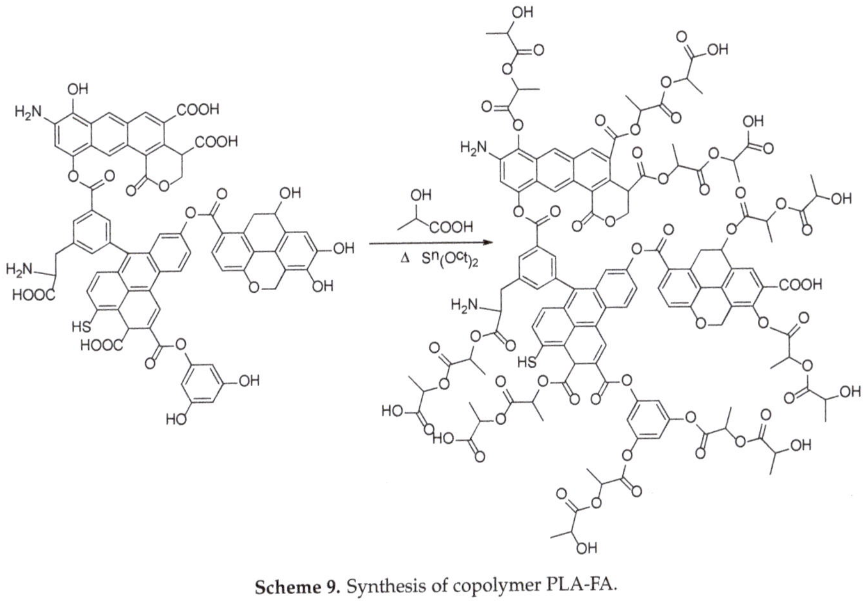

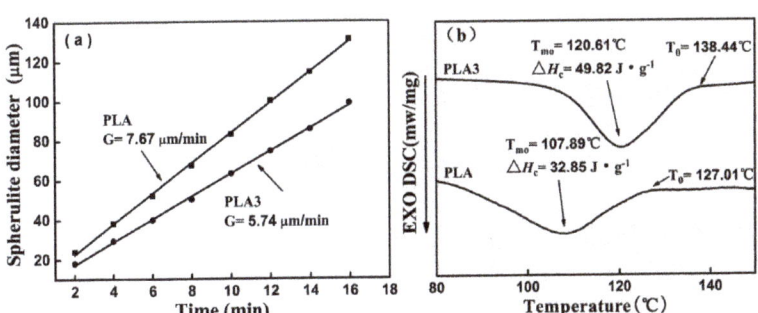

Scheme 9. Synthesis of copolymer PLA-FA.

Figure 15. (a) The spherulite growth rate of PLA and 0.5 wt% of PLA-FA (PLA3) and (b) cooling DSC curve of PLA and PLA3 [72]. Reprinted by permission of the publisher Taylor & Francis Ltd. Copyright (2019).

The effect of PLA-FA on the properties of polylactide blends can be related to the amount of the macromolecular additive. It affected not only the crystallization process but also the thermal and mechanical properties of the composite. The toughness of the PLA matrix was improved, while its strength and rigidity were enhanced. The sample containing 0.5 wt% of PLA-FA (PLA3) had the best properties (Figure 16). A significant increase in ductility and flexibility was noted for PLA3. The impact strength of this sample was improved by almost 200% compared with pure PLA. This is larger than the increase in impact strength induced by amidated FA (FAA, FA-BH, or MFA). However, when the amount of PLA-FA exceeded 0.5%, the mechanical properties of the blends gradually decreased. The effect was attributed to the agglomeration of the additive in the polylactide matrix, which led to stress concentration.

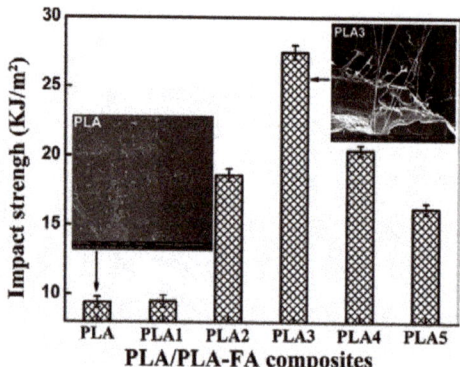

Figure 16. The impact strength (inset: impact fracture SEM images) of PLA/PLA-FA composites (PLA1: 0.1 wt%; PLA2: 0.3 wt%; PLA3: 0.5 wt%; PLA4: 0.7 wt%; PLA5: 1 wt% of PLA-FA) [72]. Reprinted by permission of the publisher Taylor & Francis Ltd. (http://www.tandfonline.com).

3. PLA/Nanocellulose Composites

Nanocellulose (NC) is an organic homopolymer, constituted of 1,4-anhydroglucopyranose repeating units (Figure 17). Cellulose nanomaterials are cost-effective, renewable, thermally stable up to 200 °C, and lightweight and provide high strength and stiffness. As a biomaterial of anisotropic shape, good biocompatibility, excellent mechanical properties, and tailorable surface chemistry, it is of high interest for material science and biomedical engineering (wound dressing, nanocarriers for drug delivery, and scaffolds for tissue engineering) [73–76]. Nanocellulose of defined nano-scale structural dimensions is derived from cellulosic extracts or processed materials. The term covers cellulose nanocrystals (CNCs; also called nanocrystalline cellulose (NCC) or cellulose nanowhiskers (CNWs)), cellulose nanofibrils (CNFs; also known as nano-fibrillated cellulose (NFC)), and bacterial cellulose (BC). They have different crystallinities, surface chemistries, and mechanical properties. Moreover, the rod-like CNCs show concentration-dependent liquid crystalline self-assembly behaviour. Micrometre-long entangled fibrils of CNFs contain both amorphous and crystalline cellulose domains, and their entanglement results in highly viscous aqueous suspensions even at relatively low concentrations (<1 wt%). BC is pure cellulose produced extracellularly by microorganisms.

The development of green nanocomposites based on polylactide and bio-based cellulose nanofillers for different applications, particularly for packaging and biomedical materials, has attracted significant attention [77]. The excellent properties of renewable and bio-based PLA/NC nanocomposites, their cytocompatibility and biodegradability, and their relatively low cost make those materials suitable for biomedical devices with enhanced mechanical performance. The earliest reports can be found in the cited reviews concerning polylactide containing various nanocellulose fillers. The most recent trends, illustrated by the examples given below, include porous and fibrous materials.

Most frequently, the preparation of PLA/nanocellulose composites involves melt processing, wet processing, or their combination. However, the hydrophilic nature of the surface hydroxyl groups on NC particles results in the poor dispersion of nanocellulose in nonpolar media, poor interface adhesion, and agglomeration in the PLA matrix. Thus, the surface modification of reinforcing nanocellulose fillers is often employed to improve their interfacial compatibility with the hydrophobic polylactide matrix. The -OH residues provide abundant active sites for covalent bond formation (oxidation, esterification, etherification, acetylation, carboxymethylation, silylation, and polymer grafting onto the polysaccharide backbone) as well as noncovalent binding. The efficiency of the surface modification strongly depends on the nature of the nanoparticles. Surface treatment may improve both the interfacial adhesion and dispersibility of NC. A significant increase in both mechanical and thermomechanical properties was observed with the addition of surface-treated cellulose. Moreover, hydrophobic nanocellulose can act as a barrier agent. Interfacial bonding and adhesion mechanisms determine the final properties of the polymer composites. Various factors should be taken into account to predict the overall effect. Molecular dynamics simulations characterized the interfacial structure and adhesion behaviour of crystalline nanocellulose in contact with polylactic acid [78]. It was found that adhesion between PLA and the surface of cellulose nanofibers is affected not only by the polarity of the functional groups and hydrogen bonds between CNC and PLA but also the surface roughness. PLA macromolecules can adopt conformations that accommodate the cellulose surface. The rougher the surface of the cellulose planes, the stronger the adhesion. Interestingly, among the four available interfaces, the best adhesion can be achieved at the (010) one due to the greater number of hydrogen bonds formed between the polar (010) surface and the polymer chain. However, the polar (110) and (1–10) surfaces produced similar adhesion work to the nonpolar (100) one (van der Waals effect), which illustrates the amphiphilicity of cellulose.

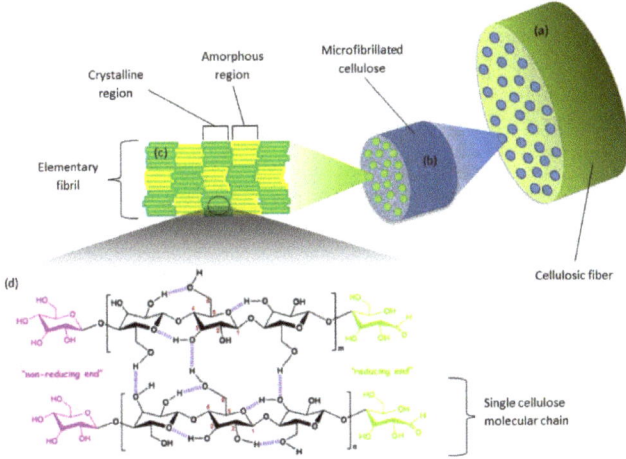

Figure 17. Structural characteristics of cellulosic fibre [75]. Reprinted from International Journal of Biological Macromolecules. Copyright (2019), with permission from Elsevier.

The functionalization not only enhanced the dispersion of modified NC in polymer composites but also may improve their mechanical properties, supporting specific interactions at the interface between the particles and the matrix. The improved association between the polymer matrix and the filler facilitates stress-transfer and enhances the nanocomposite strength. The amount of cellulose nanofillers determines the crystallinity and mechanical properties of the composite. Well-dispersed nanocellulose fillers act as nucleation sites in PLA, improving the degree of crystallinity and increasing the temperatures of phase transitions. The changes depend on the filler content and its surface chemistry, as well as on the technique used for the preparation of the nanocomposites. For example, ultra-strong PLA/CNC nanocomposites (an ultimate strength of 353 MPa and a toughness of 107 MJ/m^3) of increased T_g (93–95 °C) can be manufactured at a large scale through the surface modification of the nanocrystals, liquid-assisted extrusion, and solid-state drawing [79]. Silylation was applied to improve the surface compatibility between hydrophobic PLA and hydrophilic bamboo cellulose nanowhiskers

(BCNWs) [80,81]. The glass transition temperature, crystallinity, tensile strength, and tensile modulus decreased irrespectively of the silane used. However, the maximum elongation at break for the samples containing silanized cellulose increased to 213.8% (16 wt% triethoxyvinylsilane), 111.3% (8 wt% aminopropyltriethoxysilane), 255.3% (8 wt% methacryloxypropyltrimethoxysilane), and 209.8% (8 wt% mercaptopropyltrimethoxysilane) with respect to 12.4% of the untreated PLA/BCNW composite. The results suggest interactions between the ester groups of the PLA backbones and both the functional organic groups and silanol residues of the coupling agents. Many crazes were produced and absorbed most of the tensile energy during mechanical tests (wire-drawing and formation of thin necks) (Figure 18).

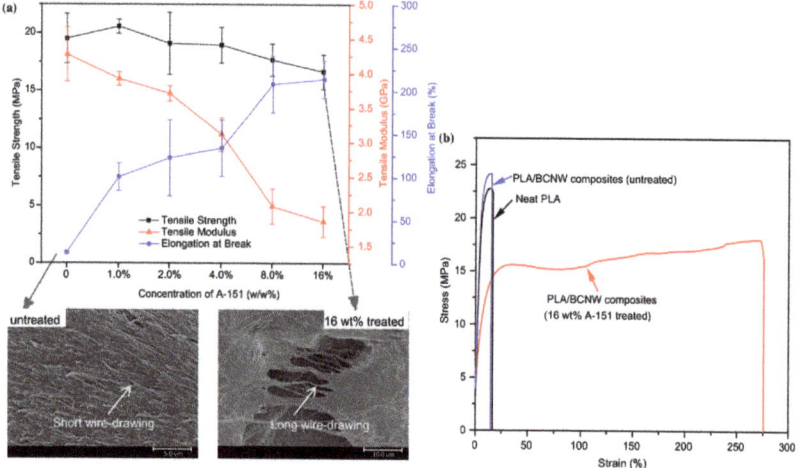

Figure 18. Tensile properties of PLA/composites containing bamboo cellulose nanowhiskers (BCNWs) treated with vinyltriethoxysilane (**a**), and exemplary stress-strain curves (**b**) [80]. Reprinted with permission from Springer Nature: Journal of Materials Science. Copyright (2018).

It was also shown that the dispersion of the cellulose nanocrystals and formation of a percolation network can be influenced by the molecular weight of the PLA and its crystallizability [76]. Lower CNC percolation concentrations could be obtained in PLA matrices of low molecular weight. CNCs more easily interpenetrate shorter PLA chains dissolved in a good solvent. The CNC percolation concentration could be lowered even more, providing high enantiomeric purity of polylactide chains. Upon solvent evaporation, PLA chains crystallized around the dispersed CNCs, which could prevent the further re-agglomeration of the latter. The rheological properties and thermal stability of such blends were improved.

3-Methacryloxypropyltrimethoxysilane was also used to modify the surface of cellulose nanofibrils (CNF) that were then used as fillers in poly(lactic acid) [82]. It was found that the silanization slightly decreased the thermal stability of the CNF, but their morphological integrity and rod-like morphology were retained. The treated nanofibrils dispersed well and crossed with each other as a percolated network in the PLA matrix. The tensile strength and elongation of the blends changed with the content of modified CNF. The highest tensile strength was shown for PLA with 1 wt% of the nanocellulose filler (Figure 19). The results were much better than those obtained with BCNW, which confirms the role of percolation for the properties of the composites.

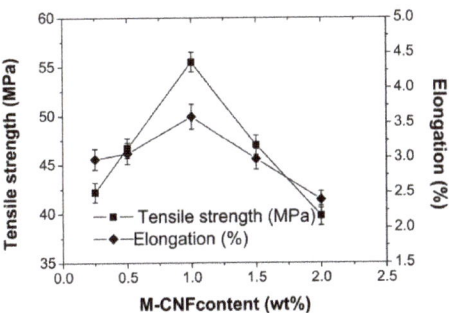

Figure 19. Tensile properties of PLA containing cellulose nanofibrils (CNF) modified with 3-methacryloxypropyltrimethoxysilane [82]. Reproduced with permission of John Wiley & Sons Ltd. Copyright (2012).

A similar concentration dependence was observed when epoxidized microfibrillated cellulose (MFC-EPI) was employed as an interfacial compatibilizer as well as a reinforcement filler in PLA/polybutylene succinate (PBS) blends [83]. The tensile strength and elongation at break of the composite containing 2% MFC-EPI reached 71.4 MPa and 273.6%, respectively. The toughening mechanism was explained by a "bridge" effect of the filler that contributes to energy transfer and dissipation during deformation (Figure 20). The entanglement of long nanofibers of MFC may enhance the load transfer between MFC and the polymer matrix during crack propagation. The interactions were enhanced by chemical cross-linking between the epoxy groups of MFC-EPI and hydroxyl/carboxyl groups of PLA and PBS. MFC-EPI acted as both the compatibilizer and reinforcement filler and endowed the PLA-based materials with high tensile strength and toughness. However, larger amounts of MFC-EPI agglomerated in the polymer matrix, deteriorating the properties of the blend.

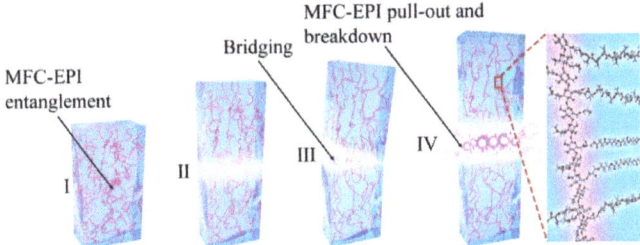

Figure 20. The fracture mechanism explaining the behaviour of PLA/polybutylene succinate (PBS)/epoxidized microfibrillated cellulose (MFC-EPI) composites [83]. Reprinted with permission from ACS Sustainable Chemistry & Engineering. Copyright (2020), American Chemical Society.

The surface of nanocellulose particles can be also rendered with a high content of carboxylic groups through the esterification of the native hydroxyl moieties with citric acid in a solid phase reaction [84]. The modified cellulose nanoparticles were further fibrillated via friction grinding. It strengthened the hydrogen bonding between NC and PLA. The resulting cellulose/PLA composites exhibited a desirable filler-polymer compatibility. The dispersion of the modified cellulose in PLA matrix was good, and the flexural properties of the prepared PLA composites were improved with regard to those of the pristine PLA resin. Maleic anhydride-grafted poly(lactic acid) (PLA-g-AMS/MAH) was used as a compatibilizer for the microcrystalline cellulose (MCC)/poly(lactic acid) (PLA) composites [85].

The modification of the nanocellulose filler may also change the gas barrier properties of the PLA nanocomposites [86]. PLA containing well-dispersed lauryl-functionalized cellulose nanocrystals (LNC) exhibited lower gas permeability due to the formation of a rigidified PLA interfacial region with a size comparable to the LNC diameter. The solvent molecules trapped in those regions were released at much higher temperatures than those locked in phase-separated blends.

Biocompatible PLA/nanocellulose composites with good absorbent properties and high mechanical strength can be also used for the preparation of artificial networks and three-dimensional scaffolds with a structure mimicking that of extracellular matrix (ECM). The high porosity of such 3D networks is desirable and beneficial for cell migration as well as nutrient input and metabolite output. Electrospinning technology can be used to create nanofibrous networks, although it is not possible to control their pore size and shape. Well-defined 3D scaffolds made of poly(lactic acid)/regenerated cellulose (PLA/RC) were thus formed using a method that combined freeze-drying and crosslinking [87]. Citric acid was applied as a non-toxic chemical cross-linker to RC nanofibers through esterification with their -OH groups. The resulting bioactive PLA/RC nanofiber-crosslinked scaffolds exhibited a dual pore structure and dimensional stability. Their high water absorption, hierarchical cellular structure, fast recovery from 80% strain, and apatite nucleating capacity indicate good osteogenic potential for bone tissue engineering. Other PLA-based open-pore porous blends suitable for this purpose, with high porosity and interconnectivity as well as superabsorbent ability, were prepared by melt-blending using crosslinked superabsorbent sodium polyacrylate particles (SAP) as a porogen [88]. The SAP particles were leached out from the PLA matrix in an aqueous environment. This generated high and tuneable porosity. The scaffolds allowed the good cell adhesion and proliferation of mouse embryo fibroblasts. It was also shown that similar 3D cellulose/PLA porous bio-composite templates can significantly facilitate PLLA/PDLA stereocomplex crystallization by accelerating the nucleation of SC crystals without the suppression of their growth (Figure 21) [89]. The interfacial hydrogen bonding between cellulose templates and polylactide molecules promoted the formation of precursor racemic helical pairs.

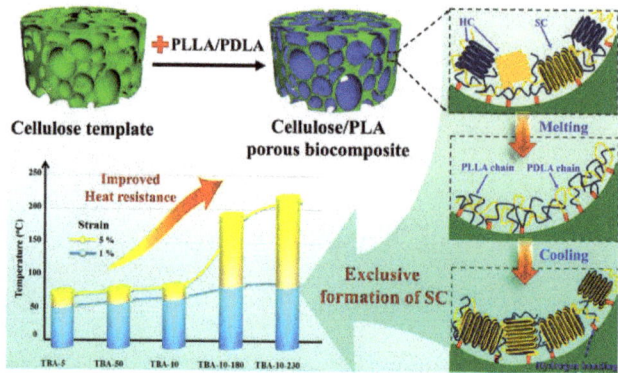

Figure 21. Formation of 3D cellulose/stereocomplex structure (SC) PLA porous biocomposites with improved thermal and mechanical properties [89]. Reprinted with permission from Springer Nature: Cellulose. Copyright (2020).

PLA-based blends containing nanocellulose as a reinforcing phase can be also used as an excellent feed material for 3D printing by means of fused deposition moulding (FDM). Unfortunately, the molecular weight of PLA typically becomes reduced after recycling, which limits the reuse of PLA in FDM-based 3D printing. The addition of an epoxy-based chain extender and a reinforcing phase of microcrystalline cellulose (MCC) to regenerated PLA improved its processability and mechanical performance [90]. The addition of 1 wt% nanocellulose enhanced the thermal stability of PLA, increased its crystallization rate, and shortened the crystallization half-time, which is vital for the solidification of a 3D printed object [91]. Polylactic acid/cellulose acetate (PLA/CA) mixtures loaded with antiseptic 1-chloro-2,2,5,5-tetramethyl-4-imidazolidinone were used as biodegradable, low-cost, antibacterial scaffolds prepared by a direct ink writing (DIW) technique [92]. The printability of the DIW inks was improved with the addition of CA due to the formation of a hydrogen-bonded 3D network between PLA and CA.

4. PLA Composites with Cyclodextrins

Cyclodextrins (CD) are natural cyclic oligosaccharides produced from starch and consisting of six (α-CD), seven (β-CD), or eight (γ-CD) D(+)-glucose units joined by α-1,4-linkages (Scheme 10) [93]. They have a shallow truncated cone shape with a hydrophobic cavity and hydrophilic outer surface. γ-cyclodextrin is the most flexible of the CD molecules, whereas α-cyclodextrin is the most rigid one. The solubility of β-CD in water (18.4 g/L) is low when compared to that of α- and γ-CD (129.5 and 249.2 g/L, respectively) [94]. It appears that intramolecular hydrogen bonding in β-CD results in a tighter crystal structure. Cyclodextrin molecules are amphiphilic, with -CH$_2$OH groups linked to the narrower rim and the wider rim displaying -OH groups connected to the glucose rings. The hydrophilic groups are situated on the outside of the molecular cavity whereas the hydrophobic inner surface is lined with ether-like anomeric oxygen atoms and methine units. Thus, cyclodextrin molecules may bind non-polar and suitably-sized aliphatic or aromatic compounds. The binding is driven by the enthalpic and entropic gain related to the reduction in the hydrophobic-aqueous surface and the release of water molecules from the CD cavities. Those specific features make CD effective hosts for a variety of guest molecules (small organic species and macromolecules), resulting in the formation of inclusion complexes (IC) [95–97]. The IC formation mechanism involves fitting the guest molecules in the CD cavity through non-covalent interactions (van der Waals interactions and the hydrophobic effect).

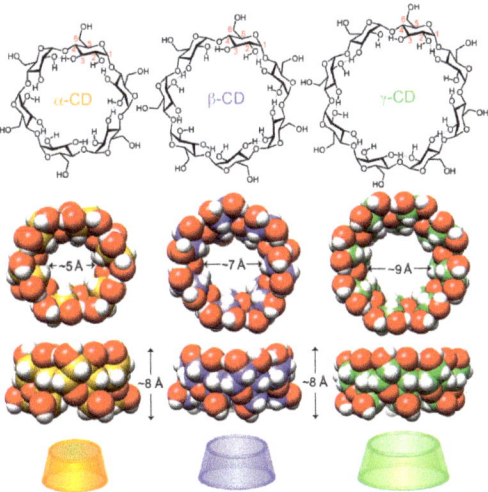

Scheme 10. Structural, space-filling, and graphical representations of cyclodextrin molecules: α-CD, β-CD, and γ-CD [96]. Reprinted with permission from Symmetry (MDPI) 2019.

Cyclodextrins of all types were used for the formation of "green" IC with polylactide chains. Such polymeric IC may have an important role for constructing supramolecular architectures (molecular tubes or poly(pseudo)rotaxanes). The formed nanometre-scale ordered structures can also be used as nucleating agents to enhance the crystallization of the guest PLA due to the restriction of macromolecular chain motions inside the CD cavities. The introduction of water-soluble cyclodextrins threaded on polymer chains not only affects the crystallization but also modifies the hydrophilicity of polymer surfaces, degradation, and thermal performances. Recently, a significant interest may be also observed in PLA/CD-based materials for the delivery of encapsulated drugs [98–104] and other medical applications [105–108].

4.1. PLA/α-Cyclodextrins

Threading α-cyclodextrin molecules onto PLA chains results in the formation of supramolecular inclusion complexes, organized by non-covalent interactions [109–111]. The formation of stable poly(pseudo)rotaxanes was reported when α-CD was threaded onto PLLA chains and PLLA-PEG-PLLA triblock copolymers of different molar ratios of [LA]/[PEG] (Tri-1: 23, Tri-2: 31, Tri-3: 45, Tri-4: 54) [110]. For poly(pseudo)rotaxanes of both kinds, threaded onto triblock copolymers or on PLLA, the stoichiometric number of [LA] monomer units and [CD] was found to be 2:1. The complexation strongly reduced the solubility of the polymers.

^{13}C CP/MAS NMR studies in the solid state indicated that α-CD in such polymeric IC adopted more symmetric cyclic conformations, which influenced the splitting of all the C1-6 glucose carbon resonances of the cyclodextrin moieties (Figure 22) [110]. The signals characteristic of the glycoside linkage (98 ppm and 80 ppm) disappeared in the spectra of channel-structured IC. The most symmetric cyclic conformations of α-CD were adopted in the inclusion complex with a 2:1 feed ratio. More evidence of IC formation may be obtained by FTIR spectroscopy. A shift of the ν (O-H) mode to higher frequencies can be observed (3389 cm^{-1} for pure α-CD and 3420 cm^{-1} for IC) due to the association of OH groups with the guest polymeric chains [110]. The stretching band ν (C-H) (300–2950 cm^{-1}) in α-CD was also shifted.

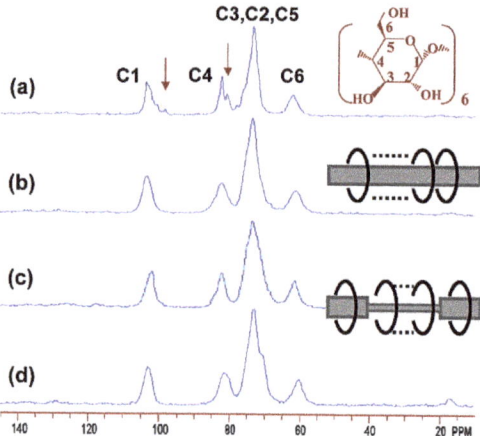

Figure 22. Solid-state ^{13}C CP/MAS NMR spectra of R-CD (**a**) and inclusion complexes of PLLA/α-CD (**b**), Tri-1/α-CD (**c**), and Tri-4/α-CD (**d**) [110]. Reprinted with permission from Macromolecules. Copyright (2003), American Chemical Society.

The formation of IC with α-CD was confirmed by wide-angle X-ray diffraction measurements (exemplary diffractograms are shown in Figure 23). The WAXS patterns of PLA-IC are different from those of PLA and α-CD. A new channel-type crystal structure was formed with the copolymers confined to CD, and the components lost their original crystalline features. The X-ray diffractograms of PLLA/α-CD and the Tri-1/α-CD inclusion complexes showed patterns different from those of host CD and guest macromolecules [110]. A set of reflections at 2θ = 7.60° (d = 11.6 Å), 13.0° (d = 6.80 Å), and 20.0° (d = 4.44 Å) indicated the formation of a columnar crystalline structure (a hexagonal unit cell with the lateral dimension a = 13.6 Å). The strong 210 reflection (2θ = 20.0°) is characteristic of the channel structure of IC crystals containing α-CD (the electron density distribution of the core of α-CD molecules with a radius of about 5 Å) [110,111].

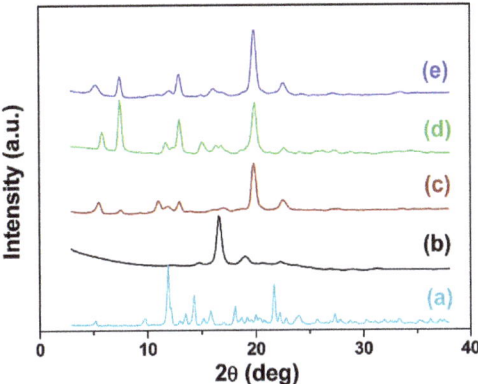

Figure 23. X-ray diffraction patterns for α-CD (**a**), PLLA (**b**), poly(ethylene glycol) (PEG)/α-CD inclusion complex (**c**), PLLA/α-CD inclusion complex (**d**), and Tri-1/α-CD inclusion complex (**e**) [110]. Reprinted with permission from Macromolecules. Copyright (2003), American Chemical Society.

Dielectric relaxation spectroscopy (DRS) studies on poly(D,L-lactic acid) (PDLLA) and its inclusion complexes in α-CD of various ratios of incorporated/initial PDLLA revealed a distinguishably different dynamical response of the PDLLA chains constrained between α-CD from the fraction of macromolecules incorporated inside the channels (Figure 24) [112]. The presence of α-CD molecules depletes the segmental α-process and resolves two sub-T_g relaxations. The cooperative motions (α-relaxation) are supressed for PDLLA hosted inside the channels whereas the relaxation of macromolecules situated between α-CD channels is reduced and shifts to higher temperatures (~4.5 °C). A secondary relaxation—the Johari–Goldstein process (βJG-process)—that can be related to the PDLLA confinement effect was also observed in the IC. An additional secondary γ process of length scale inferior to inter- or intra-channel dimensions was detected only in the inclusion complexes.

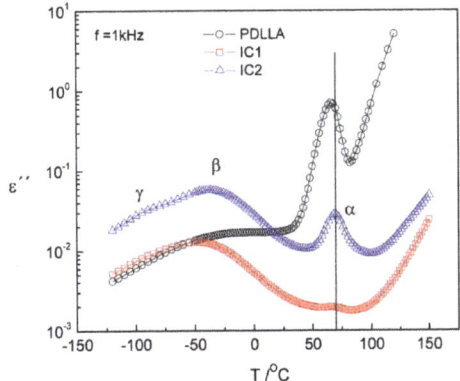

Figure 24. Isochronal (IC) plot of the imaginary part of the complex permittivity at 1 kHz for poly(D,L-lactic acid) (PDLLA) (●), IC1 (■), and IC2 (▲). The ratios of incorporated/initial PDLLA by weight percentage (%, w/w) were 10/24 (IC1) and 15/46 (IC2) [112]. Reprinted with permission from The Journal of Physical Chemistry B. Copyright (2014), American Chemical Society.

The optically active CDs are known for their ability to recognize chiral molecules [113–115]. Interestingly, Ohya et al. reported that the inclusion complex of α-cyclodextrin with PLLA is preferentially formed compared to that with poly(D-lactide) (PDLA) (Scheme 11) [116]. The recognition of PDLA chains was significant, and they were almost excluded by α-CD. The phenomenon was confirmed by tests with (LL)- and (DD)-lactic acid dimer methyl esters. Only (LL) species formed IC with α-CD. The crystalline structures of PLLA/α-CD and PDLA/α-CD were investigated by DSC and X-ray diffraction (Figure 25). The WAXS diagrams of PLLA, PDLA, and PDLA/α-CD featured intensive peaks at 2θ = 17° and 19° (typical for α-crystals of polylactide). PDLA/α-CD showed a melting point around 160 °C (similarly to neat PLLA or PDLA), indicating a crystalline nature of the sample. The diffraction pattern for PLLA/α-CD was characteristic for IC with a columnar structure (2θ = 20°).

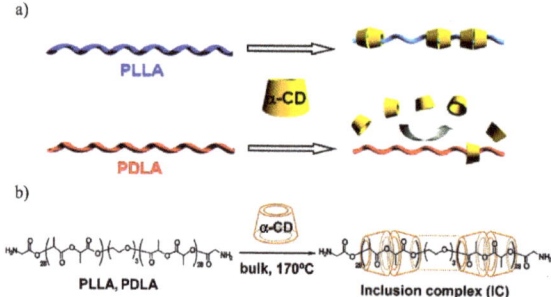

Scheme 11. (a) Chiral recognition of polylactides by α-CD, and (b) preparation of PLA/α-CD inclusion complex [116]. Reprinted with permission from Macromolecules. Copyright (2007), American Chemical Society.

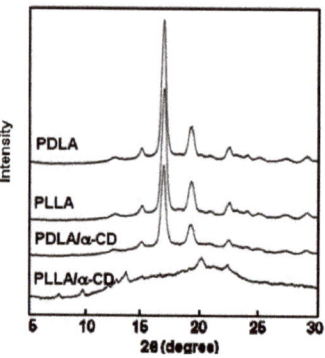

Figure 25. Wide-angle X-ray diffraction patterns of PLLA, PLLA/R-CD poly(D-lactide) (PDLA) and PDLA/R-CD [116]. Reprinted with permission from Macromolecules. Copyright (2007), American Chemical Society.

Experiments with triblock copolymers—PLLA-PEG-PLLA and PDLA-PEG-PDLA—showed that α-CD molecules were hosting PEG segments irrespectively of the enantiomeric form of the polylactide blocks. It suggests that α-CD can freely slide along the PDLA segments and that the fitting between the polymer width and diameter of the CD cavity is not the only factor that governs the IC formation. It was therefore concluded that the reason for the chiral recognition is not the polymer chain steric hindrance (similar for PLLA and PDLA) but the thermodynamic stability of the diastereomeric IC. Chiral fitting with the α-CD cavity is more favourable to PLLA and allows the linking of several cyclodextrine molecules through hydrogen bonds. The molecules of α-CD have to rotate unidirectionally on sliding down the helical chain of PLLA. This phenomenon may be exploited for the design of molecular motors.

The α-CD host polylactide guest IC with a channel-type structure had different thermal characteristics to the parent polylactide [109–111]. No cold crystallization and melting can be detected if the PLA chains are included inside the α-CD voids (Figure 26). Small amounts of IC may exert a nucleation effect and promote the crystallization of the PLA matrix during both the non-isothermal and isothermal crystallization experiments [111]. Upon cooling from melt (5 °C/min), an exotherm was observed at about 94 °C, suggesting that IC particles can accelerate the nucleation. If the PLA contained free α-CD (not IC) then the nucleation effect was small. It was also noted that α-CD may act as a nucleating agent for the crystallization of PLLA from the solution with the formation of spherulitic structures [117]. The size of the spherulites in the cast film depended on the amount of α-CD and the casting rate. Very small crystallites were formed in the presence of IC and on fast casting.

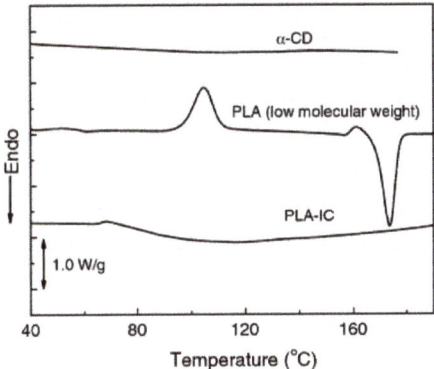

Figure 26. DSC traces of α-CD, PLA, and PLA-IC during heating at 10 °C/min [111]. Reprinted with permission from Springer Nature: Polymer Bulletin. Copyright (2013).

4.2. PLA/β-Cyclodextrins

The feeding ratio between β-CD and PLA, as well as the molecular weight of the latter, has a significant effect on the formation of IC prepared by the solution ultrasonic technique [118]. IC with PLA of Mn ≥ 500 kg/mol agglomerated in water, but if the Mn was <100 kg/mol then the IC was soluble in H_2O. Both IC had lower T_g and were more hydrophilic than those of pure PLA. FTIR spectra showed a clear difference between the IC and the components β-CD and PLA. The ν (O-H) stretching modes of β-CD shifted to higher frequencies when the inclusion complexes were formed, indicating physical interactions between the components. The ν (C=O) stretching band of PLA (1750 cm^{-1}) shifted to 1757 cm^{-1} and weakened in the spectrum of IC. Wide angle X-ray diffraction patterns revealed a specific crystal structure of IC that differed from those characteristic of neat β-CD and PLA. The T_g of the inclusion complex was lower than that of pure PLA, and the decrease was proportional to the amount of β-CD. The drop in the glass transition temperature was explained in terms of IC supramolecular structure formation with the molecular rings of β-CD hindering interactions between the PLA chains. The thermal stability of the IC was slightly lower than that of pure PLA.

The hydrogen bonds between polylactide and β-CD were confirmed by a shift of the ν (O-H) stretching IR mode of β-CD (3386 cm^{-1}) towards a lower wavelength in the IC spectrum (3304 cm^{-1}) [119]. Additionally, a shift in the ν (C-H) band of β-CD was noted. The association behaviour can be explained by the presence of an ordered phase with different chain conformation, as indicated by the ν (C=O) stretching band of PLLA. The second derivative of this band changes markedly in IC, with a decrease in the trans-gauche conformation component. Those changes may be related to the formation of hydrogen bonding between the blended molecules.

The absence of the hydrogen bonding hinders the propensity towards IC formation in polylactide/cyclodextrin systems. The addition of methyl-β-cyclodextrin (Me-β-CD) lowered the crystallinity of the PLLA and enhanced the mobility of the polyester chains [120]. The addition of Me-β-CD increased the amorphous component of the PLLA. The plasticizing of PLLA by Me-β-CD resulted in an improvement of the drawability of the composite (1000% at 60 °C in the presence of 17% of Me-β-CD). In situ measurements using combined differential scanning calorimetry (DSC) and Raman spectroscopy revealed a lowering of the glass transition temperature, cold crystallization temperature, and melting point. As no IC was formed and the amount of the amorphous phase increased, the thermal stability of the PLLA/Me-β-CD blend was lower than that of neat PLLA.

The mechanism of polylactide/β-cyclodextrin inclusion complex formation and the effects of its incorporation in the PLA matrix were investigated [121]. The changes in the thermal stability, surface morphology, and barrier and mechanical properties were studied for the composites at varying IC and β-CD concentrations. Samples admixed with neat β-CD featured agglomerates of the β-cyclodextrin molecules while PLA/IC composites had uniform structures. The reason was poor interfacial interactions between the polyester macromolecules and β-CD moieties. The structural variations were reflected in different thermal stabilities. The T_g and T_c of the studied PLA composites were shifted to higher temperatures upon increasing the concentration of IC or β-CD. The composite films containing larger amounts of IC or β-CD exhibited higher oxygen and water vapour permeability but were less flexible and had lower tensile strength than neat PLA.

High molecular weight polymer chains of stronger steric hindrance cannot easily penetrate the cavities of cyclodextrins; thus, the macromolecules tend to form partial inclusion complexes with small stoichiometric ratios. Nevertheless, the partial inclusion complexes formed between β-CD and high molecular weight PLLA crystallized better than neat PLLA. An improved mechanical performance and thermal stability were also demonstrated [122].

4.3. PLA/γ-Cyclodextrins

A poly(lactic acid) and γ-cyclodextrin inclusion complex was obtained by ultrasonic co-precipitation and used as a precursor for the preparation of its benzyl-hydrazide derivative that was subsequently applied as a nucleating agent for PLA composites [123]. PLA chains penetrated into the cavity of the γ-CD cone, which resulted in a tunnel-shaped crystalline structure. For PLA/γ-CD IC, two prominent X-ray diffraction peaks were formed (2θ = 7.5° and 19.6°). The latter indicated the channel-like structure characteristic of crystalline necklace-like IC with a fingerprint diffraction feature at 2θ = 7.5°. The comparative FTIR analysis of PLA-IC and neat PLA and γ-CD showed that the interactions between those two components were strong. The peak corresponding to symmetric and antisymmetric O-H stretching modes of γ-CD (3405 cm^{-1}) was shifted to 3415 cm^{-1} in PLA-IC. This confirmed a non-covalent nature of the interactions between the γ-CD and PLA backbones in the inclusion complex. Moreover, the carbonyl stretching (v, C=O) appeared at 1760 cm^{-1} in the neat PLA and at 1768 cm^{-1} in the IC (noticeably weakened). Calorimetric studies of PLA/PLA-IC showed that T_c and T_m trended towards lower-temperature regions. It suggested that PLA-IC could form hydrogen bonds with PLA and break the intermolecular interactions in PLA. PLA/γ-CD IC was also modified via surface-initiated atom transfer radical polymerization to obtain poly(lactic acid)-γ-cyclodextrin inclusion complex-poly(glycidyl methacrylate) (PLA-IC-PGMA) [124].

5. Conclusions

Polylactide nanocomposites, composites, and blends containing specific organic compounds (arylamides and arylhydrazides, 1,3:2,4-dibenzylidene-D-sorbitol, or orotic acid), humic and fulvic acids, nanocellulose, and cyclodextrins have been reviewed. Supramolecular interactions operating in those blends play a very important role in the properties of those novel hybrid materials. They differ from other PLA-based hybrids containing inorganic additives or composites of PLA and graphene or carbon nanotubes.

Aryl nucleators containing amide or hydrazide linkages are capable of crystallization in the polylactide melt. The formation of nanofibrils/nano-objects upon the self-organization of these moieties exploits two recognition motifs: hydrogen bonds and π-π interactions between the molecules taking part in the formation of supramolecular structures and intermolecular hydrogen bonds linking those formations and the polyester backbones. Macromolecular nucleators (humic/fulvic acids and nanocellulose) not only help in the formation of crystal nuclei but also improve the mechanical properties of the PLA. Host–guest effects in blends containing inclusion complexes of PLA and (α/β/γ)-cyclodextrins change the mobility of the polyester chains. The formed structures may also enhance the polylactide crystal growth, but they have much larger potential. Treading cyclodextrins onto PLLA backbones may be used for the design of biocompatible molecular rotors. Interestingly, chiral recognition was observed for the PLLA/PDLA mixtures and α-CD.

The performance of PLA can be improved because of those specific interactions that modify the organization of the polymer matrix. The "soft templating" may change the conformation of segments in PLA chains and influence the nucleation of polymer crystals, thus enormously enhancing the crystallization process. Despite the templating, the type of crystal structure is not changed, except that the well-organized α-crystals are more easily formed in such systems. The influence of supramolecular nucleators is of exceptional importance, as the crystallinity of PLA plays a significant role in its mechanical and barrier properties, biodegradability, and thermal stability. Many supramolecular systems described in this review are biocompatible and can be used for biomedical purposes.

Author Contributions: Conceptualisation, writing—original draft preparation, and funding acquisition, A.K.; writing—review and editing, M.N. All authors have read and agreed to the published version of the manuscript.

Funding: The APC was funded by Polish National Centre of Sciences (NCN) grant No 2016/21/B/ST5/03070.

Conflicts of Interest: The authors declare no conflict of interest.

References

1. Gross, R.A.; Kalra, B. Biodegradable polymers for the environment. *Science* **2002**, *297*, 803–807. [CrossRef] [PubMed]
2. Nair, L.S.; Laurencin, C.T. Biodegradable polymers as biomaterials. *Progr. Polym. Sci.* **2007**, *32*, 762–798. [CrossRef]
3. Lasprilla, A.J.R.; Martinez, G.A.R.; Lunelli, B.H.; Jardini, A.L.; Maciel Filho, R. Poly-lactic acid synthesis for application in biomedical devices—A review. *Biotechn. Adv.* **2012**, *30*, 321–328. [CrossRef] [PubMed]
4. Tyler, B.; Gullotti, D.; Mangraviti, A.; Utsuki, T.; Brem, H. Polylactic acid (PLA) controlled delivery carriers for biomedical applications. *Adv. Drug Deliv. Rev.* **2016**, *107*, 163–175. [CrossRef] [PubMed]
5. Poh, P.S.P.; Chhaya, M.P.; Wunner, F.M.; De-Juan-Pardo, E.M.; Schilling, A.F.; Schantz, J.-T.; van Griensven, M.; Hutmacher, D.W. Polylactides in additive biomanufacturing. *Adv. Dru Deliv. Rev.* **2016**, *107*, 228–246. [CrossRef] [PubMed]
6. Saeidlou, S.; Huneault, M.A.; Li, H.; Park, C.B. Poly(lactic acid) crystallization. *Progr. Polym. Sci.* **2012**, *37*, 1657–1677. [CrossRef]
7. Liu, G.; Zhang, X.; Wang, D. Tailoring crystallization: Towards high-performance poly(lactic acid). *Adv. Mater.* **2014**, *26*, 6905–6911. [CrossRef]
8. Lan, Q.; Li, Y. Mesophase-Mediated Crystallization of Poly(L-lactide): Deterministic pathways to nanostructured morphology and superstructure control. *Macromolecules* **2016**, *49*, 7387–7399. [CrossRef]
9. Paspali, A.; Bao, Y.; Gawne, D.T.; Piestert, F.; Reinelt, S. The influence of nanostructure on the mechanical properties of 3D printed polylactide/nanoclay composites. *Compos. B Eng.* **2018**, *152*, 160–168. [CrossRef]
10. Mihai, M.; Ton-That, M.-T. Novel bio-nanocomposite hybrids made from polylactide/nanoclay nanocomposites and short flax fibers. *J. Compos. Mater.* **2019**, *32*, 3–28. [CrossRef]
11. Rojas-Lema, S.; Quiles-Carrillo, L.; Garcia-Garcia, D.; Melendez-Rodriguez, B.; Balart, R.; Torres-Giner, S. Tailoring the properties of thermo-compressed polylactide films for food packaging applications by individual and combined additions of lactic acid oligomer and halloysite nanotubes. *Molecules* **2020**, *25*, 1976. [CrossRef]
12. Odent, J.; Raquez, J.-M.; Thomassin, J.-M.; Gloaguen, J.-M.; Lauro, F.; Jérôme, C.; Lefebvre, J.-M.; Dubois, P. Mechanistic insights on nanosilica self-networking inducing ultra-toughness of rubber-modified polylactide-based materials. *Nanocomposites* **2015**, *1*, 113–125. [CrossRef]
13. Hao, X.; Kaschta, J.; Schubert, D.W. Viscous and elastic properties of polylactide melts filled with silica particles: Effect of particle size and concentration. *Composites B Eng* **2016**, *89*, 44–53. [CrossRef]
14. Lee, J.H.; Park, S.H.; Kim, S.H.; Ito, H. Replication and surface properties of micro injection molded PLA/MWCNT nanocomposites. *Polym. Test.* **2020**, *83*, 106321. [CrossRef]

15. He, S.; Bai, H.; Bai, D.; Ju, Y.; Zhang, Q.; Fu, Q. A promising strategy for fabricating high-performance stereocomplex-type polylactide products via carbon nanotubes-assisted low-temperature sintering. *Polymer* **2019**, *162*, 50–57. [CrossRef]
16. Prasitnok, K.; In-noi, O. Functionalized graphenes as nanofillers for polylactide: Molecular dynamics simulation study. *Polym. Composites* **2020**, *41*, 294–305. [CrossRef]
17. Tsui, H. Poly(lactide) stereocomplexes: Formation, structure, properties, degradation, and applications. *Macromol. Biosci.* **2005**, *5*, 569–597.
18. Tsuji, H. Poly(lactic acid) stereocomplexes: A decade of progress. *Adv. Drug Deliv. Rev.* **2016**, *107*, 97–135. [CrossRef]
19. Jiang, L.; Shen, T.; Xu, P.; Zhao, X.; Li, X.; Dong, W.; Ma, P.; Chen, M. Crystallization modification of poly(lactide) by using nucleating agents and stereocomplexation. *e-Polymers* **2016**, *16*, 1–13. [CrossRef]
20. Michalski, A.; Makowski, T.; Biedroń, T.; Brzeziński, M.; Biela, T. Controlling polylactide stereocomplex (sc-PLA) self-assembly: From microspheres to nanoparticles. *Polymer* **2016**, *90*, 242–248. [CrossRef]
21. Yang, C.-F.; Huang, Y.-F.; Ruan, J.; Su, A.-C. Extensive development of precursory helical pairs prior to formation of stereocomplex crystals in racemic polylactide melt mixture. *Macromolecules* **2012**, *45*, 872–878. [CrossRef]
22. Carbone, M.J.; Vanhalle, M.; Goderis, B.; Van Puyvelde, P. Amino acids and poly(amino acids) as nucleating agents for poly(lactic acid). *J. Polym. Eng.* **2015**, *35*, 169–180. [CrossRef]
23. Zhou, Y.; Lei, L.; Yang, B.; Li, J.; Ren, J. Preparation and characterization of polylactic acid (PLA) carbon nanotube nanocomposites. *Polym. Test.* **2018**, *68*, 34–38. [CrossRef]
24. He, D.; Wang, Y.; Shao, C.; Zheng, G.; Li, Q.; Shen, C. Effect of phthalimide as an efficient nucleating agent on the crystallization kinetics of poly(lactic acid). *Polym. Test.* **2013**, *32*, 1088–1093. [CrossRef]
25. Xu, Y.; Wu, L. Synthesis of organic bisurea compounds and their roles as crystallization nucleating agents of poly(L-lactic acid). *Eur. Polym. J.* **2013**, *49*, 865–872. [CrossRef]
26. Zheng, X.; Luo, H.; Chen, S.; Zhang, B.; Sun, B.; Zhu, M.-F.; Zhang, B.; Ren, X.-K.; Song, J. Conformation variation induced crystallization enhancement of poly(l-lactic acid) by gluconic derivatives. *Cryst. Growth Des.* **2020**, *20*, 653–660. [CrossRef]
27. Herc, A.S.; Lewiński, P.; Kaźmierski, S.; Bojda, J.; Kowalewska, A. Hybrid SC-polylactide/poly(silsesquioxane) blends of improved thermal stability. *Thermochim. Acta* **2020**, *687*, 178592. [CrossRef]
28. Herc, A.S.; Bojda, J.; Nowacka, M.; Lewiński, P.; Maniukiewicz, W.; Piórkowska, E.; Kowalewska, A. Crystallization, structure and properties of polylactide/ladder poly(silsesquioxane) blends. *Polymer* **2020**, *201*, 122563. [CrossRef]
29. Kowalewska, A.; Herc, A.S.; Bojda, J.; Palusiak, M.; Markiewicz, E.; Ławniczak, P.; Nowacka, M.; Sołtysiak, J.; Różański, A.; Piórkowska, E. Supramolecular interactions involving fluoroaryl groups in hybrid blends of laddder polysilsesquioxanes and polylactide. *Polymer Testing.* (under review).
30. Herc, A.S.; Włodarska, M.; Nowacka, M.; Bojda, J.; Szymański, W.; Kowalewska, A. Supramolecular interactions between polylactide and model cyclosiloxanes with hydrogen bonding-capable functional groups. *eXPRESS Polym. Lett.* **2020**, *14*, 134–153. [CrossRef]
31. Nam, J.Y.; Okamoto, M.; Okamoto, H.; Nakano, M.; Usuki, A.; Matsuda, M. Morphology and crystallization kinetics in a mixture of low-molecular weight aliphatic amide and polylactide. *Polymer* **2006**, *47*, 1340. [CrossRef]
32. Li, J.; Chen, D.; Gui, B.; Gu, M.; Ren, J. Crystallization morphology and crystallization kinetics of poly(lactic acid): Effect of N-aminophthalimide as nucleating agent. *Polym. Bull.* **2011**, *67*, 775. [CrossRef]
33. Kawamoto, N.; Sakai, A.; Horikoshi, T.; Urushihara, T.; Tobita, E. Nucleating agent for poly(Llactide)—An optimization of chemical structure of hydrazide compound for advanced nucleation ability. *J. Appl. Polym. Sci.* **2007**, *103*, 198. [CrossRef]
34. Nakajima, H.; Takahashi, M.; Kimura, Y. Induced Crystallization of PLLA in the presence of 1,3,5-benzenetricarboxylamide derivatives as nucleators: Preparation of haze-free crystalline PLLA materials. *Macromol. Mater. Eng.* **2010**, *295*, 460. [CrossRef]
35. Song, P.; Wei, Z.; Liang, J.; Chen, G.; Zhang, W. Crystallization behavior and nucleation analysis of poly(L-lactic acid) with a multiamide nucleating agent. *Polym. Eng. Sci.* **2012**, *52*, 1058–1068. [CrossRef]
36. Xu, T.; Wang, Y.; Han, Q.; He, D.; Li, D.; Shen, C. Nonisothermal Crystallization Kinetics of Poly(lactic acid) Nucleated with a Multiamide Nucleating Agent. *J. Macromol. Sci. B* **2014**, *53*, 1680–1694. [CrossRef]

37. Bai, H.; Zhang, W.; Deng, H.; Zhang, Q.; Fu, Q. Control of crystal morphology in poly(L-lactide) by adding nucleating agent. *Macromolecules* **2011**, *44*, 1233–1237. [CrossRef]
38. Zhang, H.; Wang, S.; Zhang, S.; Ma, R.; Wang, Y.; Cao, W.; Liu, C.; Shena, C. Crystallization behavior of poly(lactic acid) with a self-assembly aryl amide nucleating agent probed by real-time infrared spectroscopy and X-ray diffraction. *Polym. Test.* **2017**, *64*, 12–19. [CrossRef]
39. Kong, W.; Zhu, B.; Su, F.; Wang, Z.; Shao, C.; Wang, Y.; Liu, C.; Shen, C. Melting temperature, concentration and cooling rate-dependent nucleating ability of a self-assembly aryl amide nucleator on poly(lactic acid) crystallization. *Polymer* **2019**, *168*, 77–85. [CrossRef]
40. Kong, W.; Tong, B.; Ye, A.; Ma, R.; Gou, J.; Wang, Y.; Liu, C.; Shen, C. Crystallization behavior and mechanical properties of poly(lactic acid)/poly(ethylene oxide) blends nucleated by a self-assembly nucleator. *J. Therm. Anal. Calorim.* **2019**, *135*, 3107–3114. [CrossRef]
41. Leoneé, N.; Roy, M.; Saidi, S.; de Kort, G.; Hermida-Merino, D.; Wilsens, C.H.R. Improving processing, crystallization, and performance of poly-L-lactide with an amide-based organic compound as both plasticizer and nucleating agent. *ACS Omega* **2019**, *4*, 10376–10387.
42. Wang, Y.; He, D.; Wang, X.; Cao, W.; Li, Q.; Shen, C. Crystallization of poly(lactic acid) enhanced by phthalhydrazide as nucleating agent. *Polym. Bull.* **2013**, *70*, 2911–2922. [CrossRef]
43. Xu, T.; Zhang, A.; Zhao, Y.; Han, Z.; Xue, L. Crystallization kinetics and morphology of biodegradable poly(lactic acid) with a hydrazide nucleating agent. *Polym. Test.* **2015**, *45*, 101–106. [CrossRef]
44. Cai, Y.; Yan, S.; Yin, J.; Fan, Y.; Chen, X. Crystallization behavior of biodegradable poly(L-lactic acid) filled with a powerful nucleating agent: N,N'-bis(benzoyl) suberic acid dihydrazide. *J. Appl. Polym. Sci.* **2011**, *121*, 1408–1416. [CrossRef]
45. Zhen, Z.; Xing, Q.; Li, R.; Dong, X. Crystallization behavior of polylactide nucleated by octamethylenedicarboxylic di(2-hydroxybenzohydrazide): Solubility influence. *Acta* **2020**, *683*, 178447. [CrossRef]
46. Li, C.; Luo, S.; Wang, J.; Wu, H.; Guo, S.; Zhang, X. Conformational regulation and crystalline manipulation of PLLA through a self-assembly nucleator. *Biomacromolecules* **2017**, *18*, 1440–1448. [CrossRef]
47. Zhao, L.-S.; Cai, Y.-H. Investigating the physical properties of poly(L-lactic acid) modified using an aromatics succinic dihydrazide derivative. *Polym. Sci. Ser. A* **2018**, *60*, 777–787. [CrossRef]
48. Zhao, L.-S.; Cai, Y.-H.; Liu, H.-L. N,N'-sebacic bis(hydrocinnamic acid) dihydrazide: A crystallization accelerator for poly(L-lactic acid). *e-Polymers* **2019**, *19*, 141–153. [CrossRef]
49. Cai, Y.-H.; Zhang, Y.-H.; Zhao, L.-S. Role of N,N'-bis(1H-benzotriazole) adipic acid acethydrazide in crystallization nucleating effect and melting behavior of poly(L-lactic acid). *J. Polym. Res.* **2015**, *22*, 246. [CrossRef]
50. Cai, Y.-H.; Tang, Y.; Zhao, L.-S. Poly(L-lactic acid) with the organic nucleating agent N,N,N'-tris(1H-benzotriazole) trimesinic acid acethydrazide: Crystallization and melting behaviour. *J. Appl. Polym. Sci.* **2015**, *132*, 42402. [CrossRef]
51. Zhao, L.-S.; Cai, Y.-H.; Liu, H.-L. Physical properties of Poly(L-lactic acid) fabricated using salicylic hydrazide derivative with tetraamide structure. *Polym.-Plast. Technol. Mater.* **2020**, *59*, 117–129. [CrossRef]
52. Liu, J.-H.; Cai, J.-H.; Tang, X.-H.; Weng, Y.-X.; Wang, M. Achieving highly crystalline rate and crystallinity in Poly(L-lactide) via *in-situ* melting reaction with diisocyanate and benzohydrazine to form nucleating agents. *Polym. Test.* **2020**, *81*, 106216. [CrossRef]
53. Brown, D.J. Chapter 2.13; Pyrimidines and their Benzo Derivatives. In *Comprehensive Heterocyclic Chemistry*; Katritzky, A.R., Rees, C.W., Eds.; Elsevier Science Ltd.: Oxford, UK, 1984; Volume 3, pp. 57–155.
54. West, T.P.; Chunduru, J.; Murahari, E.C. Orotic acid: Why it is important to understand its role in metabolism. *Biochem. Physiol.* **2017**, *6*, 1000e157. [CrossRef]
55. Qiu, Z.; Li, Z. Effect of orotic acid on the crystallization kinetics and morphology of biodegradable poly(L-lactide) as an efficient nucleating agent. *Ind. Eng. Chem. Res.* **2011**, *50*, 12299–12303. [CrossRef]
56. Song, P.; Sang, L.; Zheng, L.; Wang, C.; Liu, K.; Wei, Z. Insight into the role of bound water of a nucleating agent in polymer nucleation: A comparative study of anhydrous and monohydrated orotic acid on crystallization of poly(L-lactic acid). *RSC Adv.* **2017**, *7*, 27150–27161. [CrossRef]
57. Salač, J.; Šerá, J.; Jurča, M.; Verney, V.; Marek, A.A.; Koutný, M. Photodegradation and biodegradation of poly(lactic) acid containing orotic acid as a nucleation agent. *Materials* **2019**, *12*, 481. [CrossRef]

58. Okesola, B.; Vieira, V.M.P.; Cornwell, D.J.; Whitelaw, N.K.; Smith, D.K. 1,3:2,4-Dibenzylidene-D-sorbitol (DBS) and its derivatives—Efficient, versatile and industrially relevant low-molecular-weight gelators with over 100 years of history and a bright future. *Soft Matter* **2015**, *11*, 4768–4787. [CrossRef]
59. Lai, W.-C. Thermal behavior and crystal structure of poly(L-lactic acid) with 1,3:2,4-dibenzylidene-D-sorbitol. *J. Phys. Chem. B* **2011**, *115*, 11029–11037. [CrossRef]
60. Lai, W.-C.; Liao, J.-P. Nucleation and crystal growth kinetics of poly(L-lactic acid) with self-assembled DBS nanofibrils. *Mater. Chem. Phys.* **2013**, *139*, 161–168. [CrossRef]
61. Lai, W.-C.; Lee, Y.-C. Effects of self-assembled sorbitol-derived compounds on the structures and properties of biodegradable poly(L-lactic acid) prepared by melt blending. *J. Polym. Res.* **2019**, *26*, 10. [CrossRef]
62. You, J.; Yu, W.; Zhou, C. Accelerated crystallization of poly(lactic acid): Synergistic effect of poly(ethylene glycol), dibenzylidene sorbitol, and long-chain branching. *Ind. Eng. Chem. Res.* **2014**, *53*, 1097–1107. [CrossRef]
63. Petchwattana, N.; Naknaen, P.; Sanetuntikul, J.; Narupai, B. Crystallisation behaviour and transparency of poly(lactic acid) nucleated with dimethylbenzylidene sorbitol. *Plast. Rubber Compos.* **2018**, *47*, 147–155. [CrossRef]
64. Sun, C.; Li, C.; Tan, H.; Zhang, Y. Enhancing the durability of poly(lactic acid) composites by nucleated modification. *Polym. Int.* **2019**, *68*, 1450–1459. [CrossRef]
65. De Melo, B.A.G.; Motta, F.L.; Santana, M.H.A. Humic acids: Structural properties and multiple functionalities for novel technological developments. *Mat. Sci. Eng.* **2016**, *C62*, 967–974. [CrossRef]
66. Bishai, M.; De, S.; Adhikari, B.; Banerjee, R. A comprehensive study on enhanced characteristics of modified polylactic acid based versatile biopolymer. *Eur. Polym. J.* **2014**, *54*, 52–61. [CrossRef]
67. Xu, X.; Zhen, W.; Bian, S. Structure, performance and crystallization behavior of poly (lactic acid)/humic acid amide composites. *Polym.-Plast. Technol. Eng.* **2018**, *57*, 1858–1872. [CrossRef]
68. Xu, X.; Zhen, W. Preparation, performance and non-isothermal crystallization kinetics of poly(lactic acid)/amidated humic acid composites. *Polym. Bull.* **2018**, *75*, 3753–3780. [CrossRef]
69. Liu, P.; Zhen, W. Structure-property relationship, rheological behavior, and thermal degradability of poly(lactic acid)/fulvic acid amide composites. *Polym. Adv. Technol.* **2018**, *29*, 2192–2203. [CrossRef]
70. Liu, P.; Zhen, W.; Bian, S.; Wang, X. Preparation and performance of poly(lactic acid)/fulvic acid benzhydrazide composites. *Adv. Polym. Technol.* **2018**, *37*, 2788–2798. [CrossRef]
71. Zhang, H.; Zhen, W. Performance, rheological behavior and enzymatic degradation of poly (lactic acid)/modified fulvic acid composites. *Int. J. Biol. Macromol.* **2019**, *139*, 181–190. [CrossRef]
72. Duan, K.; Zhen, W. The synthesis of poly (lactic acid)-fulvic acid graft polymer and its effect on the crystallization and performance of poly (lactic acid). *Polym.-Plast. Technol. Mater* **2019**, *58*, 1875–1888. [CrossRef]
73. Abitbol, T.; Rivkin, A.; Cao, Y.; Nevo, Y.; Abraham, E.; Ben-Shalom, T.; Lapidot, S.; Shoseyov, O. Nanocellulose, a tiny fiber with huge applications. *Curr. Opin. Biotechnol.* **2016**, *39*, 76–88. [CrossRef] [PubMed]
74. Mokhena, T.C.; Sefadi, J.S.; Sadiku, E.R.; John, M.J.; Mochane, M.J.; Mtibe, A. Thermoplastic processing of PLA/cellulose nanomaterials composites. *Polymers* **2018**, *10*, 1363. [CrossRef] [PubMed]
75. Kian, L.K.; Saba, N.; Jawaid, M.; Sultan, M.T.H. A review on processing techniques of bast fibers nanocellulose and its polylactic acid (PLA) nanocomposites. *Int. J. Biol. Macromol.* **2019**, *121*, 1314–1328. [CrossRef]
76. Vatansever, E.; Arslan, D.; Nofar, M. Polylactide cellulose-based nanocomposites. *Int. J. Biol. Macromol.* **2019**, *137*, 912–938. [CrossRef]
77. Scaffaro, R.; Botta, L.; Lopresti, F.; Maio, A.; Sutera, F. Polysaccharide nanocrystals as fillers for PLA based nanocomposites. *Cellulose* **2017**, *24*, 447–478. [CrossRef]
78. Ren, Z.; Guo, R.; Bi, H.; Jia, X.; Xu, M.; Cai, L. Interfacial adhesion of polylactic acid on cellulose surface: A molecular dynamics study. *ACS Appl. Mater. Interfaces* **2020**, *12*, 3236–3244. [CrossRef]
79. Geng, S.; Wloch, D.; Herrera, N.; Oksman, K. Large-scale manufacturing of ultra-strong, strain-responsive poly(lactic acid)-based nanocomposites reinforced with cellulose nanocrystals. *Compos. Sci. Technol.* **2020**, *194*, 108144. [CrossRef]
80. Qian, S.; Sheng, K.; Yu, K.; Xu, L.; Lopez, C.A.F. Improved properties of PLA biocomposites toughened with bamboo cellulose nanowhiskers through silane modification. *J. Mater. Sci.* **2018**, *53*, 10920–10932. [CrossRef]

81. Ma, Y.; Qian, S.; Hu, L.; Qian, J.; Lopez, C.A.F.; Xu, L. Mechanical, thermal, and morphological properties of pla biocomposites toughened with silylated bamboo cellulose nanowhiskers. *Poly. Compos.* **2019**, *40*, 3012–3019. [CrossRef]
82. Qu, P.; Zhou, Y.; Zhang, X.; Yao, S.; Zhang, L. Surface modification of cellulose nanofibrils for poly(lactic acid) composite application. *J. Appl. Poly. Sci.* **2012**, *125*, 3084–3091. [CrossRef]
83. He, L.; Song, F.; Li, D.-F.; Zhao, X.; Wang, X.-L.; Wang, Y.-Z. Strong and Tough Polylactic Acid Based Composites Enabled by Simultaneous Reinforcement and Interfacial Compatibilization of Microfibrillated Cellulose. *ACS Sustain. Chem. Eng* **2020**, *8*, 1573–1582. [CrossRef]
84. Cui, X.; Ozaki, A.; Asoh, T.-A.; Uyama, H. Cellulose modified by citric acid reinforced Poly(lactic acid) resin as fillers. *Polym. Degr. Stab.* **2020**, *175*, 109118. [CrossRef]
85. Wang, F.Y.; Dai, L.; Ge, T.T.; Yue, C.B.; Song, Y.M. α-methylstyrene-assisted maleic anhydride grafted poly(lactic acid) as an effective compatibilizer affecting properties of microcrystalline cellulose/poly(lactic acid) composites. *Express Polym. Lett.* **2020**, *14*, 530–541. [CrossRef]
86. Rigotti, D.; Pegoretti, A.; Miotello, A.; Checchetto, R. Interfaces in biopolymer nanocomposites: Their role in the gas barrier properties and kinetics of residual solvent desorption. *Appl. Surf. Sci.* **2020**, *507*, 145066. [CrossRef]
87. Chen, J.; Zhang, T.; Hua, W.; Li, P.; Wang, X. 3D Porous poly(lactic acid)/regenerated cellulose composite scaffolds based on electrospun nanofibers for biomineralization. *Coll. Surf. A* **2020**, *585*, 124048. [CrossRef]
88. Sartore, L.; Pandini, S.; Day, K.; Bignotti, F.; Chiellini, F. A versatile cell-friendly approach to produce PLA-based 3D micro-macro-porous blends for tissue engineering scaffolds. *Materialia* **2020**, *9*, 100615. [CrossRef]
89. Zhang, L.-Q.; Yang, S.-G.; Li, Y.; Huang, H.-D.; Xu, L.; Xu, J.-Z.; Zhong, G.-J.; Li, Z.-M. Polylactide porous biocomposites with high heat resistance by utilizing cellulose template-directed construction. *Cellulose* **2020**, *27*, 3805–3819. [CrossRef]
90. Cisneros-López, E.O.; Pal, A.K.; Rodriguez, A.U.; Wu, F.; Misra, M.; Mielewski, D.F.; Kiziltas, A.; Mohanty, A.K. Recycled poly(lactic acid)ebased 3D printed sustainable biocomposites: A comparative study with injection molding. *Mater. Today Sust.* **2020**, *7–8*, 100027.
91. Wang, Q.; Ji, C.; Sun, J.; Yao, Q.; Liu, J.; Saeed, R.M.Y.; Zhu, Q. Kinetic thermal behavior of nanocellulose filled polylactic acid filament for fused filament fabrication 3D printing. *J. Appl. Polym. Sci.* **2020**, *137*, 48374. [CrossRef]
92. Zuo, M.; Pan, N.; Liu, Q.; Ren, X.; Liu, Y.; Huang, T.-S. Three-dimensionally printed polylactic acid/cellulose acetate scaffolds with antimicrobial effect. *RSC Adv.* **2020**, *10*, 2952–2958. [CrossRef]
93. Bender, M.L.; Komiyama, M. *Cyclodextrin Chemistry*; Springer: Berlin, Germany, 1978.
94. Sabadini, E.; Cosgrovea, T.; do Carmo Egídio, F. Solubility of cyclomaltooligosaccharides (cyclodextrins) in H_2O and D_2O: A comparative study. *Carbohydr. Res.* **2006**, *341*, 270–274. [CrossRef] [PubMed]
95. Wenz, G. Cyclodextrins as Building Blocks for Supramolecular Structures and Functional Units. *Angew. Chem. Int. Ed. Engl.* **1994**, *33*, 803–822. [CrossRef]
96. Bruns, C.J. Exploring and Exploiting the Symmetry-Breaking Effect of Cyclodextrins in Mechanomolecules. *Symmetry* **2019**, *11*, 1249. [CrossRef]
97. Yao, X.; Huang, P.; Nie, Z. Cyclodextrin-based polymer materials: From controlled synthesis to applications. *Progr. Polym. Sci.* **2019**, *93*, 1–35. [CrossRef]
98. Mostovaya, O.A.; Gorbachuk, V.V.; Padnya, P.L.; Vavilova, A.A.; Evtugyn, G.A.; Stoikov, I.I. Modification of Oligo- and Polylactides with Macrocyclic Fragments: Synthesis and Properties. *Front. Chem.* **2019**, *7*, 554. [CrossRef] [PubMed]
99. Aytac, Z.; Uyar, T. Core-shell nanofibers of curcumin/cyclodextrin inclusion complex and polylactic acid: Enhanced water solubility and slow release of curcumin. *Int. J. Pharm.* **2017**, *1–2*, 177–184. [CrossRef]
100. Li, W.; Fan, X.; Wang, X.; Shang, X.; Wang, Q.; Lin, J.; Hu, Z.; Li, Z. Stereocomplexed micelle formation through enantiomeric PLA-based Y-shaped copolymer for targeted drug delivery. *Mat. Sci. Eng.* **2018**, *91*, 688–695. [CrossRef]
101. Ordanini, S.; Cellesi, F. Complex Polymeric Architectures Self-Assembling in Unimolecular Micelles: Preparation, Characterization and Drug Nanoencapsulation. *Pharmaceutics* **2018**, *10*, 209. [CrossRef]

102. Nittayacharn, P.; Nasongkla, N. Development of self-forming doxorubicin-loaded polymeric depots as an injectable drug delivery system for liver cancer chemotherapy. *J. Mater. Sci. Mater. Med.* **2017**, *28*, 101. [CrossRef]
103. Hu, Y.; Wu, X.; JinRui, X. Self-assembled supramolecular hydrogels formed by biodegradable PLA/CS diblock copolymers and β-cyclodextrin for controlled dual drug delivery. *Int. J. Biol. Macromol.* **2018**, *108*, 18–23. [CrossRef] [PubMed]
104. Kost, B.; Brzeziński, M.; Cieślak, M.; Królewska-Golińska, K.; Makowski, T.; Socka, M.; Biela, T. Stereocomplexed micelles based on polylactides with β-cyclodextrin core as anti-cancer drug carriers. *Eur. Polym. J.* **2019**, *120*, 109271. [CrossRef]
105. Vermet, G.; Degoutin, S.; Chai, F.; Maton, M.; Flores, C.; Neut, C.; Danjou, P.E.; Martel, B.; Blanchemain, N. Cyclodextrin modified PLLA parietal reinforcement implant with prolonged antibacterial activity. *Acta Biomater.* **2017**, *53*, 222–232. [CrossRef] [PubMed]
106. Liu, Y.; Liang, X.; Zhang, R.; Lan, W.; Qin, W. Fabrication of Electrospun Polylactic Acid/Cinnamaldehyde/_-Cyclodextrin Fibers as an Antimicrobial Wound Dressing. *Polymers* **2017**, *9*, 464. [CrossRef] [PubMed]
107. Yi, W.-J.; Li, L.-J.; He, H.; Hao, Z.; Liu, B.; Chao, Z.-S.; Shen, Y. Synthesis of poly(l-lactide)/β-cyclodextrin/citrate network modified hydroxyapatite and its biomedical properties. *New J. Chem.* **2018**, *42*, 14729–14732. [CrossRef]
108. Kost, B.; Svyntkivska, M.; Brzeziński, M.; Makowski, T.; Piorkowska, E.; Rajkowska, K.; Kunicka-Styczyńska, A.; Biela, T. PLA/β-CD-based fibres loaded with quercetin as potential antibacterial dressing materials. *Coll. Surf. B Biointerf* **2020**, *190*, 110949. [CrossRef] [PubMed]
109. Rusa, C.C.; Tonelli, A.E. Polymer/Polymer Inclusion Compounds as a Novel Approach to Obtaining a PLLA/PCL Intimately Compatible Blend. *Macromolecules* **2000**, *33*, 5321–5324. [CrossRef]
110. Choi, H.; Ooya, T.; Sasaki, S.; Yui, N.; Ohya, Y.; Nakai, T.; Ouchi, T. Preparation and Characterization of Polypseudorotaxanes Based on Biodegradable Poly(L-lactide)/Poly(ethylene glycol) Triblock Copolymers. *Macromolecules* **2003**, *36*, 9313–9318. [CrossRef]
111. Zhang, R.; Wang, Y.; Wang, K.; Zheng, G.; Li, Q.; Shen, C. Crystallization of poly(lactic acid) accelerated by cyclodextrin complex as nucleating agent. *Polym. Bull* **2013**, *70*, 195–206. [CrossRef]
112. Viciosa, M.T.; Alves, N.M.; Oliveira, T.; Dionísio, M.; Mano, J.F. Confinement Effects on the Dynamic Behavior of Poly(d,l-lactic Acid) upon Incorporation in α-Cyclodextrin. *J. Phys. Chem. B* **2014**, *118*, 6972–6981. [CrossRef]
113. Kano, K. Mechanisms for chiral recognition by cyclodextrins. *J. Phys. Org. Chem.* **1997**, *10*, 286–291. [CrossRef]
114. Gingter, S.; Bezdushna, E.; Ritter, H. Chiral recognition of macromolecules with cyclodextrins: PH- and thermosensitive copolymers from N-isopropylacrylamide and N-acryloyl-D/L-phenylalanine and their inclusion complexes with cyclodextrins. *Beilstein J. Org. Chem.* **2011**, *7*, 204–209. [CrossRef]
115. Wang, S.-Y.; Li, L.; Xiao, Y.; Wang, Y. Recent advances in cyclodextrins-based chiral-recognizing platforms. *Trends Anal. Chem.* **2019**, *121*, 115691. [CrossRef]
116. Ohya, Y.; Takamido, S.; Nagahama, K.; Ouchi, T.; Ooya, T.; Katoono, R.; Yui, N. Molecular "Screw and Nut": α-Cyclodextrin Recognizes Polylactide Chirality. *Macromolecules* **2007**, *40*, 6441–6444. [CrossRef]
117. Saga, A.; Sasaki, T.; Sakurai, K. Influence of α-cyclodextrin and casting rate on the crystallization of poly(l-lactide) from a PLLA solution. *e-Polymers* **2014**, *14*, 151–157.
118. Xie, D.M.; Yang, K.S.; Sun, W.X. Formation and characterization of polylactide and β-cyclodextrin inclusion complex. *Curr. Appl. Phys.* **2007**, *7S1*, e15–e18. [CrossRef]
119. Lizundia, E.; Gómez-Galván, F.; Pérez-Álvarez, L.; Leóna, L.M.; Vilas, J.L. Poly(L-lactide)/branched β-cyclodextrin blends: Thermal, morphological and mechanical properties. *Carbohydr. Polym.* **2016**, *144*, 25–32. [CrossRef]
120. Suzuki, T.; Ei, A.; Takada, Y.; Uehara, H.; Yamanobe, T.; Takahashi, K. Modification of physical properties of poly(L-lactic acid) by addition of methyl-β-cyclodextrin. *Beilstein J. Org. Chem.* **2014**, *10*, 2997–3006. [CrossRef]
121. Byun, Y.; Rodriguez, K.; Han, J.H.; Kim, Y.T. Improved thermal stability of polylactic acid (PLA) composite film via PLA-β-cyclodextrin-inclusion complex systems. *Int. J. Biol. Macromol.* **2015**, *81*, 591–598. [CrossRef]

122. Zhou, Y.F.; Song, Y.N.; Zhen, W.J.; Wang, W.T. The Effects of Structure of Inclusion Complex between β-Cyclodextrin and Poly(L-lactic acid) on Its Performance. *Macromol. Res.* **2015**, *23*, 1103–1111. [CrossRef]
123. Li, Y.; Zhen, W. Preparation and characterization of benzoylhydrazide-derivatized poly(lactic acid) and γ-cyclodextrin inclusion complex and its effect on the performance of poly(lactic acid). *Polym. Adv. Technol.* **2017**, *28*, 1617–1628. [CrossRef]
124. Li, Y.; Zhen, W. Preparation, Structure and Performance of Poly(lactic acid)/Poly(lactic acid)-γ-Cyclodextrin Inclusion Complex-Poly(glycidyl methacrylate) Composites. *Macromol. Res.* **2018**, *26*, 215–225. [CrossRef]

© 2020 by the authors. Licensee MDPI, Basel, Switzerland. This article is an open access article distributed under the terms and conditions of the Creative Commons Attribution (CC BY) license (http://creativecommons.org/licenses/by/4.0/).

MDPI
St. Alban-Anlage 66
4052 Basel
Switzerland
Tel. +41 61 683 77 34
Fax +41 61 302 89 18
www.mdpi.com

Molecules Editorial Office
E-mail: molecules@mdpi.com
www.mdpi.com/journal/molecules

www.ingramcontent.com/pod-product-compliance
Lightning Source LLC
LaVergne TN
LVHW070242100526
838202LV00015B/2165